The Green World

The Green World

An Introduction to Plants and People

Second Edition

Richard M. Klein

University of Vermont, Burlington

HARPER & ROW, PUBLISHERS, New York
Cambridge, Philadelphia, San Francisco, Washington,
London, Mexico City, São Paulo, Singapore, Sydney

FOR DR. DEANA TARSON KLEIN
Professional colleague, collaborator in botanical research, trail buddy, companion, critic, and wife.

Sponsoring Editor: Claudia M. Wilson
Project Editor: Joan C. Gregory
Text and Cover Design: Lucy Zakarian
Text Art: J & R Technical Services, Inc.
Illustrator: Jane R. Ulrich and Linda E. Jones
Production Manager: Kewal Sharma
Compositor: Black Dot, Inc.
Printer and Binder: R. R. Donnelley & Sons Company

Cover Illustration: Ch'ien Husan (ca. 1235–after 1301). Yuan Dynasty. *Pear Blossoms.* Ink and Colors on Paper, 12 1/2″ × 37 1/2″ (31.1× 95.3 cm.). The Metropolitan Museum of Art, Purchase, The Dillon Fund Gift, 1977. (1077, 79).

THE GREEN WORLD: An Introduction to Plants and People
Second Edition

For information address Harper & Row, Publishers, Inc., 10 East 53d Street, New York, NY 10022

Library of Congress Cataloging in Publication Data

Klein, Richard M.
The green world.

Includes index.
1. Botany. 2. Botany, Economic. 3. Ethnobotany.
I. Title.
QK47.K63 1987 581 86-9813
ISBN 0-06-043713-8

86 87 88 89 9 8 7 6 5 4 3 2 1

Contents

CHAPTER 2 / *Filling Empty Bellies* 83

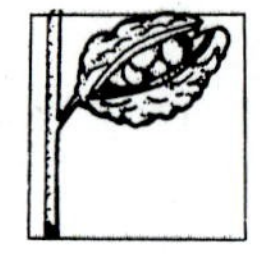

CHAPTER 3 / *The Staffs of Life* 113

CHAPTER 4 / *Vegetable Feast* 207

CHAPTER 5 / *Plants in Religion* 295

CHAPTER 10/*Plants in the Environment* 541

Preface

Botany, like any other topic, can be explored from many different points of view. At the college level, botany is usually studied as one of the biological sciences, in which the structure, function, evolution, ecology, and life histories of plants are examined in detail. Students of botany have many different career goals in botanical sciences or in plant sciences such as forestry, horticulture, or environmental science. In addition to the more strictly scientific and technological aspects of botany, all of us know that we are absolutely dependent on plants for food, shelter, and the oxygen we breathe. Our daily lives involve activities in which plants and plant products are in constant use. Our relationships with plants are so pervasive and multidimensional and so much a part of the fabric of our lives that they rarely enter our conscious thoughts. We use plant symbolisms in everyday speech: expressions like "putting down roots" or "knocking on wood" are so common that we simply don't think about them. There are religious, economic, historical, political, and other facets of our intimate relationship to the world of plants. In fact, the interactions of plants and people are co-evolutionary; both plants and the people who depend on them have been substantially altered by the interaction.

For centuries, people have lied, cheated, and killed to obtain a particular plant or to ensure its steady supply. The discovery of the Americas was motivated by a desire to obtain the plants of China, and the early exploitation of the Americas provided Europe with the potato,

Indian corn, tobacco, quinine, and other plants. The plants proved to be economically and politically more important than the gold and silver gouged from the land and its inhabitants. The introduction of sugar and cotton, together with the cultivation of the native tobacco, was responsible for the use of slave labor, whose historical, economic, and social consequences are still with us. The rise and fall of the Venetian city-state, the Western hegemony of China, the "troubles" and famines in Ireland, and the settling of the North American prairies are historical events derived from our use and misuse of plants. Thus, through an examination of the green world, we can better understand the modern world and the civilizations from which our cultures have evolved.

At the same time, plants play basic roles in our personal lives. People are born, pair off, marry, have children, and eventually die with these events memorialized by flowers and plants. From earliest times, painting, sculpture, architecture, and even such mundane items as shopping bags and fabrics have been decorated with realistic or stylized flowers, foliage, or trees. The Oriental rugs on our floors are basically garden scenes woven with yarns whose colors can be obtained from plants. We easily recognize plants carved in wood or stone in our churches, public buildings, and homes. Music has been written about plants and some of the instruments are plant stems. It is unlikely that an individual in any culture can spend a single day without coming into contact with plant designs in one form or another.

Landscapes, both natural and contrived, have been admired and even worshiped as evidences of the presence of God, and many landscapes have been dedicated as the abode of the Deity. To a large extent, private and public parks derived from religious gardens have been sources of aesthetic pleasure as well as relief from the pressures of modern civilization. Landscape and flower gardens have become important parts of our lives; gardening is listed as a hobby of more people in North America than any other avocation. The Chinese, the Japanese, and the Indian people stylize natural landscapes and through them express cultural and philosophical characteristics; to some extent, we in the West do the same thing. Gardens in the Middle East, with their emphasis on greenery and flowing water, express longings for a home far removed from the harsh reality of heat and drought; the Persian word for *garden* is *paradise*—a word that assumed new meaning in the Judeo-Christian tradition that originated in that part of the world.

I believe that botany is an intellectual activity in every way comparable to history, art, philosophy, or political science and assert that there is an intellectual excitement to be obtained by learning and investigating what plants are, what they do, and how they affect our lives. Knowing the structure of a cell or a leaf is basic to understanding how plants capture and utilize solar energy, how they move water, and how and why they produce those chemicals we use. Thus, this book includes aspects of formal botanical science. But it is not an abbreviated and simplified textbook of general botany. This would be like trying to transmit the power of Shakespeare with a shortened version of *King Lear*.

Like all other scientists, botanists are acutely conscious that the knowledge we have accumulated about plants is part of the larger, infinitely more complex world in which we live. What we know—or think we know—about our science is colored by the culture in which we grew up and by the interactions we have with people in our culture. Botany is a part of human culture and is influenced by it. Our research is, to some extent, dictated not only by perceived gaps in knowledge but also by the events that shape our lives. Hence my attempt to place our present knowledge of plants into historical and cultural frameworks.

If a book is to serve a function beyond that of helping a student pass an examination, it ought to arouse the interest and curiosity of the reader. In a book such as this, which attempts to cover a very broad range of topics, none can be presented in detail. In addition, more plant-and-people related topics have been excluded from than included in this volume. Certainly an instructor can provide further detail about a topic and can introduce other topics of interest and importance. But ultimate responsibility and intellectual pleasure must be self-generated. Each section of this book ends with an Additional Readings list designed to provide access to the topic. Some of the citations are technical; some are beautifully written expositions of historical, cultural, economic, or scientific aspects of the topic. Others blend science and art, or deal with such current subjects as the marketing of plants.

Although my name appears as the author, the expertise, critical judgments, and friendly interest of many collaborators have been invaluable. I want to thank the following persons who read all or part of the manuscript: Drs. M. J. Behan, J. D. Caponetti, J. C. Cavender, W. M. Hess, R. W. Hoshaw, R. L. Hulbary, H. W. Keller, D. T. Klein, W. J. Koch, J. C. Lockhart, R. L. Mansell, J. H. McCulloch, R. F. Raffauf, L. L. St. Clair, R. D. Schein, C. A. Schroeder, D. W. Smith, F. H. Tschirley, E. C. Weaver, and H. O. Whittier. I am indebted to Dr. Robert Raffauf who checked all chemical formulae. The epigraph for the section on Essential Oils is courtesy of Charles of the Ritz Corp. The lyrics by Cole Porter heading the section on cocaine is courtesy of Warner Bros. Music. Some of my collaborators have been dead for several centuries. I have, in the words of Isaac Newton, stood on the shoulders of giants. I must, however, assume full responsibility for any factual errors and for the many interpretations and personal opinions found herein. Since the focus of this book is on aspects of plant science that have affected the lives of people in many civilizations over thousands of years, my own biases and opinions—which are conditioned by my personal experiences and the Western civilization that molded me—cannot fail to have influenced my thinking, research, teaching, and writing. If you disagree with what I have said, so much the better!

Richard M. Klein

Chapter 1

What Plants Are and What Plants Do

The Evolution of Life on Earth

The earth was without form and void,
and darkness was upon the face of the deep.

Genesis 1:1

HOW LIFE GOT STARTED

Speculations on the origin of life and on its development have been of deep concern to the human species since we evolved. All religions have myths and legends about the origin of the earth and the origin of our species. The most familiar to us is in Genesis 1, wherein the Judeo-Christian God created the Earth and its living inhabitants. Biological and physical scientists also provide information on these important questions. The scientific theories take as their basis the concept that the universe developed and that life originated and evolved by purely physical-chemical processes. If these hypotheses are to have any value, they must be rational and subject to scientific scrutiny, they must account for the almost unimaginable size and complexity of the universe, and they must provide a basis for understanding the wonderful and awesome diversity of living things that populate the planet. Formally, cosmology and biological evolution must be based solely on physics and chemistry, without recourse to the intervention of a Prime Mover (a God). Such considerations do not necessarily rule out the possibility of a God; there are different ways of viewing the universe and its contents.

It is generally believed that the universe originated as an explosion of a mass of energy so dense that it contained all of the substance that now constitutes the universe. When the "Big Bang" occurred, some 20 billion years ago, matter was flung out so violently that it is still moving rapidly away from the point of the explosion. Matter condensed into the galaxies and into solar systems by a variety of known physical and chemical processes. Our solar system—the Sun and surrounding planets—is near the edge of a medium-sized galaxy, the Milky Way. By further condensation of matter, planets, including the one called Earth, were formed about 4.5 billion years ago—in mathematical notation, 4.5×10^{-9} before the present (B.P.). Excellent evidence supporting this time scale is found in the known rates of decay of radioactive isotopes in rocks.

Earth was, at the time of its formation, almost molten because of radioactive heat; it had no atmosphere and liquid water did not exist. No life—as we know life—could have existed. This condition lasted about 2 billion years, and, as the earth cooled, liquid water appeared as lakes and oceans. An atmosphere formed which contained water vapor and gases: hydrogen, nitrogen, ammonia, and simple carbon compounds like carbon dioxide (CO_2) and methane (CH_4). There was little or no free, molecular oxygen. Such an atmosphere is called a *reducing atmosphere,* in contrast with an oxygen-containing gas envelope called an *oxidizing atmosphere.*

For life to be initiated on earth, there had to be a supply of water in which organic (carbon-containing) building blocks of life (sugars, protein precursors, and so on) were dissolved. Indeed, during the nineteenth century, Charles Darwin suggested that life could be formed in a broth of organic substances. This idea was elaborated in 1929 by the Russian cosmologist A. I. Oparin and the English biologist J. B. S. Haldane, who wrote of a sea of rich, warm soup. If the concept of the origin of life on earth has any validity, the first task in establishing its merit was to determine whether these chemical building blocks of cells could be synthesized under the presumed abiotic (nonliving) conditions of the earth at that time.

Among the seminal experiments on abiotic synthesis of life-requiring organic molecules was the 1953 doctoral dissertation of Stanley Miller, who was working with Professor Harold Urey at the University of Chicago. In a laboratory reaction vessel, Miller introduced water vapor, hydrogen gas, small amounts of ammonia and methane (Figure 1.1). He inserted electrodes to produce sparks equivalent to bolts of lightning and then he turned on the apparatus. The next morning he found that the vessel contained a gummy brown solution which, upon analysis, proved to contain several sugars, amino acids (the building blocks of proteins), organic acids such as are found in fruit, and other compounds

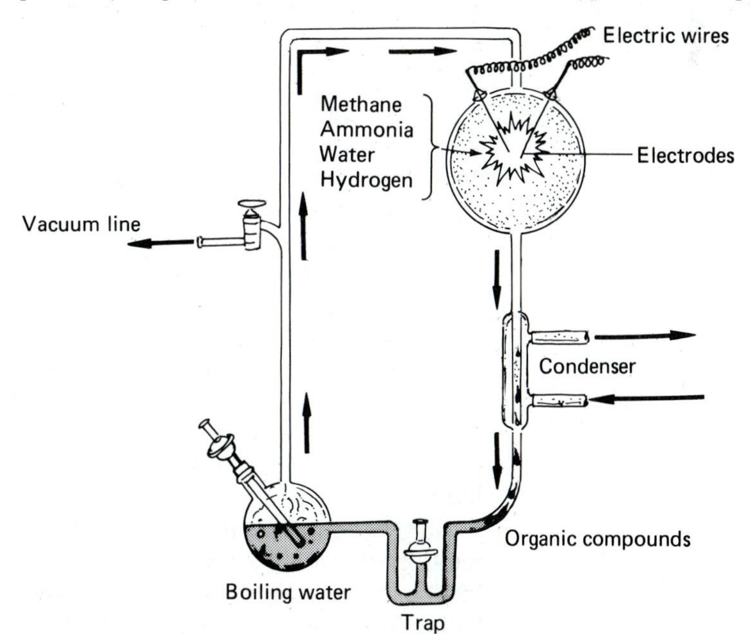

Figure 1.1 A diagram of Dr. Stanley Miller's spark-discharge apparatus used to investigate the abiotic synthesis of organic compounds. From Miller, S. L. *Journal of the American Chemical Society* 77, p. 2351. Copyright 1955 by the American Chemical Society.

common to living plants and animals. Although not the first study on abiotic synthesis of biologically important molecules, Miller's research formed the basis for many other projects to make molecules that are characteristic of living cells under primitive earth conditions. Within a few years, other investigators found that the rich, warm soup of Oparin and Haldane could easily have contained the basic molecules that are found in living creatures.

If all the compounds that can be synthesized under primitive earth conditions are brought together in a test tube, the mixture looks, smells, feels, and even tastes like living substance. But it is not alive, since it neither metabolizes nor reproduces. This jump from nonliving to living is still the greatest gap in our knowledge of biological evolution. Somehow these substances organized, developed some semblance of structure, and were transformed—perhaps, in the words of Handel's *Messiah,* "In an instant; in the twinkling of an eye"—into something that was alive. This transformation from nonliving to living might have been the act of a Prime Mover or may have occurred by pure blind, mathematically calculable chance through physical and chemical processes. Many scientists accept the latter explanation and feel reasonably confident that this immense gap in our understanding will be bridged, if not tomorrow, then in the foreseeable future. But for now, we know that the gap exists, and we can follow the evolution of this primordial blob of living stuff from which we are all descended. The precise course of biological evolutionary events leading to that gloriously colored autumnal sugar maple tree or to the joy of a Mozart or a Matisse is still not known with absolute certainty, but the broad outlines of the process are abundantly clear. Evolution is not a theory; it is a fact.

BIOLOGICAL EVOLUTION

At a minimum, archaic life had to have a place to live—a habitat—that provided liquid water, a reasonable temperature, minerals, and food. Oxygen was, for such organisms, not required; they—like their direct, modern descendants—lived anaerobically (without oxygen). Such microorganisms are in fact killed by oxygen; today they live in mud at the bottom of lakes and the oceans. These archaic organisms fed on components of the warm soup and, later, on the debris of dead organisms, as do their modern descendants.

Within a few million years, the supply of soup began to run low, and some organisms possessed the organic catalysts, the *enzymes,* that allowed them to obtain the energy they needed by transforming inorganic molecules to other states and, using this energy, fix carbon dioxide and methane into complex organic molecules that could serve as food. Some evolved the capacity to obtain energy from sulfur compounds; their descendants are responsible for the foul, rotten-egg odor that rises from some bogs and swamps. Others could obtain energy from nitrogen compounds, from hydrogen gas, or from iron compounds. These organisms, called *lithotrophic* (rock-feeders), were the dominant life forms for almost 2.5 billion years. The huge sulfur deposits in Freeport, Texas, and some of the iron ore strata in the Mesabi Range of Minnesota are the results of their activity.

THE BIOCHEMISTRY OF EVOLUTIONARY CHANGE

The physical basis for permanent alteration in plant structure or function can be analyzed. The DNA (deoxyribonucleic acid) of a cell provides the information that directs the synthesis of protein molecules within that cell. These proteins function as enzymes that regulate the formation of life molecules and life processes. If there is any change in the organization of the DNA, the protein molecules that are formed will be different and the processes controlled by these proteins will also be different. A permanent alteration of the DNA can be brought about by many environmental factors, including light, temperature, and chemicals, or it can occur spontaneously at predictable rates. Such changes are called *mutations.*

Most mutations are deleterious; the loss of genetic information or a modification of information usually leads to death. Other mutations are innocuous—the change in the hereditary material DNA that instructs a plant to make a smooth-edged leaf instead of a rough-edged leaf, for example. Only a very few mutations confer upon the organism the ability to do something new that allows that organism to get along better in its environment. But, speaking teleologically, cells or organisms cannot say, "Well, since I am confronted with a life-threatening situation, I had better alter my structure or one of my biochemical processes so that I can survive this threat."

Mutational change in DNA is undirected and occurs by chance. Therefore, the "need" for a mutation, even for the survival of that cell and hence for survival of that lineage, cannot activate a linked series of mutations that control a major life process or structure. The environment cannot instruct or imprint upon the genetic apparatus because the genetic information is self-generating. For example, some bacteria have the genetic capacity to synthesize protein enzymes that can destroy life-threatening chemicals that never existed until humans—who came along eons after the bacteria—created these chemicals in test tubes. Such mutations occurred by statistically calculable chance, perhaps millions of years ago, and were carried along in the DNA through millions of generations of that organism and of organisms that evolved from the original mutated one. The concept that mutations occur independently of need is fundamental to our understanding of evolution. If this formulation is correct, the varied and wondrous differences among organisms were pure happenstance, a series of probabilities; this is the working hypothesis of most biologists.

THE ANCESTORS OF GREEN PLANTS

The chemical energy obtained by the breakdown of inorganic compounds was used to convert carbon dioxide, dissolved in the water in which they lived, into one or more organic compounds related to sugar. It is impossible to overestimate the importance of this simple carbon dioxide fixation route. It set the biochemical stage for the eventual evolution of the photosynthetic process in green plants upon which all life on Earth now depends. In fact, some of the enzymes of these carbon-dioxide-fixing bacteria are present, virtually unchanged, in flowering plants and are used in the photosynthetic fixation of carbon dioxide into sugar. Such genetic information controlling the formation of an enzyme or of a whole biological process becomes locked into the hereditary apparatus, is perpetuated because it is efficient, and is used unchanged or is modified in newly evolved organisms. Carbon dioxide fixation, the chemical structure and function of the hereditary machinery, energy metabolism, and many other fundamental life processes have come down through several billion years. The rather precise similarities of the metabolic processes and the underlying genetic regulation of an alga, an orchid, a maple tree, and the human species provide excellent evidence for the concepts of biological evolution.

Those organisms that developed the carbon-fixing mechanism, the photosynthetic bacteria, possessed a selective advantage in survival and reproduction. Some of these organisms evolved new enzymes and compounds that allowed them to obtain the energy necessary for the fixation of a carbon dioxide from the sun. They were the first photosynthetic organisms, and some of their direct descendants are present in mud, where, if they occur in large numbers, they form a red mass.

Along with the evolution of food-making ability, photosynthetic organisms diverged in other ways from their direct ancestors and altered slightly the chemical organization of the pigment molecules used to capture sunlight. They are recognizably bacteria, but these photobacteria can survive in situations that tend to be lethal to most common bacteria.

The next major evolutionary jump can be dated with some precision. About 2.1 billion years ago, free oxygen began to accumulate in the earth's atmosphere. Although this oxygen could have been formed only by some life process, it could not have resulted from the photosynthetic activities of the photobacteria, because their photosynthesis does not produce oxygen. Some organism like a photobacterium altered slightly its light-capturing photopigment into the bright green photopigment called *chlorophyll a,* the same chlorophyll that is found in all green plants. The organisms that contained chlorophyll *a* were similar to bacteria: they possessed the same carbon-dioxide-fixing enzymes and showed the same internal organization as other bacteria. But this small chemical alteration in the photopigment (a few chemical bonds) was enough to allow the blue-green algae (the Cyanobacteria) to use water plus sunlight to make the chemical energy needed to drive carbon dioxide fixation into sugar and to release oxygen as a waste or end product.

The release of oxygen into the atmosphere drastically altered life on earth. Initially possessing a reducing atmosphere, the earth developed an oxidizing atmosphere. Most organisms—actually almost all of the organisms on earth at that time—were restricted to special anaerobic habitats, leaving much of the earth's surface available for those that could not only tolerate oxygen but actually use it for their metabolic processes. After blue-green algae evolved, certainly by 2 billion years B.P., a burst of evolutionary activity resulted in the appearance of those aerobic species of bacteria with which we are familiar. They include most of the disease-producing organisms and those involved in decay processes, as well as those that carry on activities that are of great benefit to us.

By 1.9 billion years B.P., the aerobic bacteria and the blue-green algae had colonized much of the water and damp land surfaces. Their remains have been found as fossils. We do not consider photosynthesizing bacteria to be green plants, although they certainly are plants. In contrast to green plants, their cells do not have rigid, woody walls and lack the compartmentalization of structure and function characteristic of the green plants and of all animals. Chloroplasts, nuclei, and other structures are not found in bacteria or in blue-green algae, which are therefore placed together in a subkingdom of life called the *prokaryotes* (organisms without nuclei). Of course, they contained the hereditary material DNA, but it was not well organized into a form that facilitated regularized exchange of genetic information among compatible organisms. Basically, prokaryotes lack a regular sexual process.

THE EVOLUTION OF GREEN PLANTS

The evolutionary alterations that led to the packaging of metabolic processes—respiration, photosynthesis, enzyme activity—are among the most interesting and exciting questions in biological evolution. The resulting organisms, called the *eukaryotes* (organisms with true nuclei), include all of the green plants, all of the fungi, and all of the animals that have ever lived or still live on earth. There is a good deal of discussion and several contending theories on how and exactly when this profound evolutionary event occurred. But occur it did, and we can follow the subsequent course of green plant evolution with some confidence.

The first eukaryotic plants were probably single-celled green algae with one or few large chloroplasts, a true nucleus containing the hereditary DNA, and a regularized sexual process in which DNA is exchanged between individuals. The cells of the first eukaryotic algae contained all of the packaged functions that are found today in all eukaryotes. As they diverged in structure and function, a number of different major groups of algae evolved, each adapted to different habitats and different conditions. These differences allowed them to colonize a variety of habitats; damp soil, fresh and salt water, and rocks soon had their own distinctive flora. When some algae lost their chloroplasts to become the first animals, Kingdom Animalia started on its own lines of evolution including, eventually, the primates. Other algae that also lost the photosynthetic ability formed the organisms that led to the evolution of the fungi, the third of the triumvirate of eukaryotic Kingdoms.

Evolution within the plant kingdom can be viewed as a tree with many branches arising at different points along the stem of the tree (Figure 1.2). The main branches have a number of smaller branches, each designating a line of evolution within that divergence. Groups that occupy the terminal portions of each branch are extant; they are the living descendants of their line of evolution. Determination of the temporal and spatial position of each of the branches and of its twigs is the important topic called *phylogeny,* which Charles Darwin called the origin of species. Biologists who study phylogeny attempt to determine not only what the overall relationships among plants are, but also how these relationships are established. Attempting to establish phylogenetic relationships among plants (or animals) is wonderful fun. It requires that the scientist integrate available information from many areas of scientific study, not just fossils but also patterns of biochemical processes, structural attributes including the results of electron microscopy and X-ray diffraction patterns, and many other specialties. Sherlock Holmes would find companion souls among the phylogeneticists.

Evolution within the plant kingdom is marked by successful modifications in structure and function that allows plants to cope more efficiently with the cold, cruel world. The evolution of mosses and liverworts resulted in the development of structural mechanisms that allowed them to protect and to nourish each new generation by enclosing the fertilized egg within a nest of maternal tissues, a modification that was perpetuated among the ferns. Mosses and liverworts are, however, restricted in their distribution because they have neither roots nor a system of conductive tissues that permit rapid and efficient water movement. They cannot grow up away from the soil very well. This is not true for the ferns. Not only do they have a root system; they also have a plumbing system (the vascular system) that permits water to move from the soil to the roots and up a considerable distance into the stem. Some ferns, particularly those that during the Carboniferous Period were the dominant plant forms on earth, were the size of many of our forest tree giants; their fossil remains provide much of our coal. Some tree ferns still exist today where they dominate forests in New Zealand and parts of Central America.

Algae, mosses and liverworts, and ferns reproduce sexually, but the movement of the sperm to the egg requires that the sperm swim in liquid water to the egg. This restricts sexual reproduction to moist habitats,

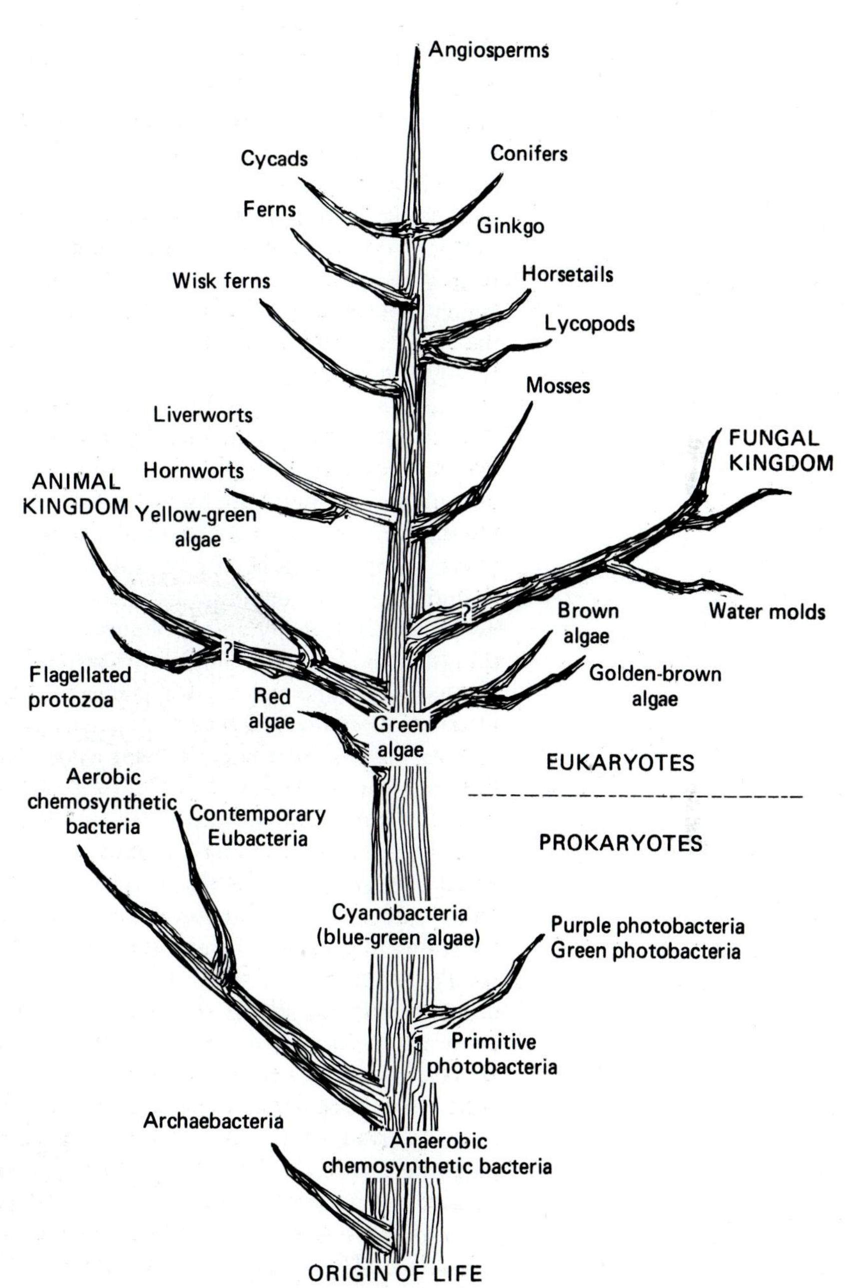

Figure 1.2 An evolutionary or phylogenetic "tree" indicating the lives of evolution of the plant kingdom. Taxa at the end of each branch are extant.

something that is not always available. The conifers (gymnosperms) and the flowering plants evolved a sexual reproductive mechanism in which the sperm nucleus, encapsulated within a pollen grain, can be transferred to the egg without the need for free water. Wind was the first vector for pollen transfers; later in time animals became important vectors. And finally, the flowering plants—the angiosperms—developed a more efficient method of bringing sperm and egg together when the flower and the fruit evolved. The selective advantage of the flower is that the production of female and male sex organs are well packaged and is sufficiently adaptable in structure and function to allow for many different types of pollen transmission. The selective advantage of the fruit is that it protects and nourishes the new generation (the seed), which itself is an important modification of structure and function since the seed contains an embryo with all of the basic parts of a new generation of plants.

The sequence of evolutionary events leading from primitive green algae to the flowering plants is well established, and the time relations of the processes are known within a relatively few thousands of years—pretty good, all things considered. There are differences of opinion as to the precise structural and functional changes whereby each of the modifications occurred, and there are even greater differences of opinion in our evaluation of the forces that determined the course of evolution. Phylogeny and evolution are not dry, sterile topics. Everything that is done in biological research expands our knowledge of the mechanisms and the consequences of evolution. Evolution is the ultimate base for biological sciences and, as such, is an intellectual activity in every way comparable to intellectual endeavors in any other area of human concern.

The flowering plants have many common attributes. There is considerable basic uniformity in structure and function and, within this unity, there is a good deal of diversity of detail. All flowering plants have roots, stems, leaves, and flowers, organs that perform basically the same functions in all plants. Reproduction, metabolism, water and mineral uptake and utilization, photosynthesis, and developmental processes are almost indistinguishable. Many of the differences are small ones, and most are relatively unimportant for survival. Variations in the leaf shapes or sizes among different species of oak trees are unlikely to have any importance in survival, although in certain environments differences in flower structure, leaf form, photosynthetic pathways, and so on may affect survival and reproductive success. Such diversity represents genetic adaptation of the plant to a particular environment, allowing it to colonize and utilize that environment to maintain itself and to reproduce. These differences were not caused by the environment, but were chance alterations in the genetic apparatus that facilitated adaptation. If the environment suddenly is altered, the organisms that survive will be those that inadvertently possessed the genetic information that permits structural or functional change.

There are, however, differences in specific structures and in a few life processes that allow the taxonomists (scientists who classify living plants) and phylogeneticists to puzzle out which groups of plants are most primitive, which are advanced, and which came from which. Such speculations can be organized in a phylogenetic "tree" that gives a two- or three-dimensional view of the evolution of plants. These constructions arrange plants so that their relationships in time and space can be determined. Figure 1.2 is an example of one such tree. It is certainly not the definitive view of plant evolution for, as more is learned about plant structure, chemistry, and function, there will be appropriate alterations in the position of the branches and the amount of their twigging. Grand syntheses of knowledge are always subject to modification.

THE DISPERSAL OF GREEN PLANTS

Based on an extensive fossil record—analysis of plant fossils in rock strata whose geological history is known—it is possible to reconstruct the course of biological evolution (Table 1.1). Primitive algae were widespread in ancient oceans by 700 million years B.P., and multicellular algae, including seaweeds, were present by 600 million B.P. Only limited evidence for land plants has been found in rocks dating to 400 million B.P. Land plants much like modern mosses were flourishing by 350 million B.P., when insects also reached the land accompanied by ancestors of frogs and other amphibians. The late Devonian and early Carboniferous periods witnessed the burst of evolution that resulted in great forests of ferns that were subsequently transformed into coal beds. Dinosaurs were kings of all they surveyed, and the small mammals that were ancestral to mammals of today led a precarious existence. By 250 million years ago (the mid-Carboniferous), the dominance of the ferns was fading and gymnosperms, like our modern cycads and conifers, successfully competed for space, water, light, and nutrients. Dinosaurs were still abundant, birds were evolving, and mammals were increasing in size and diversity. In rocks of the Cretaceous Period of the Mesozoic Era, fossil examples of flowering plants, probably much like buttercups and magnolias, have been found, and the dinosaurs were inching toward extinction. Mammals and birds were populating the habitats previously occupied by amphibians and reptiles. During the Cenozoic Era, the geological time span in which we live, most of the modern kinds of plants and animals evolved and diversified, showing characteristics that fitted them for life in habitats as different as deserts and tropical rain forests. Most of the flowering plant families were well established when the human genus, *Homo,* orginated in Africa.

As paleobotanists (scientists who unearth and study fossil plants) and taxonomists studied fossil and modern plants, they were struck by obvious discontinuities in the floras of different continents. They found fossils and modern descendants of the same groups on continents

Table 1.1 GEOLOGICAL, BIOLOGICAL, AND GEOGRAPHICAL TIME SCALES

Era	Period	Epoch	Millions of Years B.P.	Plants
Cenozoic	Quaternary	Recent	0.01	Dominance of angiosperms
		Pleistocene	2.5	
	Tertiary	Pliocene	7	Angiosperms expand and diversify
		Miocene	26	
		Oligocene	38	
		Eocene	54	
		Paleocene	64	
Mesozoic	Cretaceous		136	Angiosperms arise Gymnosperms decline
	Jurassic		190	Cycads and conifers appear Seed ferns decline
	Triassic		225	Cycads expand Ferns decline
Paleozoic	Permian		280	Seed ferns appear
	Carboniferous		345	Coal forest ferns dominate
	Devonian		395	Radiation of vascular plants
	Silurian		430	Invasion of land
	Ordovician		500	Marine algae expand
	Cambrian		570	Fungi arise Primitive marine algae appear
Precambrian			4000	Origin of life

Animals	Continental Drift*
Genus *Homo*	
Advanced primates arise Dominance of mammals, birds and insects	Modern continents form India reaches Asia Separation of N. America from Eurasia Separation of Africa from S. America
Dinosaurs extinct Mammals expand	
First birds Dinosaurs dominant	Gondwanaland/Laurasia
Dinosaurs expand First mammals appear	
Reptiles expand	Pangaea
Age of amphibians	
Age of fishes	
Arthropods invade land	
First vertebrates appear	
Marine invertebrates appear	

*See Figure 1.3 for continental drift maps.

separated by oceans or mountains that could not possibly be traversed by known methods of seed dispersal. Plants of west Africa were in the same taxonomic group as some in eastern South America. Some plants in India were much the same as some in Australia. The geographic basis for these disjunctions in distribution could not be explained until the concepts of *plate tectonics* and *continental drift* were developed.

Actually, the ideas underlying these concepts had been floating around for many years. Sir Francis Bacon (1561–1626) was the first to note that the outline of the west coast of Africa fit nicely with the east coast of South America. The Austrian geologist Eduard Suess was so struck by the similarities between the geological formations in lands of the Southern Hemisphere that, with no solid evidence, he proposed that there must have, at one time, been a land mass or supercontinent, which he named Gondwanaland. Suess's ideas were dismissed because it seemed impossible that land masses could move. In 1912, the German geologist Alfred Wegener suggested that all of the earth's land masses were plates that floated on the semisolid interior of the earth and that they could, indeed, move in response to powerful forces generated by heat and pressure in the earth. He suggested that all of the modern continents were, during the late Permian and Early Triassic periods (250 million years B.P.), bound together in one supercontinent, which he named Pangaea. Hotly debated and generally rejected at the time, Wegener's ideas were largely confirmed in the 1960s, when large-scale studies of the ocean floors were conducted by geologists of many countries, led by a group at the Lamont laboratories of Columbia University (Figure 1.3).

Actually, all of the existing continents are combinations of several plates. Part of North Carolina appears to have once been connected to west Africa, and the Indian plate collided with the Asian plate less than 100 million years ago with a force that resulted in the uplifting of the Himalayan Mountains (Figure 1.4). The Calfornia plate, marked by the San Andreas fault, will eventually drift away to become a mid-Pacific island. The Americas will move away from Africa and Europe and closer to Asia. Geologists can demark the outlines of the plates by faults that slip to cause earthquakes and by volcanic activity.

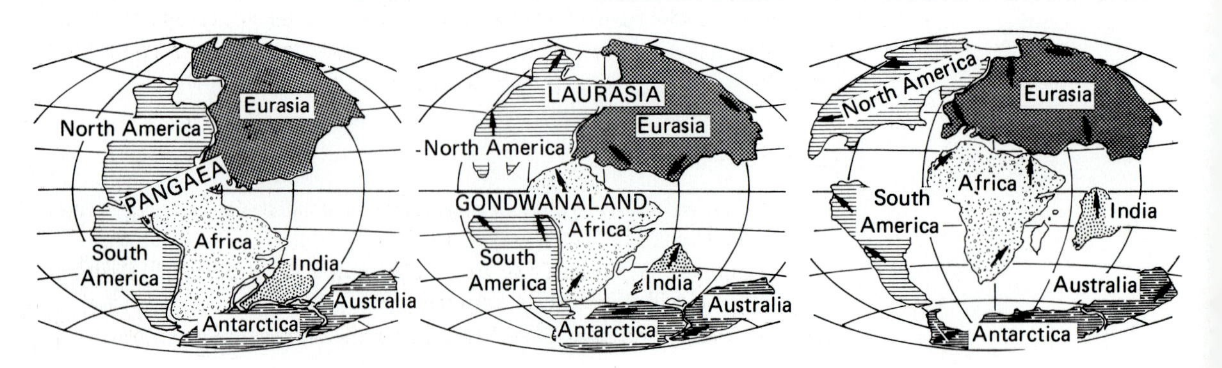

Figure 1.3 Three major stages in continental drift from the time of Pangaea (30 million years B.P.) during the late Paleozoic Era. During the Mesozoic Era (180 million years B.P.) the two supercontinents, Laurasia and Gondwanaland, had emerged and, by the end of the Mesozoic Era (110 million years B.P.) the present continents had separated. India had not yet collided with Eurasia, and Australia had not reached its present position.

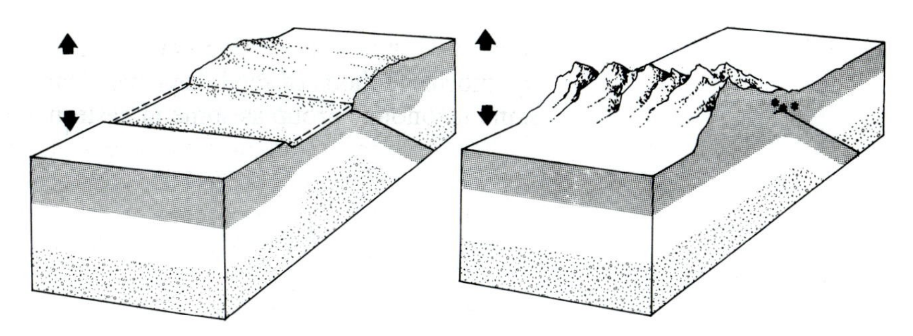

Figure 1.4 The collision of the Indian tectonic plate with the Eurasian plate during the late Mesozoic Era, which caused the uplifting of the Himalayan Mountains. The asterisks denote a region of active earthquakes along the fault line where the two plates joined.

The implications of plate tectonics and continental drift for plant distribution are still being explored, although many of the apparent discrepancies and disjunctions in plant distribution are being resolved. The drifting apart of Pangaea to form Laurasia and Gondwanaland in the Jurassic Period (180 million years B.P.) allowed cycads and conifers and many ferns to become dispersed among the land masses that eventually formed the Southern and Northern hemispheres. Some plant types in India, Australia, and southern South America would show similarities, and plants of North Africa and southern North America would also show affinities. As continental drift resulted in the separation of Africa and South America about 110 million years ago, this piece of the geological jigsaw puzzle and the disjunction of similar plants between these regions made sense. We can now also understand disjunctions among animals: the presence of one marsupial in the Americas, the oppossum, separated from the other marsupials (kangaroos and so forth) in Australia and the presence of fossil tigers in both hemispheres.

Prior to the final separation of the continents, about 110 million years ago at the end of the Cretaceous Period and the beginning of the Paleocene Epoch, the flowering plants, angiosperms, had evolved from gymnospermic ancestors. Their evolutionary radiation had resulted in the establishment of groups that we now recognize as representing various lines of evolution in flowering plants. Following continental separation, evolutionary development continued to produce genera, species, and occasional families that diverged significantly from their progenitors. Yet, within a genus—the oaks, for example—new species evolved in isolation so that there are unique floras on different continents which can, nevertheless, be traced back to ancestral types that were the same for both species constellations.

ADDITIONAL READINGS

Briggs, D., and S. M. Walter. *Plant Variation and Evolution.* Cambridge, England: Cambridge University Press, 1984.

Cairns-Smith, A. G. *Seven Clues to the Origin of Life: A Scientific Detective Story.* New York: Cambridge University Press, 1985.

Grant, V. *Plant Speciation.* New York: Columbia University Press, 1981.

Moore, D. M. (ed.). *Green Planet: The Story of Plant Life on Earth.* Cambridge, England: Cambridge University Press, 1982.

Nelson, G., and N. Platnick. *Biogeography.* Burlington, N.C.: Carolina Biological Readers, 1984.

Niklas, K. J. *Paleobotany, Paleoecology and Evolution.* New York: Praeger, 1981.

Stewart, W. N. *Paleobotany and the Evolution of Plants.* Cambridge, England: Cambridge University Press, 1983.

Sullivan, W. *Landprints: On the Magnificent American Landscape.* New York: Times Books, 1984.

Takhtajan, A. L. "Outline of the Classification of Flowering Plants (Magnoliophyta)." *Botanical Review* 46, no. 3 (1980):225–359

Woese, C. R. *The Origin of Life.* Burlington, N.C.: Carolina Biological Readers, 1984.

How Plants Get Their Names

What's in a name? That which we call a rose
By any other name would smell as sweet.

WILLIAM SHAKESPEARE
Romeo and Juliet

The human species seems to be driven to give things names and to arrange them in some orderly fashion. Witness the 500 different plants classified as trees, shrubs, undershrubs, and herbs by Theophrastus (2372–2287 B.P.) in his *History of Plants,* or the classification by Pliny the Elder (23–79) in his *Historia Naturalis,* in which categories included medicinal plants, trees, and agricultural plants. The herbalists of the Middle Ages up to the seventeenth century paid only minor attention to the biological relationships of plants. These writers did, however, categorize plants by their medical use; their work, although not then recognized as reflecting relationships, formed a foundation for later classification systems. With the Renaissance, botanists cautiously began to dissociate themselves from still powerful institutionalized traditions and began to reorganize classification into a more rational form.

NEED FOR A CLASSIFICATION SYSTEM

By the middle of the seventeenth century, Spain, Holland, France, and Britain were well embarked on their imperialistic endeavors. Tremendous voyages of exploration and conquest extended the hegemony of these nations over lands scattered about the world. The treasures of the subjugated countries, including their plant wealth, were pouring back to Europe. The Royal Society of London, organized in 1660, prevailed upon the British Admiralty to allow biologists to accompany the fleets to collect plants and animals for universities, museums, botanical gardens, and zoological parks. Plant hunters were commissioned by scientific organizations, by ministeries of agriculture, by admiralties looking for suitable plants for masts and other naval stores, by medical societies, by industries searching for new sources of raw materials, and by wealthy patrons hoping to be the first in their neighborhood to have a new plant in their gardens. In keeping with the ideas of this Age of Reason, the enthusiastic Fellows of the Royal Society believed that they could understand the world by cataloguing the contents of the earth—animal, vegetable, and mineral. They believed that by developing appropriate inventories, by documenting and preserving specimens and by working out a system of plant nomenclature, it would be possible to classify all of the plants then known and also to accommodate the plants that they knew would eventually be discovered. It was a grand dream, and their courage should be applauded.

It was apparent that certain conditions had to be met and rules had to be applied to any classification system. The language had to be in the scholar's unchanging Latin; the common names that people had used for centuries simply wouldn't do. For instance, despite its name the Jerusalem cherry isn't a cherry. It looks like a miniature tomato plant and, like the tomato, came from South America and not the Holy Land. There are Jerusalem artichokes, a weedy Jerusalem oak that looks like our common pigweed, another Jerusalem cherry that is really a hot pepper, plus another 50 or so plants that are named Jerusalem something or other. The Scotch pine isn't from Scotland, but originated in Asia Minor. Poison oak isn't an oak, club mosses aren't mosses, and eggplant isn't related to hen fruit. There are many examples of completely different plants that bear exactly the same common name in different countries. Linguistic variations had to be eliminated; your *moss rose* might be my *portulaca* and her *purslane.* Any classification system had to allow for pigeonholing so that it would be easy to put like with like. The system selected also had to permit rapid information retrieval. How can one determine whether a dried plant from Tierra del Fuego is different from a salt-incrusted specimen which had been in transit for several years from the Malay Straits? Are there similarities between plants from Brazil and those from the west coast of Africa? There had to be some standardizing procedure to ensure, first, that the plant namer, the taxonomist, could determine in which group to put a hitherto undescribed plant and, second, that the same plant wouldn't be given a different name by a different taxonomist.

LINNAEUS

It is usual to credit Linnaeus with the genius to put plant classification into the form in which it is used today. It is also appropriate to stress the fact that the structure of plant classification which he developed was based, as are almost all seminal advances in science, on the intellectual labors of people who preceded him. Certainly, the herbalists played a role: they demonstrated the need for classification of herbs used medicinally. Andrea Cesalpino (1519–1603), director of the Botanical Garden at Bologna, developed the technique of preserving plants on sheets of heavy paper so that they could be stored for comparison with other plants—the technique leading to documentation and preservation. Joachim Jung (1587–1657) publicly questioned the significance of the classification schemes of the ancient Greeks and Romans, an act requiring considerable courage, and invented names for plant parts that are still in use. John Ray of Britain (1627–1705) recognized the differences between monocots and dicots. Pierre Mangol (1638–1715) and Joseph de Tournefort (1656–1708), both of France, developed concepts of the plant genus and the plant family, important hierarchical groups in classification.

The man who put it all together was the son of a Lutheran pastor, educated in medicine at Uppsala University in Sweden and later professor of medicine and botany at Uppsala: Carl Linnaeus (1707–1778). He perfected the hierarchical classification system starting with the species and going up to the kingdom—the ultimate grouping that separates plants from the animals and the fungi. The concept of giving each new kind of plant a double name—a binomial—had been conceived in 1623 by the Swiss Caspar Bauhin, who, following the idea that people have two names, called all of the species of the rose *Rosa* (a genus name) and then delimited each rose with a species name or specific epithet. Linnaeus adopted this idea enthusiastically because it gave him the two primary levels in his hierarchy, the species and the next level, the genus.

John Ray and Joachim Jung had earlier suggested that in order to classify plants, one must have some stable characteristic of the plant. Growing conditions, climate, and other factors can easily result in variations in size and form of stems and leaves, but flowers are the plant structure least susceptible to environmental modification. Using the flower and its parts also eliminated the problem that was noted by Aristotle and Theophrastus when they used the tree-shrub-herb classification system. Plants that were undoubtedly related, on the basis of the arrangement of their flower and other plant parts, could easily be separated into categories. There are many plants whose floral organization demonstrates a clear relationship to the common rose, but apples and cherries are trees, blackberry and cinquefoil are shrubs, strawberry is an herb, and some, like spiraea, have species in all three groups.

Linnaeus's great books, *Species Plantarum* and *Genera Plantarum,* appeared between 1727 and 1753. His work adequately addressed all of the questions relating to a taxonomic scheme and was almost immediately adopted by all biologists. It still forms the basis for our classification of organisms. In Uppsala, Linnaeus built a collection of plant specimens on sheets of paper, a herbarium, which is the base for the naming of all plants. Using the Linnean system, a taxonomist who believes that she or he has discovered a species new to science will prepare a concise description of the plant, make herbarium specimens, and then compare these with published descriptions and with a previously described and properly identified herbarium specimen to determine if the plant has already been described and named. There are hundreds of herbaria throughout the world, with some being major repositories of named plants. Today, much of this information is available on computer printouts.

An example will show how the classification system works. Comparing the structure and organization of the flower of the oaks with those of other genera of plants, it was apparent that those of the beeches *(Fagus)* and those of the chestnuts *(Castanea)* are similar and all three genera can be grouped in the next higher category, the family—in this case, the family Fagaceae. Families can be further grouped into Orders, Orders into Classes, and Classes into Divisions. At the other end, species can be subdivided into subspecies or varieties or forms that reflect variations within a species that are too minor to allow species differentiation. It is now common practice to recognize genetic uniformity of cultivated plants under the name of *cultivar,* abbreviated *cv.* The cv taxonomic grouping allows the categorization of the many kinds of corn or tomato or beans which people have selected or bred (Table 1.2). The taxonomic classification system is logical and, most important, it works.

MODERN SYSTEMS: THE EVOLUTIONARY CONNECTION

The classification of flowering plants—the science of taxonomy—is an attempt to put like with like, essentially to pigeonhole all known plants into groups *(taxa)* that reflect their structural (usually floral) similarities.

Table 1.2 A COMPLETE TAXONOMIC CLASSIFICATION OF THE RED KIDNEY BEAN*

KINGDOM:	Plantae
DIVISION:	Angiosperm**ae**
CLASS:	Dicotyled**oneae**
SUPERORDER:	Ros**idae**
ORDER:	Fab**ales**
FAMILY:	Fab**aceae**
SUBFAMILY:	Papilion**oideae**
TRIBE:	Phaseol**eae**
SUBTRIBE:	Phaseol**inae**
GENUS:	*Phaseolus*
SPECIES:	*vulgaris*
VARIETY:	humulis
CULTIVAR:	'Red Kidney'

*Boldface endings are standarized indications of the ranking within the hierarchy.

Other stable traits, or characters, are also used. Phylogeny, although dependent upon taxonomy, has a different mission. Phylogeny attempts to show how plants in one taxonomic group—anywhere from subspecies up through the entire plant kingdom—are related in an evolutionary sense. An underlying assumption in modern classification systems is that when plants are placed in the same taxonomic category, they are evolutionarily related. This works well for taxa of species and genera and sometimes up to the family or the order, but then it becomes difficult to determine whether one plant family or order is derived evolutionarily from another family or order.

This was not much of a problem before Charles Darwin; most pre-Darwinian taxonomists believed that plant cataloguing reflected God's plan of special creation and that lineages didn't exist. The nineteenth century, however, saw the glimmering and then the flowering of evolutionary theory. J. B. P. de Lamarck (1744–1829), the de Jussieu family of four related botanists, the De Candolle father and son team, the German collaborators Engler and Prantl, and the British botanists George Bentham and Joseph Hooker attempted to reorganize the species-genus-family taxa into orders and classes that reflected their presumed evolutionary lineages. As Darwin's ideas gained full acceptance by the scientific and intellectual communities, renewed efforts at devising a scheme that expressed the genetic and evolutionary relationships of taxa were made. This process is still going on as information from genetics, physiology, ecology, and other scientific disciplines becomes available. The attempts to include lineages in taxonomic schemes is fundamental to our understanding of the evolutionary process, which is, after all, the foundation upon which all of biological science rests.

The species is the basis for a modern plant classification system. *Species* is a Latin word meaning "a particular kind or sort." Species are, by definition, the smallest group of individuals that are hereditarily different from any other group, are interfertile, and are perpetuated over generations in their uniqueness. Over time, a species may diverge genetically and one species may evolve into two or more, each possessing the unique characteristic of a species. There is a low but statistically determinable probability that any cell or any organism can, through genetic mutation—a permanent alteration in the genes of a chromosome—acquire or lose the ability to make any particular enzyme, compound, process, or structure. The alterations in the genetic constitution leading to structural and functional changes usually involve a series of mutations. Equally important, these mutations are undirected and occur by chance. The "need" for a particular structure or function cannot activate specific mutations; environment cannot instruct or imprint genetic information upon the genes or chromosomes, for genetic information is self-generating.

Phylogeneticists attempt to trace the relatedness of species or of higher taxonomic categories by an examination of the structure, function, and ecology of organisms that might be ancestral to the species being studied. In many instances, a coherent picture can be obtained. If, for example, several species in a particular genus are found in one geographical area, it is reasonable to assume that they are descended from a common species that had, itself, evolved in that area or invaded the area. An isolated island or a valley surrounded by high mountains may develop a unique flora or fauna because of *adaptive radiation*. That is, a few organisms reaching the new area evolved into forms that are adapted to various habitats provided by the geography and climate. Some of the most dramatic examples of adaptive radiation are seen in the variation of the Darwin finches of the Galápagos Islands. From a single species, bird phylogeneticists have been able to trace the adaptive radiation that culminated in finches that look, act, and feed like our woodpeckers, seed eaters, or insectivores. Similar evolutionary patterns have been found in other animal taxa and among groups of plants.

It is not particularly difficult to visualize how, by mutation, isolation, and environmental selection, the original, archaic species of the rose could be the direct and indirect progenitor of all members of the genus *Rosa* and of the family Rosaceae. The primordial grass—to take another example—could give rise over a long period of time to genera as diverse as wheat, lawn grasses, bamboo, sugar cane, and rice. Within a family, some genera or species may have evolved by complications of the basic structure; maize *(Zea)* cobs may be an example of this (Chapter 3). Simplifications in structure or function—internal parasites of animals, for example—can also be fitted into the evolutionary scheme. But in order to make broader phylogenetic interrelationships, many criteria must be used, including, as in the algae, the spectrum of pigments formed, the type of cell division, the presence or absence of starch or oil, or the chemical constitution of cell walls. This is a difficult job and is, as you would suspect, not free of disagreements and controversy, but, as noted earlier, is as intellectually satisfying and as worthy of study as any other intellectual activity.

FINDING A NAME

Even with a taxonomic system, choosing a name is, as any parent can testify, a difficult problem. By the time Linnaeus started his work, about 5000 plants were widely known in Europe, most of them with common names—pea, wheat, barley, birch—that were in everyday use. People distinguished between different types of willows, oaks, or beans and Linnaeus found it convenient to simply Latinize these genus names so that the willows became *Salix,* the oaks were *Quercus,* and the violets were *Viola.* Linnaeus classified, using the Latin binomial system, over 18,000 plants, but by 1790 there were still 50,000 that had not been catalogued, and the Latinization of common names was reaching a dead end. One way of handling this problem is to shuffle names around. For example, let's look at the junipers, whose genus name is *Juniperus. Juniperus chinensis* is a species from China, *J. japonica* is from Japan, *J.*

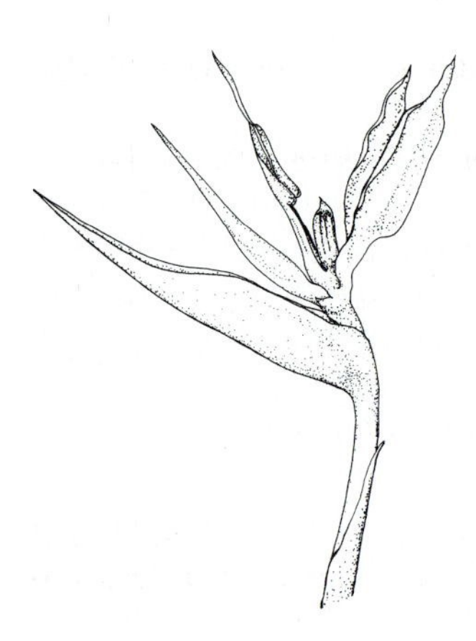

Figure 1.5 Bird-of-paradise flower *(Strelitzia reginae),* a member of the banana (Musaceae) family. On this brilliantly colored plant, the bright blue "tongue" of two fused petals encloses yellow stamens and is joined to the orange-colored third petal.

occidentalis is from California, and *J. virginiana* is from eastern North America. *Arabica, californica,* and *pennsylvanicus* are from other obvious place names. Many species names reflected the use of the plant; *vulgaris* (common or ordinary), *deliciosa, edulis (edible), cereale.* A spring flowering plant might be *vernalis,* while a plant that flowered in the fall would be *autumnalis.*

Many plant names reflect an outstanding characteristic of the plant. *Thymus* is the Latinized genus of the common thyme, and *T. serphyllum* is creeping thyme (serpent and leaf). Sandwort is *Arenaria serphyllum; Houstonia serphyllulm* is a creeping bluet. *Microphyllum* means small leaf; there is a *Thymus microphyllum* and also a big-leaved thyme *T. macrophyllum. Glaucum* means smooth, *hirsute* is hairy, and *pubescent* is fuzzy. *Rotundifolium* means round-leaved, *quinquifolium* is five-leaved, and so it goes. Color names are as common as *Ms. Brown* or *Mr. White* except that the names are Latinized. The yellow birch is *Betula lutina,* black birch is *Betula nigra,* and white birch is *B. alba. Acidus* means sour, *altus* is tall, and *barbarus* is foreign. *Denticulatus* is slightly toothed, *dentifera* is tooth-bearing, and *bidentalus* is two-toothed—not especially imaginative, but very effective. Some names roll grandly off the tongue. How about *entomophilis* (insect-loving) or *erythrocephalus* (red-headed) or *foetidissimus* (very foul-smelling)?

People's names, Latinized of course, are very common. One can honor one's professor, spouse, good friend, or lover. Names have been given plants for political and economic reasons. The South African bird-of-paradise plant (Figure 1.5), *Strelitzia,* was named for Charlotte Sophia, wife of England's George III. She was from the German house of Mechlenberg-Strelitz, and the plant was named by a man who was then given a soft, well-paying job as royal gardener. The poinsettia is a French family name once removed. Joel Roberts Poinsett was an American from South Carolina who, as a diplomat, was sent to Chile in 1809 and brought the plant back to the United States. A specimen ended up in a German botanical laboratory where it was named in Poinsett's honor.

It is also customary to append the name (abbreviated) of the person who first applied a particular scientific name to a plant. The "L." following many plant genus and species names refers to Linnaeus, while the authority of the river or black birch *Betula nigra* Michx. f.) indicates that it was first described by François Michaux, a French botanical explorer of colonial North American and the son *(fils)* of André Michaux, who also botanized in North America.

ADDITIONAL READINGS

Bailey, L. H. *How Plants Get Their Names.* New York: Macmillan, 1933.

Bold, H. C. *The Plant Kingdom.* 4th Ed. Englewood Cliffs, N.J.: Prentice-Hall, 1977.

Coombes, A. J. *Dictionary of Plant Names.* Beaverton, Ore.: ISBS/Timber Press, 1986.

Cronquist, A. *The Evolution and Classification of Flowering Plants.* Boston: Houghton Mifflin, 1968.

Gardner, E. J. *History of Biology.* 2nd Ed. Minneapolis: Burgess, 1965.

Grant, W. (ed.). *Plant Biosystematics.* New York: Academic Press, 1984.

Heywood, V. H., and D. M. Moore (eds.). *Current Concepts in Plant Taxonomy.* New York: Academic Press, 1984.

Jones, S. B., and A. E. Luchsinger. *Plant Systematics.* New York: McGraw-Hill, 1979.

Kartesz, J. T., and C. R. Bell. "A Summary of the Taxa in the Vascular Flora of the United States, Canada and Greenland." *American Journal of Botany* 67 (1980):1495–1500.

Klein, R. M., and A. Cronquist. "A Consideration of the Evolutionary and Taxonomic Significance of Some Biochemical, Micromorphological, and Physiological Characters in the Thallophytes." *Quarterly Review of Biology* 42 (1967):108–296.

Parker, S. P. (ed.). *Synopsis and Classification of Living Organisms.* New York: McGraw-Hill, 1983.

Plowden, C. C. *A Manual of Plant Names.* New York: Philosophical Library, 1970.

Scagel, R. F., R. J. Bandoni, J. R. Maze, G. E. Rouse, W. B. Schofield, and J. R. Stein. *Plants: An Evolutionary Survey.* Belmont, Calif.: Wadsworth, 1984.

Wooster, M. S. *What's in a Name: A Guide to Botanical Names.* Montclair, N.J.: Garden Clubs of Montclair, 1982.

Cells

The body is a State in which each cell is a citizen.

RUDOLPH VIRCHOW

Plant organs were obvious to fifteenth-century observers, and the general organization of roots, stems, leaves, and flowers could be made out, even if dimly. Robert Hooke looked at a thin slice of cork under a microscope in 1665 and reported in his *Micrographia* that the pores or cavities into which cork was partitioned looked like the rooms or cells in which monks lived (Figure 1.6). Hooke turned his microscope on bits of charcoal, fruits, and on vegetables and found that this same "schematisme" or organization was present. Almost in passing, he noted that fresh tissues from green plants are "fulled of nutritive juices." Almost 200 years later it was recognized that the living cell was the juice and what Hooke called a cell was the nonliving cell wall, an excretion product of the living matter in many ways comparable to fingernails and hair.

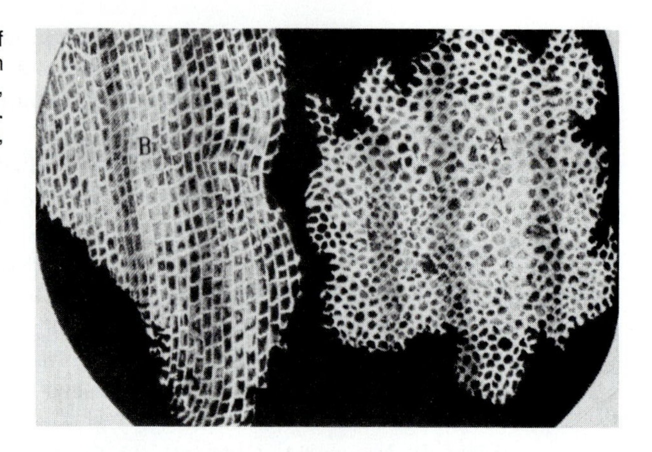

Figure 1.6 First illustration of plant cells, those of cork. From *Micrographia* by Robert Hooke, published in London in 1665. Courtesy Dr. Arthur Cronquist, New York Botanical Garden.

Mathias Schleiden, professor of botany at the University of Jena, announced in 1838 to a most skeptical scientific world that the cell, the fluid stuff, was the fundamental living unit of all plant structures, and that leaves, stems, and roots were made of similar units. As an additional heresy, he suggested that botany was more than classifying plants and that the examination of how these units came into being and how they were organized would yield much useful information. He pointed out that each cell was a self-contained unit and that each new cell added to the structure of organs must be derived from a preexisting cell. These assertions were previously made by several others. Within a year, Theodore Schwann, professor of anatomy at Louvain University, extended Schleiden's cell theory to animals. "Growth," he wrote, "is

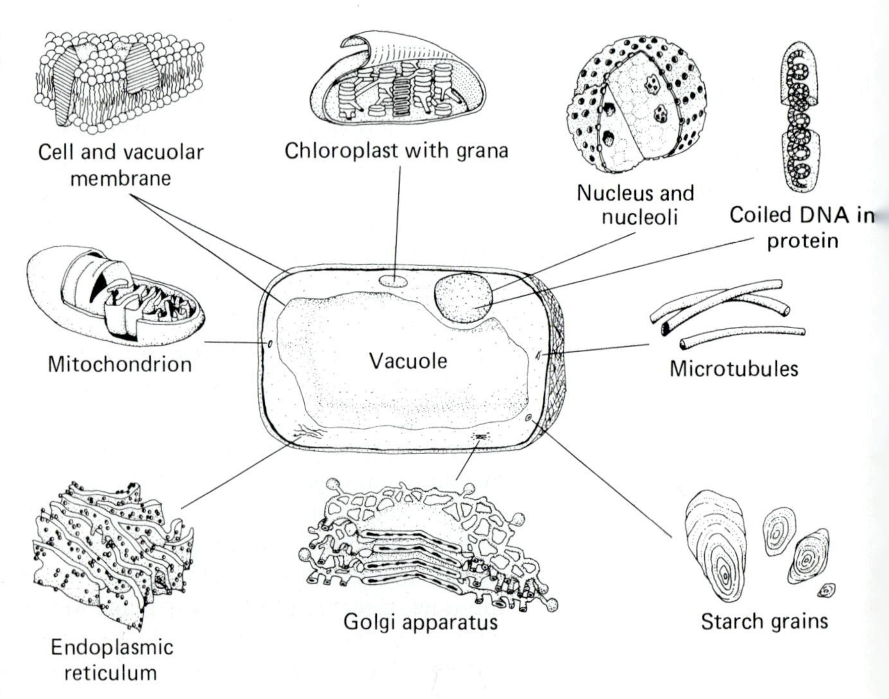

Figure 1.7 Diagram of the organization of a plant cell and the fine structure of cell organelles.

not the result of a force having its ground in the organism as a whole, but each of the elementary parts possesses a force of its own, a life of its own, if you will . . . the totality of the organism can indeed be a condition, but on this view, it cannot be a cause."

As microscopes improved in the late nineteenth century, botanists turned them on all sorts of living tissues, building up a picture of what Hooke's "nutritive juices" actually contained. The plant cell could be viewed as a box. Indeed, the wall about the living cell is made up of several layers of cellulose fibers impregnated with other chemicals, the whole box being essentially wood, for wood in the modern sense is composed of the dead cell walls of specialized plant cells.

Directly inside the cell wall is a thin membrane, the *cell membrane,* which encloses the rest of the living cell (Figure 1.7). Composed of a sandwich of fat and protein, this bounding membrane is part of the living cell that serves to regulate the movement of water in and out of the cell, excluding some chemicals and facilitating the passage of others. It is extremely sensitive, losing its functional integrity with heat (the release of red juice when beets are boiled), with cold (when frozen strawberries are thawed), or with chemicals (when lettuce wilts as vinegar is added to a tossed salad). There is a second, structurally similar membrane that serves as the other boundary of the living cell—the *vacuolar membrane*—which has the functions of regulating the movement of dissolved material into the *vacuole.* Plant cells are different from animal cells in the possession of a vacuole, and this difference bears directly on how plant cells enlarge.

THE VACUOLE

Within any animal organ—the liver, for example—there is little variation in the sizes of cells of a given type; a cell divides and its progeny reaches mature size. Plant cells of a given type—cortical cells in a stem, for example—may vary by a factor of three to five in size, starting out as small units and growing in volume as water accumulates in the vacuole, much as a balloon increases in size as air is blown into it. There is, however, one important difference between the balloon and the cell. As the plant cell enlarges through water uptake, its cell wall also grows. Thus, for a plant cell to grow in length and/or diameter, not only must water accumulate in the vacuole by expedited passage through the cell and vacuolar membranes, but the cell wall must loosen and new cellulose fibrils be added. Both processes involve growth-regulating chemicals produced primarily in the stem and root tips and in the young leaves.

Vacuoles are of variable size. At one extreme are individual cells in which each cell has a huge vacuole containing water, and at the other are newly formed cells of meristems which lack a vacuole, forming one only when the cell enlarges. Although as cells mature their cell walls become thicker and stronger, a plant like a coleus or a geranium does not have sufficient wall strength to maintain its upright position without the water pressure that develops as stem cell vacuoles accumulate water. If water is limited, individual cells lose their vacuolar water, water pressure is reduced, and the plant wilts. Cells with high vacuolar water pressures are said to be *turgid*; the opposite condition is termed *flaccid.* A plant under *water stress,* when water is suboptimal, will do poorly and will grow slowly, with individual cells failing to achieve their maximum potential for elongation. Water stress can occur from many causes: inadequate soil moisture, a damaged root system, derangement of water translocation, and excessive water loss through the leaf stomata. Moreover, the vacuole is not merely a passive bag of water, but is a participant in the metabolic economy of the cell. For instance, in orange fruits citric acid and sugars accumulate in the vacuole as side products of cell metabolism. The brilliant red pigments (anthocyanins) in geranium petals or the equally bright betacyanin pigments of red beets are dissolved in the vacuolar sap. Tannins used in leathermaking and a variety of crystals are also found in the vacuole.

THE CYTOPLASM

The living material of a cell is bounded by the two membrane systems with the *cytoplasm,* consisting of an emulsion of protein dispersed in water, much like Jello without artificial colorings and flavorings, between the two membranes. These proteins are not inert, but are enzymes, each functioning to synthesize, transform, or degrade one of the multitude of compounds that exist within cells. To some extent, cytoplasm can be simulated by adding appropriate enzyme proteins, fats, sugars, and so on to water, but it doesn't behave like a living system because the reactions that occur in the test tube are neither regulated nor integrated, and life is both. We don't know all the constituents of living stuff, how it is organized and how it is integrated. Perhaps some day. . . .

THE CHLOROPLAST

Within the cytoplasmic fluid-gel are discrete bodies, each bounded by the same fat-protein membrane system that delimits the cell and the vacuole. A few of these *organelles* (little organs) were large enough to have been seen rather early; Robert Brown observed in 1831 a large, roundish organelle he called the nuclear body, although he didn't think it was important. Green bodies, somewhat football-shaped, were seen even earlier and given the name *chloroplastids*—green bodies. It was obvious that they had something to do with the self-feeding (autotrophy) of plant cells.

By 1850–1860 it was agreed that all plant and animal cells contained a nucleus and that only plant cells contained chloroplastids. This was as far as observations could go until the latter part of the nineteenth century when light microscopes became powerful enough to magnify a thousand-fold with good resolution. With this new tool botanists were able to make out smaller bodies. Individual starch grains could be stained with tincture of iodine and visualized as small blue-black dots in the cytoplasm. Small ovoid bodies, about the size of bacteria, were named *mitochondria* (a Greek construction: *mitos* = thread; *chondrus* = cartilage), although they were frequently confused with bacteria and their function was unknown. An Italian microscopist, Camillo Golgi, saw what appeared to be roundish blobs with rough edges, and this organelle, the *Golgi apparatus,* was named for him. A series of threads and tubes, the *endoplasmic reticulum,* could be seen if the microscopist strained very hard and had a good imagination.

The invention of the electron microscope in Germany in the 1920s and its perfection in the late 1940s allowed magnifications of over 400,000 times, permitting confirmation of earlier findings and descriptions of the fine structure of small organelles. As high-speed centrifuges and new biophysical techniques became available, organelles could be removed from the cells for detailed examination under the electron microscope and used for experimental study of function. Biochemists began to probe their functions, ushering in an era of macromolecular biochemistry that is still an active research area.

These cell organelles are found in both animal and plant cells, and although there are structural and functional differences between those in the cells of tigers and of tiger lilies, they are sufficiently alike so that researchers who work at the cell level can study either plants or animals with a reasonable assurance that the information obtained is applicable to both. The chloroplast, however, is unique to plants.

In spite of tremendous diversity in life styles, in habitats, and in form among various plants, the internal structure and the function of the cholorplasts is virtually identical. Surrounded by a double protein-fat membrane, the chloroplast contains a semiliquid gel in which an array of enzymes active in photosynthesis are found. The chloroplast pigments are contained on and in stacks of flattened hollow disks, each disk called a *thylakoid,* each stack called a *granum;* the granum resembles a stack of coins. Thylakoid membranes are similar to other plant membranes in that they are composed of protein and fatty substances to which the pigments are attached. The chemical composition of individual pigments is known with some precision. There are two chlorophylls, one bright green (chlorophyll *a)* and the other a blueish-green (chlorophyll *b).* Several chemically related yellow xanthophyll pigments, orange-red carotene pigments, and bright red cytochrome pigments are also present. The pigments and nonpigmented chemicals are spatially arranged to facilitate the capturing of light energy and the transfer of chemical energy that occurs in the process of photosynthesis.

Photosynthesis Photosynthesis is the most efficient energy-converting process on earth; the transformation of the physical energy of light into

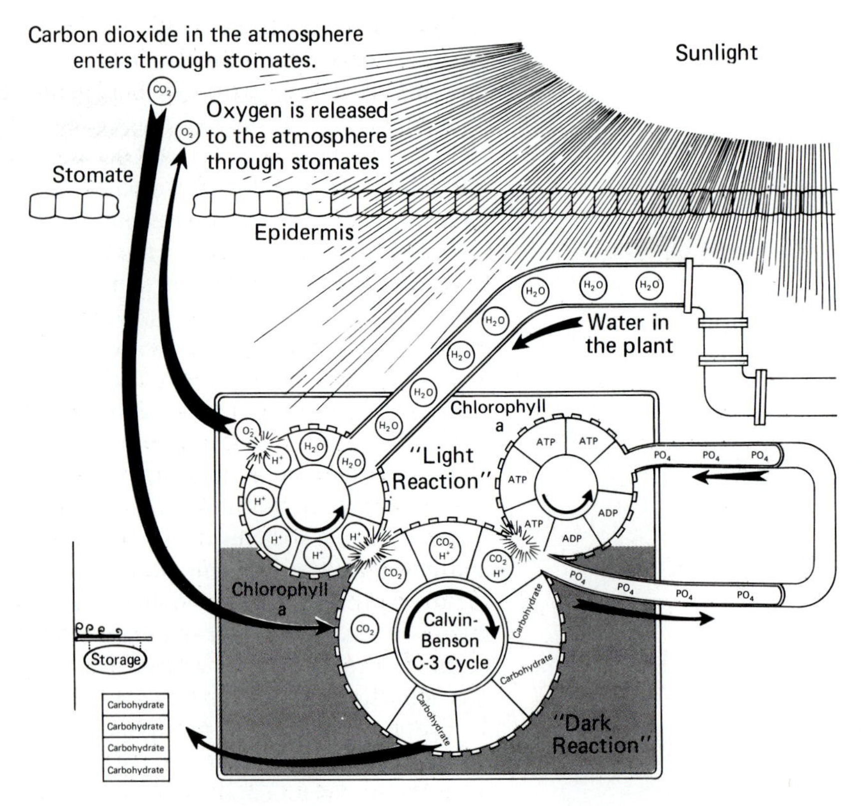

Figure 1.8 Photosynthesis as a chemical factory. In chloroplasts (the rectangular box), light energy is used by chlorophyll to split water to form molecular oxygen and hydrogen ions (H*). In a second set of light reactions, chlorophyll is excited, and the excitation energy is used to link phosphate ions to form the high energy compound ATP. Both the H* and the ATP are then used in the dark reactions to convert atmospheric carbon dioxide into carbohydrate. Reproduced from *The Garden Journal* of the New York Botanical Garden, October 1974, with permission of the author.

the chemical energy bound into a sugar molecule can approach 20 percent. The overall process can conveniently be divided into two phases, the first requiring light—the light reactions—and the second phase independent of light—the dark reactions, which occur either in light or in darkness (Figure 1.8). Light, primarily red and blue wavelengths, is captured by the green chlorophyll molecules contained in the membrane system of the chloroplast (which is why leaves look green; the red and blue wavelengths of sunlight have been absorbed by chlorophylls, and the green wavelengths in white light are either transmitted through the leaf or reflected from its surface). Light absorption can also occur in other pigments in the chloroplasts, especially the orange carotenes and yellow xanthophylls, and the energy is transmitted to the chlorophylls. When radiant energy is absorbed by

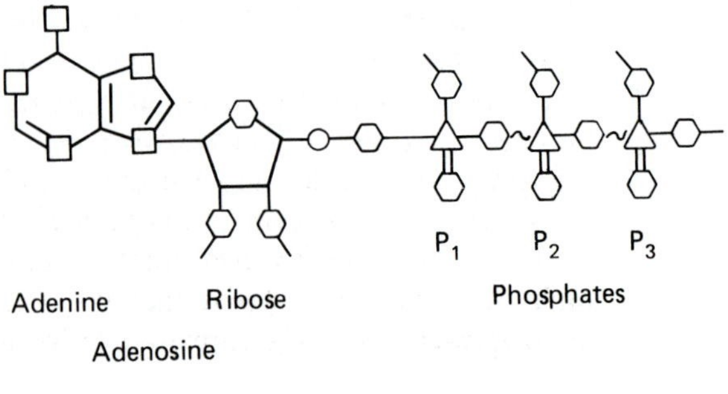

P_1 P_2 P_3

Adenine Ribose Phosphates

Adenosine

Figure 1.9 Molecular structure of adenosine triphosphate (ATP), the primary biological energy source. Hydrogen atoms and carbon atoms in the rings are not shown. Carbon atoms are represented as circles, oxygen atoms as hexagons, nitrogen atoms as squares, and phosphorus atoms as triangles.

chlorophylls, electrons in the molecules are activated and shift to a higher energy state, roughly analogous to lifting a rock from the ground into the air. The electrons are held in the "excited state" for a fraction of a second and, like the rock suddenly released, can fall back down into the chlorophyll molecule with the excitation energy released. If, however, the electrons trickle through a series of compounds, much as water cascades through the paddles on a water wheel, the potential energy can be captured into chemical bonds of a compound, abbreviated as ATP (adenosine triphosphate), required for the photosynthetic dark reactions (Figure 1.9). In another set of light-driven reactions, excited electrons activate another compound, PN (pyridine nucleotide), which can hold the hydrogen ions also needed for the dark reactions. These hydrogen ions are obtained by a chlorophyll-mediated splitting of water with the oxygen from the water released from the plant as the molecular oxygen we and most other organisms must have. The speed with which these reactions occur is measured in milliseconds, and, when each complete cycle is finished, the chlorophyll molecules are again ready to absorb light and repeat the process. Since there are thousands of active, light-reaction centers in each granum of a chloroplast, thousands of grana in a chloroplast, and thousands of chloroplasts in a leaf, the number of ATP and PN molecules produced per minute is numbered in the tens of millions per plant.

The raw material for formation of sugar through the dark reaction is the gaseous compound carbon dioxide (see Figure 1.8). Composed of two oxygen atoms connected to a single carbon atom, CO_2 is normally present in air at the low concentration of 0.035 percent. During the day, when the "little mouths," the stomata, are open, CO_2 diffuses into the leaf cells and moves to chloroplasts. There it is attached to a carrier molecule, a 5-carbon compound called "RuBP (ribulose bisphosphate). The carbon atom in CO_2, being attached to two oxygen atoms, is in a fully oxidized state in which it cannot link up with other carbons to form the 6-carbon chain of a simple hexose sugar such as glucose. To permit chain formation, not only must the oxidation level of carbon be reduced by shifting carbon-oxygen bonds, but hydrogen atoms must also be added. Both of these reactions are accomplished by the actions of the two compounds produced in the photosynthetic light reactions; the hydrogen atoms carried by PN are added to the carbon with required energy supplied by the ATP molecules. This "reductive fixation" of CO_2 into a carbohydrate is a cyclic process, with CO_2 entering the chloroplastid, linking to the RuBP carrier, being reduced with energy from ATP, and then being released in solution into the cell with the carrier freed to accept another CO_2. The analogy with a factory taking in raw material, putting it on an assembly line, fabricating the end product, and shipping it to the consumer is quite apt.

There are, however, built-in inefficiencies in the photosynthesis factory. One is the low concentration of CO_2 in air. The efficiency of the light reactions is so great that there are excesses of both ATP and hydrogen-carrying PN formed relative to the CO_2 available. In some plants like the jade plant *(Crassula)* and the stonecrops *(Sedum)*, stomata remain open at night, with the incoming CO_2 held as an acid which can release the CO_2 again when the light reactions begin in the morning. This is called the CAM (crassulacean acid metabolism) pathway. In other plants, including corn, sorghum, bamboos, and sugar cane, the limitation of low CO_2 is overcome by a biochemical mechanism in which CO_2 is taken up by special cells and held until it can be used in the light reactions. This is called the C_4 pathway. Another saving is achieved by the repression of photorespiration, a process in which sugars are metabolized rapidly in light but not in darkness. By increasing the amount of CO_2 available for photosynthetic sugar formation and by preventing the loss of sugar by photorespiration, more sugar is made. Some of this sugar is used as the energy source for the rapid growth of the sugar cane plant, some is used as units which are polymerized into cellulosic plant cell walls, and some is used to synthesize amino acids and other compounds that comprise a plant cell, but so much sugar is formed that it accumulates in the vacuoles of cells, available for human exploitation.

Autumnal Coloration In the eastern provinces of Canada and adjacent states of Maine, New Hampshire, Vermont, and northern New York, October is known as "leaf season," when these areas are invaded by people looking at the colors of leaves on maples, oaks, sumacs, and other deciduous (leaf-dropping) trees. Leaf season is of economic importance to the tourist industry, and a good year may bring a million people and several million dollars into the state of Vermont alone. It isn't a case of "Well, if you've seen one tree, you've seen them all," since natives of Sherbrooke, Quebec, or Burlington, Vermont, also drive or walk through the woods with as much pleasure as the visitor from British Columbia or Florida. The brilliance of the foliage is unique not only because of the tree species involved but also because the fall weather in the Northeast is nearly perfect for the development of the colors.

Careful analysis of the colors that, together, provide the beauty of an autumn sugar maple leaf indicates that there are five basic colors. The maple leaf will show green, orange-red, yellow, bright red, and brown; and from these five colors can be obtained the delicate shadings and mixtures that bring dollars to the economy and pleasure to the viewer. Three of these colors are present in the leaf throughout the spring and summer.

The summer leaf is seen as bright green because the chlorophyll molecules preferentially absorb blue and red wavelengths of sunlight, allowing green wavelengths to be reflected from the leaf or transmitted through it. The high concentration of chlorophylls in chloroplasts masks or covers up the orange-red pigments of carotene (seen in its full hue in the carrot) and the yellow pigments of the xanthophylls (seen in yellow Delicious apples and in daffodil petals). As days get shorter and cool nights occur, chlorophyll is broken down faster than it is synthesized, and oranges and yellows become visible. Hues of brown are the result of breakdown products of chlorophylls and other leaf chemicals. Just these four colors give the leaves a good deal of color; the wonderful yellow-golds of poplars gleaming through the deep green of conifers in the Rockies represent the amalgam of these four colors. But a fifth pigment can bring Staghorn sumacs into flame and add that special touch to sugar maples on a hillside. The red pigment is anthocyanin, the pigment of geranium petals, Macintosh apple skin, and rhubarb petioles.

As both day and night temperatures continue to decline, all pigments fade as leaf enzymes degrade them. As cells lose their capacity for biosynthesis, they no longer synthesize growth substances that can move down the leaf into the petioles, which connect the leaf blade to the stem. With reduction in the concentration of these substances, special cells at the base of the petiole become corky and die, and the leaves fall (abscise) from the trees. Winter is well on its way. Skiing, anyone?

THE NUCLEUS

The most prominent organelle is the nucleus, a spheroid-shaped body enclosed in a double membrane with many fine pores and containing the *chromosomes* (the hereditary apparatus of the cell) and one or more smaller bodies, the *nucleoli.* Chromosomes are made up of the nucleic acid DNA, arranged as a coiled helix or spiral surrounded by a protein overcoat. The number of chromosomes per nucleus is constant for any particular plant or plant species, although the number may be as few as four or as many as several hundred. With certain exceptions, all cells in a plant have the same number of chromosomes and each cell contains exactly the same genetic information, which raises exciting research questions of how and why cells become functionally and structurally specialized. Based on the arrangement of four different chemicals in the DNA subunits (two purines and two pyrimidines), the genetic information is coded in such a way that it can be transcribed onto the other type of nucleic acid—RNA (ribonucleic acid)—which is synthesized in the nucleoli. These RNAs move out into the cytoplasm through nuclear pores and direct the synthesis of protein enzymes which, in turn, control all cellular activities. Protein synthesis itself takes place on the *ribosomes,* small spheres attached to the endoplasmic reticulum.

Multiplication of cells occurs by division, a process called *mitosis.* Typically, the dual (diploid) set of chromosomes in the nucleus of a dividing cell duplicates itself, giving four sets and then transferring a diploid set to the new cell formed by cytoplasmic division. The evolutionary logic of this process is impeccable; it ensures not only that each new cell is the product of a preexisting cell but also that the new cell has exactly the same hereditary complement as its progenitor. Stability and uniformity are maintained. Variation, with the consequent possibility of giving rise to a cell better adapted to the totality of the environment, can occur by a change in the ordering of the genetic chemicals and a mutation, or *sport,* may arise. The navel orange, most of our apple varieties, and many useful economic plants arose by somatic mutations and have been propagated by cuttings or grafts.

THE MITOCHONDRIA

All cells with nuclei, both plant and animal, contain the cigar-shaped bodies called mitochondria. These organelles, like nuclei and chloroplasts, have a wrapper or outer membrane; mitochondria also possess a folded inner membrane system called the *cristae.* On and in the cristae are bundles of protein enzymes that are responsible for major portions of the energy metabolism of cells. This concept of cellular energy is one of the most important ideas in biology.

A physical chemist, looking at living cells, would suggest that life is thermodynamically impossible. Living systems are so complex, so dynamic, and so delicately poised that they should lose their integration and "decay" to reach the lowest possible energy level. Entropy, a measure of this "decay," should be high. In order for this not to occur, there must be a continual input of energy into the system. As an analogy, consider a house. If neglected, it will weather, decay, and collapse. The repair and maintenance of a house requires the input of materials and of energy to stave off the inroads of entropic chaos. So it is with a cell. Formation of new cell walls, replacement of worn out protein, and revitalization of structure and function requires not only the synthesis of the building materials but also the availability of energy needed for the requisite construction. The biochemical processes of respiration and fermentation provide the cell with both energy and building supplies.

Respiration and Fermentation If one sets a piece of wood on fire, the wood is burned fully to form water and carbon dioxide and the chemical energy of the wood is converted into light and heat. Burning, or oxidation, can be expressed as a simple chemical equation:

$$\text{Wood (cellulose)} + O_2 \rightarrow H_2O + CO_2 + \text{energy (heat and light)}$$

The difficulty with obtaining the energy needed for cellular processes is that a fire is too hot for the cells. Thus a process was required in which the energy of oxidation is given off or made available in smaller amounts so that heat buildup does not occur and the cell can handle the energy that is developed. The solution found in evolutionary time is what is called the *cellular respiratory process* (Figure 1.10). Essentially, the equation given above is fulfilled once water and carbon dioxide are formed from the combustion of a compound, a simple 6-carbon sugar called glucose (which, in fact, is the basic unit of cellulose). Instead of a hot flame, the oxidation of the sugar occurs in a series of steps mediated by specific enzymes, and instead of the energy simply being given off as

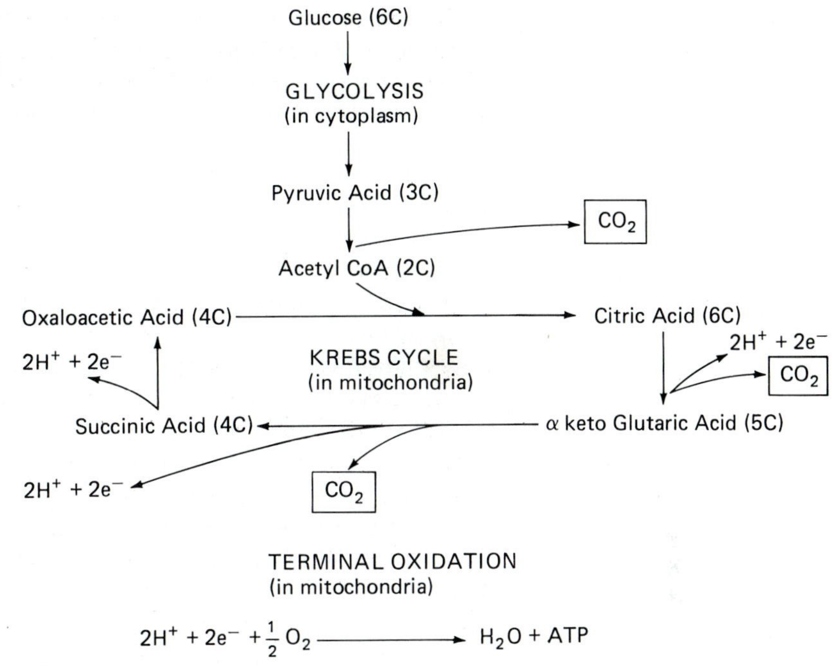

Figure 1.10 A summary of the enzymatic reactions of glycolysis, the Krebs Cyle and the terminal oxidation steps in the cellular respiratory pathway.

heat and light, a portion of it is transferred into a chemical bond that has the potential of being utilized for the cell processes requiring energy. The chemical is ATP (see Figure 1.9), and when the third phosphate bond in ATP is split away from the rest of the molecule, energy is released for cellular work.

The respiratory process involves a linked sequence of enzymatic reactions which have been studied in great detail. The starting material is the simple 6-carbon sugar, glucose, or the glucose derived from starch or from table sugar (sucrose). The first portion of the process involves 11 enzymes that convert the 6-carbon compound into two 3-carbon compounds called *pyruvic acid.* This portion of the process is known as *glycolysis* (the *lysis,* or dissolving, of glucose). Only a little energy is released, with considerable energy still residing in the pyruvic acid. All of these glycolysis reactions occur freely in the cytoplasm of the cell.

When the supply of oxygen is adequate, each pyruvic acid molecule is converted into a 2-carbon molecule, with the third carbon being oxidized to carbon dioxide. Thus we have accounted for two of the six carbons in the initial glucose molecule.

The 2-carbon molecule then enters the mitochondrion, where it enters into a cyclic series of enzymatic reactions called the *Krebs cycle,* to honor the German Nobel Laureate Hans Krebs, who worked out its details. The 2-carbon compound combines with a 4-carbon compound to form a 6-carbon compound, citric acid, the acidic component prominent in citrus fruits. The citric acid is enzymatically degraded into a 5-carbon compound (accounting for two more of the carbons in a glucose molecule) and then into a 4-carbon compound (accounting for the remaining two carbons in each glucose). During these enzymatic transformations, hydrogen ions (H^+) and electrons are removed from the molecules and are captured by a hydrogen ion and electron-carrying compound. The hydrogen ions and electrons are moved through a third enzymatic process, during which they combine with oxygen to form water and simultaneously make ATP.

But there are times and circumstances when not enough oxygen is available for the full respiratory process to take place. One such time occurs when you are actively exercising and your muscle cells are respiring at rates faster than your lungs and blood can supply oxygen to the muscle. When this situation happens, cells can initiate another type of respiratory activity, called *anaerobic* (without oxygen) *respiration,* or, more usually, *fermentation.* Under these conditions, the pyruvic acid formed in glycolysis is converted into carbon dioxide plus ethyl alcohol, or the pyruvic acid is converted into lactic acid. It is the alcohol and lactic acid that make your muscles burn and ache. When the exercise stress is over, the alcohol or lactic acid can be metabolized by routes that funnel them into the more usual aerobic respiratory pathway. The fermentation pathway can be summarzied as:

$$\text{Pyruvic acid (3-C)} \rightarrow CO_2 + \text{ethyl alcohol (2-C)}$$
$$\text{Pyruvic acid (3-C)} \rightarrow \text{lactic acid (3-C)}$$

A few organisms regularly use the fermentation pathway even in the presence of normal amounts of oxygen. The most important of these are the yeasts. When glucose substrate is available, they will actively convert it into carbon dioxide and ethyl alcohol. And a good thing they can! The carbon dioxide formed from yeast by fermentation is what makes bread rise, with the alcohol giving a lovely smell as it evaporates during baking (Chaper 3). When the sugar of grapes or of malted barley or of any other sugar-containing plant material is fermented by yeasts, the carbon dioxide is released and the alcohol accumulates to form wine, beer, or, through distillation, the brandies, whiskies, and other high-alcohol beverages (see Chapter 8).

Respiration provides not only the ATP energy required by cells but also the compounds needed for cellular repair, maintenance, and growth. For example, α-keto glutaric acid or pyruvic acid can be converted into amino acids needed for protein synthesis, and from these amino acids others can be made. Pyruvic acid can be converted into glycerol, the backbone of fats and oils, and the fatty acids can be made from acetyl CoA. There are hundreds of these reaction pathways, all utilizing one or more of the intermediates of the glycolytic, Krebs cycle, and fermentative routes. The accumulation of compounds in plants also involves the respiratory route. If the enzymes which convert citric acid into α-keto glutaric acid work slower than those converting acetyl CoA and oxaloacetic acid into citric acid, citric acid will accumulate in cells, as it does in citrus fruits.

Functional roles of the other cell organelles, aside from the chloroplast, are less well understood and active research is still underway on these problems. We know that Golgi and microtubules accumulate where new cell membranes and walls are being formed, and there is considerable evidence that these organelles play roles in wall formation. Still something of a mystery is the structure and function of very minute structures that even with the resolution of the electron microscope are seen as blobs or tubules—the *spherosomes.*

CELL GROWTH

All cells in an individual plant start out as a product of mitotic division of a preexisting cell and hence have exactly the same *genome*—the same genetic complement. As these cells enlarge, each may begin to differentiate into one of a variety of possible cell types, their final form keying with their final function in the plant. Similar differentiations lead to the formation of a tissue or tissue system which may be made of one or a few cell types, just as a mammalian organ like a kidney contains several cell types each performing certain functions. The existence of plant tissue systems was recognized in the late seventeenth century by Marcello Malpighi, John Ray, and especially Nehemiah Grew. It wasn't until the nineteenth century that the arrangement of cell types into tissue systems and thence into organs was fully appreciated, and it wasn't until the twentieth century that the functional significance of tissue systems began to be understood with any precision.

The study of the alteration—the differentiation—of a simple cell, as described above, is the study of growth and development. Some scholars call it the dismal science, because as soon as we think we have an answer, someone comes along and demolishes the theory. At least theoretically, we are fairly sure that the forces that bring about cell differentiation are chemical and physical, and they operate through the genetic machinery.

Cell growth in plants can be divided into three categories. Cells can divide—growth by increase in number. Cells can enlarge in one, two, or three dimensions—growth by increase in volume or size. They can also grow by differentiation—changes in their structure or function. Ultimately controlled by the genetic constitution of the cell, the direction and scope of cell growth is regulated both physically and chemically. Physical regulations include stresses and strains resulting from the position of the cell in the plant body as well as from temperature and light.

Chemical regulations of growth are diverse. Photosynthesis contributes the sugars and is regulated by carbon dioxide, water, and light. The supply of ATP and the building blocks for the synthesis of cell constituents are regulated by the rates and patterns of respiration. The uptake and distribution of water and minerals from the soil are also important factors in growth.

Precise control of cellular and, ultimately, of whole plant growth is provided by a number of growth-regulating substances. Five major and

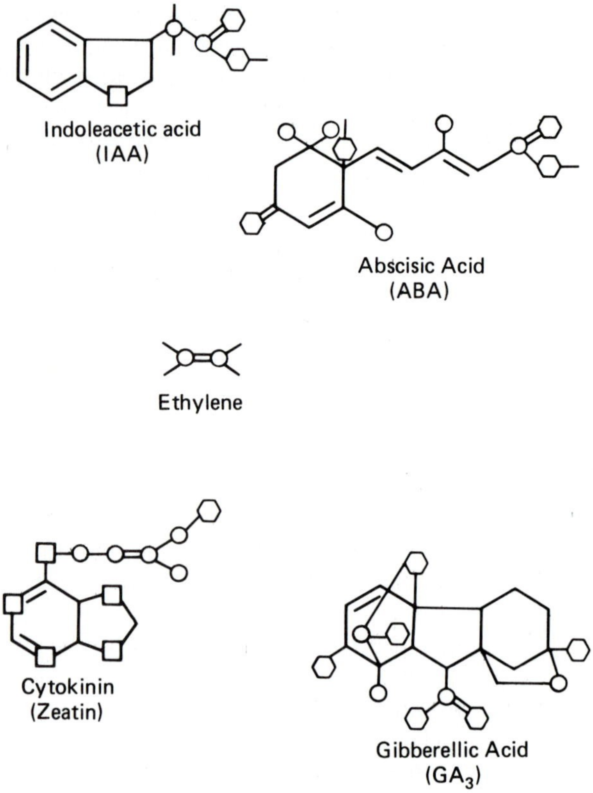

Figure 1.11 Molecular structures of major plant growth-regulating substances. Hydrogen atoms and carbon atoms in the rings are not shown. Carbon atoms are represented as circles, oxygen atoms as hexagons, and nitrogen atoms as squares.

Indoleacetic acid (IAA)

Abscisic Acid (ABA)

Ethylene

Cytokinin (Zeatin)

Gibberellic Acid (GA_3)

at least a dozen minor classes of compounds have been characterized (Figure 1.11). There are one or possibly several auxins, the most important being indoleacetic acid (IAA). Over 50 chemically distinct gibberellins, including the well-known gibberellic acid (GA_3), have been isolated from various plants. Four or five cytokinins have been studied. Abscisic acid is found in most plants, and the simple gaseous compound ethylene is produced by all flowering plants.

Although the details of exactly how these growth-regulating compounds do what they do is still not well understood, a great deal of empirical information has accumulated. The cellular response to each of the compounds depends on its chemical composition, the concentration available, and the readiness of the cell to respond in a particular way. Growth substances also work in conjunction with each other. Thus auxin alone promotes water uptake into the vacuole of a plant cell and causes cell enlargement. Cytokinins alone have very little effect on cells. Yet, when the two work together, cells both enlarge and undergo rapid division. The concentrations and ratios of auxins, gibberellins, and cytokinins can regulate the differentation of cells that will, ultimately, have a special structure that is adapted to particular functions. In addition, there are opposing effects of the gibberellins, abscisic acid, and ethylene.

The interactions among the growth-regulating substances and between growth substances, other chemical and physical forces, and the genetic machinery are subtle and wonderfully orchestrated. Indeed, these interactions have to occur if there is to be any pattern of growth and development at the cellular, tissue, organ, or whole plant levels of bioorganization. One can predict, with a high degree of probability, where particular cellular structure and function will be developed in a plant and can, with equal assurance, know what a particular species or even cultivar will look like and will do under a given set of conditions. It is the huge question of why and how we can have such confidence in predicting plant form and function that is at the cutting edge of plant sciences.

ADDITIONAL READINGS

Fahn, A. *Plant Anatomy.* 3rd Ed. Elmsford, N.Y.: Pergamon Press, 1982.

Galston, A. W., P. J. Davies, and R. L. Satter. *The Life of the Green Plant.* 3rd Ed. Englewood Cliffs, N.J.: Prentice-Hall, 1980.

Salisbury, F. B., and C. W. Ross. *Plant Physiology.* 3rd Ed. Belmont, Calif.: Wadsworth, 1985.

Thompsom, D. W. *On Growth and Form.* 2nd Ed. New York: Cambridge University Press, 1942.

Wardlaw, C. W. *Morphogenesis in Plants: A Contemporary Study.* London: Methuen, 1968

Wareing, P. F., and I. D. G. Phillips. *Growth and Differentiation in Plants.* 3rd Ed. Elmsford, N.Y.: Pergamon Press, 1981.

What Is a Plant?

Flower in the crannied wall
I pluck you out of the crannies
I hold you here, root and all, in my hand
Little flower—but if I could understand
What you are, root and all and all in all
I should know what God and man is.

ALFRED LORD TENNYSON

If one looks up the word *plant* in a dictionary, one is told that "any member of the vegetable group of organisms" is a plant. Other definitions include putting seeds into the ground, an industrial building, or to defraud, as by planting evidence. Common sense tells us that the botanical plant is a living organism, that it is green, and that it grows in the ground. But a tree in winter may not be green, and many plants grow in water or, like orchids and bromeliads, grow in trees or even on telephone wires. A person in a green sweater isn't a plant by any definition, but we correctly associate greenness with plants, and we know that this color is related to the process of photosynthesis.

Any living system has to have provided for it, or must provide for itself, several fundamental conditions for its very existence. It must be able to reproduce itself. It requires a place to live, formally called a habitat. It must have appropriate mechanisms for obtaining minerals, water, and energy for maintenance and growth—it must have food. Organisms can obtain food by one of three methods. They can absorb food in liquid form from their habitats—bacteria and fungi use this method. Some organisms can passively or actively ingest particulate matter—animals use this method. Finally, some organisms can make their own food—and only green plants can do this. We can now define a plant as a living organism that can make its own food from water, inorganic minerals, and a simple source of carbon such as carbon dioxide. This definition is essentially a definition of photosynthesis and needs only the additional qualifier that the energy needed to convert carbon dioxide into food is derived from light.

A few additional attributes must be included. Not all of the organism need carry on photosynthesis, and the organism need not carry on photosynthesis all the time. This allows for roots, for wood in the center of an oak tree, for deciduous trees in winter and for plants at night. Some structural attributes we associate with plants separate them from animals. We think of animals as moving about and plants as staying in one place. While this is certainly true of evolutionarily advanced organisms like lions and dandelions, there are algae that swim and marine invertebrates that are bound firmly to rocks. As research tools and research concepts become more sophisticated, subtle differences between the cells of plants and animals show up, but these exciting developments in molecular biology are not part of our mind's-eye definition.

THE PLANT AS AN ORGANISM

The gross form of a plant is very different from that of animals. With some important exceptions, animals have appendages—heads, legs and arms, wings, tails, antennae. Plants, notably the flowering plants with which we are most familiar, have three major vegetative organ systems: roots, stems, and leaves. Flowers, fruits, and seeds may be formed at specific times of the year. As in animals, plants show a division of labor and of responsibility in function. Each plant organ is a structural entity with functional roles necessary for the survival of the plant as an organism.

Except for some marine invertebrates that look a great deal like plants, we have no difficulty in recognizing a plant. Indeed, we are able to identify many plants by their body plan or form as far away as we can see them. This, too, is part of our concept of a plant. A sturdy oak silhouetted against the sky or the tall spire of a redwood pulling our eyes upward to its summit evokes images that are translated into common phrases: sturdy as an oak, willowy, reaching for the heavens were orginally developed from the idea of plant form as translated into human attributes.

Although each plant species has its own unique form, it is convenient to give botanical names to each of the various organs. The terms *roots, stems,* and *leaves* are obvious and have been used for centuries. So are the words *flower, fruit,* and *seed.*

THE ROOT—WATER AND MINERAL UPTAKE

Clichés are trite and overused statements, but they are generally true. One biological cliché is that form and function are but two sides of the same coin. Nowhere in the biological world is this more clearly seen than in the root systems of plants. Indeed, the root system's form and function has given rise to anthropomorphic clichés about *putting down* or *discovering one's roots, rooting around,* or *rooting out something*—preferably something evil. These expressions parallel roles of the root in the economy of plants. Roots serve to anchor a plant to one place, they ramify through the soil to absorb water and nutrient minerals, and they store substances needed for the growth and development of the plant as a whole.

Most plants are fixed in one spot; their mobility is nil. If the soil around roots becomes depleted of water and nutrients, the plant can obtain these materials only by continuing to form new roots that increase the surface area in contact with the soil. A single winter rye plant *(Secale cereale)* grown for four months had 13 million roots with a total length of 600 kilometers (380 miles). The surface area for water and mineral uptake was 200 square meters (1800 square feet); root hairs added another 400 square meters (36,000 square feet) of surface area. Depending on the species, root systems can extend laterally for 30 meters (about 100 feet) from the stem axis and may extend vertically for the same distance to tap underground water supplies. The food storage function of roots has been exploited by selecting plants with enlarged, food-containing root systems: carrots, both table beets and sugar beets, cassava, radishes, true yams, and the rutabaga. Starch, the primary storage substance in roots, is compact, relatively inert and stable, and readily transformed back into sugar for recycling through the plant when growth processes are resumed. In some storage roots, smaller but nutritionally significant quantities of amino acids, proteins, and vitamins are present (Chapter 4).

There are minor differences in the structure of roots of different plant species, but the basic structural plan is surprisingly uniform, especially at the tip of the root (Figure 1.12). The root tip, or apex, is where most of the action is, with the young rootlet growing in length, taking up water and minerals, and developing a vascular system for the transmission of the water to the rest of the plant. The primary root is formed during the development of the embryo within the seed and consists of a

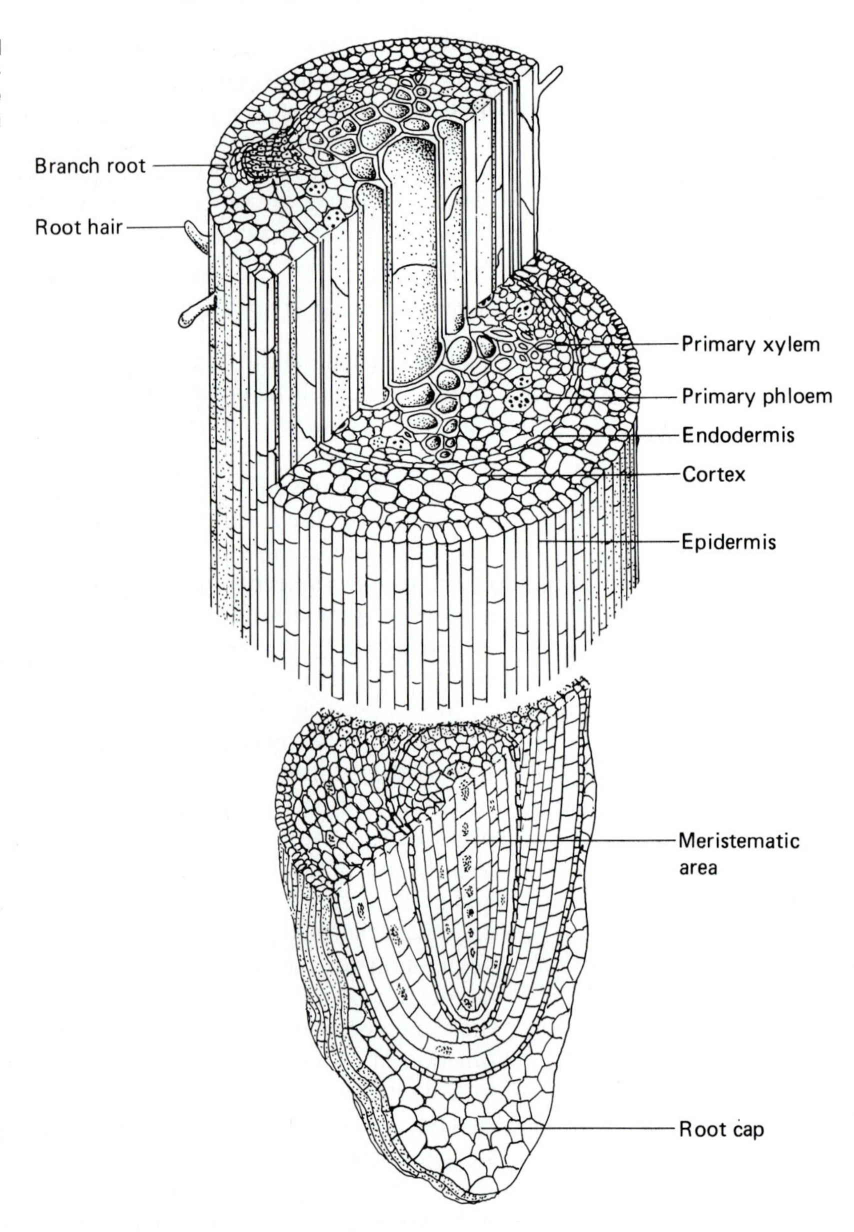

Figure 1.12 Diagram of a typical plant root showing the meristematic (cell division) zone and the more mature region in which branch roots originate.

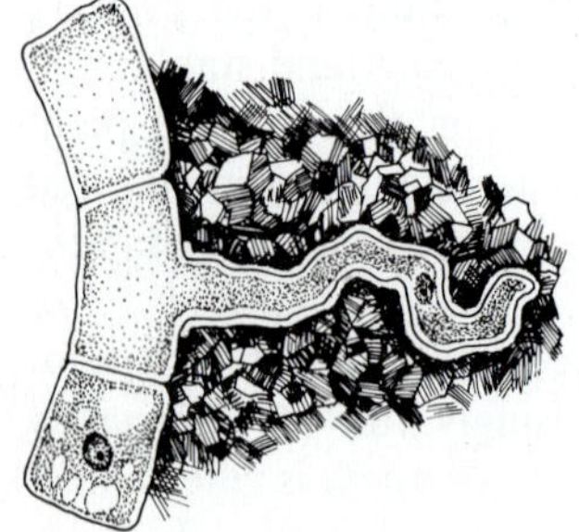

Figure 1.13 Epidermal cells of the root with root hairs growing among particles of soil.

root tip attached to a short axis. As the seed germinates, this embryonic root (the radicle) is usually the first organ to elongate through the seed coat and to develop its final form.

The root is very well adapted for the uptake of both water and minerals. Surrounding each young feeder root is a film of water containing dissolved minerals. This zone, the *rhizosphere* (root environment), is in direct contact with the epidermal cells of the root and their root hairs so that there is a large surface area for water and mineral uptake (Figure 1.13). Where water supplies are adequate, the root rapidly takes up the water, which is then replaced by soil water in close proximity to the rhizosphere. Since water moves relatively slowly through soil, continued uptake depends on root growth into undepleted soil. If soil is dry and soil rhizosphere solutions are absent, water may move out of the root and severe wilting or death of the plant may occur. Once within the root epidermal cells, water and minerals move to the vascular cylinder, where they move quickly to the root xylem and are pulled up the plant by the process called transpiration (see section on the leaf).

Mineral Nutrition Minerals are dissolved in the rhizosphere soil solution, but their uptake is not necessarily at the same rate as water uptake. Mineral uptake is, at least to some extent, regulated by different chemical, biological, and physical forces, and some minerals are taken up more rapidly than are others. This is the science of plant mineral nutrition.

Plant physiologists learn the memory aid "See Hopkins Café; mighty good with a pinch of salt," which is a listing of the major inorganic minerals—as their elements—needed by plants: C, H, O, P, K, I, N, S, Ca, Fe, Mg, Na, Cl. The first three, carbon, hydrogen, and oxygen, enter plants as carbon dioxide and water. All the other required mineral elements are taken up as inorganic ions dissolved in the soil solution surrounding the root (Table 1.3). These minerals are present in the soil as it is formed from rock or are deposited in rain or as dry fallout (dust). As plants and animals die and decay, their mineral content is restored to the soil. Physiologists and ecologists have worked out the mineral cycles for each of these nutrients, determining their journey from plant to animal and back to the soil. The cycle is broken when plants are harvested for human use, when erosion sweeps away soil, or when water percolating through soil carries dissolved minerals below the root zone and eventually to rivers and oceans. The inherent fertility of soil is always decreasing and where high plant productivity is desired, it is usually necessary to supplement soils with minerals supplied as fertilizer.

In the hundred or so years in which plant scientists have studied the mineral requirements of plants, we now have a fairly complete picture of what minerals are needed, in what quantities they are required, and how they are used (see Table 1.3). Plants can be grown to maturity in the absence of soil, using water solution (hydroponic) culture in which the quantity of each mineral is supplied in the amount necessary for optimum growth and development. Hydroponic culture of plants is commercially feasible, although its economic value is limited to specialty crops.

Some of the minerals are usually present in soils in amounts that are not limiting for plant growth and development. Calcium, sulfur, sodium, and chlorine may, in fact, be present in excessive amounts, as when salts accumulate from irrigation (Chapter 10), near the oceans, or as a

Table 1.3 ROLE OF ESSENTIAL NUTRIENTS IN PLANT CELL METABOLISM

Nutrient	Role	Deficiency symptoms
Carbon (C)	All organic compounds	Death
Oxygen (O)	Most organic compounds, respiratory acceptor	Death
Hydrogen (H)	All organic compounds	Death
Nitrogen (N)	All proteins	Pale green leaves, old leaves yellow and dry, short stems
Phosphorus (P)	Many vital chemicals	Dark green leaves, old leaves yellow, slender stems
Sulfur (S)	Most proteins, synthesis of pigments	Young leaves light green
Calcium (Ca)	In cell walls, cell membranes, cofactor for enzymes	Stem tip death, wilting
Potassium (K)	Cofactor in enzymes, photosynthesis, respiration, nucleic acids, water movement	Pale leaves, stem tip death, dead spots on leaves between veins
Magnesium (Mg)	Chlorophyll molecule, cofactor in enzymes	Pale leaves, curling of leaves, weak growth
Iron (Fe)	Cytochrome molecule, chlorophyll formation	Short stems, pale leaves, weak plants
Boron (B)	Role uncertain	Stem tip death
Copper (Cu)	Enzyme action, synthesis of vitamin C	Drooping of stem tip, wilting of young leaves
Manganese (Mn)	Chlorophyll synthesis, photosynsynthesis, enzyme activation	Leaves spotted with dead areas Plant pale and weak
Molybdenum (Mo)	Nitrogen fixation and metabolism	Nitrogen deficiency symptoms
Zinc (Zn)	Auxin formation, enzyme cofactor	Large dead spots on leaf, leaves excessively thick
Chlorine (Cl)	Oxygen formation in photosynthesis	Poor growth

result of industrial pollutions. Too much of any of the mineral nutrients can be toxic and cause poor plant growth.

Of the required minerals, nitrogen is frequently in limiting supply. Although 78 percent of air is nitrogen gas (N_2), this elemental nitrogen must be converted into an ionic form as ammonium ion (NH_4^+) or the nitrate ion (NO_3^-) before plant roots can absorb it and before it can be used by the plant. The conversion of nitrogen gas into the ammonium ion occurs to a limited extent with the energy of lightning, but the bulk of the nitrogen fixed into inorganic ions is mediated by the biochemical activities of free-living bacteria and blue-green algae and by bacteria in the root nodules of legumes and a few other plants (Chapter 3). Most of the nitrate taken up by plant roots is enzymatically converted into ammonium, which can be built into amino acids, proteins, and other biologically essential compounds. In order for a plant or animal cell to make a single protein macromolecule, several thousand nitrogen-containing amino acids are linked together; 16 percent of protein is nitrogen. Complex rings of carbon and nitrogen atoms form the purine and pyrimidine compounds of nucleic acids, which are part of the genetic machinery of cells and are part of the chlorophyll molecule. Alkaloids (Chapter 6), vitamins (Chapter 4), and many other constituents of life either are nitrogen-containing or have their synthesis and degradation regulated by nitrogen-containing enzymes. Cellular growth and development is thus directly limited by the supply of inorganic nitrogen compounds in soils. Nitrogen-deficient plants will look stunted and have thin, woody stems and small, pale green leaves that cannot conduct photosynthesis efficiently (Figure 1.14).

The two other elements in what is called the NPK triad are also required in large amounts. Phosphorus, as the phosphate ion ($-PO_4^{-3}$), is built into nucleic acids and the phosphate-phosphate bond in adenosine triphosphate (ATP) bears the chemical-physical energy needed for every energy-requiring process in cells. Potassium (K) is the third letter in the NPK ratio. A 10:10:10 fertilizer ratio simply means that every pound of fertilizer has 10 percent each of NPK (nitrogen, phosphorus, and potassium) and a 5:10:5 means 5 percent nitrogen, 10 percent phosphorus, and 5 percent potassium. Potassium plays many different roles. It is the primary regulator of the movement of water into and out of plant cells, and it forms complexes with organic acids (citric, malic, and so on) within the cells.

Figure 1.14 Erosion and mineral deficiencies in tobacco. Courtesy U.S. Department of Agriculture.

Few soils are deficient in calcium, magnesium, or sulfur, and few fertilizers contain them, but this does not reduce their importance to the plant. Calcium (Ca) and magnesium (Mg) are enzyme activators, serve as carriers for other ions through plant cell membranes, and reduce the toxicity of other ions which may be in excess. Magnesium, as part of the chlorophyll molecule, has a vital role in photosynthesis. Sulfur's (S) main roles are as constituents of certain amino acids, cysteine and methionine, that determine the three-dimensional shape of proteins and as parts of compounds that regulate the oxidation-reduction state of cells. Sodium (Na), needed in higher concentrations in animal cells than in plant cells, acts much like potassium, and the only known role of the chloride ion (Cl) is as part of the enzyme complex that breaks water in photosynthesis, releasing molecular oxygen into the atmosphere. Iron (Fe) is an ionic constituent of the cytochromes that serve as links in the chain of compounds through which electrons flow to provide cellular energy (ATP) in both respiration and photosynthesis. Because iron is also required for the synthesis of chlorophyll, iron-deficient plants are pale or yellow and grow poorly.

In addition to these minerals that make up Hopkins Café, at least five others are required in very small or trace amounts. These are called *micronutrients.* Zinc (Zn) and copper (Cu), very poisonous in high concentrations, are absolutely required by all plant cells as parts of certain enzymes; the blackening of a cut potato involves the enzyme tyrosinase which includes copper in its active site. Manganese (Mn), like chlorine, is part of the enzyme complex that releases oxygen in photosynthesis. We know virtually nothing about the role of boron (B) except that plant cells will not divide without it, and actively growing tissues of the root and shoot meristem will die without continuous supplies of borate ions. We also know that the difference between adequate boron levels and toxic ones is very small and that even slightly elevated concentrations will result in plant cell death within a few days. All plants require molybdenum (Mo) in order to carry on normal nitrogen metabolism, and many plants also require silicon (Si), germanium (Ge), vanadium (V), or gallium (Ga), for reasons which are obscure.

It is a general biological and thermodynamic rule that the efficiency with which nutrients are used by plants is about 10 percent. This means that of the nitrate ion incorporated into a plant, only about 10 percent of it is made available to plants when the original plant decays and returns nitrogen to the soil. The relevance of this recycling loss to many aspects of plant growth is not always recognized. When plant residues are composted and the compost is spread on a garden plot, only a tenth of the nitrogen originally present in the initial compost is available to the new plants. Some of the nitrogen has been converted to ammonia and other volatile compounds, some has been incorporated into microorganisms present in the compost pile, and some has been changed into chemicals not available to the plant. This same order of magnitude loss occurs at every turn of the nitrogen cycle. Additional losses of minerals take place when erosion sweeps away topsoil or, as noted earlier, when water percolates through soil carrying dissolved minerals below the root zone and eventually into rivers and oceans. When parts of the crop are eaten by people or animals, the minerals in those plant parts are removed from the original soil and may be deposited long distances away. Merely returning to the soil all of the residues of harvested crops will not restore the soil to its original fertility, and adequate plant growth depends on the availability of new sources of nitrogen from rock particles and from precipitation.

Obviously, this does not mean that the recycling of organic residues should not be done. Wasting of potentially useful sewage and other organic materials is nothing short of stupid. It does mean that even if all of the potentially useful plant and animal wastes were efficiently recycled, we would still have to supply soils with additional sources of fertilizer if we wish to continue to produce adequate levels of food and other plant products. Because of the need to grow more plants to feed the results of enthusiastic human overbreeding, the fertilizer industry is a vital factor in the world's economy.

That organic additives such as manures and composts can supply plants with minerals needed for growth and development is unquestioned. The organic gardener and the thrifty farmer restore some of these minerals by adding plant materials or animal wastes to the soil. But serious questions can be raised about the assertion that organic additives supply plants with other substances such as vitamins. Not only are intact plants fully capable of synthesizing these organic substances themselves, but there is good evidence that root systems cannot absorb most organic substances present in composts.

Plant physiologists and organic gardeners disagree about whether plants can distinguish between a phosphate ion derived from compost and one from a bag of chemical fertilizer. All the available evidence indicates that it matters not one whit to plants whether the mineral nutrient addition is from a sack of industrially produced fertilizer or from manure or compost. There is no chemical or biological difference between molecules of nutrient minerals from the two sources. Be it noted, however, that organic additions do improve soil by modifying their structure and texture to permit good root aeration and to facilitate adequate regulation of soil moisture that, in turn, permits the efficient absorption of inorganic nutrients by roots. Whether organically grown plants are healthier, disease-free, or better-tasting has, surprisingly, received little study under controlled experimental conditions with adequate statistical evaluation of the data.

THE STEM

Because early land plants were generally short, the movement of water from the soil to the above-ground parts was not a limiting factor for growth. As plants increased in size and as the distance between the

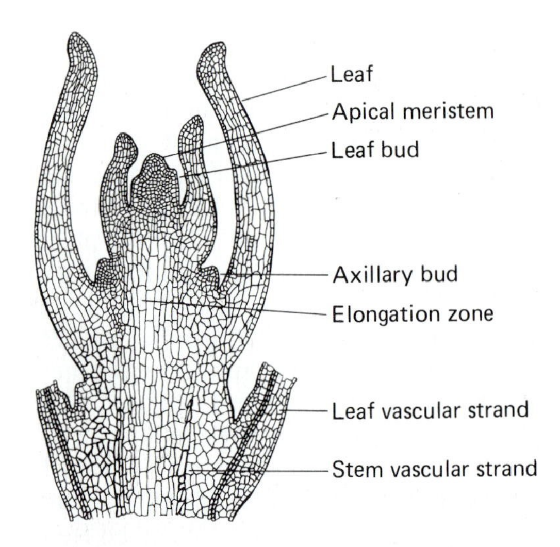

Figure 1.15 The apical meristem. At the left is a diagramatic representation of a typical apical meristem *(Elodea)*. The photograph on the right is of a head of cabbage sliced lengthwise.

source of water and the leaves increased, stems developed. Plant stems serve several functions. They maintain a plant erect so that the leaves are exposed to sunlight. Stems also contain the vascular or plumbing system through which water and minerals move up from the soil. The vascular system also permits the flow of food made in the leaves to be transported to the lower parts, including the roots. In flowering plants, the stem originates as part of the embryo within the seed. This stem can be seen by opening up the two halves (cotyledons) of a peanut; the embryonic stem is a short axis terminating in a pair of tiny leaves. During germination, this embryonic axis elongates, and, upon completion of germination, the young stem together with the seed leaves assumes its typical erect position.

The elongation of the stem involves the activity of the *apical meristem,* a dome-shaped structure whose outer layer covers a mass of rapidly dividing cells (Figure 1.15). Continued cell division in the upper portion of the apical meristem results in some elongation of the stem; additional stem extension occurs as these cells become organized into files or columns of cells that elongate rapidly. During this elongation, cells in the central part of the young stem begin to differentiate into the beginnings of the vascular system under the influence of plant growth substances synthesized in the apical meristem. Differentiation is a well-orchestrated process so that, within a few millimeters of the stem tip, the primary vascular system is formed.

In stems of monocots—plants with one cotyledon—like maize (corn) or palm trees, further differentiation and development of the stem and its vascular system does not occur; the mature stem becomes no thicker than the young stem (Figure 1.16). In most conifers and dicots, however, a circle of cells, the vascular *cambium,* forms between the primary *xylem* and the primary *phloem* (Figure 1.17). This tissue is capable of dividing laterally to give rise to a new cambium cell and to either a secondary xylem vessel or xylem tracheid or to a phloem cell (Figure 1.18). In general, the vascular cambium forms many more

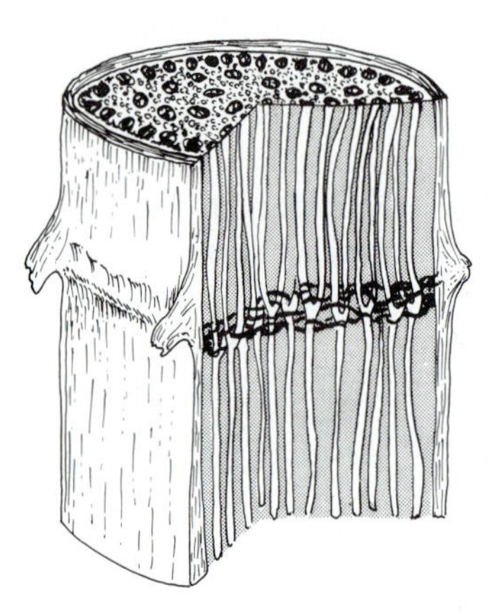

Figure 1.16 Structure of the stem of a typical monocot showing the scattered vascular bundles.

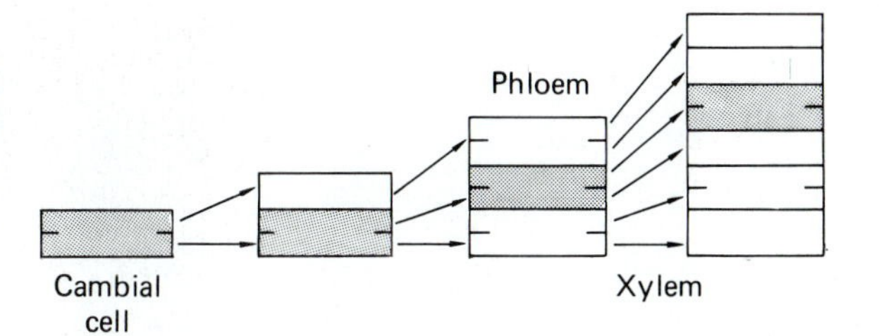

Figure 1.17 Diagram of the formation of xylem (wood) to the inside of a stem and of phloem (bark) to the outside of a stem through the activity of the cambium layer. It is by cambial activity that stems and roots increase in girth.

xylem than phloem cells, so that as secondary xylem accumulates, the girth or diameter of the stem increases. This growth in stem diameter confers a good deal of strength to the plant, enabling the stem to support branches and a large crown of leaves. The stem of a tree is mostly xylem. Not all stems are erect and not all plants have a single stem. Shrubs are woody perennials in which shoot buds are produced from the rootstock. Some plants, like the strawberry *(Fragaria)* have runners that are actually horizontal stems. Bulbs of onion *(Allium),* corms of crocus and gladiolus, the underground tubers of potato *(Solanum tuberosum),* or the lotus and the rhizomes of ginger or iris are all stems.

In many perennial woody plants, activity of the cambium is seasonal. Most of the larger xylem cells are formed in the spring; fewer and smaller xylem cells are formed later in the growing season. This difference in the size of xylem cells accounts for the annual rings in trees. The ring represents the boundary between the very small cells formed at the end of a growing season and the large xylem cells formed by the reactivation of cambial activity the following spring (Figure 1.19). As initially described by Leonardo da Vinci, the size of an annual ring provides information on the rate of growth of a tree. If growing conditions are optimum, the annual ring will be wide with many cells,

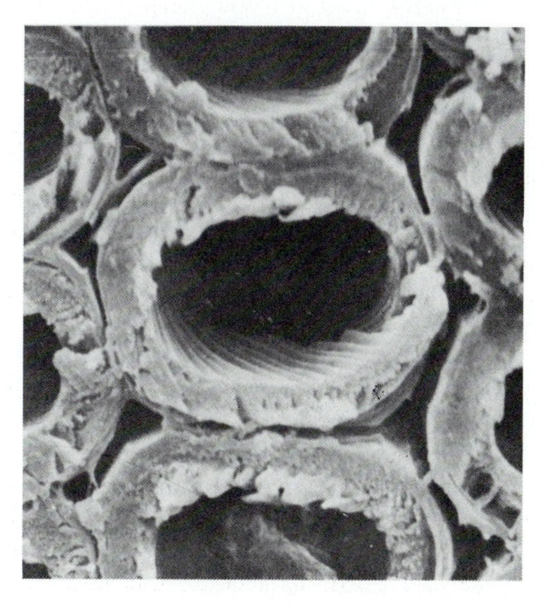

Figure 1.18 Scanning electron micrograph of xylem tracheids in pine wood. Courtesy Barry S. Faigel.

Figure 1.19 Five annual rings of the wood of the southern yellow pine *(Pinus palustris).* Round holes are resin ducts. Courtesy U.S. Department of Agriculture, Forest Service, Forest Products Laboratory.

while a poor growing year results in a narrow ring with fewer cells. Examination of a freshly cut stump will reveal the pattern of annual growth rates. It is not necessary to cut down a tree to evaluate its growth history; a small diameter core can be taken from the tree (Figure 1.19). By detailed examination of such tree cores, it is possible to determine the growth pattern and to infer the conditions, including the climate and the possibility of damaging pollutions, that the tree experienced over its lifetime. *Dendrochronology,* as this topic is called, has been used in anthropological research to date the beams in cliff dwellings, using established chronologies obtained from the area of interest. By comparing series of narrow or wide rings from tree cores with pieces of logs or lumber, it is possible to determine when a tree was cut down and used in construction. It is also possible to establish the pattern of climatic changes over centuries by examining the tree cores of long-lived species such as the bristlecone pine *(Pinus aristata),* whose life span extends back for over four thousand years.

The Forest Products Industry In addition to their roles in the structure and function of plants, woody plant stems are extremely important in human activity (Figure 1.20). The word *xylem* is derived from the Greek *xylon,* which means *wood.* In addition to many specialty products obtained from tree stems (rubber, turpentine, dyes, and tanning compounds), we can divide wood products into three utilization classes: pulp for paper, energy sources, and lumber products. The forest products industry utilizes about a quarter of the land area of the United States (Tables 1.4 and 1.5) and close to 40 percent of the Canadian land

Figure 1.20 Anatomy of a tree trunk. From the outside, the tissue systems are: bark, phloem, cambium, and xylem. Courtesy St. Regis Paper Company.

mass. Although the Soviet Union has the greatest area of potentially exploitable forested lands (Figure 1.21), the Soviets have not become a major factor in the international forest products market (Table 1.6). The United States produces the greatest volume of forest products, but also imports large amounts of pulpwood and construction lumber from Canada (Table 1.7). Both the United States and Canada have extensive overseas markets, and forest products are significant factors in the international economies of both countries. Nevertheless, 60 percent of the world's wooded lands are in the tropics (Chapter 10).

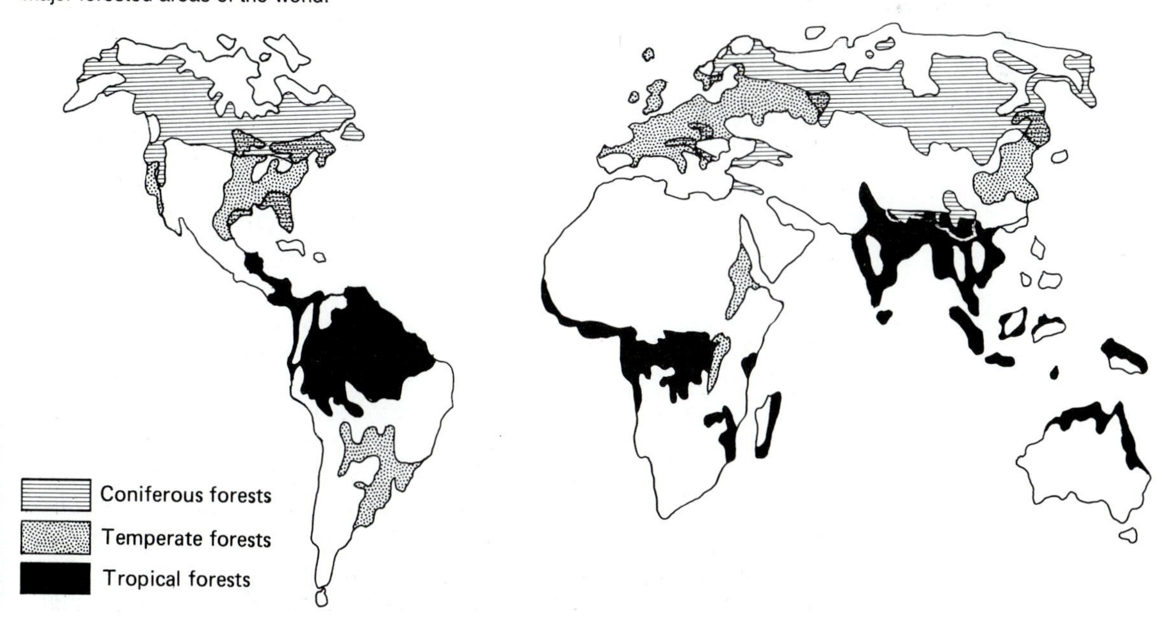

Figure 1.21 Distribution of the major forested areas of the world.

Table 1.4 FORESTED LANDS OF THE UNITED STATES

	Billions of acres	Percent
Land use		
Rangeland, mountains, deserts	0.81	36
Farms, roads, structures	0.70	31
Commercial forests	0.49	22
Noncommercial forests	0.25	11
Total U.S. land area	2.25	100
Forest distribution		
Commercial forests	0.49	66
Unproductive forests	0.23	31
Reserve forests	0.02	3
Total U.S. forested area	0.74	100
Forest ownership		
Privately owned	0.28	58
Federally owned	0.10	20
Industry-owned	0.07	14
Public (state, etc.) owned	0.04	8
Total U.S. commercial forest area	0.49	100

Table 1.5 TOTAL FORESTLAND ACREAGE BY STATES

	Total area	Forest	Percent
Alabama	32,678	21,770	67
Alaska	365,481	119,051	33
Arizona	72,688	1,883	26
Arkansas	33,324	18,227	55
California	100,091	42,408	42
Colorado	66,485	22,535	34
Connecticut	3,117	2,186	70
Delaware	1,269	391	31
Florida	35,179	17,932	51
Georgia	37,295	25,545	69
Hawaii	4,106	1,974	48
Idaho	52,933	21,591	41
Illinois	35,761	3,789	11
Indiana	23,161	3,908	17
Iowa	35,867	2,455	7
Kansas	52,515	1,344	3
Kentucky	25,504	11,968	47
Louisiana	28,867	15,380	53
Maine	19,797	17,749	90
Maryland	6,369	2,960	42
Massachusetts	5,013	3,520	70
Michigan	36,492	19,273	53
Minnesota	50,745	18,984	37

Table 1.5 (*continued*) TOTAL FORESTLAND ACREAGE BY STATES

	Total area	Forest	Percent
Mississippi	30,290	16,913	56
Missouri	44,189	14,919	34
Montana	93,258	22,777	24
Nebraska	48,974	1,045	2
Nevada	70,264	7,660	11
New Hampshire	5,781	5,132	89
New Jersey	4,821	2,463	51
New Mexico	77,766	18,313	24
New York	30,636	17,378	57
North Carolina	31,367	20,613	66
North Dakota	44,339	421	1
Ohio	26,251	6,498	25
Oklahoma	44,149	9,340	21
Oregon	61,574	30,404	49
Pennsylvania	28,816	17,832	62
Rhode Island	671	433	65
South Carolina	19,366	12,493	65
South Dakota	48,606	1,734	6
Tennessee	26,474	13,136	50
Texas	168,300	24,091	14
Utah	52,697	15,288	29
Vermont	5,935	4,391	74
Virginia	25,496	16,389	64
Washington	42,665	23,098	54
West Virginia	15,414	12,172	79
Wisconsin	34,858	14,945	43
Wyoming	62,342	10,085	16

Wood as an Energy Source Wood has been the primary source of heat energy since people discovered and harnessed fire. For uncounted centuries, open fires were used for warmth, cooking, and protection from marauding animals. Open fires have efficiencies of only 10 to 15

Table 1.6 THE SOFTWOOD INDUSTRY, 1981
(Millions of cubic meters)

Country	Softwood reserves	Softwood harvest	Softwood exports
Soviet Union	61,000	141	8
United States	13,000	112	6
Canada	15,000	73	25
Europe	10,000	107	10
Other	8,000	120	7
	107,000	562	56

Table 1.7 PRODUCTION AND CONSUMPTION OF WOOD (INCLUDING BARK) IN THE UNITED STATES.
(Data in millions of tons for 1982)

	Softwood	Hardwood	Total
Domestic production	150.0	80.5	230.5
Imports	32.5	7.0	39.5
Total available	182.5	87.5	270.0
Domestic consumption	160.0	82.0	242.0
Exports	18.5	4.0	22.5
Total used	178.5	86.0	264.5
Surplus and waste	4.0	1.5	5.0

percent for converting potential energy of the wood into utilizable energy. Over centuries, various kinds of fireplaces and stoves were invented which more than doubled the efficiency. Nevertheless, control of burning and the focusing of heat have continued to challenge engineers.

The amount of wood used throughout the world for energy production is a significant percentage of the available timber resources. With adjustments made for the type of wood burned, the effective heat value today is close to 8500 British Thermal Units (BTUs) per pound, an efficiency of close to 60 percent; most of the remaining energy is dissipated as light and unburned gases (Table 1.8). Colonists in North America, with simple stoves and drafty, poorly insulated homes, burned 40 to 60 cords of wood a year with an efficiency of only about 30 percent. (A *cord* is a pile of wood 8 feet long, 4 feet wide, and 4 feet high; a ton of air-dry wood is 0.6 cord.) By 1740, the Rhode Island colony had used up all its forests and was importing wood from Long Island. Boston was getting its firewood from Maine and New Hampshire. President Washington worried about where the young nation was going to get wood for ships' timbers and masts for the developing navy and merchant marine.

Table 1.8 HEAT UNITS OF VARIOUS HARDWOOD AND SOFTWOOD SPECIES*

Hardwoods			Softwoods		
	Millions of BTU/cord	Pounds in 1 cord		Millions of BTU/cord	Pounds in 1 cord
Hickory	25	4200	Tamarack	18	2400
Oaks	22	3800	Spruces	15	2300
Beech	22	3800	Firs	13	2200
Hard maples	21	3700	Pines	13	2200
Yellow birch	21	3700			
Ashes	20	3400			
Aspens	12	2200			

*Data based on a 50 percent efficiency in a modern heating stove and on air-dry wood with a 20 percent moisture content ("green" wood averages 60 percent moisture). One British Thermal Unit (BTU) is the amount of heat energy needed to raise 1 pound of water by 1 degree Fahrenheit. A standard cord of wood is a pile 4 feet by 4 feet by 8 feet and contains 80 cubic feet. For comparison, one gallon of fuel oil contains 16,660 BTU at a 60 percent efficiency: 1 gallon of L.P. gas ("bottled gas") contains almost 120,000 BTU at a 65 percent efficiency.

To some extent, the fuel shortage was eased by the invention in 1720 of the Franklin stove, which reduced fuel requirements by 15 to 20 cords per year. By 1900, wood became too valuable in North America to continue as the major energy source and was supplanted by fossil fuels, primarily coal and later oil, and from electricity produced by nuclear energy.

The North American experience of fuel shortages was short-lived and minor compared with the situation in much of the rest of the world. In Europe and Asia, major destruction of forests occurred before the Americas were colonized. Not only was wood used as energy sources at rates in excess of reforestation, but huge amounts were consumed in the form of charcoal. Produced by burning wood with limiting oxygen, wood is converted into almost pure carbon. Because the heat value of charcoal is greater than that of wood, the combustion process can attain the higher temperatures required for the smelting of ores into metals. Charcoal smelting was invented at least 7000 years ago to obtain the copper that ushered in the age of metals. By 5000 years ago, the technology of alloying copper and tin into bronze was developed independently in China and the Mediterranean. Shortly thereafter, charcoal-fired furnaces were developed to smelt iron and to make glass, pottery, and porcelain. The deforestation of Europe by the fourteenth century was primarily the result of ore smelting; 14 tons of charcoal were required to produce a ton of iron. England experienced a wood shortage by 1600, but within 50 to 75 years coal replaced charcoal. In central Europe, particularly in Germany, the profession of forestry was developed by 1700 to ensure a steady volume of wood for internal use and for export. The industrial revolution was fired by coal rather than wood, particularly as coke from coal produced the even higher temperatures needed to obtain steel.

Wood combustion is, at least theoretically, an environmentally sound utilization of a low-sulfur, renewable natural resource. Sweden expects to generate 12 percent of its electric power needs from wood by the end of the century. In Vermont, a wood-burning electric power generating plant is developing 62 megawatts of power, using 1500 tons of wood chips daily. This amounts to 25 percent of the state's annual growth of trees, since, on the average, tree growth in the northern temperate zone is 0.6 tons per acre per year. The United States now uses 130 million tons of wood as fuel each year, close to 4 percent of energy production. This is double the amount used in 1970 and represents 15 percent of the total consumption of wood for all purposes.

These figures suggest that wood is being used at a faster rate than forests can regenerate. More serious, however, are the consequences of extensive deforestation in tropical countries and semiarid lands (see Chapter 10). Central and South America harvest 11 percent of the world's wood but use over 80 percent of it for fuel. As populations increase, the pressures on forests become overwhelming. Large areas of forested lands have been mismanaged to the point where they have become useless, erosion-prone, and barren landscapes that resist refor-

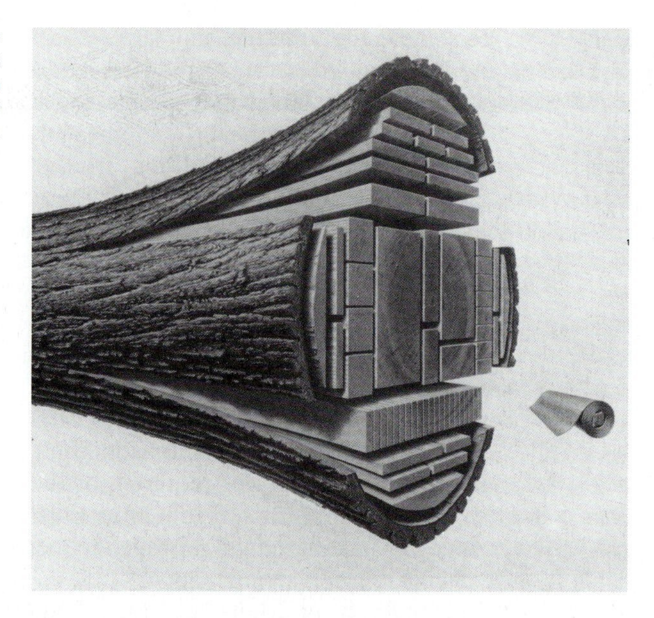

Figure 1.22 Tree stems are used to make lumber. The various cuts of the log are designed to provide different types of lumber and to utilize the tree efficiently. Courtesy St. Regis Paper Co.

estation. India's 730 million people include millions of rural residents who use wood for fuel and construction, requiring up to 200 pounds of wood per month per family. Massive efforts to replant trees have encountered severe problems, although some progress has been made.

The ability of modern machinery to uproot an entire tree and convert trunk, roots, and foliage into chips has resulted in the loss of scarce nutrients from soil. Subsequent tree growth is repressed unless the forest is fertilized. The accumulation of large heaps of alkaline wood ashes resulting from wood combustion has created disposal problems that have yet to be satisfactorily solved.

Lumber For many purposes, wood is an ideal construction material. It has good dimensional stability, is easily worked into intricate shapes, can be joined with a variety of fasteners, has fair insulating and water-repellent qualities, and has considerable inherent beauty and

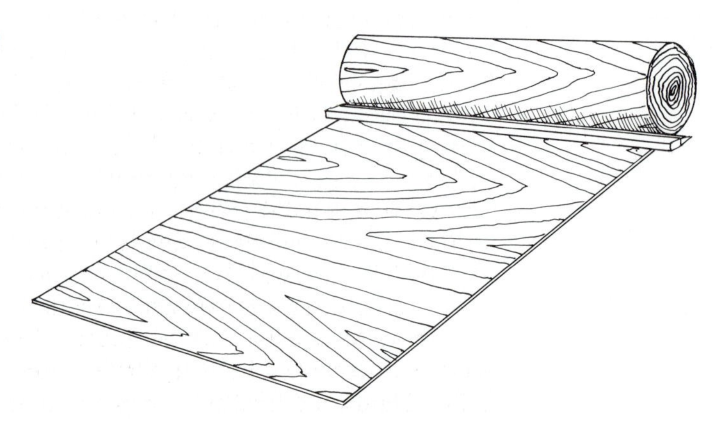

Figure 1.23 Cutting plywood from a log.

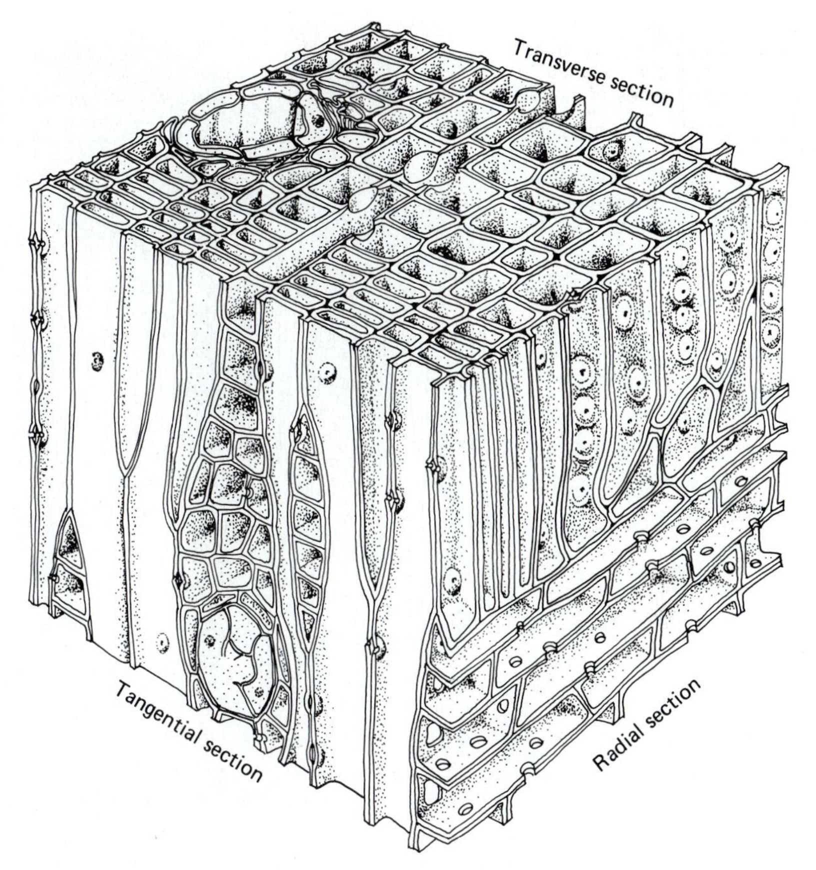

Figure 1.24 Diagram of the cellular organization of a block of pine wood, an important softwood tree. Redrawn from Dean, H. L. *Biology of Plants. Laboratory Exercises.* 5th Ed. ©1968, 1973, 1978, 1982 Wm. C. Brown Publishers, Dubuque, Iowa. All rights reserved. Reprinted by permission.

warmth. Wood has traditionally been inexpensive, generally available, and a renewable resource so long as supply and demand are in balance. Logs have been used for centuries for walls, foundations, and vertical supports. The stone or wooden columns that grace churches, public buildings, and mansions are stylized tree trunks: Corinthian, Doric, and Ionic columns were derived from the major timber tree species of the ancient world. The conversion of the tree trunks into lumber awaited the invention of sharp tools. Splitting axes gave way to hand-powered saws, and finally to powered saws that make many cuts simultaneously. With the increasing cost of logs, methods have been developed to allow the utilization of virtually the entire tree (Figure 1.22).

The pattern of grain (annual rings) is determined by the method of slicing the log. For furniture and construction lumber, dimensionally stable, large sheets are prepared by slicing the log into thin sheets of veneer (Figure 1.23) that are glued together in laminae of threes, fives, or sevens, with the grain of each layer running in a different direction. Plywood lumber may have a surface layer of valuable woods such as teak, walnut, oak, or mahagony used as paneling and for furniture.

Lumber has traditionally been divided into two major categories. The softwoods include all of the conifer species, the gymnosperms (Figure 1.24). Hardwood lumber is primarily from broad-leaved species, both

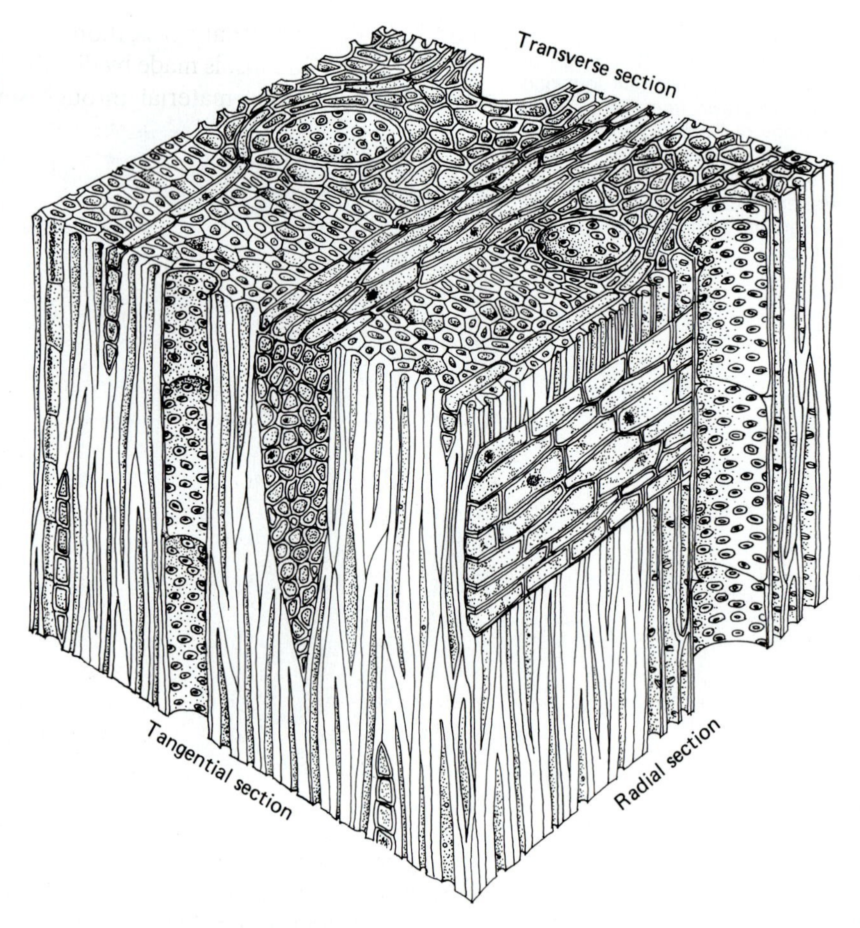

Figure 1.25 Diagram of the cellular organization of a block of oak wood, an important hardwood tree. Redrawn from Dean, H. L. *Biology of Plants. Laboratory Exercises.* 5th Ed. ©1968, 1973, 1978, 1982 Wm. C. Brown Publishers, Dubuque, Iowa. All rights reserved. Reprinted by permission.

evergreen and deciduous dicots (Figure 1.25). The distinction between the types is arbitrary, because some conifers have denser and harder wood than some of the so-called hardwoods like poplar (aspen) or balsawood. However, most softwood conifer species tend to have xylem cells with thinner cell walls and are easier to cut and work. Most construction lumber in North America is obtained from softwoods—pine, spruce, Douglas fir. Almost 80 percent of the lumber is from conifer species. The United States cuts 400 million board feet of softwood lumber per year. (A *board foot* is a plank 1 inch thick and 1 foot square or its equivalent.) Less than 4 percent of the wood harvest is used for telephone poles, pilings, mine timbers, and posts, but the trees used for these purposes must meet exacting specifications for strength and durability. Both hardwood and softwood species are used for pulp and paper; many papers are composed of a mixture of pulp from both types of wood. Hardwood, more expensive and frequently more durable, is used for flooring, furniture, and specialty purposes (wine barrels, handles, baseball bats).

In addition to its role in the production of energy, paper, and lumber, wood has other industrial applications. Rayon, the first of the modern synthetic fabric threads, is made by dissolving wood pulp in solvents and extruding the viscous material through small holes into a bath that hardens the fibers. Cellulose acetate is made by treating solutions of wood cellulose with acetic acid and acetic anhydride, after which it can be rolled into sheets or spun into fibers. Wood can be distilled to provide liquid fuels and gases; wood-derived "gasoline" is used in many countries.

There are necessary tradeoffs between the forest products industry and the environmentalist ethic. At stake are jobs and the conservation of the land. The wood needed for pulp can be obtained from plantations of specially bred soft- and hardwood species and from second- or third-growth forests that reach harvesting size in 25 to 40 years, but large trees make the best lumber. It takes at least 90 and usually more than 150 years to grow such trees. It can be done; there are few natural forests in Europe, and the forestry systems in Germany and Scandinavia are models of controlled, long-term tree cultivation that yield top-grade lumber. But these forested lands are plantations of even-aged trees of a single species and are very different from the lands that we in North America think of as forests. In North America and in some tropical countries, most of the original forests are on public lands, and confrontations between the forest products industry and conservation-minded groups have become adversarial. Automation of tree harvesting and of lumber preparation has allowed large tracts of land to be quickly cut and processed. The degradation of water quality, increased erosion, and loss of prime recreation land that are the consequences of frequently unsound logging practices have affected other segments of the economy (tourism, agriculture, fisheries) and reduced the quality of life for the people of the affected regions. This is now particularly pronounced in tropical lands, whose original forests provide some of the fine cabinet woods and whose soils and climate make natural or controlled replanting difficult (Chapter 10).

Cork As plants age, the thin epidermal cell layer of the stem is replaced by a cork or bark layer that, in trees, may become thick and furrowed. Cork—*phellem* in botanical terminology—is derived from a layer of living cells, the cork cambium or phellogen (Figure 1.26). This generative layer produces a file of cells that, as they die, have their cell walls impregnated with a waterproof material called suberin. This effectively prevents water loss through the stem. Bark or cork is not restricted to aerial stems; the skin of a potato tuber, the peelable outer layers of a carrot, and most other roots have the same developmental sequence and structure. If the continuity of cork or bark is disrupted by wounds, exposed living cells can differentiate into a cork cambium that forms new cork cells. This happens as a tree increases in diameter: bark splits and new cork cambium cells are formed at the bottom of the

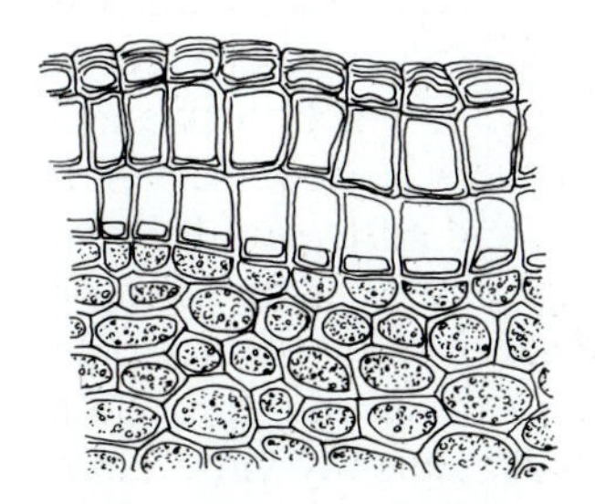

Figure 1.26 Young cork or bark on a plant stem. In trees, the cork or bark is much thicker.

furrow. As cork cells die, the center of the cell that contained the living protoplasm fills with air. With several million cork cells to a cubic inch, and with half this volume being air, cork is lightweight, impervious to moisture, decay-resistant, springy, and of high insulating value.

The physical and chemical properties of cork have been exploited for at least 3000 years. Mediterraneans removed sheets of the unusually thick bark of the cork oak *(Quercus Suber)* to make fishing floats, sandals, and bottle stoppers that have survived to the present. The tree is native to lands surrounding the Mediterranean Sea, with southern Portugal, Spain, Morocco, Algeria, and the islands of Corsica, Sardinia, and Sicily producing commercial quantities of cork. The cork oak is one of the few trees whose bark can be stripped repeatedly without significant injury to the tree (Figure 1.27). Trees known to be several hundred years old are stripped at ten-year intervals.

The sheets of cork, frequently a foot thick, can be punched into bottle stoppers. The residue and cork from less desirable trees are ground into granules of suitable size and glued into sheets for floorings, gaskets, tiles, and many other products for which no synthetic substitute has been developed.

Figure 1.27 Workers stripping bark from a cork oak tree. Photograph courtesy of Armstrong World Industries, Inc.

THE LEAF

Long before flowering plants evolved, plants carried on photosynthesis in their green stems. The gradual evolutionary trend toward expanded and flattened stem tips that captured light efficiently can be followed by examining plants preserved in rocks as fossils. The broad leaf with its flattened blade is reflected in human speech: the term *leaves* (of a book) in English is equivalent to the German *blatt,* the Spanish *hoja,* and the French *feuillet.* So, too, the annual cycle of leaves on deciduous plants from small, light green, delicate leaves in spring to the large, sturdy leaves in summer and the change in color and eventual dropping of leaves in the autumn, signaling the onset of winter, conveys, respectively, images of resurrection or birth, vigorous adulthood, decline and death.

Leaves originate as small bumps on the sides of the shoot tip—the apical meristem (see Figure 1.15). The position of these leaf primordia on the meristem is quite precise, so that as the stem elongates, each leaf or leaf pair is offset in a spiral pattern from the ones above or below. The spiral pattern *(phyllotaxy)* prevents leaves above from shading completely those lower down on the stem. After initiation of the primordium and its development into a somewhat elongated projection, an axillary bud is formed by a second bulge of cells that organize into a tightly furled stem bud, capable of developing into a shoot under appropriate conditions. This axillary bud has the structure of an apical meristem, but its development into a shoot—a stem—is normally repressed by plant growth substances formed in the apical meristem and diffusing down the stem. If the apical meristem is injured or ceases to function, the axillary bud is derepressed and is capable of forming a shoot. It is by this method that side branches form on stems. Dominance of the stem apex over the development of axillary buds may be complete, as in sunflower plants, which usually do not branch at all, or it may be partial, as in a coleus plant, where side branches form on the lower part of the stem. Release of apical dominance to increase the bushiness of a plant is usually accomplished by pinching out the apical meristem. This procedure is widely used in horticulture.

The leaf blade and its stalk (the petiole) attain their species-specific size and shape by a linked series of cell divisions, enlargements, and differentiations. Veins are formed from preexisting cells that differentiate into the xylem and phloem cells that, in turn, connect the leaf blade to the vascular system of the stem.

Like stems, leaves vary greatly in size, shape, and ornamentation. Many of these variations reflect genetic programming for adaptation to specific environmental conditions. Plants adapted to dry *(xeric)* habitats tend to have small or highly modified leaves, like the cacti and the crown of thorns *(Euphorbia milii).* Plants adapted to arid lands frequently bear hairs that reflect heat, may secrete a thick, waxy cuticle to reduce water loss, or may have fleshy water-storing organs. The needles of conifers, the leaves of the common jade plant *(Crassula argentea),* and the narrow leaves of grasses illustrate such xeromorphic characteristics. Leaves of plants adapted to humid or moist habitats also are structurally adapted to the conditions prevailing in their environment.

At maturity, leaves are well adapted for their two major functions—as the producer of food by photosynthesis (Table 1.9) and as the terminus for the flow of water through a plant (Figure 1.28). The upper and lower layers, composed of epidermal cells, are covered with a layer of a water-impervious substance called *cutin;* the leaf may also have additional wax coatings that reduce water loss and protect the more delicate inner tissues. Both epidermal layers, particularly the lower layer, are dotted with small pores, the stomata (little mouths), surrounded by specialized epidermal cells, the guard cells. Depending on the amount of water vapor, the concentration of carbon dioxide needed for photosynthesis, and other internal and external micro-environmental factors, guard cells can change shape (Figure 1.29). The stomatal pores can open or close to permit or to restrict the flow of water vapor into the atmosphere or the uptake of carbon dioxide from the atmosphere for photosynthesis.

The bulk of the typical leaf consists of two major cell types. Beneath the upper epidermis is a tightly packed array of elongated palisade cells that contain chloroplasts which conduct photosynthesis. Below the palisade layer is a more loosely arranged cell type, the spongy mesophyll, with large intercellular spaces. Ramifying through the mesophyll

Table 1.9 GLOBAL EXTENT OF PHOTOSYNTHESIS
As Determined by the Amount of Atmospheric Carbon Fixed into Photosynthate and Its Human Utilization

Plant Biome	Area in $Km^2 \times 10^6$	Carbon Fixed Kg/yr $\times 10^{12}$	Carbon Fixed 10^6 tons/yr.	Productivity (10^6 tons/area)
Oceans	360	2.93	150	0.42
Fresh water	0.2	0.002	1	5.0
Land				
Forest	44	6.2	126	2.9
Arable land	23	2.1	4	0.17
Grassland	27	2.5	3	0.11
Desert	33	.15	<1	0.03
Tundra/ice	22	.05	<1	0.04

Direct Utilization of Carbon by People (Billions of tons per year)

Use	Billions of tons per year
Human food	0.4
Loss to pests	0.3
Animal food	0.06
Waste products	3.0
Loss to microoganisms	0.24
Other direct use (fiber, shelter, wood, etc.)	0.8

Figure 1.28 Diagram of a leaf.

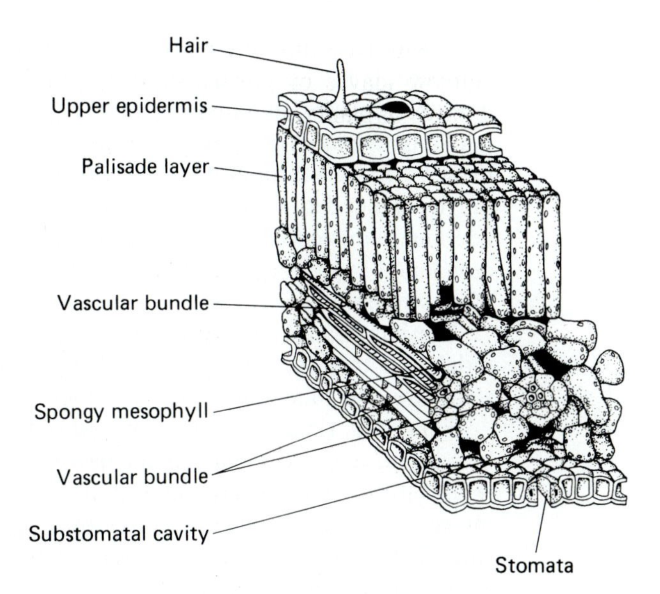

are veins composed of both xylem and phloem. In most dicotyledenous plants, veins are usually arranged in a netlike pattern; in monocots like grasses and lilies, they tend to be parallel.

Some plants, maize for example, have a somewhat different organization. The veins are surrounded by a ring of highly chlorophyllous cells called a *bundle sheath.* Such plants can store carbon dioxide in their mesophyll cells; it moves into the bundle sheath cells to supplement the carbon dioxide received directly from the atmosphere. Plants that possess these structural and functional adaptations include maize, sorghum, and sugar cane, plants that are highly efficient in photosynthesis, grow rapidly, and store sugars. They play an important role in feeding people (see Chapter 3 and Chapter 4).

Water Movement Through the Plant Water is a predominant factor in plant life. Most actively metabolizing plant tissue is about 70 percent water; inactive seeds have close to 6 percent water, and the flesh of the watermelon is over 93 percent water. The roles of water in the metabolism of all living cells include its lysis to yield oxygen during

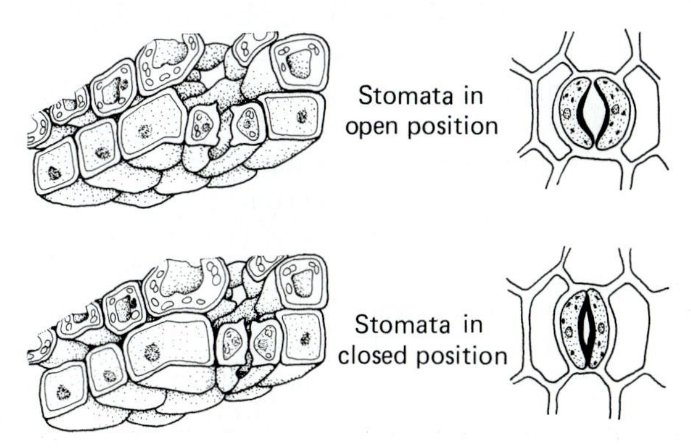

Figure 1.29 Diagram showing stomata in the opened and closed positions.

photosynthesis, its ability to keep the many enzymes and reacting compounds in cell biochemical activity in solution, its participation in growth processes, and many others. The minerals dissolved in the water taken up by plant roots are in low concentrations and considerable water must be passed through the plant to concentrate the amounts of these nutrients needed by the plant. Water plays an indispensable role in keeping plants cool; its evaporation from leaves removes the heat load developed from solar radiation in the same way that you can get chilled after swimming.

The huge range of water requirements among different plants is an indication that the water relations of plants are closely bound to the structure and function of plants in their environments.

From the time a seed is put into the ground until the crop is harvested, a single tomato plant moves over 30 gallons of water from the soil through the plant and back into the air. In one growing season a maize *(Zea mays)* plant transpires over 100 gallons of water; an apple tree, 2000 gallons; and a date palm, 35,000 gallons. On a daily basis, a palm can lose 100 gallons and ragweed *(Ambrosia)* can lose two gallons, while a cactus plant will transpire little more than a spoonful. The huge range of water requirements of different plants is an indication that the water relations of plants are closely bound to the structure and the function of a plant in its environment. Thus the ability to adapt to varying supplies of water is basic to plant survival.

Water Stress A student leaves the dormitory on Friday afternoon, comes back from a wonderful weekend late Sunday night, and finds the geranium on the windowsill looking as bad as the student feels. Leaves are flaccid, shriveled, and some are beginning to turn yellow-brown along their margins. Stems are drooping and flowers have fallen. Rushing to the rescue, the student pours water onto the bone-dry soil until it flows from the bottom of the pot—usually all over the floor. If the plant has not been wilted too long and if the room hasn't been too hot, the plant will begin to recover within a few hours. Leaves will again become succulent and turgid, and the stems will again become upright. Since many house plants are fairly rugged, recovery is frequently complete, although the death toll of dorm plants is high.

Wilting is a cellular phenomenon, the loss of water from the vacuoles of living cells. Vacuolar water has moved through the cytoplasm, cell membrane, and cell wall into the spaces between the cells from where it is pulled up to the stomata in the leaves and thence out into the atmosphere. As water pressure within the cells decreases, vacuoles collapse and cells become flaccid, much like a balloon with most of the air let out. The flaccid cells are no longer capable of maintaining the rigidity of stem and leaf tissues, and the plant wilts. Wilting is a biological example of the economists' Law of Supply and Demand; transpirational losses of water from the leaves exceed the supply of water available to the roots. As with economic supply and demand, the consequences reverberate in the plant long after the plant appears to have recovered.

Wilting is one extreme of a continuum between the fully water-saturated, or turgid, cellular state to the dehydrated cell. Although the wilted state can result in cellular or plant death, most plants under field or natural conditions are rarely at either end of the continuum, but are usually somewhere in between; plants live perpetually in a state of water stress. Those adapted to dry *(xeric)* habitats have been selected for one or more responses that lead to survival. A drought-resistant species such as white pine *(Pinus strobus)* will maintain cellular turgidity and exhibit little evidence of stress under conditions of available soil moisture that would cause the permanent wilting and death of most house plants. The pine survives because its needles have a smaller surface area and fewer stomata per unit area than does a geranium and hence loses water at a proportionally slower rate. Most cacti do not have leaves, and their green, photosynthetic stems have few, if any, stomata. Some desert plants lose their leaves during the dry season, growing a new set only when water becomes available during the rainy season. Other plants, especially the desert annual flowers, are capable of going through seed germination, vegetative growth, flowering, fruiting, and seed maturation in short periods of time when spring rains wet the soil. These are drought-avoiding adaptations, and, like drought resistance, are programmed into the cellular hereditary material. Knowing this, geneticists can choose individuals or races of economically important plants possessing drought-resistance or drought-avoidance characteristics as crop plants. They may use these plants as breeding stock to improve existing crops and extend their growing range. Green Revolution cereal grains (Chapter 3) now include drought-resistant characters and were specifically engineered for drought-prone climates.

In order to develop such improved crops, an understanding of the cellular consequences of water stress is necessary. When asked how severe water stress must be in order to reduce growth and yield, the plant physiologist gives a typical scientific answer: "It depends." It depends on what cellular function or cellular system is limited for growth and yield, on the species under study, on the expected severity of water deficit, and on when during the growing season this deficit is expected. With appropriate instruments and under controlled conditions, test plants are placed under conditions of zero water stress and the level of stress can be raised to any point up to permanent wilting. Various components of the physiological, biochemical, and growth activity of the stressed plants can then be evaluated to determine whether they are, relative to appropriate controls, functioning efficiently. Different cell types are not injured to the same extent by a given level of stress, nor does injury start at the same stress level for all species. In general, cell division and cell enlargment are the most sensitive; this is why that windowsill geranium grows poorly for a while after it has been severely water-stressed.

Protein and chlorophyll synthesis, activities of growth-regulating chemicals, photosynthesis, respiration, and the accumulation of metabolites within cells are injured as stress becomes more and more severe, but the sequence is variable with species, stage of development, and even the previous water status of the plant. Seedlings and young sprouts are usually very sensitive to even slight reductions in water supply, for they are in a period of rapid cell division and cell enlargement. Plants grown with high levels of water—most houseplants are in this category—are ill-adapted to any but short-term water stress. A tree forming its winter buds and preparing for winter dormancy suffers little from water stress, and, indeed, high levels of available water may suppress normal winter-hardiness. Corn and soybeans, both adapted to hot, dry summers, may wilt at the end of a hot, midwest "scorcher" but will suffer little permanent damage, while a rice plant at the same stage of development will show reduced yields after just a few evening wiltings.

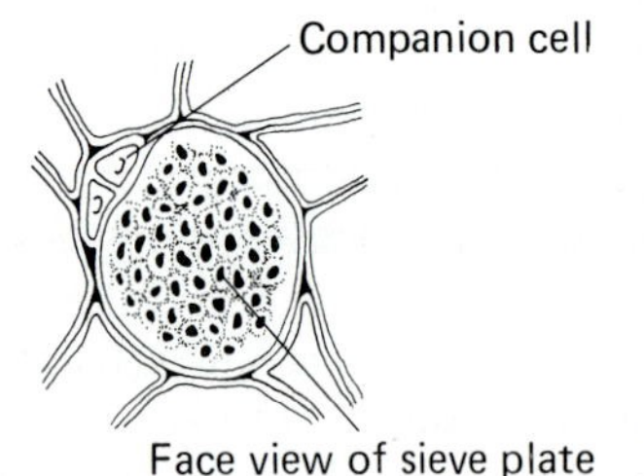

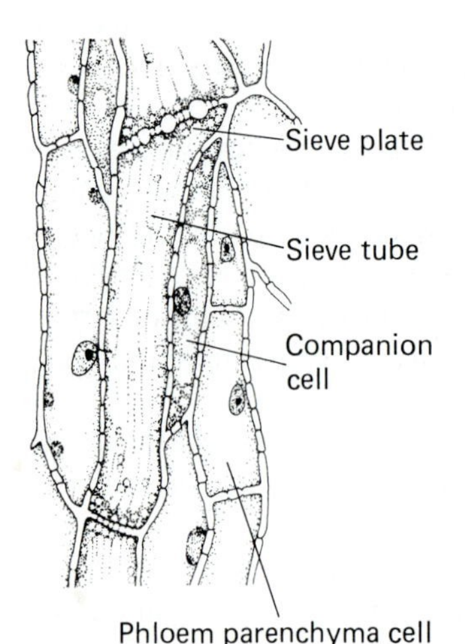

Figure 1.30 Diagram of the conducting cells of the phloem.

With much of the world's available agricultural land subjected to drought, investigation of water stress in the laboratory and in the field is basic to food production. Physiologists, ecologists, taxonomists, anatomists, cell biochemists, geobotanists, and horticultural and agricultural scientsts are cooperating in this vitally important study.

Translocation of Photosynthate As sugars accumulate in a leaf as a result of photosynthesis, they must be transported to the rest of the plant for its continued development. This translocation of photosynthate occurs through the phloem tissues of leaf veins, petioles, and the stem and roots (Figure 1.30). The photosynthate is primarily in the form of sucrose (table sugar), which can be converted into starch at the site of storage. Sugar cane and sugar beets, sorghum, and other crop plants have been selected for their high sugar content. The sucrose made in sugar maple leaves is stored in the stem as starch which is converted into sucrose when the stem warms up on sunny, early spring days (Chapter 4). People have also selected plants that store large amounts of starch; our root crop vegetables (potatoes, sweet potatoes and yams, and the cassava) owe their high starch concentrations to actively photosynthesizing leaves and an efficient translocation mechanism (Chapter 3).

ADDITIONAL READINGS

Dale, J. E. *The Growth of Leaves.* London: Edward Arnold, 1982.

Earl, D. W. *Forest Energy and Economic Development.* Oxford: Clarendon Press, 1975.

Edlin, H. L. *The Illustrated Encyclopedia of Trees, Timber and Forests of the World.* New York: Harmony Books, 1978.

Fahn, A. *Plant Anatomy.* 3rd Ed. Elmsford, N.Y.: Pergamon Press, 1982.

Fritts, H. C. *Tree Rings and Climate.* New York: Academic Press, 1977.

Hall, D. O., and K. K. Rao. *Photosynthesis.* 3rd Ed. Baltimore: University Park Press, 1981.

Heath, O. V. S. *Stomata.* Oxford Biology Readers No. 37. Burlington, N.C.: Carolina Biological Supply, 1975.

Hewitt, E. J., and T. A. Smith. *Plant Mineral Nutrition.* New York: Wiley, 1975.

Marden, L. "Bamboo: The Giant Grass." *National Geographic* 158, no. 4 (1980).

Meidner, H. *Plants and Water.* New York: Halsted Press, 1976.

Meiggs, R. *Trees and Timber in the Ancient Mediterranean World.* Oxford: Clarendon Press, 1982.

Pike, R. E. *Tall Trees, Tough Men.* New York: Norton, 1967.

Prance, G. T., and K. B. Sandveg. *The Formation, Characteristics and Uses of Hundreds of Leaves Found in All Parts of the World.* New York: Crown, 1984.

Rost, T. L., M. G. Barbour, R. M. Thornton, T. E. Weier, and C. R. Stocking. *Botany: A Brief Introduction.* New York: Wiley, 1979.

Saigo, R. H., and B. W. Saigo. *Botany: Principles and Applications.* Englewood Cliffs, N.J.: Prentice-Hall, 1983.

Shepherd, J. *The Forest Killers: The Destruction of the American Wilderness.* New York: Weybright and Talley, 1975.

Stokes, M. A., and T. L. Smiley. *An Introduction to Tree-Ring Dating.* Chicago: University of Chicago Press, 1968.

Sutcliffe, J. F., and D. A. Baker. *Plants and Mineral Salts.* Studies in Biology No. 48. Baltimore: University Park Press, 1974.

Tippo, O., and W. L. Stern, *Humanistic Botany.* New York: Norton, 1977.

Youngquist, W. *Wood in American Life.* Madison, Wis.: Forest Products Research Society, 1977.

The Flower and Its Consequences

Beloved, Thou hast brought me many flowers
Plucked in the garden, all the summer through.

ELIZABETH BARRETT BROWNING
Sonnets from the Portuguese

The flower, in its many forms, has provided us with a wide range of images. The archetypical concept of the flower is a source of verbal and cultural images that can be traced for thousands of years in the art and speech of many cultures. Above all, the flower is a sexual symbol of great power and antiquity. The corsage, wedding flowers, and floral gifts are used because of their obvious symbolism of fertility. Floral offerings to the dead and those who mourn them are appropriate because of the fundamental nature of the short-lived flower, with its development from a tiny bud, its mature beauty, its promise of generational continuity, and its eventual fading. Floral symbols are found in all religions (Chapter 5) and appear, accurately portrayed or in abstract form, on everyday items from fabrics to wallpaper. The odors of flowers, evolved to attract insects, have been transformed into scents that evoke the image and many of the symbolic attributes of the flower itself (Chapter 9).

The flower is, evolutionarily, the solution to several major problems faced by plants. Prior to the evolution of the flowering plants, sexual reproduction in algae, mosses and liverworts, and the ferns required that the coming together (fusion) of the motile sperm and the egg (fertilization) occur in free, liquid water. As plants developed vascular systems to transport water up to the top of elongated structures, free water became less available. Sequences of structural and functional changes occurred that replaced the naked sperm cell with a specialized entity, the pollen grain. Pollen could be transferred to structurally adapted female organs without the need for free water. Thus originated the cone of gymnosperms and, eventually, the flower of the angiosperms. The flower also developed the fruit, a structure which protected the developing embryo and provided the nourishment for the next generation—the seed.

There is an enormous range of shapes, colors, and sizes in flowers and in their patterns of development. Flower initiation may require signals from the plant itself (nutrients, growth substances) and from the environment (light, temperature, water, and minerals). In spite of variations on a common theme, not only is flower function fairly constant, but there is an underlying similarity among the parts of a flower. This allows generalization about floral structure (Figure 1.31).

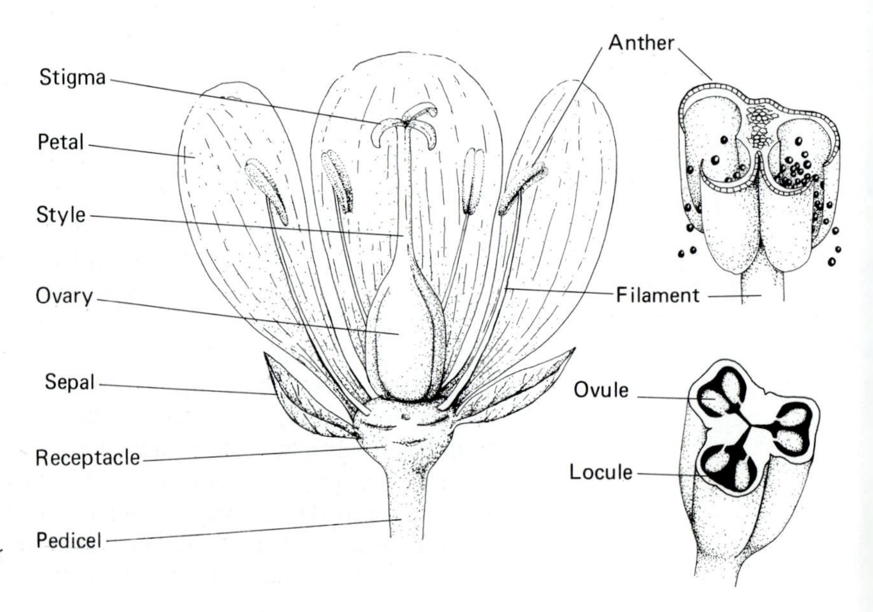

Figure 1.31 Diagram of a flower and its parts.

The typical flower has four major parts: *sepals,* which form part of the protective bud; *petals,* which may be attractive signaling devices for insects; *stamens,* which produce the male entity, the pollen grain, in which the sperm nuclei are formed; and the *pistil,* in which the egg forms and the seed develops. The sepals and the petals are obviously modified leaves, possessing veins and even stomata. The stamen has a long stalk, the *filament,* which is essentially a modified main leaf vein. The enlarged *anther* contains cells that undergo a special cell division (reduction division) in which the chromosome number is reduced to one-half that of the cells that comprise the vegetative plant body. From this pollen mother cell, four pollen grains develop which contain sperm nuclei. The pistil has a receptive surface, the *stigma,* upon which pollen is deposited; a *style,* through which a pollen tube carries the sperm nuclei; and one or more *carpels,* which each contain an ovary. As in the formation of pollen, cells in the ovary undergo reduction division to form egg cells. The carpel forms the fruit in which the fertilized egg develops into a seed.

Plants may form flowers that lack sepals or petals; this is seen in many of the grasses. Sepals or petals may be exaggerated or brightly colored. In many of our cultivated flowers, the numbers of sepals or petals may be multiplied many times over the normal numbers; the American Beauty rose has several times as many petals as the five of the wild rose. Not all plants have both anthers and pistils. An "imperfect" flower has one but not the other. Both anther-bearing (male) and ovary-bearing (female) flowers may be on the same plant (*monoecious* plants) or on different plants (*dioecious* plants).

POLLINATION—BIRDS AND BEES

The buzzin' of the bees
In the sycamore trees . . .

The Big Rock Candy Mountain

A flower's biological role can be performed only if pollen from the anthers reaches the stigma at the top of the female organ, the pistil, and the pollen's tube grows down to the ovary where fertilization occurs. This is a dull scientific fact that fails to give even a hint of the mind-boggling diversity of methods by which pollination takes place. The obvious way for pollination to occur is for the pollen produced in the same flower to travel the small distance between anther and stigma. The problem with this method is that it precludes introduction of new genetic characteristics into the offspring, since both male and female parts of the flower are derived from the same vegetative plant. Where self-pollination does occur, as in garden peas, the offspring are genetically identical—as alike as two peas in a pod—and while this is advantageous for canners, it means that adaptability of the plants to environmental stress is reduced. Self-pollination in many plants is prevented by one or more processes. Pollen may be self-incompatible, incapable of causing fertilization of the egg of the same plant, but effective when it reaches the pistil of the flower on another plant. Some plants separate the male and female flowers, restricting them to different locations on the same plant (the silks and tassels of corn, for example), or having male and female flowers on different plants (holly is a good example). Pollen may mature earlier or later than the pistil, or anthers may hang out and away from female organs.

If self-pollination cannot occur, pollen must reach the flowers of another plant—that is, cross-pollination must occur. The most obvious movement of pollen through space occurs by its being passively carried by wind or water. This is a chancy proposition, and, in order to secure a reasonable amount of pollination, the number of pollen grains produced is enormous. If you have ever parked an automobile under a pine tree in late spring, you have some idea of the amount of pollen produced by a wind-pollinated plant. All grasses, including cereal grains, are wind-pollinated, and so are most of our forest trees like oaks, maples, birches, and beeches.

Air-borne pollen can travel long distances; 30 miles is not unusual. Pollen has been captured by balloons more than ten miles above the earth, and just a few high-pollen-producing plants can blanket a

Figure 1.32 Flowering head of giant ragweed *(Ambrosia trifida)*.

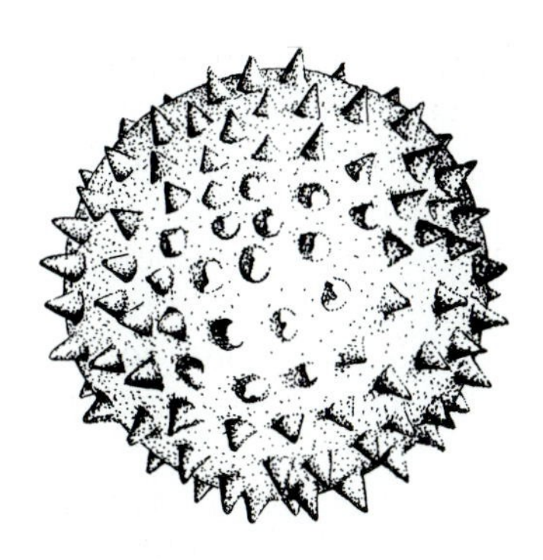

Figure 1.33 Pollen grain of giant ragweed *(Ambrosia trifida).*

ten-mile radius. Wind-pollinated flowers are the cause of considerable human discomfort, as any hay fever victim can testify. A hyperallergic person may react in spring to pollen from maples, willows, and oaks; in midsummer to grass pollen; in early fall to pollen from several weeds; to breathe freely again only when plants die back in early winter. The most common of the pollen allergies is due to the ragweeds, which by some strange quirk of botanical sadism were given the genus name *Ambrosia* —"food of the gods" (Figure 1.32). There are several species of ragweed with wide, overlapping geographical distributions, and their pollen production is prodigious. Some people are so exquisitely sensitive that merely a few pollen grains per milliliter of air will cause severe allergic reactions (Figure 1.33). In some cases, allergies to specific pollens may be inherited; allergic children frequently have allergic parents or grandparents.

The allergic reactions are caused by proteins in the outer walls of the pollen grains. These proteins move from the pollen to the sensitive lining of the nose, where they react with antigens already in the victim's system. Desensitization by a series of inoculations is frequently successful in lessening the severity of the reactions.

Pollination by Animals It is, however, pollination mechanisms involving animals that attract the most attention. Evolutionary evidence suggests that the flower evolved about 150 million years ago (Jurassic Period) and that flowering plants began to dominate plant life forms about a million years ago (Cretaceous Period). Insects evolved about 400 million years ago, and most of the different kinds of insects were evolutionarily well advanced about 300 million years ago (Carboniferous Period). Bees, wasps, flies, and butterflies—all insect pollinators—are, today, obligatorily dependent upon plants for their food, and yet they were on earth 200 million years before there were flowers. The obvious question, "What did they eat before flowers evolved?" simply cannot be answered.

This question aside, we do know a good deal about the interaction of insects and plants. Pollen is a fine nutritional source, containing proteins, fats, carbohydrates, and minerals. It, and the sugar-rich nectar of flowers, constitutes virtually all the food of bees and a significant part of the food of other insect pollinators. The association is obligatory for a large number of wild plants, and many of our food crops are absolutely dependent on insects for pollination. This means that over 60 million years or so, both the flower and its pollinator adapted structurally and functionally for mutual survival. If, as modern scientific thinking suggests, evolution is not directed but is a random process that occurs by pure blind chance, the statistical probability of so many specific adaptations occurring is very small—and yet they did occur.

Many of these adaptations are so precise that they boggle our minds. Bees, for example, are attracted to plants for the nectar—the classical "drink of the gods"—secreted by cells at the base of the showy petals (Figure 1.34). Since bees work hard, requiring sugar for energy, nectar

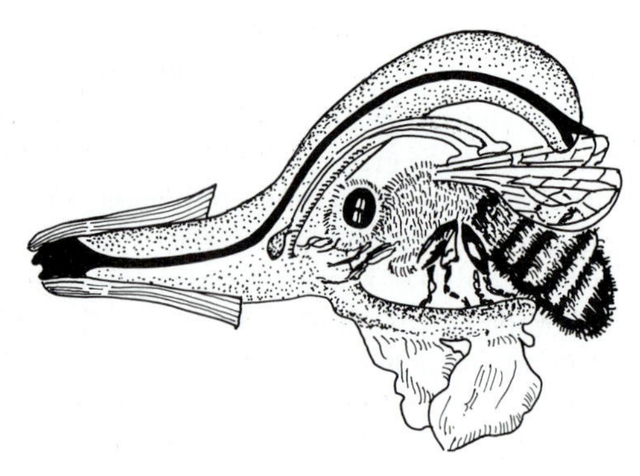

Figure 1.34 A bee inside the flower of *Salvia*. Pressure of the bee's proboscis pushes the anthers down to deposit pollen on the bee's back. This pollen can be transferred to the stigma of the next flower visited.

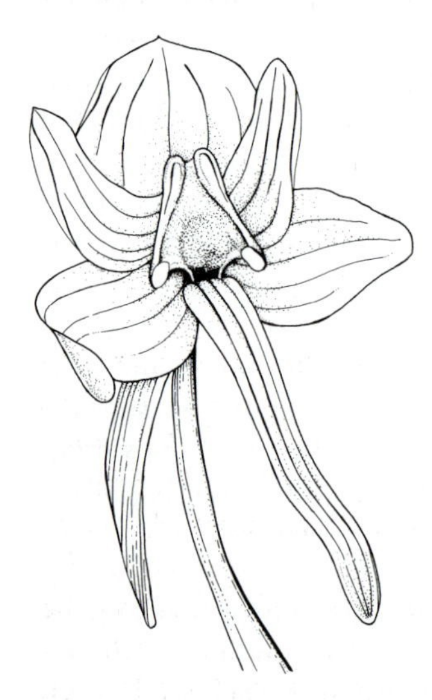

Figure 1.35 Flower of the orchid *(Habenaria orbiculata)*, with a grossly elongated petal serving as a landing strip for pollinating insects.

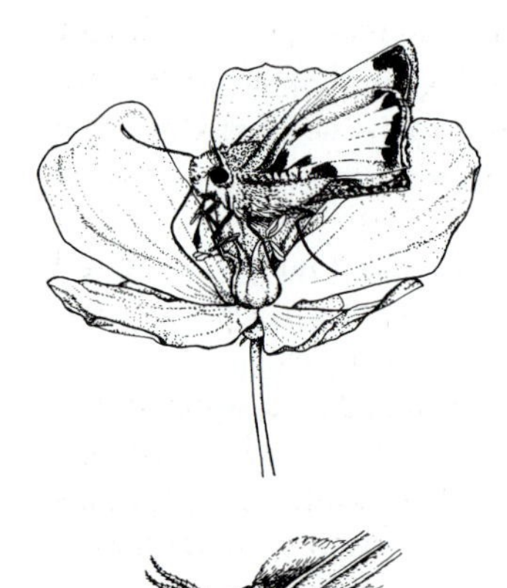

Figure 1.36 Pollination of wild geranium *(Geranium maculatum)* by a skipper moth *(Poanes spp.)*. The nectar-sipping tongue of a moth is shown below.

must be at least 18 percent sugar to make its collection efficient. Most bee-visited flowers have about 20-25 percent sugar in their nectars, and some, like the horse chestnut, have so much dissolved sugar that the nectar is syrupy. As long as the flowers of a species produce nectar, bees will remain exclusively with that species, thereby ensuring adequate pollination. Apple flowers produce nectar daily, just before bees leave the hives. Bees have color vision, seeing best in the blue and near-ultraviolet ranges, and bee flowers tend to be just those colors. Botanists used to worry about the fact that many bee flowers are red or even pale yellow or white and thus should not be seen by the insects. Only recently was it discovered that red and white flowers reflect solar ultraviolet rays and, to a bee's eye, appear as bright spots against a dull gray background of leaves and grass. In some plants, the ultraviolet-reflecting areas of petals are arranged in streaks or a series of dots that, like painted stripes on an airplane landing field, guide the bee directly into the center of the flower (Figure 1.35). The perfumes of flowers are not designed to attract humans, but are keyed to the organs of smell of the insect pollinators. The smell of rotting meat, offensive as it is in skunk cabbage, is irresistible to flies. Pollinating insects are structurally and functionally adapted for pollination. Their long tongues are usually precisely the right length to tap the nectar pool at the base of their host flowers. Hairs on their bodies collect pollen as they feed, and the sex organs of the plants are keyed in length and in structure for their task.

Bees are not the only insect pollinators; moths and wasps are also important pollinators of many flowers (Figure 1.36). The flowers visited by these insects are usually different species than those visited by bees. The mouth parts of these insects are different from those of bees and are structurally adapted to the flowers they visit. Flower shape and strong scents are the main attractors for wasps. Certain orchids have flowers that so nearly mimic the shape of the body of the female wasp that the not-too-bright male wasp attempts copulation with the flowers, one after another (Figure 1.37). Beetles, although attracted to flowers by their odors, are not important pollinators. Generally, beetles eat the soft parts of flowers and rarely carry pollen from one plant to another.

Flies which pollinate plants (Figure 1.38) can be separated into two groups. Those with long tongues compete with bees for the same flowers; their mouth parts are very much the same size as those of bees, and the flowers are not specifically adapted to the flies. The short-tongued flies are attracted to flowers that give off odors that we perceive as decay. The short-tongued flies are fairly crude, flying about and bumping into the anthers; one scientist described them as "unindustrious, unskilled and stupid." Some flowers show specific adaptations to take advantage of these attributes. The flowers of the Dutchman's Pipe *(Aristolochia macrophylla)* attract flies into tubular flowers, and they become imprisoned there in a floral trap where the flies buzz around madly trying to escape. When the flowers release them, the flies, not having learned a thing, enter another flower of the same species.

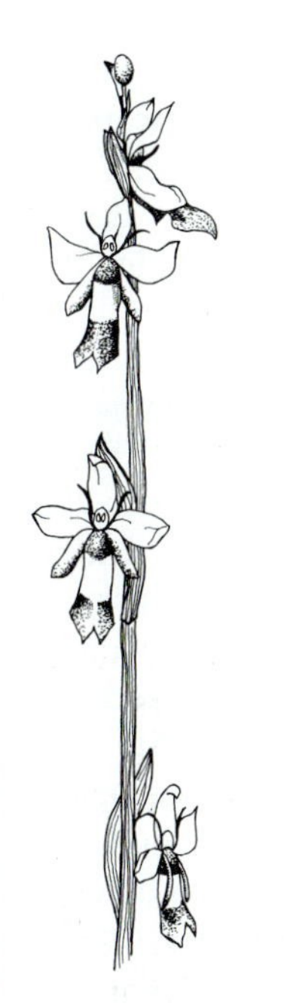

Figure 1.37 The wasp-mimicking orchid *(Ophyrs muscifera).*

Birds are very important pollinators. In the tropics, birds may be more important than bees. In the Americas, the bird pollinators are the jewels of the avian world—the hummingbirds (Figure 1.39). They are replaced by the sunbirds of Africa and Asia, the honey creepers of the Pacific islands, or the honey eaters and lorikeets of Australia. Most birds have excellent vision and a poor sense of smell; flowers visited by birds are usually red or yellow, colors birds see best. Hummingbirds rarely perch on flowers. They suck nectar while hovering with their wings beating so rapidly that they appear to us as a blur. Flowers visited by hummers are usually tubular and hang out and down, facilitating the insertion of the long slender bill and even longer tongue. African sunbirds cannot hover, however, and the flowers they visit are formed with a landing platform next to the flower. Since sunbirds (Figure 1.40) probe for nectar with pointed bills, the delicate parts of the plant are placed out of harm's way, but the anthers are where the bird cannot fail to brush its head against them. Somewhat different, but equally appropriate, floral adapations are found in flowers visited by honey creepers (Figure 1.41) and honey eaters. Even though these bird pollinators are not closely related, their heads and bills are very much alike. Lorikeets and other parrotlike birds are not very selective feeders and use their strong beaks to cut holes in the sides of the flowers to get at the nectar. In spite of extensive damage to petals, pollination is ensured because anther stalks are short and next to the nectar sacs so that the lorikeets have their heads dusted with pollen as they feed.

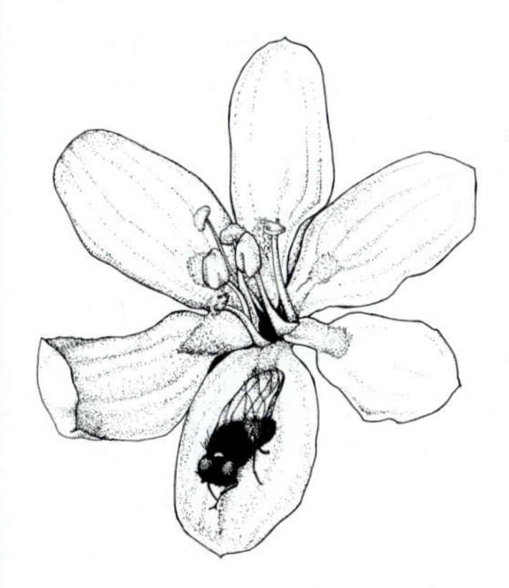

Figure 1.38 Pollination of a lily *(Zigadenus fremontii)* by a fly in the Syrphidae group.

Some bats (Figure 1.42) feed on nectar and on small insects present in the pools of nectar. Since bats are nocturnal, the flowers they visit are usually night-bloomers and their petals are usually white. The muzzles of pollinating bats are slender, their tongues are long and extendable,

Figure 1.39 Female ruby-throated hummingbird.

Figure 1.40 African sunbird feeding on heath flowers.

and their front teeth, which would otherwise be in the way, are either short or, in some species, undeveloped. Bats are guided to flowers by sight and by fruity odors. Of assistance to the bat is the fact that the flowers are usually large and raised above the foliage. In a few cases, bat-visited plants drop their leaves at the time of flowering, making the flower more accessible.

Pollen Archeology Because of their thick outer walls, pollen grains are resistant to decomposition and may last for long periods of time. Since they are also sculptured, identification of a pollen grain down to the genus level is possible. These facts have been the basis of a scientific specialty, the field called *pollen archeology,* started in Denmark by Johannes Iverson. Iverson reasoned that since we can identify pollen of specific plants, and since pollen can remain intact for thousands of years, we should be able to study the ebb and flow of plant life in a particular location by analyzing the pollen obtained in archeological sites. During the last ice age, Denmark was essentially scraped down to rock by the ice. As soil formed, the land was covered by forests, mainly spruce and fir as determined by pollen resting on rock. As forests became dense, game decreased and Man, then a hunter, retreated to the coasts and fished for a living. Sometime later, during the Neolithic Period, people returned and began to clear the land and burn the logs, as evidenced by a layer of spruce and fir charcoal just above the pollen of these species. For a period of time, pollens were mainly of herbs and of barley and wheat, proof that this land was being farmed. New weeds, those from the Danish coast, were mixed with local weeds. The farmers moved on, and their fields began to return to their natural state, first to weeds and then to species of trees that typically follow forest clearings; willows, aspen, and birch. The fact that birch was found is also an indication of clearing by fire, since birch requires high levels of soil phosphorus, which is found in wood ashes.

Apparently the people didn't completely abandon this land, because pollen of clover and other pasture plants was found, an indication that the land was in pasturage for goats and sheep. Later, the land was completely abandoned, and birch gave way to hazel and then to elm,

Figure 1.41 Hawaiian honey creeper.

Figure 1.42 A bat *(Leptonycteris)* feeding on flowers of the organ-pipe cactus *(Lemaireocereus thurberi).*

linden, and finally to oak. During this transition period, there was some renewed cutting and burning (charcoal again), and the land grew small crops and blueberries, a plant that seeds in after fires. Some climatic shifts occurred, for pollen of sedges and spores of sphagnum moss show that this area became marshy. A shift in topography allowed the land to drain and forests of spruce and fir developed. Again, people moved back, cleared the forests, and planted barley and wheat, but pollen of weeds native to Russia was found mixed with grain pollen. This suggested that the new farmers weren't Danes, but were likely to have been Finns, driven west by Russian invaders of their land. This prompted a search for evidence of human artifacts which, when found, were unmistakably from eastern Europe. In addition, pollen of pasture grasses was mixed with the cereal pollen, an indication that these people practiced some kind of crop rotation or left land in pasture for several years between plantings of grain. The eastern invaders interbred with the natives, and the land became permanently agricultural by the end of the Neolithic Period.

THE SEX LIFE OF FLOWERS

The flowers that bloom in the spring
Tra La.

GILBERT AND SULLIVAN

Sometimes when a child asks, "Where did I come from?" parents embark on a lecture on sex when the kid merely wants to know whether he or she came from Chicago or Toronto. In many instances, this discussion takes the classical birds-bees-flowers route, which is fine as far as it goes, but it is basically about how reproductive entities get together and not about sex as biologists understand it. Biologically, sex is the process by which two reproductive nuclei, each bearing one-half the complement of chromosomes, fuse to form a fertilized cell containing the full complement of hereditary material. All the rest, including the information in How-To books, is recreation, not procreation.

The correlation between intercourse and initiation of the next generation was obscure for many centuries. That there was a relationship was obvious; when men and women live together, children are born. But the relationship itself? In 2458 B.P. Eumenides of Aeschylus said: "The mother of what is called her child is no parent of it, but nurse only to the young life that is sown in her. The parent is the male and she but a stranger, a friend, who if fate spares his plant, preserves it till it puts forth." That the male serves as the procreator is seen in the biblical story of Onan, who masturbated and "spilled his seed on the ground." Aristotle concluded that males are warm and complete, while females, being colder or more passive, are less complete. She contributes matter and a place for development, and the male contributes his seed, which is the future child and its soul. This concept of the female as a mere receptacle held sway in Europe for 1500 years. What it did for the position of women can only be surmised, but it certainly didn't advance their dignity. The Koran states: "Have we not created you of a repulsive drop of seed which we placed in a sure depository until the fixed time of delivery?"

Identification of the child as the product of a seed implanted into the soil of the womb was accepted throughout most of the world, and the parallelism between copulation in humans and sowing crops was also widespread. Woman is the field, the furrow is the vulva, and the seed is the fructifying principle. So, too, did Oedipus "sow his seed in the sacred furrows," and Sophocles referred to the furrows of paternity. The act of planting seeds in the earth was homologous to human copulation, and this parallelism was fervently made manifest in spring rites.

Plants Do Not Have Sex! Dating from about the second century, people were concerned with the concept of intercourse as sin. Since the relationship between copulation and child-making was obscure, many believed that intercourse was unnecessary and a part of original sin. Because it was also pleasurable, it was even more sinful. And the intercourse of animals? . . . Were they also contaminated through Adam and Eve, and was copulation in animals necessary for procreation? This was a hard question, but since animals didn't have souls—or had lesser ones—it wasn't too immediate a problem, and it was pushed under the rug of scholasticism. Plants, on the other hand, were pure entities by definition and were clearly innocent of any taints (Figure 1.43). Their reproduction could not involve sex, for if even the lilies of the field reproduced by sexual means, every creature on earth was drowned in sin, and this couldn't be rationalized with the idea of a benign Creator. It didn't make sense that plants were sexual and hence sinful, and if this notion were promulgated by theologians, maybe these doctors of the Church were wrong about original sin and, perhaps, wrong about other things the priests thundered from their pulpits. It was incumbent upon learned theologians to insist that plants were without sex. And insist they did!

Figure 1.43 Fifteenth-century concept of the birth of the souls of little children in flowers. Taken from a woodcut by Meydenbeck, made in 1491.

Theophrastus, a pupil of Aristotle, described in detail the ancient North African method of initiating fruit formation in the date palm by dusting the female plant with a frond of the male flower. St. Augustine had read Theophrastus and lived in North Africa, and yet he explicitly denied that plants had any trace of a sex life. St. Thomas wriggled out of the problem during the thirteenth century by insisting that crude cross-breeding of plants was unnatural and thus contrary to moral law. Asians didn't have this hangup; the Chinese *Classic of the Golden Emperor* stated that the fine dust of plants was the male entity, and

Hindus of the Vedic period of 3000 B.P. were aware that "the golden powder" was needed to ensure fruit development.

Denial of sex in plants was still dogma by the end of the sixteenth century. Andrea Cesalpino, an influential professor of botany and medicine, stated in 1590 (following Galen, who followed Aristotle) that plants could not have sexual phase. The fruit, he insisted, was formed from buds produced by the pith and inner bark. In 1583 Dodoens wrote, "The flower we call the joy of trees and plants. It is the hope of fruit to come. . . ." William Harvey, newly famous for his discovery of the circulation of the blood, stated that the male principle conferred upon the passive human egg some vital force that caused it to develop into an embryo. The male principle of humans and other animals was discovered in 1667 by Antony van Leeuwenhoek and was identified as such, but either his report to the Royal Society of London was ignored, or it was assumed that the little "beasties" were parasites. Another Hollander, Jan Swammerdam, agreed with his friend that the things Leeuwenhoek saw were sperm and reported that Aristotle was right after all, because one could see a little man, a *homunculus,* coiled tightly within the head of the sperm. Yet another Dutchman, Reinier de Graf, found the human egg cell—actually he saw the ovarian follicle—and asked whether, perhaps, and with all due respect to Galen and Aristotle, this female cell might not play a more important role in embryo development than had previously been believed.

Well, Maybe They Do! Marcello Malpighi, born in Bologna in 1628, was one of the first of the enlightened breed of scientists to focus a microscope on plants. He published detailed, accurate drawings of flower structure, but concluded that the flower itself served only to purify the juices of plants, thus allowing seeds to develop. "The pollen dust," he said, "is a mere secretion and may be compared with the menstrual discharge of women." At the same time, the English plant anatomist Nehemiah Grew advanced the most appalling idea that the pistil of a flower is the female part and that stamens and their pollen were male. He spoke of male and female "juices" but copped out on the question of how seeds were formed except to suggest that plants might reproduce "like snails." This was essentially hermaphroditism, a neat idea since Grew shrewdly noted that stamens and pistils were frequently found in the same flower. Grew's ideas were, however, demolished by the pompous Joseph de Tournefort, professor of medicine and director of the Jardin des Plantes in Paris, who insisted that Aristotle could not be denied, that the scholastics were correct, and that this heretic Protestant, this country parson, this Grew, was dealing in superstition. John Ray came to his fellow Englishman's defense, stating that plants must have sex, since all other living things on earth did, but "for the life of himself," he didn't understand the mechanism.

Yet the stability of the form of flowers and the precision of their timing in spite of variations in the vigor and growth habit of the plants within a species suggested that there was something fundamental about flowers. Fifty years before Linnaeus, Joachim Jung used the flower as the main criterion for classifying plants, although he refused to comment on its function. In 1694 Rudolph Jacob Camerarius wrote *A Letter on the Sex of Plants* in which he discussed the ideas of the ancients and concluded that they really didn't know what they were talking about. He reported experimental studies done in the gardens of the University of Tübingen, where he removed the female parts of complete flowers and obtained no fruit. He also grew plants in which the pollen-containing flowers were separate from pistil-containing flowers and found that fruits developed only when they had been dusted with appropriate pollen. Being aware of what happened to heretics even at the end of the seventeenth century, Camerarius simply left off at this point without reaching any conclusions on whether what he had reported represented sex in plants.

In 1730, at age 23, Linnaeus published his *Introduction to Floral Nuptials.* In spite of the deliberately provocative title, the volume was entirely on plant anatomy and created a stir only in the limited botanical world. Linnaeus read Camerarius's writings and decided that plants could best be classified by comparing and contrasting the numbers and arrangements of their sexual parts. His views on sexuality in plants were most literal. Thus he described the condition of *monandism* as "one husband in a marriage," and used the term *polyandry* when a plant showed "twenty or more males in the same bed as the female." For the composite family—sunflowers and others with sterile ray and fertile disk flowers—he said that "the beds of the married occupy the disc and those of the concubines, the circumference." In 1760 he suggested a "new employment for botanists to attempt the production of new species of vegetables by scattering the pollen of various plants over various widowed females." Such shocking words made life difficult for botanists of the day. Many botanists were medical men whose interest in plants had been sparked by medicinal plants, or they were dedicated amateurs—gentle souls all—with a high proportion of ministers in their ranks. Linnaeus's taxonomic system had many sexual implications that were abhorrent, but the system worked—and worked beautifully. There ensued a series of intellectual juggling acts in which the implications were thundered down while the method continued to be used. The Rev. Samuel Goodenough, bishop of Carlisle, took indignant pen in hand to write a letter to a friend "to tell you that nothing could equal the gross prurience of Linnaeus' mind," and the celebrated Goethe worried publicly about the potential embarrassment of botany texts to chaste young minds.

Yes, They Do! The topic was, however, ripe for clarification. Father Spallanzini made "trousers of waxed fabric" for male frogs and prevented fertilization; the collected semen was used for artificial insemination of frogs. Thomas Fairchild cross-pollinated a carnation with a pink in

1717 and got a hybrid. Showing mixed characters, "Fairchild's mule" was famous throughout Europe. Gardeners on royal estates were reporting to their lords of the hybridizations of fruit trees, flowers, and medicinal plants. Buffon in France, von Haller in Switzerland, von Baer in Germany, and others had overthrown the male seed idea, isolated animal eggs, and worked out the fundamentals of fertilization and early embryonic development by the middle of the eighteenth century. Two botanists gave the death knell to Aristotelianism.

Joseph Koelreuter, curator of the botanical gardens in Karlsruhe, speculated in 1761 on sex in plants as a result of his extensive cross-breeding of medicinal herbs. His experiments involved pollen transfer, selective removal of floral parts, and an attempt to keep records on his crosses. His thoughts on plant sex are, upon modern reading, quite accurate, but they were so contrary to Church doctrine, and thus so abhorrent, that his contemporaries ignored him and his ideas.

In 1793, Christian Sprengel published *Nature's Secret Revealed: The Structure and Fertilization of Flowers.* Sprengel was the son of a Brandenburg Calvinist minister and was convinced that since nature is preconceived by the wisdom of God down unto the smallest thing, there was a purpose for everything on earth, even flowers. Sprengel had a lot of time on his hands. Although he had studied theology, he was a mere schoolteacher, a recluse, and a most irascible citizen without a social life. So, as he spent his days looking at the interactions between bees and flowers, he discovered that plants with flowers having both male and female parts could have the parts develop at different times to mandate cross-fertilization, and he saw that some plants were specifically designed for wind pollination. We can now be a bit condescending about Sprengel's failure to distinguish between pollination—the transfer of pollen to the pistil—and fertilization—the fusion of male and female nuclei—but he specifically used the word *fertilization* with the clear implication that plants did have sex. The philosophers of the day, churchmen all, had little patience with the concept of detailed research and the laborious reporting thereof. They couldn't understand Sprengel's book and, therefore, could safely ignore or denouce it. They did both, to the point where poor Sprengel became more of a recluse and even more irascible and eventually faded away.

By the early years of the nineteenth century, the last bastions of scholastic belief were falling. If the common people could have read the learned discussions denying sex in plants, they would not have believed that educated people could be so blind. Of course plants had sex. Why were they dusting pollen on flowers to get better crops or being so concerned that the bees worked in blooming apple orchards? The governor of Pennsylvania, James Logan, published experiments on pollinating Indian corn. Gregor Mendel was fully aware that pollen was necessary for the development of peas. But Victorian England and its American counterpart stood firm. Alfred, Lord Tennyson could write of "the wild flowers of a blameless life," while others, like Croisat, noted that the sex organs of plants were "blatantly displayed and act like husbands and wives in unconcerned freedom." Jamie Alston, professor of botany at Edinburgh, attacked the whole idea of sex in plants, although he used Linnaeus's system to reorganize the university herbarium. In 1829 a Mrs. Almira Lincoln of The Troy Female Seminary, in upstate New York, wrote a famous botany text, *Familiar Lectures in Botany,* for young ladies. It went through ten editions until 1849 and sold 275,000 copies. Botanical language contained several unmentionable words, and so she substituted *germ* for *ovary,* and the word *placenta* was never defined, although it had to be used. Mrs. Lincoln, in fact, never discussed reproduction in plants at all.

The Victorian lady who diligently cultivated orchids in her conservatory was either unaware of or repressed the knowledge that the name is derived from the greek *Orchis,* meaning testicles. In rural England, however, native orchids were called dog's stones or bull's bags. A careful look at the Unicorn Tapestry in the Metropolitan Museum of Art clearly shows the orchid as a sexual symbol. In his book *On the Various Contrivances by which British and Foreign Orchids are Fertilized by Insects and On the Good Effects of Crossing,* Charles Darwin demonstrated that insect pollination was a prelude to a sexual fertilization in every way comparable to that of animals. This really struck home because the orchids, of all nature's creations, were held up to the youth of England as a perfect example of botanical art for art's sake—pure, innocent, chaste. That orchids have traps, nectars, perfumes, and structures specifically to attract insects that ensure pollination and hence fertilization was as distasteful to a Victorian curate as the thought that the Queen's children were conceived by the same procedure as were those of the least of his congregation. Since Victorian young ladies received instruction in botany as a suitable avocation, and the curate was frequently the local botanist, how was he to explain to these girls the function of the pollen grain, the discharge of the sperm nuclei through the micropyle, and the union of sperm and egg nuclei? No, this he could not do. By 1870 botany was no longer part of social instruction. Painting, music, poetry, and needlework were more appropriate; our legacy is some charming petitpoint, some horrible watercolors, and a tradition of genteel poets. It wasn't until after World War I that many women entered botany departments.

If one wishes to be stuffy about it, the word sex has gotten out of hand. Biologically, sex is the fusion of two gametes (sex cells or nuclei), each bearing one-half of the total hereditary apparatus (genome) of the animal or the flowering plant. This fusion restores the complete genome of the organism. Biologically, sex allows the variation in the genome that comes of receiving genes from organisms with different hereditary characteristics and thus may permit variation in structure and function of the resulting organism.

When a pollen grain lands on the stigma, sugars and other chemicals stimulate it to germinate with the formation of a pollen tube that grows down the style, directed by substances diffusing from the ovary (Figure 1.44). During growth, the reproductive pollen nucleus divides again, resulting in the formation of two sperm nuclei. When the pollen tube reaches the ovule, it bursts at the tip, releasing the two generative nuclei, one of which fuses with the egg nucleus to reinitiate the normal, vegetative, diploid number of chromosomes. The other sperm nucleus combines with the two endosperm nuclei to initiate a special tissue, the *endosperm,* which has one and a half times the diploid number of chromosomes per cell and serves to nourish the developing embryo derived from the sexual fusion of sperm and egg nuclei. Sometimes, as in young coconuts or the "milk" of sweet corn kernels, the endosperm may be liquid.

FRUIT AND SEEDS

Scientists have observed the development of the embryo from the fertilized egg (the *zygote*) in many different species. While there are several distinct patterns of development, the result is the formation of

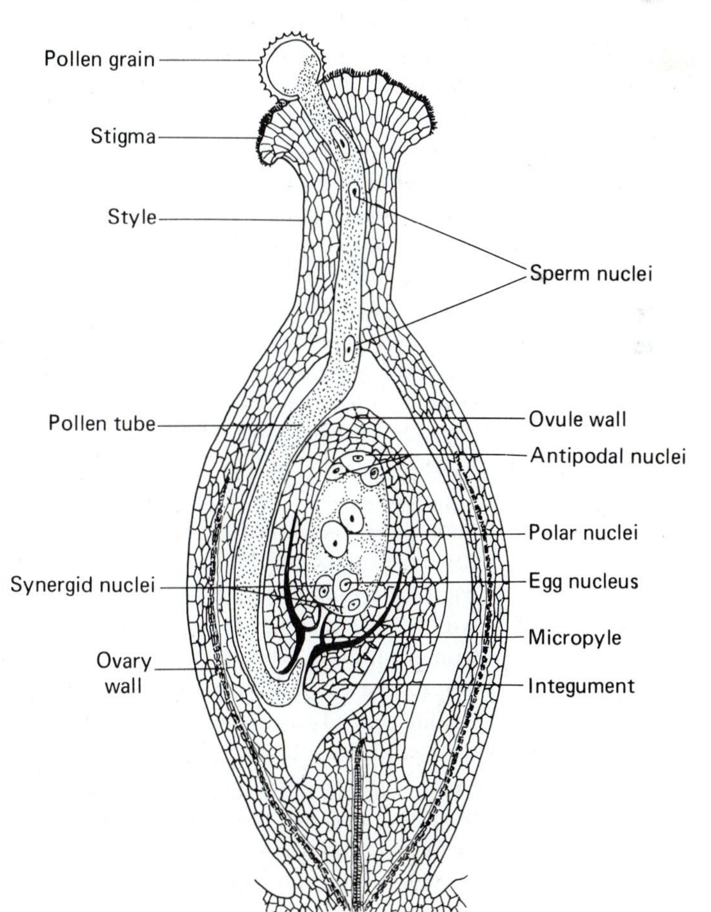

Figure 1.44 Representation of a carpel of a flowering plant showing the development of the pollen tube to the ovule. The bursting of the tip of the pollen tube releases the two male nuclei through the micropyle to the ovule, where one nucleus fuses with the egg (female) nucleus to reestablish the diploid (2N) chromosome number. The result is the formation of the zygote—the first cell of the new plant generation. The other male nucleus fuses with the pair of polar nuclei to form the 3N endosperm, which forms a tissue that nourishes the developing embryo.

an embryo possessing rudimentary root, shoot, and leaves surrounded by tissues that provide the nourishment for the germination and early growth of the new plant when it germinates. The ovule walls (see Figure 1.44) become seed coats that, at maturity, may be dry and papery, as in the peanut, or hard and stony, as in the peach or the cherry.

Fruit Development The fruit is derived from the ovary wall and, at maturity, may be fleshy, woody, or dry depending on the species. The conifers and related plants are called *gymnosperms,* or "naked-seed plants." In the flowering plants, the *angiosperms,* the seeds receive additional protection by being enclosed within the fruit, which is defined as the ripened ovary plus contained, mature seeds. It is believed that the major signals for the development of the fruit come from the growing embryo or embryos. At least some of the signals are growth substances synthesized by the embryo. In some plants, treatment of unfertilized carpels with growth substances will induce fruit growth without seed growth. Such fruits are referred to as *parthenocarpic.* In other plants, the seeds abort early in development, but the stimuli for fruit growth has been received by the carpel wall and it will still go ahead to form a fruit. The banana and some of the seedless citrus fruits are examples.

Seeds possess many characteristics that are useful or even essential for the continuity of generations and the reproductive survival of the species. Each seed can be defined as a young plant of the species in a box (the seed coats) with its lunch (the food stored in cotyledons or endosperm). In pines and other gymnosperms, the seeds are borne on cone scales and are exposed to the elements. In some species, the seed can germinate and form a new plant as soon as it is mature, but in other species, there must be a period before germination can begin. For some plants, particularly those that grow where winters are harsh, a dormancy period enables the seed to avoid germinating in the middle of the winter. In some desert plants, germination will not occur until the seed coat has been washed free of germination inhibitors by heavy rains. These and other protective mechanisms are of considerable ecological significance.

How Seeds Get Around If a young plant is to have a reasonable chance to develop, competition for light, water, and nutrients by well-established plants should be minimal. Where seed production is heavy, widely dispersed seeds can develop more rapidly when sibling competition is reduced. Dispersal also allows seeds to be deposited in habitats where the variation consequent upon the genetic recombination of sexual reproduction may be expressed more fully. Parent plants may also produce substances that repress the growth of other plants of the same and other species. This phenomenon is termed *allelopathy* and is assumed to result in a lowered competition for the producing plant.

Most of the dispersal mechanisms used by plants are fairly obvious, although little is really known of the evolutionary forces that shaped the

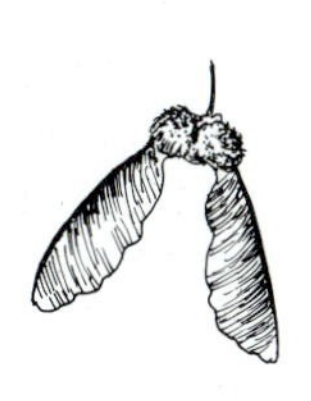

Figure 1.45 Adaptations for seed dispersal. Maple seeds and dandelion fruits (achenes) are dispersed by wind, while the fruits of burdock are moved by animals.

mechanisms (Figure 1.45). Dispersal by wind is seen in those plants whose seeds (or fruits) have plumes of hairs that can waft the seeds for long distances. The fruit of the dandelion, and seeds of thistle, firewood *(Epilobium),* poplar, and milkweed *(Asclepias)* are examples. Winged seeds of maple and ash spin like miniature propellers, those of elms soar like Frisbees for some distance. In a few plants—the tumbleweeds *(Salsola)* and Jim Hill mustard *(Sisymbrium)*—the whole plant may catch the wind and roll for miles, scattering its seeds across the prairie. The tiny seeds of many orchids are carried like pollen over land and oceans.

Visitors to islands of the Pacific have seen coconuts sprouting that may have matured on coconut palms thousands of miles away and took months on their sea voyages. So, too, can seeds of water lilies be carried, kept afloat by an air-filled sac (aril) surrounding the embryo.

The presence of oak and nut-tree seedlings growing in lawns attests to the efficiency of animal dispersal by squirrels and other rodents. Attractive fleshy fruits of cherry and related plants and of some seeds are eaten by birds, who may deposit the seed hundreds of miles from the parent plant. Seeds and fruits of cocklebur, burdock, and Queen Anne's lace bind tenaciously to hair, fur, feathers, and clothing. Some animal dispersal mechanisms are less obvious but no less effective. Seeds of marsh plants can be carried in the mud plastered to the feet of wading birds or the fur of water animals. As long-distance voyages became common in the fourteenth and fifteenth centuries, people became important vectors of dispersal. In many instances, such transport was deliberate (see Chapter 3), while in other cases the plants were unnoticed stowaways in baggage or in ship's ballast, or as contaminants in batches of desirable seeds. Not only have many of our economically desirable plants been dispersed throughout the world, but some of the most noxious and persistent weeds are now ubiquitous because of inadvertent human distributions.

ADDITIONAL READINGS

Barth, F. G. *Insects and Flowers: The Biology of a Partnership.* Princeton, N.J.: Princeton University Press, 1985.

Dimbleby, G. W. *Plants and Archeology.* New York: Humanities Press, 1967.

Faegri, K., and L. van den Pijl. *The Principles of Pollination Ecology.* 3rd Ed. Elmsford, N.Y.: Pergamon Press, 1979.

Farley, J. *Gametes and Spores: Ideas About Sexual Reproduction.* Baltimore: Johns Hopkins University Press, 1982.

Fritsch, K. von. *Bees.* Ithaca, N.Y.: Cornell University Press, 1972.

Gilbert, L. E., and P. H. Raven, *Coevolution of Animals and Plants.* Austin: University of Texas Press, 1975.

Holm, E. *The Biology of Flowers.* New York: Penguin Books, 1979.

Iverson, J. "Forest Clearance in the Stone Age." *Scientific American* 194 (1956):36–41.

Jaeger, P. *The Wonderful Life of Flowers.* New York: Dutton, 1961.

Meusse, B., and S. Morris. *The Sex Life of Flowers.* New York: Facts on File, 1984.

Percival, M. S. *Floral Biology.* Elmsford, N.Y.: Pergamon Press, 1965.

Pilj, L. van den. *Principles of Dispersal in Higher Plants.* New York: Springer-Verlag, 1982.

Stanley, R. G., and H. F. Linskins. *Pollen.* New York: Springer-Verlag, 1973.

West, R. G. *Studying the Past by Pollen Analysis.* Oxford Biology Reader No. 10. New York: Oxford University Press, 1971.

Whitehead, D. R. "Wind Pollination: Some Ecological and Evolutionary Perspectives." In L. Real (ed.). *Pollination Biology.* New York: Academic Press, 1983.

Chapter 2

Filling Empty Bellies

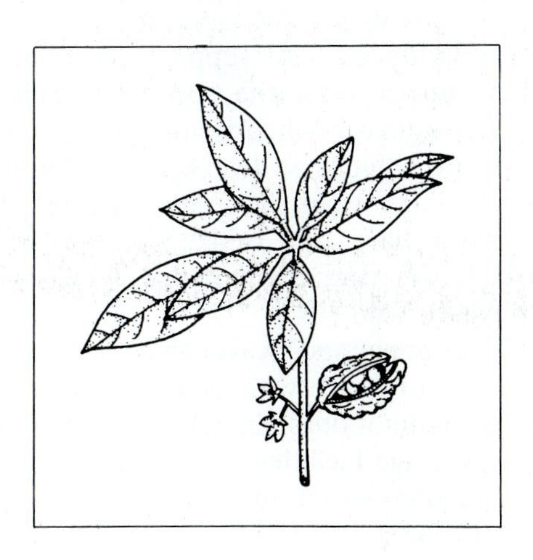

Agricultural Beginnings

And you shall eat the plants of the field.

Genesis 3:18

The origin of agriculture can only be inferred, since no completely satisfactory anthropological data have emerged. However, certain reasonable assumptions can be made. About 250,000 years ago in Africa, where the human genus evolved, small family bands of *Homo erectus* and *H. habilis* obtained their food by scavenging and foraging—collecting plants and capturing small animals in their wanderings. Gradual dispersion of these early peoples from Africa to Europe, the Middle East, and Asia, some evolution in the species, and slow increases in population occurred. By 50,000 years ago, when modern Man *(Homo sapiens)* appeared, units of people had established home bases from which they could range over a modestly sized area to hunt, gather wild edible plants, and return to the base. Fire had been discovered, stone tool construction was well advanced, and vessels of gathered gourds, hollow logs, and holes in rock permitted storage and some stability of food supplies. Because population could increase rapidly, dispersal of the species became necessary as the hunter-gatherers exhausted the resources of an area. Their movements ranged widely into habitats as diverse as tropical forests and arctic tundra. By perhaps 30,000 years ago, the migration of people to the New World took place via a land bridge over the Bering Straits that connected Alaska to the Asiatic mainland. Australia was colonized somewhat later from south Asia.

The hunting and scavenging way of life continued for several thousands of years with increasing sophistication in tools, possible domestication of dogs, and elaboration of life styles—clothing of skins, pottery storage facilities, the development of ritual, and regularization of interaction within and among groups of people. Our assumptions about the life styles of preagricultural peoples are, to some extent, based on analogy with preagricultural societies of our time, including the Inuit and Eskimo peoples, some tribes of North and South American Indians, and central African groups.

The transition from gathering to agriculture took a long time and occurred independently in different parts of the world. For the European–Middle East region, the threshold of deliberate food production took place between 9000 and 11,000 B.P. in the area of the Middle East named the Fertile Crescent, an arc extending from modern Lebanon to the Persian Gulf (Figure 2.1). Excavations in Jarmō, an ancient settlement in modern Iraqi Kurdistan, and in Tepe Sarab in Iran

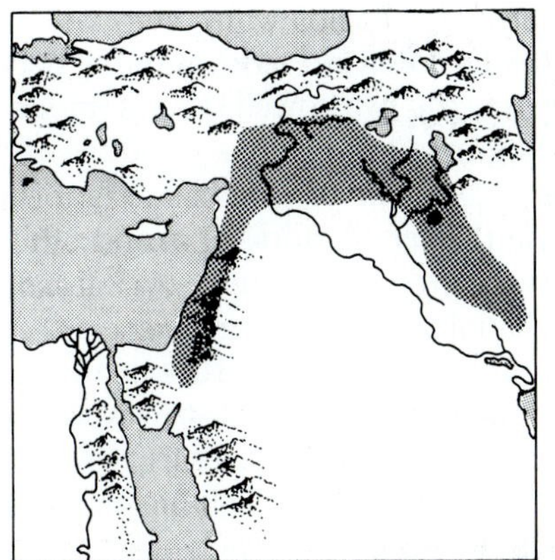

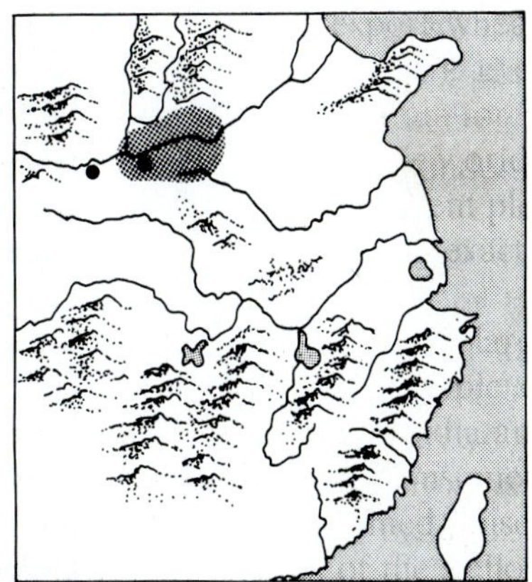

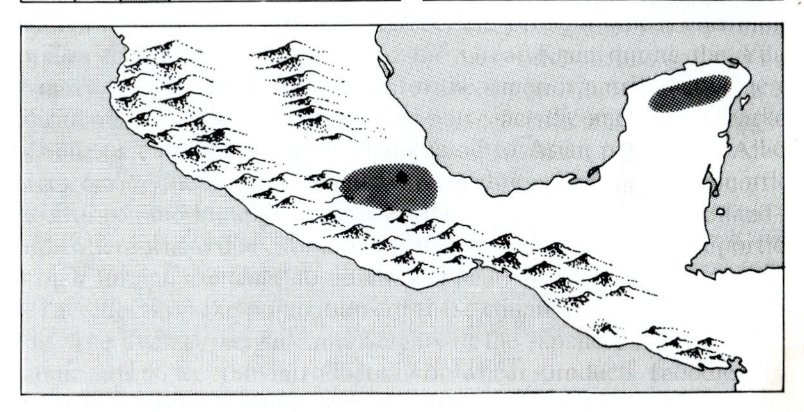

Figure 2.1 Three major sites where agriculture was established. The Fertile Crescent embraces modern Israel, western Syria, Afghanistan, and the region down to the Persian Gulf. The Chinese area centers at Yangshao on the Yellow River in north China. There may have been two Meso-American sites, one in Tehuacán in the Valley of Mexico and the other in the Mayan lowlands near Yucatán.

have yielded sickles, pottery, and bones of dogs and goats as well as carbonized grains of wheat *(Triticum)*, barley *(Hordeum)*, and the millets *(Panicum* and *Pennisetum)* dating to 10,000 years B.P. These lands, recently freed from the burden of the last ice age, were grassy with forest areas, were well supplied with water, and had a climate that included winter and spring rains. Population increase was probably a major impetus for food production, but factors of convenience or plain laziness were involved. How cultivation began is highly speculative; observations of food plant seed sprouting in refuse heaps, the germination of seed in storage pits, and other possibilities have been suggested. Whether these first cultivated plants were deliberately grown from those already in use or were plants that were difficult to gather is unknown; both hypotheses are probably correct. In either case, the domestication process—the origin of agriculture in the Fertile Crescent—took close to 1500 years for cereal grains. One might say that crops and Man coevolved.

Major attention has been paid to the domestication of the cereal grains because these grasses provide people with their basic carbohydrate and protein foods (Chapter 3). The cereals supply nourishing food, but because they must be sown and harvested at definite times of the year, early agriculturists had to be settled, with at least the rudiments of village life. In the Fertile Crescent, both barley and wheat were under intensive cultivation by 8000 years B.P. The first cultivated barley was the two-rowed species (modern barleys have either two or six rows of grain on each spike), and the earliest wheat was the wild emmer species, *Triticum dicoccoides.* This wheat was minimally satisfactory because the seed-holding spike becomes brittle as the plant ripens and the grains drop off. Plants with a gene that controls the production of tough, nonbrittle spikes are more difficult to grow, but a reaper would collect more grain from such plants. Over time, such variants, now believed to include several species with different chromosome numbers, came to dominate the seed source for future planting. Such selection of mutants has long been the basis for advances in agricultural productivity and is now supplemented (but not replaced) by deliberate breeding and genetic engineering.

The domestication of animals occurred concomitant with the development of Middle East agriculture. Perhaps the human interest in pets was an initiating factor, as was the "imprinting" tendency of newly born animals to attach themselves to the first moving creature they see or hear. Animals were used as hunting decoys and as readily available sources of meat, milk, and skins. Selection toward rapid maturation and against aggressive behavior produced submissive, manageable herds of goats, sheep, and cattle.

The dispersion of agriculture from the Fertile Crescent to Europe, North Africa, and eastward, apparently was accomplished by 6000 B.P. In the new areas, native crops supplemented wheat, barley, millet, and legumes. The stomach contents of an Iron Age man found in Denmark contained over 50 species of plants. Other crop plants were inadvertent travelers, accompanying the main crop. Rye and oats, mildly troublesome weeds of wheat in the Mediterranean, became dominant cereals in northern Europe because they germinate and develop at lower temperatures than wheat.

Equally important for the evolution of civilization was the continued development of the technical arts of toolmaking and the sophistication of fine arts. Leisure time allowed people to make their lives more pleasant and comfortable. Community life facilitated, indeed required, the development of government, a legal framework, organized ritual, and other good and bad consequences of a sedentary life. All these attributes of civilization are based on agricultural surpluses over immediate food needs. This abundance allowed some individuals, freed from the pressures of obtaining food, to devote their energies to other activities, including toolmaking, art, religion, and, most important, thinking.

There is increasing evidence that plant and animal domestication occurred independently in Southeast Asia. Excavations of Non Nok Tha in northern Thailand and Spirit Cave in Kampuchea show that rice cultivation, the husbandry of humped cattle, and bronze metallurgy flourished at least 7000 B.P., suggesting that the origin of these complex agricultural civilizations predated those of the Fertile Crescent by 1000 years. Contemporaneous sites have been found on Taiwan. Sophisticated storage pottery in India and China has been dated to 6000 B.P., certainly matching the initiation of civilization in the Middle East and predating that of Europe.

When Europeans reached Mexico and South America in the sixteenth century, they destroyed several advanced civilizations. These had their origin somewhat later than those of the Old World. The Indians brought a large number of plants into agricultural production. Staple carbohydrate crops included corn as the cereal as well as potatoes, squashes, and beans. Since corn (maize) is a species developed by and dependent upon agricultural practice, the excavation of corncobs and corn pollen effectively dates the origin of sophisticated agriculture to 5000 years B.P. in Mexico, where people had been living for 12,000 years. In contrast to the Old World, the Indians domesticated few animals (dogs and the llama); all New World cultures depended on hunting for meat, furs, and other animal products.

The tropics played a seminal role in the origin of agriculture. The ecosystems of tropical lands, with their great diversity of species, inherent structural and functional stability, and complex food chains, differ greatly from temperate lands of Southeast Asia, the highlands of South America, or the Mediterranean. Early agricultural practices in temperate zones evolved toward monoculture (fields containing a single crop species), consequent simplicity, and an emphasis on continued use of fixed-plot, clearly delineated fields. The agricultural integration of crop plants and animals was, and in many cases still is, restricted to temperate regions. Such practices are economically and agriculturally efficient; the assured diet increased the opportunities for an intensive selection of efficient species with emphasis on storable grain, root, and legume crops. Many Old and New World tropical peoples practice shifting cultivation (swidden or slash-and-burn) and intercropping of several agricultural products. The limited ability of tropical people to store crops for long periods of time, rapid maturation of vegetable crops under the uniform climatic conditions, year-round agriculture, and the inability of the soil to maintain fertility over a period of years resulted in swidden techniques that are efficient so long as population size remains small (see Chapter 10).

Although seemingly a prosaic consideration, the development of agriculture depended heavily on weed control. Because thrifty farmers recognized that weeds robbed their crops of water and nutrients, they pulled weeds by hand. The Romans sought the intercession of the goddess Spiniensis, whose aid was invoked to keep thorns out of their fields. Rooster blood was sprinkled in grain fields to kill dodder and bindweed, and virgins unbound their hair and carried yellow and black chickens around the fields. Farmers in the Mideast used mummy dust mixed with sparrow's blood, and Chinese rice farmers burned incense whose smoke wafted across the rice paddies.

The hoe, sickle, weeding hook, and plow were employed for centuries to weed between rows, a task (like housework) that is never finished. Methods for freeing seed of weed seeds were used as early as the tenth century. Seeds were soaked in lime or salt brine to allow the weed seeds to float to the top of the container. Denmark, France, and other European countries recognized that an unweeded field was a source of contamination for the entire area and required farmers to use weed control. Chemical weed control—the use of herbicides—was long a dream of agriculturists. Democritis recommended lupine flowers steeped in hemlock to kill persistent rootstocks. Varro (b. 2027 B.P.) noted that the soil around olive trees was weed-free and suggested that farmers sprinkle their fields with the water skimmed from the oil resulting from pressing olives. Salting of beets and asparagus killed weeds and promoted the growth of these salt-loving plants. Arsenic compounds and other herbicides were employed from the fourteenth century but were known to poison the land and the people. The modern era of the development of selective herbicides dates only from the 1940s. Many are now used for different crops, although the problem of land injury is not yet satisfactorily settled.

ADDITIONAL READINGS

Bender, B. *Farming in Prehistory: From Hunter-Gatherer to Food-Producer.* New York: St. Martin's Press, 1975.

Ebeling, W. *The Fruited Plain: The Story of American Agriculture.* Berkeley: University of California Press, 1979.

Harlan, J. R. *Crops and Man.* Madison, Wis.: American Society of Agronomy and Crop Science Society of America, 1975.

Harris, J. R. "The Origins of Agriculture in the Tropics." *American Scientist* 60 (1972):180–193.

Hawkes, J. G. *The Diversity of Crop Plants.* Cambridge, Mass.: Harvard University Press, 1983.

Heiser, C. B., Jr. *Seed to Civilization: The Story of Man's Food.* San Francisco: Freeman, 1973.

Hyams, E. *Plants in the Service of Man: 10,000 Years of Domestication.* London: J. M. Dent, 1971.

Langer, R. H. M. *Agricultural Plants.* New York: Cambridge Univ. Press, 1982.

Leonard, J. N. *The First Farmers.* New York: Time-Life Books, 1973.

Needham, J., and F. Bray. *Sciences and Civilization in China.* Vol. 6, Part 2, *Agriculture.* Cambridge, England: Cambridge University Press, 1984.

Renfrew, J. M. *Paleoethnobotany: The Prehistorical Food Plants of the Near East and Europe.* New York: Columbia University Press, 1983.

Richardson, W. N., and T. Stubbs. *Plants, Agriculture and Human Society.* Menlo Park, Calif.: Benjamin, 1978.

Rimdos, D. *The Origin of Agriculture: An Evolutionary Perspective.* New York: Academic Press, 1984.

Simmonds, N. W. (ed.). *Evolution of Crop Plants.* London: Longmans, 1976.

Verril A. H. *Foods America Gave the World.* Boston: Page, 1937.

Origin of Our Food Crops

Who among our ancestors first had the courage to taste rhubarb?

A visit to a supermarket will show that our plant foods are varied and abundant. Surprisingly few, however, are native to North America, but were brought into cultivation in many different parts of the world. North America has provided the sunflower, cranberries, bramble fruits, maple syrup, and wild rice (*Zizania*) (Table 2.1). Some staple crops that we think of as native originated in Central and South America. Corn, one of the three primary cereals, came from the highlands of Central America and was dispersed throughout the Americas long before Columbus arrived. The same is true of potatoes, peppers, tomatoes, beans, and the squashes.

Table 2.1 PLANT PRODUCTS ORIGINATING IN NORTH AMERICA

Common name	Scientific name	Use
Avocado	*Persea americana*	Food
Black walnut	*Juglans nigra*	Food
Butternut	*Juglans cinerea*	Food
Canada balsam	*Abies balsamea*	Laboratory use
Cascara	*Rhamnus Purshiana*	Medicine
Ceriman	*Monstera deliciosa*	Food
Chayote	*Sechium edule*	Food
Cranberry	*Vaccinium macrocarpon*	Food
Guayule	*Parthenium argentatum*	Rubber
Henequen	*Agave fourcroydes*	Fiber
Hickory	*Carya ovata*	Food
Jalap root	*Ipemoa purga*	Medicine
Jerusalem artichoke	*Helianthus tuberosus*	Food
Juniper	*Juniperus communis*	Flavoring
Maple sugar	*Acer saccharum*	Sweetening
Milkweed	*Asclepias syriaca*	Fiber
Mushroom	*Agaricus campestris*	Food
Pecan	*Carya Pecan*	Food
Piñon nut	*Pinus edulis*	Food
Redwood fiber	*Sequoia sempervirens*	Insulation
Sarsaparilla	*Smilax officinalis ornatus*	Flavoring for beverage
Sassafras	*Sassafras albidum*	Flavoring, medicine
Senega root	*Polygala senega*	Medicine
Sisal	*Agave sisalana*	Fiber
Spanish moss	*Tillandsia usneoides*	Fiber
Turpentine	*Pinus palustris*	Paint industry
Vanilla	*Vanilla planifolia*	Flavoring
Wild rice	*Zizania aquatica*	Food

Among the first to conduct a systematic study of the centers of origin of cultivated plants was Nikolai I. Vavilov, a Russian economic botanist. Vavilov believed that the plants that people adapted for use were initially dissimilar to their immediate wild relatives and hence must have been easy to identify. He suggested that they arose in the wild by mutation and by isolation in eight independent centers of the world. Although there have been some modifications of Vavilov's concepts and conclusions, his ideas have been generally confirmed (Figure 2.2). Vavilov ran afoul of the Soviet authorities and died in 1943 in Saratov prison. In some cases it has been possible, using genetic and anthropological techniques, to trace the course of domestication of plants from their wild ancestors and to follow the dispersal of specific plants from their centers of origin.

The dispersal of food plants probably started as people migrated, taking their crops with them. Few crop plants have been dispersed without human intervention: coconuts and sweet potato capsules may have floated across the Pacific Ocean, but human travel may also have been involved. The interchange of plants among African, European, and Asian peoples is associated with both migration and trade. The Sabeans maintained a sea route along the shore bordering the Indian Ocean and the Persian Gulf. The Fifth Dynasty Egyptian Queen

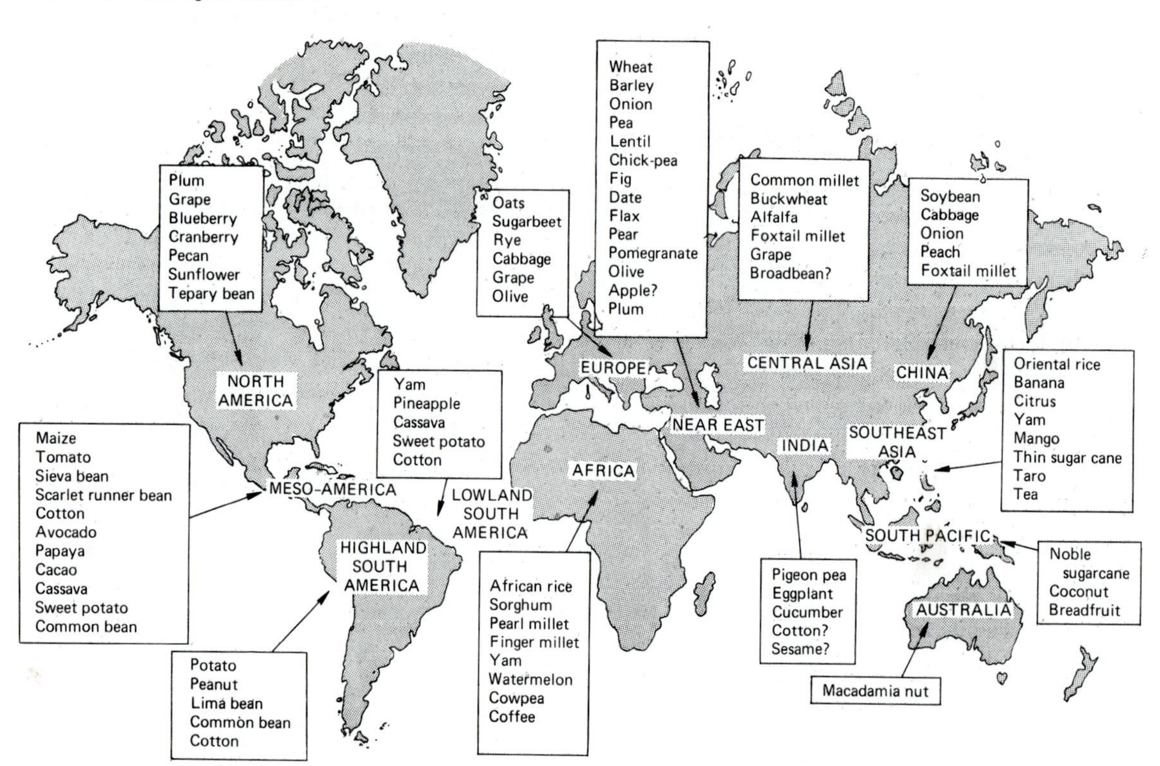

Figure 2.2 Presumed locations where major crop plants were domesticated. From J. R. Harlan. *Scientific American* 235 (3):88–7. Copyright 1976 by Scientific American, Inc. All rights reserved.

Hatshepsut (2500 B.P.) is reported to have brought tree seeds from the Land of Punt for her gardens in Luxor. Movement of plants overland from India and China to the Middle East was common long before the conquests of Alexander the Great, who brought many plants back to Europe. As Europeans embarked on bolder voyages at the end of the Dark Ages, trade with the coastal regions of Africa, the Middle East, and into India, Indonesia, and Japan was initiated. By the end of the fifteenth century, plants had been introduced into the Americas, and native plants of Central and South America had reached Europe. When Europeans settled in North America, they brought with them the plants they had been growing.

Benjamin Franklin and Thomas Jefferson made it a point to arrange for cuttings and seeds of promising plants to be sent to the colonies, a practice that was formalized after the American Revolution. Consuls and ships' captains were instructed to engage in plant accession by President John Quincy Adams. By the end of the nineteenth century, the United States and Canada had plant introduction sections in their Agricultural Departments whose mission was to conduct plant explorations throughout the world and to determine the potential use of the plants in North America.

ADDITIONAL READINGS

Cobley, L. S., and W. M. Steele. *An Introduction to the Botany of Tropical Crops*. 2nd Ed. New York: Longman, 1977.

Harlan, J. R. "The Plants and Animals That Nourish Man." *Scientific American* 235, no. 3 (1976):88–97.

Heiser, C. B., Jr. "Origin of Some Cultivated New World Plants." *Annual Review of Ecological Systems*. 10 (1979):309–326.

Sauer, C. O. *Agricultural Origins and Dispersals*. Cambridge, Mass.: M.I.T. Press, 1969.

Schery, R. W. *Plants for Man*. 2nd Ed. Englewood Cliffs, N.J.: Prentice-Hall, 1972.

Zevin, A. C., and P. M. Zhukovsky. *Dictionary of Cultivated Plants and their Centres of Origin*. Wageningen, Netherlands: Centre for Agricultural Publishing, 1975.

Food Production

The race between the stork and the plow

Agricultural productivity is necessarily a somewhat imprecise concept, and its definition depends on the criteria used to measure it. At a fundamental level, it is determined by the amount of usable product generated during the life cycle of a crop per unit of land cultivated (Table 2.2). Bushels of apples or metric tons of wheat, board feet of lumber or the dollar value of the crop per acre or hectare are frequently used. But other considerations must enter the calculations. A bushel of wormy apples or a metric ton of shriveled, protein-poor wheat has less nutritional or monetary value than superior apples or plump grains.

Table 2.2 WORLD PRODUCTION OF MAJOR FOOD CROPS

Food group	Major crop	Production (Millions of metric tons)
Cereal grains	Wheat	390
	Rice	370
	Corn (maize)	350
	Barley	180
Roots and tubers	White potato	300
	Sweet potato and yam	160
	Cassava (manioc)	110
Sugar	Sugar cane	740
	Sugar beets	260
Legumes	Soybean	80
	Beans and peas	40
	Peanut	18
Oil plants	Cotton seed	27
	Sunflower	12
	Olive	8
	Kanola (mustard)	8
	Oil palm	5
	Sesame	2
Tree nuts	Coconut	30
	Walnuts, etc.	3
Fruits	Wine grapes	60
	Banana	40
	Orange	35
	Melons	28
	Plantain	20
	Apples	20
	Pears, peaches, etc.	20
	Mango	12
	Pineapple	7
	Date	2
Vegetables	Tomato	45
	Cabbage family	30
	Onions and garlic	20
	Cucumber, squashes	15
	Bell and chili peppers	6

Table 2.2 (continued) **WORLD PRODUCTION OF MAJOR FOOD CROPS**

Vegetables	Peas and beans	8
	Carrots	8
Beverage crops	Coffee beans	5
	Tea leaves	2
	Cocoa beans	2

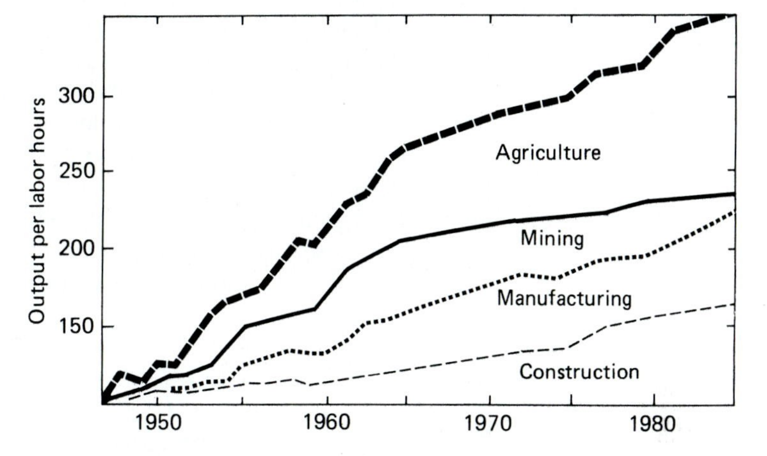

Figure 2.3 A comparison of the productivity of American agriculture with the productivity of other basic industries. Data from the U.S. Department of Agriculture.

Productivity can also be measured in physical energy units, usually as kilocalories used to produce a food crop relative to the kilocalories of food energy in the crop. This can be viewed as a return on investment (Figures 2.3 and 2.4). Economic production will, of course, vary with the crop plant or animal produced and the type of agriculture practiced. The investment in the harvesting of wild rice, for example, is low: some human labor, a canoe, packaging, and transportation. Because its dollar value is high and its nutritional value is moderate, return/investment ratios are high. The ratio for a crop like tobacco, on the other hand, is negative from both the nutritional and the health standpoint, but it is quite high if the return is measured in dollars.

Agriculture in industrialized countries is based on production in which energy inputs from human labor have been replaced by sophisticated machinery whose energy costs (manufacture, maintenance, fuel) are high. Fertilizers, herbicides, and pesticides, absolutely necessary for continued cropping of land and for high production, also require large investments for fabrication and application. In terms of nutritional value the return/investment ratio of cereal grains is high, although the ratio in

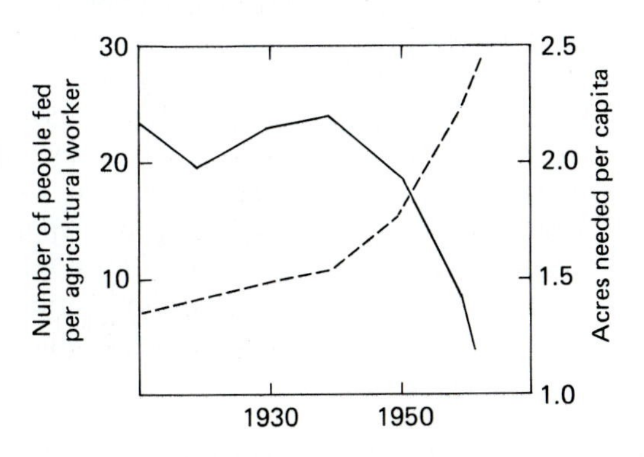

Figure 2.4 Productivity of U.S. agriculture. Introduction of efficient machinery and intensive cultivation practices has decreased the amount of land needed and the number of agricultural workers.

terms of dollar value will vary with market factors and may be less than 1 when farmers receive less per bushel then their production costs.

Similar calculations can be made for production of food animals. Wheat or corn used directly as human food will have a high human nutritional return/investment ratio, but a low ratio when the wheat or corn is used to feed cattle we then eat. In large part, this is due to an almost inflexible biological rule that states that at each move of energy in a food chain from a producer (a plant) to the final consumer (a steak inside your stomach), there is only about a 10 percent retention of nutritional value. Thus it requires close to 10 pounds of wheat protein to add a pound of protein to a steer's body, and we build less than 0.1 pound of steer protein into our proteins. Of the common domestic animal food sources, swine and chickens are the best converters, with efficiencies closer to 15 percent—one reason why Asians and African societies favor these animals over cattle.

RESOURCES FOR PRODUCTION

The resources available for agricultural productivity can be considered in several major categories. Primary among these are physical resources, including soil, climate, water, and energy—virtually the four basic elements of the Greeks: earth, air, water, and fire (one type of energy). Soil is finite. The area of land surface on earth is fixed, and virtually all of the best-quality land is already in production or is unavailable because it bears roads, cities, and other human constructions.Some lands are completely unsuitable for crops. Marginal lands, including semiarid regions, have been brought into production; others may, with sufficiently high investments, become productive. Soil can be modified by fertilizers and conditioners, again with known costs.

We can do little about climate or its short-term component, weather, except to complain or enjoy it. The exploitation of a particular crop plant is limited climatically by many factors. Obviously, bananas are not a viable fruit crop in Minnesota and rice can't be grown in the high and dry plateaus of Wyoming. The limits for productivity are largely set by prevailing climatic factors, including temperature, availability of water, and soil quality. Wheat is more productive in regions with more water than the Great Plains, but economically adequate crops can be grown there; this semiarid land is unsuitable for many other crop plants.

Temperature is a key limiting factor. Potatoes, a major carbohydrate food crop, will form tubers only when nighttime temperature falls to 50–57° F (10–14° C), effectively restricting the plant to the temperate zone (Chapter 3). On the other hand, manihot (manioc, or cassava), the potato's tropical nutritional equivalent, cannot withstand summer night temperatures in the temperate zone. Apples will not open their buds in the spring, nor will winter wheat form flowers, unless there has been a period of eight to ten weeks of temperatures close to freezing. Conversely, most citrus crops are damaged or sometimes killed when day or night temperatures fall to freezing for even a few hours.

The modifications of plants to become adjusted or adapted to particular climatic conditions have occupied the attention of agriculturists for millennia. Natural selection is usually ineffective; plants rarely acclimatize to conditions for which they are genetically unsuitable. For centuries, better-adapted crop plants were obtained by conscious or unconscious selection from among the harvested plants. Such time-tested selections form the genetic base for the plant breeding that has been practiced for close to 150 years and is still the foundation upon which modern genetic plant engineering depends.

Although the earth's water is a finite resource, its supply to a particular area can be modified. For centuries, irrigation, drainage, and diking have been practiced, but such modifications are energy-intensive and may have undesirable consequences in the long run. As irrigation water evaporates, it leaves behind in the soil the salts that it had carried, so that the land becomes more and more saline. Since most agricultural plants (except beets, asparagus, and a few other minor crops) are not especially salt-tolerant, production is decreased. The irrigated valleys in the American West, a major source of many of our vegetables and fruits, have shown a 25 percent reduction in yield since large-scale irrigation started in the 1920s. Withdrawing water for irrigation to grow wheat on the Great Plains has seriously reduced the level of a major underground aquifer that provides water for the American Southwest and Mexico (Chapter 10).

The fourth of the physical resources in agricultural productivity, energy, is capable of considerable human modification. The reduction of human labor in modern agriculture by the use of machinery has allowed an ever-decreasing population of farmers not only to feed the growing urban population but to produce food for export (Tables 2.3 and 2.4). At present, this energy is largely derived from fossil fuels including coal, oil, and natural gas, all of which are finite resources whose extraction and preparation for use are themselves energy-consuming processes.

Table 2.3 CHRONOLOGY OF FARMING POPULATIONS IN THE UNITED STATES

Year	U.S. population (in millions)	Farm population (in millions)	Percentage of total
1800	4.0	3.8	95.0
1820	9.6	6.9	72.0
1850	23.2	11.7	50.0
1880	50.1	24.5	49.0
1910	92.0	28.5	31.0
1940	131.8	30.8	24.0
1960	180.0	15.6	9.0
1980	220.1	7.7	3.5

There is a broad range of biological resources available for the production of agricultural products. The human resources are large and their limits have yet to be determined. We are, socially and biologically, the most adaptable of living creatures, as evidenced by our geographical

Table 2.4 PERCENTAGE OF WORLD'S PRODUCTION OF WHEAT, RICE, AND CORN AMONG MAJOR PRODUCING COUNTRIES

Country	Wheat	Rice	Corn
Soviet Union	30		4
United States	13	2	46
China	8	33	8
India	7	21	
France	5		4
Canada	5		
Turkey	2		
Pakistan	2		
Bangladesh		8	
Indonesia		7	
Japan		5	
Sri Lanka		5	
Brazil		2	5
South Africa			4
Mexico			4

and cultural diversity. The development of machinery to reduce human toil, the invention of chemical aids to increase yields and reduce the ravages of insects and diseases, and the progress made in storage, transportation, and marketing of agricultural products have resulted in the availability of high-quality foods undreamed of even a century ago. Even without alteration in the cultivars used, plant productivity has almost doubled for cereal grains in the past half-century and has more than doubled for other crops. Modern genetic theory, initiated close to a century ago and greatly modified in recent times, has allowed the development of new crop plants with greatly enhanced productivity. The use of new techniques, including gene splicing and genetic recombination, has already resulted in plants with superior disease resistance, adaptability to different climatic and cultural conditions, and enhanced growth potential. Fertilizers and pesticides can be tailored for specific situations with high selectivity. At present, however, slow applications of social planning and practice in many developing countries limit the exploitation of these new concepts in agriculture.

There is an increasing search for new food plants. In some cases, plants used by different cultures have been ignored by Western societies simply because we usually had surpluses of those plants with which we already were familiar (Chapter 3). The amaranth genus, known to us primarily by the garden plant love-lies-bleeding (*Amaranthus caudatus)* or the tumbleweed (*A. albus*), contains several species that have long been staple green vegetables or edible seed plants. The millets are rarely grown in North America today, although they are high-yielding cereal grains with good nutritional value. Several legumes, including the winged pea (*Lotus tetragonolobus*), pigeon pea (*Cajanus cajan*), Goa bean (*Psophacarpus tetragonolobus*), and Bambara groundnut (*Voandzeia subterranea*), may play increasing roles as protein and oil sources, particularly since they can grow well without nitrogen fertilizers.

Although developed primarily for tropical lands, the techniques for extracting proteins from foliage that would otherwise be discarded have proved to be so effective that the end products are being used in North America as supplements to animal food. The protein of the soybean, commonly used in its compressed form as tofu, can be processed as textured vegetable protein (TVP) for use as a nutritional adjunct in sausages, hamburgers, and baby formulas. Microorganisms, particularly several yeast species, are grown in special media in huge fermentation tanks and their proteins extracted to produce single-cell protein (SCP) as additives for human and animal consumption.

ADDITIONAL READINGS

Hedrick, U. P. (ed.). *Sturdevant's Edible Plants of the World.* New York: Dover, 1972.

Janick, J., R. W. Schery, F. W. Woods, and V. W. Ruttan. *Plant Science: An Introduction to World Crops.* 3rd Ed. San Francisco: Freeman, 1981.

Moore, D. M. (ed.). *Green Planet: The Story of Plant Life on Earth.* Cambridge, England: Cambridge University Press, 1982.

Pimental, D., and C. W. Hall (eds.). *Foods and Energy Resources.* New York: Academic Press, 1984.

Reichcigl, M. (ed.). *Handbook of Agricultural Productivity.* Vol. 1, *Plant Productivity.* Boca Raton, Fla.: CRC Press, 1982.

Rosenblum, J. W. *Agriculture in the Twenty-first Century.* New York: Wiley, 1983.

Seigler, D. (ed.). *Crop Resources.* New York: Academic Press, 1977.

Malnutrition and Starvation

A man doubtful of his dinner or trembling at a creditor is not disposed to abstracted meditation or remote inquiries.

DR. SAMUEL JOHNSON

The relationship between population and food was clearly enunciated in 1798 by the Rev. Thomas Malthus in his *Principle of Population.* He concluded that population is ultimately kept in balance with the available resources by either conscious regulation of numbers or catastrophic reductions through war, disease, or famine. Malthus calculated that food production is capable of increasing arithmetically, while unchecked population growth is geometric. Since Malthus's time, many people have been insisting that he has been proved wrong, but now we are not so sure that his formulation is incorrect.

The world's population increased in 1984 by 85 million—about as many people as now live in Mexico and Austria combined. There are almost 5 billion people on earth, double the number alive in 1950. If the rate of population increase (now 1.9 percent per year) remains constant (Figure 2.5), there will be 6 billion souls by 2000 (Figure 2.6). The average life span, thanks to preventive medicine and a generally more adequate diet, is 61 years, with Icelanders averaging 77 years and a newborn infant in Ethiopia or Chad having a life expectancy of less than 40 years—assuming that she or he survives the first year of life. Infant mortality is high among malnourished people. The average life expectancy in most tropical and semitropical lands—which includes most of the underdeveloped countries—is less than 58 years.

Worldwide, virtually all of the highly fertile, well-watered lands in the temperate zone is in use. One may quarrel with decisions on land use; considerable acreage is covered by buildings and roads or is used to grow grasses and grains to feed cattle. Excellent land is set aside for tobacco, other narcotics, and crops that do not provide needed products. Some lands are poorly managed for sustained agriculture, and crops that could be grown better elsewhere may be produced because of national pride or for foreign exchange.

Marginal and less desirable lands are also under cultivation. Some are arid or semiarid; others are subject to recurrent flooding. These and other lands may have intrinsic low productivity. Their use is, however, mandatory; without them there is no way that even·minimal food production for the present population can be obtained (Figure 2.7). But

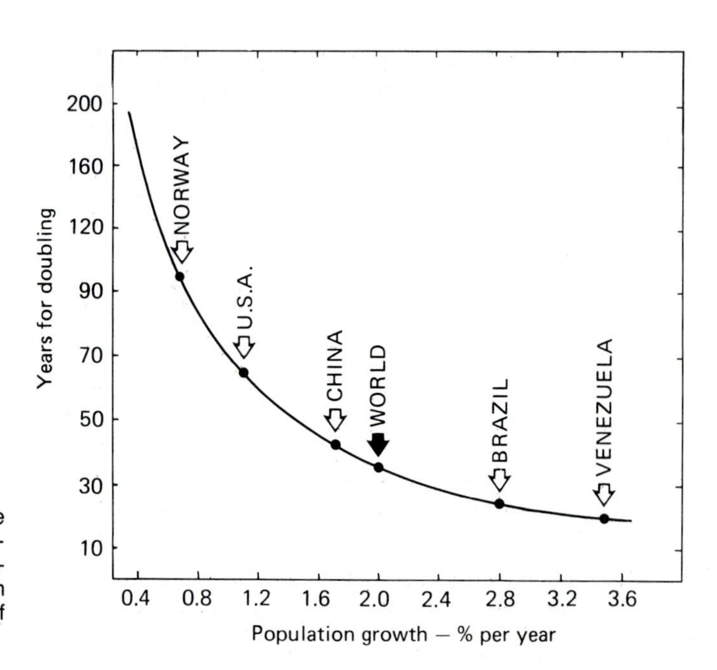

Figure 2.5 Curve showing the doubling time of a population relative to the number of years needed for that doubling. The growth rates of various countries (data of 1975) are shown as arrows.

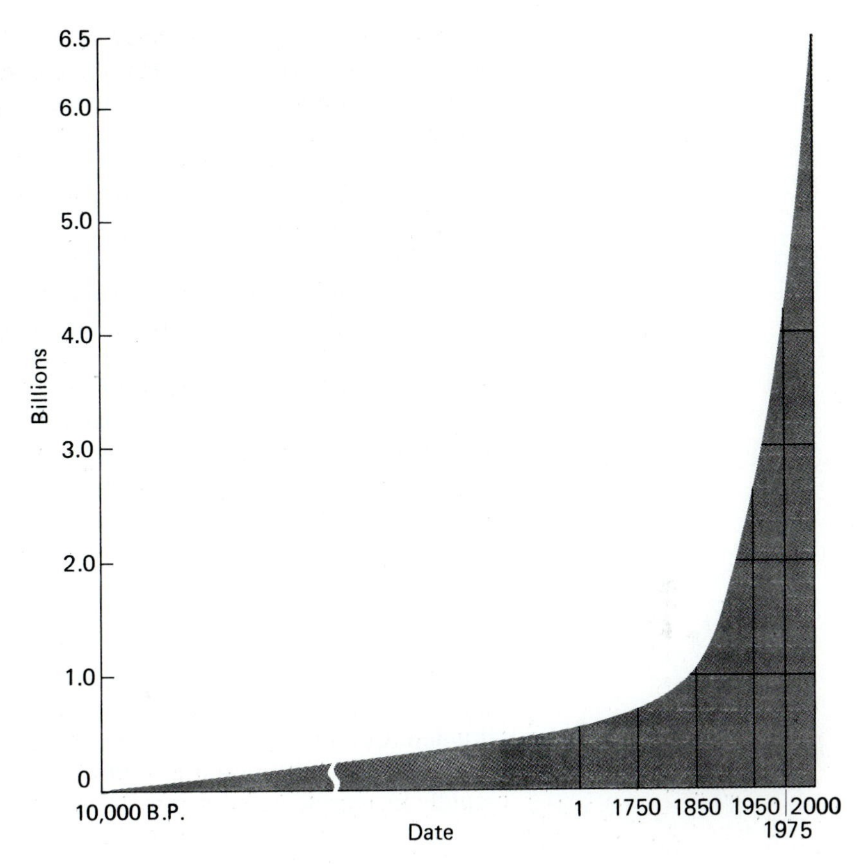

Figure 2.6 Exponential growth of the world's population. The period from 10,000 B.P. to 1750 has been compressed. Data from the United Nations.

even to maintain current productivity, large inputs of energy are needed, whether human or mechanical-chemical. Significant advances in productivity on such marginal lands depend on additional energy inputs that have diminishing returns, since productivity rates increase

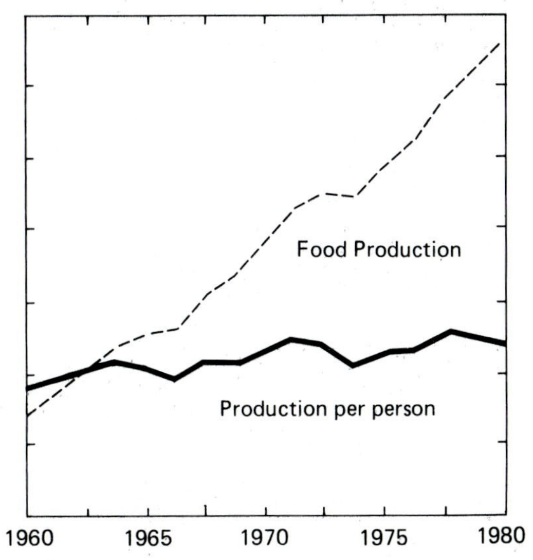

Figure 2.7 Although food production in underdeveloped countries has increased greatly, there has been almost no change in the production per person because of rises in population.

only slowly. Fragmentation of family holdings by inheritance, problems of land tenure, and lack of funds for purchases of seed and fertilizer have limited the productivity of many regions—and are likely to continue to do so.

The disparity between producers of food and consumers is widening (see Table 2.3). Before 1850, no industrialized nation was predominantly urban, but in North America today, over 50 percent of our population is concentrated in urbanized areas that occupy less than 1 percent of the land mass. Over 75 percent of our population no longer participates directly or indirectly in the production of the major food crops or meat animals. Even in less well endowed nations, the desperately poor and dispossessed are moving to cities. If this trend continues, by the end of the century the North American experience will be a worldwide reality; by the year 2000, half of the Third World's population will be urban. Because of massive inputs of energy, farms in North America can, even with reduced populations, not only support the urban population but provide food for export. The loss of farmers in less energy-dependent societies has created huge problems in distribution of food, in housing, employment, education, and, above all, in feeding these new city dwellers.

Population increases are most dramatic in the tropical and subtropical regions where a good deal of agriculturally marginal lands are situated. Countries in these regions are generally poor, with inadequate funds for the necessities of agricultural self-sufficiency, and many have rapidly increasing populations. Bringing tropical lands into production is time-consuming and expensive, may run contrary to established law and custom, and may, as in tropical forest lands, be futile (Chapter 10).

FOOD NEEDS

In spite of the great advances in agricultural efficiency and some attempts to alleviate the lot of the deprived, the inexorable march of malnutrition has not been significantly slowed. Major food-producing countries, especially the United States and Canada, have expanded their use of marginal lands as the world's demands for cereal grains has increased. Such lands are drought-prone and can be productive only with investments of irrigation and fertilizer that not only raise the price of the food beyond the means of those who need it but will continue to damage the land itself.

The affluent countries of Europe and North America, the Soviet Union, Australia, and Argentina are diverting a significant portion of grain to feed domestic animals for meat. North Americans use close to 5 percent of our wheat for this purpose, a large proportion of our corn, and most of our soybeans. But even were we to put these supplies into the export market, the problem of food distribution would not be solved. Strong cultural forces are involved; corn is not considered to be suitable food by people in some cultures, and rice-eating people are

Table 2.5 PRODUCERS AND IMPORTERS OF CEREAL GRAINS

Cereal grain	Producer	Yield	Importer	Import
	(In millions of metric tons per year)			
Wheat	Soviet Union	60–70	Soviet Union	12–15
	United States	55–65	Japan	5–6
	China	30–35	India	5–6
	India	25–28	Britain	3–4
	Canada	17–20	Egypt	3–4
Rice	China	75–85	Indonesia	1–2
	India	44–48	South Korea	0.5–0.7
	Indonesia	15–17	Iran	0.4–0.6
	Bangladesh	10–13	Sri Lanka	0.3–0.5
	Japan	12–13	Hong Kong	0.3–0.4

unlikely to switch to wheat even when they are close to starving. There is not enough international exchange to allow grain-producing countries to feed the world without at least receiving their costs of production and distribution (Table 2.5). In 1980, the total available plant food supply was enough to provide every person on earth with 2600 Calories per day, including 68 grams of protein—a complete and adequate diet (Table 2.6). But the undernourished peoples of Africa, Asia, and South America do not have the foreign exchange to cover one-third of the cost of producing, transporting, and distributing this food, and the economies of the producing countries cannot, in the long run, be maintained if food were to be given away on a permanent basis. Yet we cannot ignore the fact that an increasing percentage of humanity has become dependent upon food exported by a few countries (Figure 2.8).

While food production has, up until now, slightly exceeded Malthus's calculations, it has not attained—nor is it likely to attain—the geometric rate of increase that population has. The yearly world's grain production in 1961 was enough to feed the world's population for three months, but by 1980 the figure was down to 29 days. Thus, merely ensuring equitable distribution of food has, because of population increases, become no longer adequate. It now takes 1.2 billion metric tons of basic cereal grain (wheat, rice, corn, millets, and barley) to feed all of us for a year. Because of population growth, this must be increased by 0.25 billion

Table 2.6 DIETARY CALORIES AND PROTEINS FOR THE HUMAN POPULATION

Region	Percent of Calories from plant foods	Percent of protein from plant foods
World	88	80
Asia	95	78
Africa	93	78
Europe	81	53
Latin America	85	76
Oceania	65	30
North America	68	30

Source: U.S. Department of Agriculture.

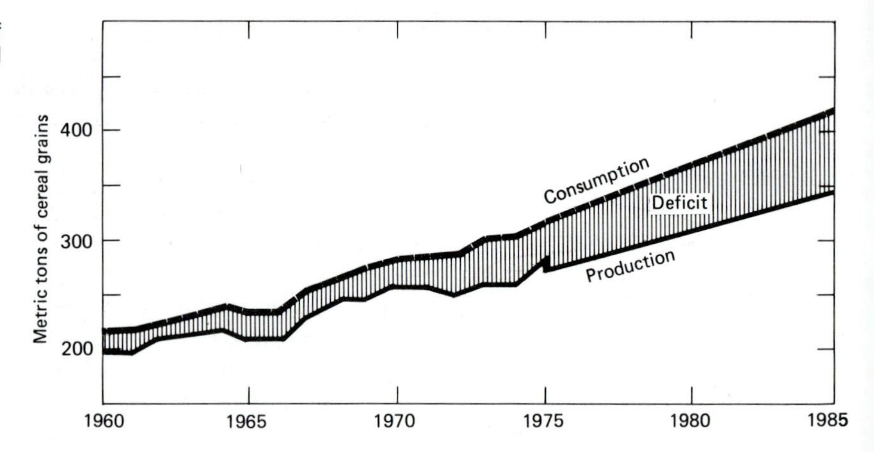

Figure 2.8 Projected deficits of cereal grains in underdeveloped lands.

metric tons each year—by a 6 percent rise in land use and/or an 8 percent increase in productivity each year. Put another way, China's population is increasing by 10 million per year, and its rice crop must triple by the end of this century.

Mankind has, in this century, come close to the precipice of mass starvation and, in the Sahal region of the southern Sahara, in Bangladesh, Ethiopia, and West Bengal, has plunged over the edge. Only massive gifts of grain from those countries producing food surpluses have limited this horror to a few places.

Certainly agricultural modernization in underdeveloped countries will significantly alleviate the situation. Innovative methods, including marine farming, formulation of synthetic, highly nutritious foods, and isolation of proteins from leaves and microorganisms, must be expanded. But without population controls, standards of living cannot even be maintained, nor can people look forward to improving their conditions of life.

If food self-sufficiency is to be achieved, many practices will have to be modified. Much of Africa, where population increase has already exceeded food production, is naturally semiarid or actually desert. Gambia cut down 60 percent of its forests between 1920 and 1966 for firewood, to produce lumber for export, and to increase agricultural land area. These forests created wind patterns that brought rain, the forest soils soaked up water that otherwise would have caused erosion, and the forest lands were fire breaks that limited the spread of bush fires that raged through West Africa. The organic matter in Gambian and other African soils is almost gone and the topsoil is blown on the wind. Low-lying lands along the rim of the Indian Ocean are unprotected and massive losses of soil and lives occur regularly during monsoon seasons. The peoples of Africa and Southeast Asia have made more changes in agricultural techniques in the past 30 years than have most societies, but the methods used by their farmers are still inadequate. The introduction of new methods will require inputs of seed, fertilizers, machinery, and irrigation equipment that cannot readily be obtained with available foreign exchange.

STARVATION

As a bare minimum, adult humans require close to 2600 Calories per day. People receiving less than this die young and can expect a painful history of decline in health and vigor. Their children, if they are born at all, will usually not live beyond a year. Total caloric intake is, however, an inadequate measure of nutritional status. Cereal grains are fine carbohydrate sources, but processed flour and polished rice have insufficient levels of fats and of certain vitamins; their proteins are incomplete, with suboptimum amounts of several required amino acids.

Large numbers of people today know intimately the Four Horsemen of the Apocalypse; the one named Famine thunders nightly through their dreams and walks with them during the day. Expressed in starkest terms, famine results when large numbers of people take in fewer Calories and fewer carbohydrate, protein, fat, and other essential nutrients than they require for their daily activities. An active adult can lose up to a kilogram of body weight per week on a 900-Calorie diet; if the diet is also deficient in protein, fats, or other nutrients, weight loss will be more rapid (Table 2.7). Two months of fewer than 900 Calories per day can reduce weight to dangerous levels, and three months will result in the death of almost 50 percent of the affected people. Children and the aged will die first, soon followed by the sick, pregnant women, and those adults with disabilities. Survivors are physically and emotionally unable to work. Sanitation breaks down, rapid increases in communicable diseases occur, and eventually the structure of the entire society is destroyed.

Less obvious are the effects of chronic malnutrition (Figure 2.9). Controlled studies have demonstrated that when children were clinically malnourished during gestation and the first year of life, they would fail to show normal growth even after they were given an adequate diet. Head circumference—a measure of brain development—was reduced, IQs were substantially below average, and permanent abnormalities in

Table 2.7 EXTENT OF BODY RESERVES OF NUTRIENTS

Nutrient	Time required to deplete reserves
Amino acids	Few hours
Carbohydrate	13 hours
Sodium	2–3 days
Water	4 days
Fat	20–40 days
Thiamin	30–60 days
Ascorbic acid	60–120 days
Niacin	60–180 days
Riboflavin	60–180 days
Vitamin A	90–365 days
Iron	125 days (women) 750 days (men)
Iodine	1000 days
Calcium	2500 days

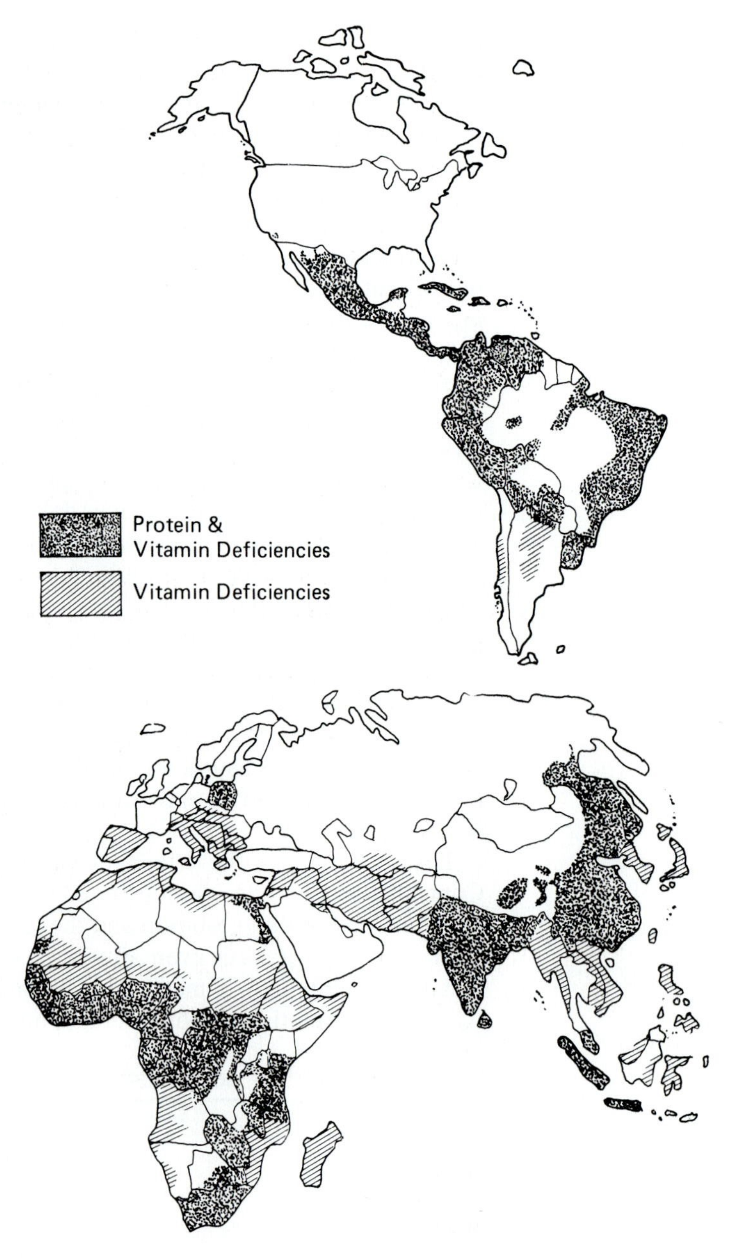

Figure 2.9 Areas of the world in which protein and vitamin deficiencies are commonly found.

body chemistry were present. Those children born of mothers who were themselves malnourished had a much higher mortality rate; the survivors were smaller, had lower brain weights, were apathetic, and had a shorter life expectancy.

In young children up to about the age of 4, brain damage is caused by inadequate protein supplies and not by inadequate caloric intake. The human body requires several amino acids that cannot be synthesized by our bodies but must be taken in as part of our diet. Since most plant proteins are low in these essential amino acids, the diet of the malnourished is inadequate for proper development. *Kwashiorkor,* grossly manifested by swollen bellies and a reddish cast to skin and hair (Figure 2.10), is an African word translated as "taken too early from the breast," because the supply of the nutritionally complete milk proteins is replaced by an inadequate plant diet. A young child given adequate protein may recover physically, but most victims never recover mentally. The skin-and-bones type of starvation called *marasmus* (see Figure 2.10) also causes brain damage.

We need not travel to other countries to find nutritionally related brain maldevelopment. An estimate in 1979 of malnourished pregnant women in the United States was 900,000, with over a million infants and children in physical or mental jeopardy. A proportion of the difficulties experienced by children in school and in their subsequent work lives can be traced directly to undernutrition in the womb and during the first years of life. And, as is obvious, this need not and should not happen.

The World Bank has estimated that a quarter of humanity lives in "absolute poverty" characterized by poor health, undernutrition, chronic deprivation, and reduced life expectancy. Human suffering, waste of talent, and repression of creativity are among the tragic consequences.

The litany of mass starvation—famine—is long and unpleasant. On the average, 50,000 people starve to death each year, and at least twice that number die from diseases they would not have had with adequate food. On the average, three children die of malnutrition every 10 seconds. Pestilence—whose horse's hooves stir up rats with plague-bearing ticks and fleas, cholera and typhoid bacteria in water supplies, and dust carrying the bacteria of tuberculosis—is a companion rider

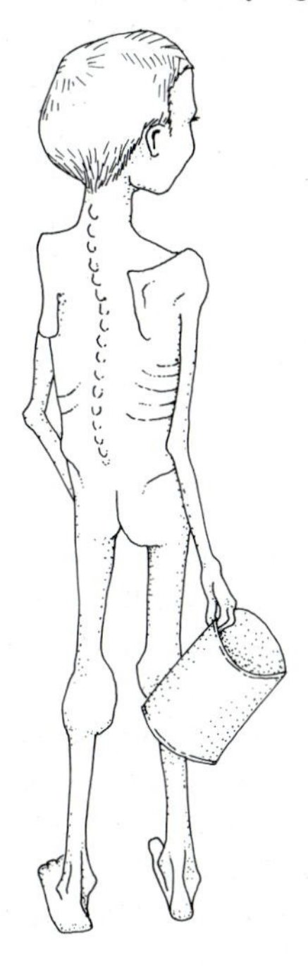

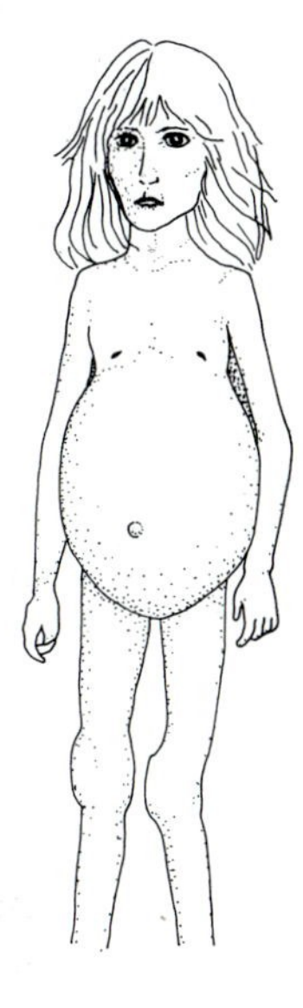

Figure 2.10 The two major types of starvation in children. The boy shows the skin-and-bones syndrome of *marasmus,* and the girl has the swollen belly and staring eyes of *kwashiorkor.* Unless protein feeding is started within a week, such children would die in less than a month.

with Famine, War, and Death. With the irreplaceable loss of humanity and the unfulfilled potential of individuals, mass starvation is a matter of concern to all of us.

ADDITIONAL READINGS

Broek, J. *Geography of Mankind.* New York: McGraw-Hill, 1973.

Crispeels, M. J., and D. Sadava. *Plants, Food and People.* San Francisco: Freeman, 1977.

Gupta, P. *The Crowded Earth.* New York: Norton, 1984.

Loomis, R. S. "Agricultural Systems." *Scientific American* 234, no. 9 (1976): 98–127.

McGee, H. *On Food and Cooking.* New York: Scribner, 1984.

Nicholson, H. J. *Distant Hunger.* West Lafayette, Ind.: Purdue University Press, 1979.

Prentice, P. *Hunger and History.* Caldwell, Idaho: Caxton Printers, 1951.

New York Times Staff. *Give Us This Day . . . A Report on the World Food Crisis.* New York: Arno Press, 1976.

Reichcigl, M. (ed.). *World Food Problem: A Selective Bibliography of Reviews.* Boca Raton, Fla.: CRC Press, 1975.

Revelle, R. "Food and Population." *Scientific American* 231, no. 3 (1974): 161–170.

U.S. National Academy of Sciences. *World Food and Nutrition Study.* Washington, D.C., 1977.

The Green Revolution

And he gave it for his opinion that whoever could make two ears of corn or two blades of grass to grow upon a spot of ground where only one grew before, would deserve better of mankind and do more essential service than the whole race of politicians put together.

JONATHAN SWIFT
Gulliver's Travels

If Swift's words symbolize the gloomy and perhaps overly pessimistic view that societies have had about the difficulty of feeding the world's peoples, is it any wonder that the Green Revolution was greeted with so much joy and hope? The possibility of doubling or redoubling the production of cereal grains galvanized the scientific, political, and economic communities. When Dr. Norman Borlaug was awarded the Nobel Peace Prize in 1970 for his encouragement of governments to provide grain to poverty-stricken peoples and for the development of the first of the miracle grains, the whole world applauded. Nothing is more important for world peace than a decent standard of nutrition and the assurance of an adequate food supply for one's present and future family.

GREEN REVOLUTION GRAIN

For thousands of years, wheats planted in different parts of the world were unstandardized local strains, moderately well adapted to the climate and growing conditions of the region. After wheat was introduced into North America, the immigrants selected seed from those plants that did well under the new conditions. Most of the selections were adequate producers in good years, but few had the potential for resistance to the diseases found in the new country, and even fewer had the drought resistance that was needed for wheat production on the beckoning Great Plains.

Nevertheless, some of the European introductions did have desirable characteristics. During the late nineteenth century, the famous 'Turkey Red' wheat was the preferred strain on the North American prairies because it was moderately disease- and drought-resistant. The introduction of 'Turkey Red' can be dated with precision. German-speaking people, Mennonites, emigrated from Holland to the Russian Crimea in 1568 to escape religious persecution. Catherine the Great annexed the Crimea in 1773, and to promote the Crimea as the breadbasket of Europe, gave these thrifty Mennonite farmers tax relief and exemption from military service. Because Czar Alexander II refused to renew these privileges, the Mennonites immigrated in 1873–1875 to the Canadian and American prairies, taking with them the hard red winter wheat they had named 'Turkey Red.' They were attracted to Kansas because land could be obtained cheaply through the Homestead Act of 1862. The Atchinson, Topeka and Santa Fe Railroad offered passage to Topeka, Kansas, for $11 to get good farmers to grow wheat for shipment back to the eastern flour mills.

Using 'Turkey Red' as breeding stock, David Fife of Ontario developed 'Red Fife' in 1880. 'Red Fife' was crossed with 'Calcutta' by Charles Saunders of Canada to yield 'Marquis,' and, by 1900, these plus 'Blue Stem' were the standard wheats of North America. William J. Spillman at Washington State University bred a very hardy winter wheat from these varieties in 1902, and discovered that his crosses followed with precision the 1866 predictions of Gregor Mendel, the Austrian founder of modern genetics. After Dr. Mark Carlton of the U.S. Department of Agriculture found 'Kharkov' wheat in Russia, it soon was as popular as 'Marquis.' Carlton also introduced a hard durum wheat, 'Kubanka,' for making spaghetti and macaroni. Fewer than ten varieties formed the solid foundation for virtually all North American wheats until the 1940s.

The recognizable drawbacks of these wheats seemed to make them unsuitable to breeding techniques of the time. They were all tall plants (120–140 centimeters; 48–56 inches in height). This meant that a considerable proportion of water and nutrients taken from the soil was used to develop the straw and not the grain. Because they were tall, they had a great tendency to lodge—that is, to be knocked over by high winds and hail. A shorter-stemmed plant was desirable. The semidwarf character was first found in Japan in a wheat called 'Daruma.' There is a toy with a heavy, rounded bottom that allows it to rock when hit by a child, but not fall over. It is called a *roly-poly* in the West and a *daruma* in Japan. Crossed with a tall, drought- and disease-resistant wheat, the semidwarf progeny was named 'Norin.' 'Norin' was bred with descendants of 'Marquis' and 'Fife,' to give a semidwarf (90–120 centimeters; 36–48 inches) variety, 'Gaines,' which was the first of the Green Revolution wheats.

IS THE GREEN REVOLUTION WITHERING?

There are problems with the implementation of the Green Revolution. For rice yields to be high, growing conditions must be exactly right. Water is critical even for drought-resistant varieties, for a short drought seriously reduces yields. Fertilizers, particularly nitrogen and phosphorus, must be provided at high levels; the new wheats and rices require two to three times more fertilizer. In 1982, new strains of rice required $42 worth of fertilizer per hectare per year, while traditional rices needed only $28 for fertilizer. Three times more pesticides must be used. Lacking foreign exchange to purchase fertilizers and pesticides, whose costs skyrocketed during the oil shortage of the 1970s, farmers and whole nations were helpless when yields decreased. Irrigation, chemicals, and advanced machinery became limiting factors for a successful Green Revolution.

There were other problems, unforeseen by the scientists and virtually beyond the control of political leaders. Most of these problems arose from the failure to consider cultural, social, religious, and economic factors in underdeveloped countries. In Thailand, for example, it is believed that rice grows from the womb of the Rice Mother, *Mae Phosop,* whose body, the earth, is impregnated by the seed inserted by the farmer. Since it is "well-known" that machinery or a man can frighten a pregnant woman, mechanical cultivators are rarely used in the paddies, and even weed-pulling is done by women.

Planting, cultivating, and harvesting are, throughout the world, inextricably bound to religion. The natural flow of the seasons and of the crops is marked and highlighted by planting and harvest ceremonies. Celebrating Thanksgiving in June or the Easter renewal of the earth in August would require large readjustments in Western thought and action, and the same is true for other cultures. Techniques for planting, cultivating, and harvesting the new types of grain differ from those used for centuries, and people do not lightly discard tradition. The earlier miracle rices looked and tasted different from traditional strains to the point where people wouldn't eat them, but tried to sell them on glutted local markets or fed them to their animals. This almost negated the advantages gained by increased yields. The taste problem has largely been solved by newer strains whose consumer acceptance is much greater, although traditional strains still command premium prices.

Requirements for massive pesticide use to control insects and diseases to which some of the Green Revolution crops were highly sensitive has, in itself, created problems the farmers were ill-equipped to handle. Because many were not told that these chemicals could kill fish, they soon discovered that their fish ponds, which had supplied a significant portion of the family's protein, had been poisoned. Bees that had pollinated their fruit trees and their legume crops were killed, and water supplies were contaminated. To some extent, these problems have been solved, but reports of pesticide pollution are increasing in regions that had been free of these substances.

Some of the countries that wholeheartedly adopted the practices inherent in the Green Revolution developed large surpluses they hoped to use for foreign exchange. The Philippines and the United States (California and Louisiana in particular) have available large stocks of rice for which markets are limited because other countries, formerly importers, either lack funds to purchase the excess or are much closer to self-sufficiency in this grain. Wheat, a cereal grain with a higher protein level than rice, is still in short supply on a worldwide basis even though prices paid to North American farmers may be depressed. Here, again, value is determined at least as much by funds available for purchase as by the need for the crop.

BIOTECHNOLOGY

Development of cereal grains that sparked the Green Revolution utilized classical breeding techniques in which crosses among plants with desirable characteristics were made. After the progeny were evaluated in rigorous field trials, the few plants that possesssed desirable traits were again crossed. This is a slow, painstaking process, requiring years for the emergence of a new variety. The revolution in genetic theory and practice during the 1970s has dramatically altered the pace and scope of tailoring crop plants to meet specific site and nutritional requirements. This is particularly true for the tropical or semiarid parts of the world.

The propagation by cell and tissue culture has had a tremendous impact on agriculture and horticulture. It is now possible to obtain thousands of genetically identical plants (the *homozygous* genetic condition) from a single plant or even a single cell. In the past, standard methods of asexual (vegetative) propagation through cuttings and grafts took up to a year to produce a relatively few plants, and costs were very high. Cell propagation methods frequently require only a few months. The identical plants are invaluable in pure research and in the evaluation of herbicides, insecticides, and growth-controlling chemicals. Research is underway to use tissue and cell culture techniques to grow the plant cells that synthesize flavorings, medicinal chemicals, and vitamins.

The cell propagation industry was foreseen as early as 1902 by the prescient German plant physiologist Gottlieb Haberlandt and was developed in the 1930s by Philip White, who worked out the basic methods of growing sterile plant cells under controlled conditions. In the 1950s, Carlos Miller and Folke Skoog at the University of Wisconsin discovered chemicals that allowed precise control of plant cell development.

The range of plants amenable to micropropagation is large and the list is being extended rapidly. The small, colorful Vanda orchid was among the first. Banana, citrus, maize, rice, potatoes, tomatoes, tobacco, berry fruits, ornamental flowers, and pulpwood trees are being micropropagated on commercial scales. Vegetable and flower crop breeding and the selection necessary to develop improved cultivars can be accomplished in half the time it used to take.

First developed to study the mechanisms of inheritance, gene splicing has been adapted for genetic engineering. Genes that control metabolic or growth processes of a plant are transferred to bacteria, yeasts, or plant cells that lack cell walls (protoplasts), where they multiply rapidly and can then be transferred to new host plants. Here they fuse with the hereditary apparatus and can direct the formation of enzymes that control biochemical and growth processes (Figure 2.11). New varieties can be made in a much shorter time and with a greater probability of successs. It is now at least theoretically possible to make a blueprint of a desirable plant for particular needs and to plan the methods needed to form such a plant. Improving photosynthetic capacity to produce more vigorous growth, promoting the dwarf growth habit to allow more energy to go into seed production, enhancing earliness of maturity to allow double cropping, and other projects are under intensive study. One goal, far from realized, is to confer the ability of a plant to make its own nitrogen fertilizer; another is to add resistance to salt or drought; a third is to increase the nutritional value of a crop.

Most of these new plants are at least 10–15 years away; many social and technical problems and theoretical considerations remain, but the foundation has been laid and many people are working on aspects of the total genetic engineering problem.

Along with genetic engineering, a quieter revolution in agricultural practice has been going on. There is increasing emphasis on the search for plants used primarily by small, isolated groups of people, usually in less sophisticated societies, that show promise in modern agriculture. Such plants, long adapted and selected for the conditions in their native

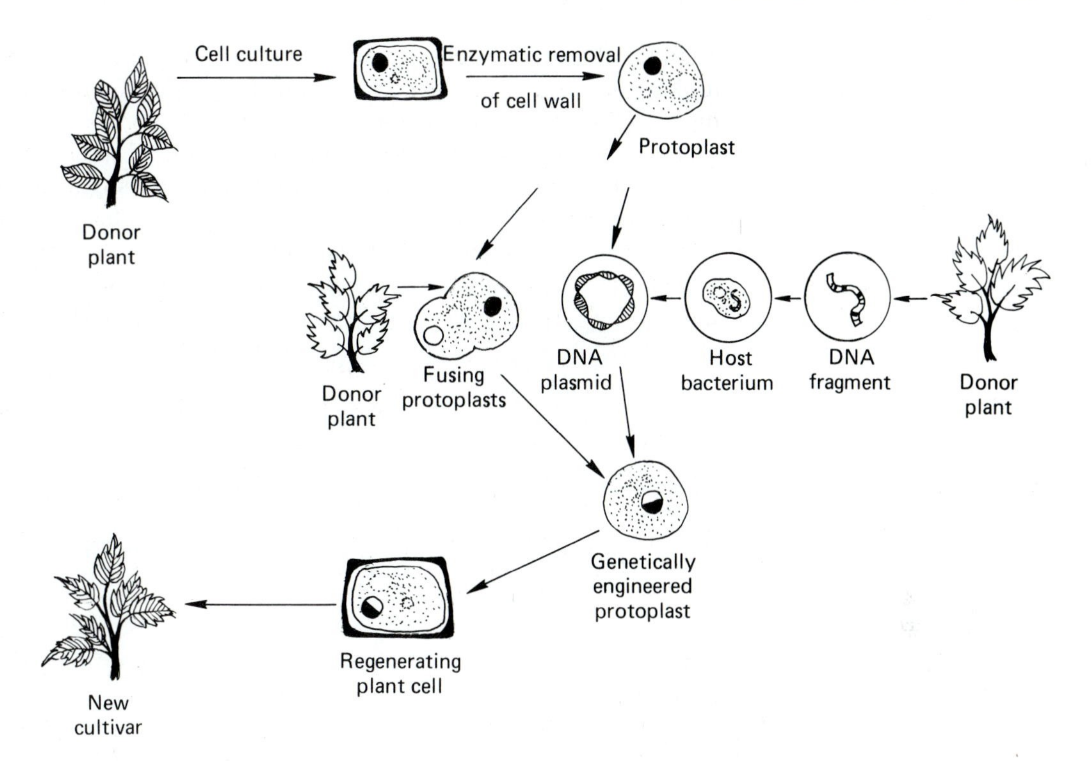

Figure 2.11 Diagram of techniques used to obtain plants possessing desirable genetic characteristics. Techniques include fusing protoplasts (cells without walls) from genetically different plants and gene splicing. In this process, desired genetic characters are transferred from plant cells to bacteria for incorporation into DNA plasmids, which can then be transferred into protoplasts. The altered protoplasts form cell walls, multiply in culture, and eventually form a new plant showing the added genetic components.

habitats, are being tested in North America as potential crops for marginal lands here and abroad. Beans of the !Kung people of the Kalahari Desert, vetches from Asia, cabbages from Inner Mongolia, and root crops from the mountains of central Africa are now being grown far from their original homes. Sometimes they have been genetically modified to extend the range in which they will grow, or they have been crossed with similar species to enhance a desired characteristic.

Agriculturalists are beginning to ask whether the farming methods that have become standard are really the best way of doing things. This questioning developed as farm populations became smaller, as energy costs rose, and as the need to feed more people became obvious. Must fields be plowed? Can manuring be improved? Should we consider ways to store and transport food plants in a less energy-intensive way? Are there crop plants that were once used but have fallen out of favor? The U.S. Department of Agriculture has found that the lowly turnip—a vegetable almost universally disliked by the young—is a treat for cattle and sheep. A forage crop in Europe during the eighteenth and nineteenth centuries, it was used only sporadically in North America because forage lands were so available. The prolific turnip does well in poor land; since the roots are above the soil line, there is no need to harvest them at all—livestock are simply turned out to feed in the fields. Other "old ways" are, with suitable modifications, becoming part of the agriculturist's business of feeding people.

ADDITIONAL READINGS

Borlaug, N. "Genetic Improvement of Food Crops." *Nutrition Today* 7 (1972): 20–25.

Dahlberg, K. A. *Beyond the Green Revolution: The Ecology and Politics of Global Agricultural Development.* New York: Plenum Press, 1979.

Griffen, K. *The Political Economy of Agrarian Change: An Essay on the Green Revolution.* Cambridge, Mass.: Harvard University Press, 1974.

Hardin, G. "Living on a Lifeboat." *BioScience* 24 (1974):561–568.

Katz, R. *Giant in the Earth.* Briarcliff Manor, N.Y.: Stein & Day, 1973.

Panopoulos, N. J. (ed.). *Genetic Engineering in the Plant Sciences.* New York: Praeger, 1981.

Pearson, C. (ed.). *Control of Crop Productivity.* New York: Academic Press, 1984.

Rosenblum, J. W. *Agriculture in the Twenty-first Century.* New York: Wiley, 1983.

Whitcomb, J. R., and W. Erskine. *Genetic Resources and Their Exploitation.* Hingham, Mass.: Kluwer, 1984.

Chapter 3

The Staffs of Life

Wheat

Oh, beautiful for spacious skies For amber waves of grain.

KATHERINE L. BATES
"America the Beautiful"

EVOLUTION OF WHEAT

The plant that has had the greatest impact on civilization is undoubtedly wheat. The plant has the ability to make a crop under a wide range of soil, water, and climatic conditions. It is conveniently harvested, cleaned, and stored for several years. The kernels have among the highest nutritional value of any plant and can be prepared in a variety of ways. These and other factors were highly significant in the cultural evolution of settled communities. Wheat must be planted, cultivated, and harvested in a set seasonal pattern. In short, the crop must be tended, which almost demands a settled life style of its farmers. The tale of the seven fat and the seven lean years in the Bible (Genesis 41) is more than a legend; it describes a basic fact about the development of agricultural civilizations: It is possible to provide for the future. The concept and practice of maintaining an "ever-normal granary" was also employed in China during the Han Dynasty (2200 B.P.) as well as in other ancient civilizations. Concomitant with a secured food supply was the opportunity for accumulation of wealth and the leisure that permits and even encourages the arts, government, organized religion, and the cultivation of the human intellect.

The importance of wheat led inexorably to it becoming of great cultural and religious significance. Every one of the agricultural deities has been credited with giving wheat to people. Most of the Earth Mothers of the Middle East, Isis of Egypt, Demeter of Greece, Ceres of Rome (Figure 3.1), and gods of the Indian pantheon are epitomized by sheaves of golden wheat, by the kernels, or by bread baked from the wheat. Priests of Egypt and other wheat-growing countries invoked the gods to provide weather conditions favorable to wheat growing and developed the astronomy and mathematics necessary to make their prophecies reasonably accurate.

Figure 3.1 Ceres, Goddess of Grain. Taken from a Pompeian wall painting.

The history of the origin and development of the bread wheat *(Triticum aestivum = T. vulgare)* has long been a mystery, with participants in the search—including archeologists, geneticists, cell biologists, and anthropologists—each contributing to the solution. Yet the puzzle is still not completely solved. Archeological and anthropological records indicate that bread wheat was cultivated in the Nile Valley by at least 7000 B.P., in the Middle East (Euphrates and Indus River valleys) before 6000 B.P., and in north China by 4000–3500 B.P. Other wheat species were grown even earlier. Einkorn wheat has been found at Jarmō dating to 9000 B.P.; emmer wheat grains dated to 8000 B.P. have been unearthed in modern Lebanon, Jordan, and Israel and have been found in Bronze Age Swiss Lake Dwellings. Spelt, another wheat, was in cultivation 2000 B.P. in Mesopotamia.

There is still some question as to whether bread wheat is derived from one species in the genus or whether it is the result of deliberate selection of individual plants from cultivated or wild populations of several species. To evaluate this question, scientists have used the shapes and numbers of chromosomes in the nucleus to evaluate possible parentage (Figure 3.2; Table 3.1). Einkorn *(T. monococcum),* the *kissemeth* of the Scriptures, is a single-grained wheat with a basic chromosome number (1N) of 7, but it does not seem to have been a major contributor to modern bread wheat. A related species, *T. boeoticum,* also has 7 chromosomes that match up well with 7 of those in bread wheat. It apparently crossed (naturally ?) with a wild goat grass, *T. speltoides,* to give rise to the doubled chromosome (1N = 14) species of emmer wheat *(T. dicoccum)* and durum wheat *(T. durum)*. Another emmer wheat, *T. dicoccoides,* interbred with the wild grass *T. squarrosa* to give rise to the hexaploid (1N = 21) bread wheats. Several of these crosses are still hypothetical.

By 1500 B.P. four major types of bread wheats had been selected for different growth and cultural characteristics. Based on husk color, kernel hardness, amounts and types of protein and starch, and cultural characteristics, we now distinguish hard red winter, hard red spring, soft red winter, and white lines of wheat. The hard red wheats all have close to 14 percent protein; a proportion of the protein is glutenin, which forms the elastic framework of leavening. Soft red winter wheats, which have less protein and reduced glutenin, are used to prepare cracker

Figure 3.2 Four important species of wheat. From the left: *Triticum monococcum* (einkorn wheat), *T. dicoccum* (emmer wheat), *T. durum* (durum wheat), and *T. aestivum* (bread wheat). Einkorn has 7 chromosomes, emmer and durum have 14 chromosomes, and bread wheat has 21 chromosomes.

Table 3.1 CHROMOSOME COMPLEMENTS OF THE WHEATS

Species	Common name
Diploid wheats (2N = 14)	
Triticum aegilopoides	Wild einkorn
Triticum monococcum	Cultivated einkorn
Aegilops speltoides	Wild wheatlike plant
Aegilops caudata	Wild wheatlike plant
Aegilops squarrosa	Wild wheatlike plant
Tetraploid wheats (2N = 28)	
Triticum dicoccoides	Wild emmer
Triticum dicoccum	Cultivated emmer
Triticum durum	Durum
Triticum persicum	Cultivated Persian
Triticum polonicum	Cultivated Polish
Triticum turgidum	Solid stem
Triticum timopheevi	Wild timopheevi
Aegilops cylindrica	Wild goat grass
Hexaploid wheats (2N = 42)	
Triticum compactum	Cultivated club
Triticum spelta	Cultivated spelt
Triticum aestivum	Bread wheat

flours and porridges. White or club wheat has the lowest protein concentration and mills to a silky flour used primarily for cake flours that are leavened with air, baking powder, or egg whites. Durum wheat has even lower protein, very little glutenin, and a hard starch ideal for making pasta and seminola flours. Several of the other ancient species are still grown in small amounts in regions that do not support the development of bread wheats.

The statistics on wheat production are staggering. Worldwide, 400 million metric tons (14 billion bushels of 60 pounds each) are produced annually on 600 million acres; 20 percent of the world's croplands are in wheat. Yields under favorable conditions can exceed 100 bushels per acre, with the world's record being close to 150 bushels per acre. The bushel measure is an Anglo-Saxon volume, based on a cylinder now in Winchester, England, and is equal to four pecks that, in turn, equal eight quarts. As a bit of trivia, there are about 15,000 kernels per pound, 700,000 kernels in a bushel, and an astronomical 33 million kernels per metric ton. Close to 90 percent of wheat production is in the Northern Hemisphere, with the Soviet Union producing the largest amount, followed in descending order by the United States, China, India, Canada, and France. The United States produces 1.8 billion bushels per year on 80 million acres (16 percent of world production). Canada plants over 30 million acres to wheat. In the Southern Hemisphere, 10 percent of the world's supply comes from Argentina, Australia, and South Africa. Most of the wheat produced is consumed locally; only a few countries produce surpluses and can export wheat. Many wheat-producing countries do not grow all they require even under the best growing conditions. Wheat is the staff-of-life grain for 35 percent of the people of the world, who obtain 20 percent of the world's Calories and almost 50 percent of their protein from it. The fervent plea in the Lord's Prayer, "Give us this day our daily bread," means exactly what it says.

The staple cereal of the Orient is rice, and this is true for large numbers of the people in this part of the world. Yet the people of northern China, and many of the inhabitants of what is now India and Pakistan, Tibet, and northern Korea, have traditionally been consumers of wheat and millet rather than rice. Noodle dishes and steamed, raised wheat breads and millet have been eaten in China north of the Yellow River for thousands of years. Marco Polo brought wheat and millet noodles to Italy from the land of the Great Khan during the Yüan Dynasty (1260–1368). Recognition of the superior nutritional value of wheat compared to rice has occurred only since the mid-1940s, sparked to some extent the introduction of bread to Asian peoples by Allied forces during the Second World War. Traditional rice-eating countries like Burma, the Philippines, Sri Lanka, South Korea, and Thailand—lands where wheat does not flourish—are spending a significant portion of their foreign exchange to purchase wheat.

The effects on the populations of these countries have been remarkable. The average heights and weights of the Japanese have increased significantly since the introduction of wheat products (noodles and breads) with higher protein levels compared with rice. Prior to the occupation of Japan in 1945, Japan's wheat consumption was under 50 pounds per person per year, but by 1983 this had increased by a factor of almost five. A concerted effort has been made in the People's Republic of China to boost wheat production in the northern part of the country and to introduce north China's noodles and breads into the diets of the southern Chinese, who formerly ate only rice. China's goal is to produce as much wheat as rice by the end of this century.

WHEAT LANDS

The movement of people throughout history has been at least in part a response to the need for wheat. The changing political fortunes of the farmers of the Middle East with dynastic and political successions for thousands of years were a result of the quest for grain. Alexander, and before him Xerxes and others, moved to obtain grain and secure the trade routes for grain. Rome's wealth was based on the wheat trade and its military might was developed to acquire wheat and to secure trade routes for wheat. The conquest of Egypt was initiated to provide access to the bountiful wheat of the Nile Valley and North Africa. Carthage was invaded and Egypt conquered to obtain the wheat of the Delta and

North Africa. The Roman legions invaded Gaul to increase the lands upon which wheat could be grown. Palestine was secured to serve as a barrier to the movements of Phoenicians and Persians, whose armies and naval forces threatened Roman hegemony in the Mediterranean. The Roman fleet needed protection while carrying grain from North Africa to Ostia, the seaport for Rome. By the first century, Rome was importing 14 million bushels per year from North Africa, where it held all land north of the Sahara. And all this grain was needed to feed Rome's burgeoning population and to placate the barbarians to the north. The "bread and circuses" deplored by Juvenal were the spoils of war. In the time of Augustus Caesar, about a kilogram of grain per person per day was being given to 320,000 people, a third of Rome's population. And a few years later, bread itself was doled out because the citizens complained about having to mill and bake their own bread.

Wheat reached France during the period of hegemony of the Celts, but apparently was not grown in Britain during Roman times; barley was the primary cereal grain. The Venerable Bede noted that spring wheat was planted by Anglo-Saxons in the eighth century, but wheat became important only during the reign of Elizabeth I. The Spanish brought wheat to their American possessions by 1520, but it did poorly in Mesoamerica or northern South America. Only in the late nineteenth century were the excellent wheat-growing areas of Argentina exploited. In North America, wheat was cultivated on Elizabeth Island off the Massachusetts coast and at the Jamestown Colony in 1602, at Port Royal, Nova Scotia, in 1606, and in the Plymouth Colony by 1621. It is possible that by the twelfth century the Norse grew wheat in what is now the Canadian Maritimes, but no records exist. Missionaries brought wheat to California by 1770.

Settlement by Europeans in the lands of the Atlantic seaboard of upper North America saw farms carved from thick forests. Wheat was the primary crop for home consumption and was the major if not sole cash crop. The 20–30 bushels per acre produced on newly cleared land would repay the farm family for clearing, seed, fencing, and labor and it increased the value of the land holdings almost tenfold within two years. Indian corn—maize—was new to the immigrants and was, for over 50 years, a less desirable secondary crop grown for emergencies and for domestic animals. The settlers in the southern colonies found to their dismay that wheat did not develop well and they turned to maize as their primary cereal.

The lack of grist mills and nearly impassable roads were major deterrents to movement away from settled communities. New Englanders soon found that the harsh climatic conditions were generally unfavorable for wheat production on a commercial scale. As transportation became more reliable in the early nineteenth century, a second but short-lived burst of wheat farming occurred in northern New England and adjacent Canada, but wheat farming declined once more, never again to be a significant agricultural enterprise. Instead, the settlers, anxious to grow a valuable cash crop, began the westward movement that culminated, in the middle and late nineteenth century, in the opening of the prairie lands bordering the Mississippi River. Securing the western prairies was an act of "manifest destiny," and Indian lands were taken with scarcely a backward glance. Deliberate slaughter of the buffalo and the murder or deportation of the hunting tribes followed inexorably. The development of facilities to store and move grain was an economic necessity. Government and the robber barons connived to subsidize railroad lines with huge grants of money and land. Immigrant Irish refugees from the potato famines of 1845–1848 formed construction crews working from east to west, while imported Chinese "coolies" built from west to east. In 1840, there were only 2500 miles of track in the United States, but this increased to 31,000 miles in 1851 and to 160,000 miles by 1871.

In Canada, John A. MacDonald dreamed of linking the Dominion from Atlantic to the Pacific with a railroad—the Canadian Pacific. In order to do so, he forced the Hudson's Bay Company to relinquish its grip on the prairies and to have the land surveyed, a move the French Canadians opposed, since they believed—correctly, as it turned out—that they would lose the land. In 1869 these actions and other accumulated grievances led to the Riel Rebellion, which was ruthlessly crushed. The Canadian Pacific reached Winnipeg by 1878 and British Columbia a short time later. Spurs from the main lines connected grain elevators to the primary routes, and villages arose in their shadows. Villages became towns and some towns became cities: St. Paul, Duluth, Winnipeg, and Grand Forks.

As the grain supplies began to flood the eastern markets, it became obvious that there was more than enough to feed the people of North America. Both Canada and the United States entered the export market then dominated by Hungary, the Ukraine, and Scandinavia. Bankrupt Swedish, German, and Ukrainian farmers came to the prairies by the thousands and were joined by other nationalities. Horace Greeley of New York was advising, "Go west, young man, go west," and thousands heeded him. Canadian land salesmen were advertising: "Buy land in the Canadian West. You can leave home after Easter, sow your seed and take in a harvest and come home with your pockets full of money in time for Thanksgiving." By 1900, the international economies of both Canada and the United States were tied irrevocably to wheat.

In the United States, a large proportion of the hard wheats for export leave from the Texas Gulf ports, with some moving out from ports on the delta of the Mississippi River. The construction of locks through the Great Lakes, a joint venture of Canada and the United States, profoundly altered the economies of states and provinces with access to the lakes route, since a cheap and direct flow of grain to eastern ports became possible. Movement across the Pacific of white wheats of the Northwest and hard wheats from Idaho, Montana, and the Dakotas is primarily through ports in the Columbia River district, while most of the westward export of grain grown in the prairie Provinces embarks from British Columbia.

WHEAT ECONOMICS

The economics and geopolitics of the import-export commerce in wheat and other grains on an international level is one of the more arcane areas of knowledge, but some indication of their importance must be given because they have become significant factors in the economies of almost all nations. U.S. income from wheat exports constitutes close to 65 percent of the total international market ($7 billion in 1983); it is the major segment of food grain trade and hence is a major factor in balance of payments. Canada, another of the major granaries for the world, depends heavily on wheat for its domestic economic health. Grain sales in North America are handled as a free enterprise undertaking, with major trading activity controlled by a relatively few multinational corporations whose power extends into legislative bodies and international organizations. In Europe, by contrast, the European Economic Community adjudicates the economics of wheat sales both within and without the community.

Prices for wheat in North America are set through commodity exchanges, where buyers and sellers meet. These markets are regulated in the United States by Commodity Exchange legislation and an equivalent series of laws in Canada. The United States has over 30 exchanges, the most active being in Chicago (Figure 3.3), Kansas City, and Minneapolis. The major Canadian exchange is in Winnipeg. The cash grain market serves immediate delivery needs (delivery within ten days) and is usually for grain or flour at railhead storage facilities. In the grain futures markets, contracts are bought and sold that require delivery of an agreed-upon amount of a specified wheat at some future date. At least in theory, futures markets provide a mechanism for year-round stability by smoothing the flow of grain prices on the basis of known or predicted supply and demand. By hedging (balancing future's buy and sell contracts) producers, consumers, and speculators attempt to minimize financial risks caused by fluctuations in supply and hence

Figure 3.3 The Chicago Board of Trade, the largest grain market in the world. Courtesy U.S. Department of Agriculture.

price. In contrast to other countries, the United States government does not directly engage in marketing wheat, although price supports and restrictions of acreages to be planted do regulate the supply. Grades and standards are also set by most nations. Developing countries and lands experiencing food crises (crop failures, wars) are aided by grain purchased on the open market by governments using international balance of credit adjustments.

At present, the world's grain reserves are low and the needs are increasing as populations expand and land deteriorates. Even if the exporting nations were to raise production by an almost impossible 5 percent a year, the shortfalls would still exist in those lands, mostly tropical, whose needs are growing most rapidly. Increasing life expectancies and reduced infant mortalities are major factors influencing the reduced grain availability per capita. Developed countries have faced economic problems too. Upgrading life styles by increasing meat supplies through use of feed grain is a major factor in the importation of wheat and other cereals by the Soviet Union, usually the largest wheat-producing country. Disastrous harvests and equally disastrous central planning have also contributed to the Soviet Republic's change from an exporter to an importer of grain. Throughout the world, distribution—including the costs of producing, storing, and shipping the grain—and access to funds or credits to purchase wheat are significant components in the inequitable availability of wheat and other basic cereals. Long-term projections suggest that means of facilitating local production and reducing the population pressures on food supplies are basic for a future that can be contemplated with any equanimity.

PRODUCTION OF WHEAT

Tilling the Soil For centuries after wheat was domesticated, the methods used to grow grain changed very little. It was likely that a simple, sharpened stick was used to scratch the soil, although we have little solid evidence for this belief. The Sumarian scratch plow was a major advance, since it could be pulled by the husband or wife and allowed more land to be prepared for sowing. Scratch plows merely score the soil surface, and, since the grain is broadcast on the bare fields, birds and other animals can eat a fair share of it. Dry weather or a sudden rainstorm may cause massive losses. The Egyptians substituted domestic animals for human muscle (Figure 3.4). The Chinese independently had a more sophisticated, animal-drawn plow before the Han Dynasty, about 2500 B.P. (Figure 3.5). The Chinese plow, which required less effort to operate than the Egyptian plow, did not reach the West.

The scratch plow was inefficient in the light soils of the Middle East, North Africa, and northern China, but it was almost useless in wet, heavy soils of Europe, North Africa, and parts of China. In Europe, the

Figure 3.4 The scratch plow used throughout the Middle East and North Africa. Taken from an Egyptian tomb painting.

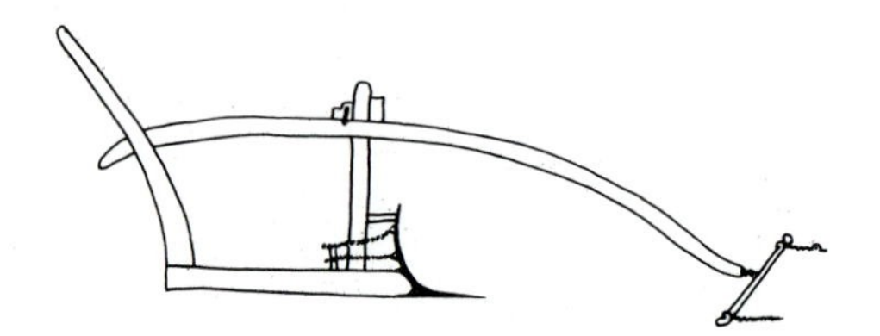

Figure 3.5 The plow in China. The upper figure was taken from a Han Dynasty stone relief from Kansu province. The lower figure of a steel shared moldboard plow pulled by two draft animals is redrawn from a figure in a fourteenth-century agricultural text, the *Lei Ssu Ching*.

scratch plow was used throughout most of the Dark Ages (Figure 3.6). During the late Dark Ages, people in eastern Europe invented a simple plow which had a sharp wooden point (a plowshare) that cut into the soil; they also invented a moldboard that turned over the clods (Figure 3.7). Even earlier, about the mid-Han period, the Chinese had moldboard plows (Figure 3.5). The sharp plowshare and moldboard were important inventions, since the plow had to be engineered to pull at an angle through heavy soils. Such plows significantly increased the efficiency of the plowing operation; yields could easily be doubled (Figure 3.8). Moldboards were, at first, merely slabs of wood which turned over the soil poorly. The curved moldboard, invented in China, reached Europe when the Dutch found such plows in their Indonesian colonies, brought there by Chinese immigrants. The Chinese animal harness, in which the strain of pulling was transferred from the animal's neck to its chest and legs, did not reach Europe until the thirteenth or fourteenth centuries, but once available, it allowed horses to be substituted for the oxen that were then in use.

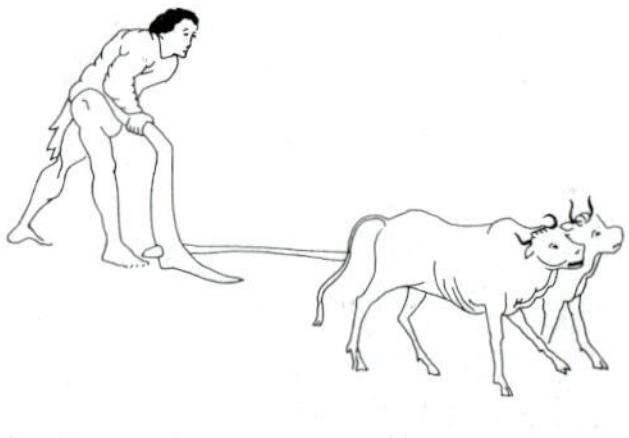

The all-steel plow was invented by John Deere in the nineteenth century (Figure 3.9) and has been replaced in this century by tractor-drawn plows and disk harrows that allow very efficient preparation of the soils (Figure 3.10).

Figure 3.6 The scratch plow used in Europe throughout the Dark Ages was essentially the same as the Egyptian plow.

Figure 3.7 The scratch plow was modified during the Dark Ages by the addition of a wooden moldboard.

Figure 3.8 During the late Middle Ages, the plow was modified by the iron plowshare attached to a wooden or sheet metal moldboard. Taken from a fourteenth-century woodcut.

Figure 3.9 John Deere's plow of 1838 had an integrated share and moldboard of steel and was capable of breaking the tough sod of the North American prairie. Courtesy Deere & Co.

Figure 3.10 Modern disk harrow. Courtesy International Harvester Co.

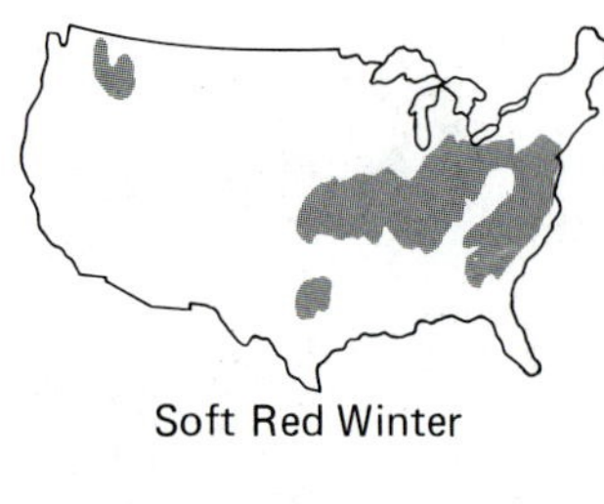
Soft Red Winter

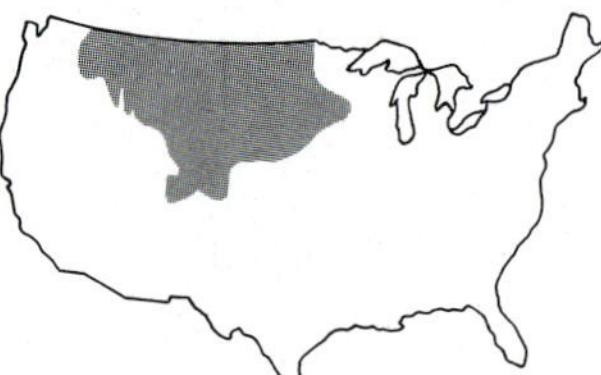
Hard Red Spring

Hard Red Winter

Durum

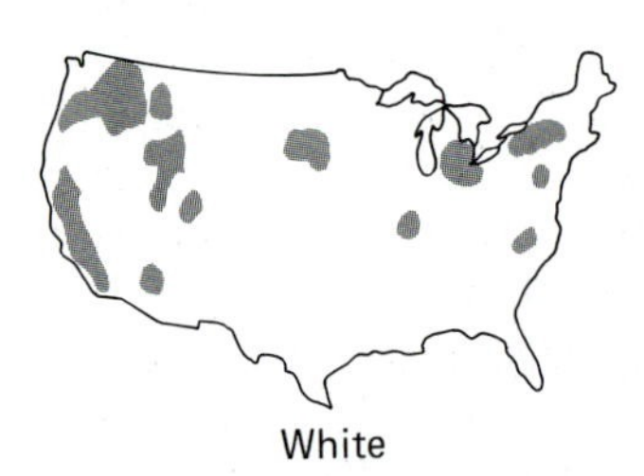
White

Figure 3.11 Distribution of growing areas of the major types of wheat in the United States.

Growing the Grain In North America, winter wheats are planted in the fall, they germinate, and, because of the ability of the seedling to resist low temperatures, they overwinter and start growing again when frost is out of the ground. Winter wheats are ready for harvest in late spring. Late fall planting is obligatory for winter wheats because the young seedlings require a cold period for rapid growth and maturation of the grain. This is called *vernalization,* which means to "make springlike" and is a translation of a word coined by a Russian scientist who studied the phenomenon. Many plants require a cold period for the growth of buds, root tips, and other plant parts.

Spring wheats are planted in the spring and mature in the fall. They are, in contrast to the winter wheats, *photoperiodic* in that they will not form flower buds until daylengths have decreased from their peaks in midsummer. Since the winter temperatures in the Dakotas and on the Canadian prairies are too low for the survival of the seedlings of winter wheat, these lands are planted to hard red spring wheats (Figure 3.11), which are as nutritious as the winter wheats. The winter wheats in North America are generally grown south of Nebraska. White wheat is planted on the West Coast of North America, and some soft winter wheat is grown in Ohio, Pennsylvania, and Maryland. Durum wheat, used for pasta, is grown in the Dakotas, Saskatchewan, and Manitoba. Durum wheat accounts for about 5 percent of the total North American production, white wheat for 10 percent, soft red winter wheat for 25 percent, and hard red winter and spring wheat—by far the best for bread—for 60 percent of the harvest.

Most of the modern wheats in North America and Europe are cool season crops that will give good yields with 10–15 inches (25–38 centimeters) of rain, although optimum yields require 25–30 inches (63–76 centimeters) or extensive irrigation, as has been practiced in the Middle East, in India, and in China for centuries. Irrigation of the high prairies of the North American West has allowed these lands to become producers of large amounts of excellent hard wheats, but the diminishing supply of ground water in aquifers has not been given adequate consideration.

The major fungus scourge of wheat is called stem rust. In early spring, spores of the fungus that overwintered in the soil (teliospores) form smaller spores that float on the wind and can infect young wheat plants. As the fungus develops on the stems of the wheat, it forms other spores (urediospores), red in color, that make leaves and stems look as though they had been streaked with red iron rust. Billions of these red spores are released into the air where they can travel for hundreds of miles before infecting other wheat plants. This cycle can be repeated many times during a single growing season. Infected plants grow poorly; the kernels shrivel and are of poor nutritional quality (Figure 3.12).

Wheat rust has been known since wheat was domesticated. In Roman times, a ceremony was held on April 25 to placate Robigus, a grain deity. At His temple, red wine was poured on the altar and a red dog

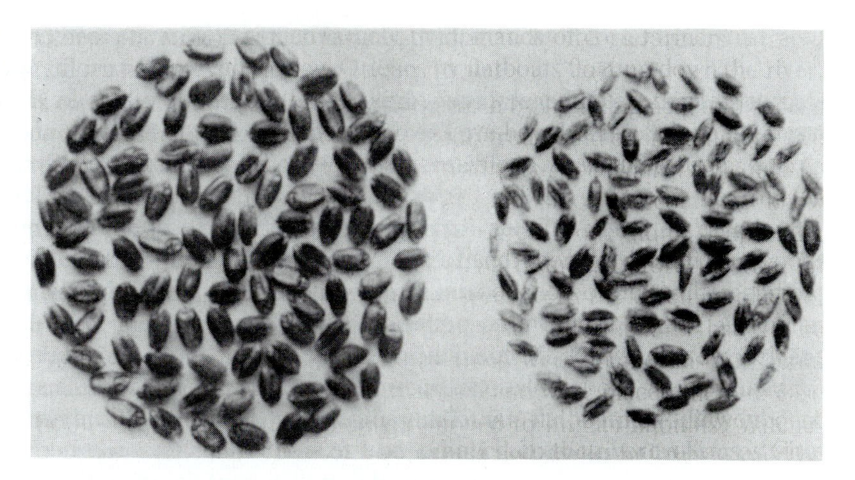

Figure 3.12 Effects of stem rust on wheat yield. Grains from healthy plants are on left; shriveled grains from diseased plants are on right. Courtesy U.S. Department of Agriculture.

was sacrificed in the hope that the Deity would protect the grain crop. Agricultural workers were forbidden to wear red clothing during the growing season. The Romans also had the vague notion that the barberry plant *(Berberis spp.),* because of its red berries and red autumnal foliage, was somehow related to the rust. This idea eventually resulted in a French law of 1660 requiring that all barberry plants growing near wheat fields had to be destroyed. Other European countries had barberry eradication laws by the end of the seventeenth century. The Connecticut Colony outlawed barberry in 1726 and the other North American colonies followed before 1750.

Proof that a specific fungus caused wheat rust was provided by Pier Michele in France in 1729, and the infective red spores were seen by the Tuslane brothers in 1845. But because this information gave no clue as to why barberry plants seemed to have some connection to the rust, people scoffed at this old superstition. It wasn't until 1927 that the Canadian mycologist (one who studies fungi) Dr. J. H. Craigie demonstrated that the life cycle of the wheat rust fungus was very complex, involving the red summer spores and black autumn spores on the wheat plant and three different kinds of spores on the barberry plant. Equally upsetting was the discovery, actually made in 1894 by Dr. Jacob Eriksson in Sweden, that there were different races of the stem rust capable of attacking different genetic lines of wheat. In 1914, Dr. E. C. Stakman of the University of Minnesota extended Eriksson's work and found that there were at least several hundred genetic races of the fungus and that none of the cultivated wheats could possibly be immune to all of them. It takes between three to five years for one genetic type of wheat rust fungus to build up to sufficient numbers so that it becomes a serious threat to the crop, but it takes several more years to develop a genetic strain of wheat that is resistant to the mutated fungus. Modern wheat breeders still face the never-ending task of developing disease-resistant strains of wheat that still retain the desirable characteristics needed for high yields. Micropropagation methods show promise in reducing this time lag. Continuous research is also needed in controlling insect pests, particularly the Hessian fly, which can destroy a crop in just a few days.

The short-stemmed progenitor of most of the Green Revolution wheats can be traced back to a 1917 cross made in Japan between 'Daruma' and the American cultivar 'Glassy Fultz,' which, bred to 'Turkey Red' in 1925, gave rise to 'Norin 10.' Breeders at Washington State University crossed 'Norin 10' to a domestic winter wheat. Further breeding by Dr. Norman Borlaug began in Mexico in 1953. Borlaug's work was tremendously influenced by the 1951 discovery by Dr. H. Kihari in Japan, who was able to induce male sterility in wheat so that normal self-pollination could not occur. This advance allowed controlled pollination and fertilization with pollen of desired wild or cultivated races of wheat and reduced the tedious job of having to remove the pollen-containing anthers by hand. The work at the International Maize and Wheat Improvement Center (CIMMYT is the acronymn for the Spanish name) has permitted the development of new breeding methods that have facilitated the crossing of many different wheats with enhanced vigor, drought resistance, tolerance to specific climatic conditions, and enhanced grain production. There are now over 200 cultivars of bread wheat grown in North America alone, each well adapted to site and climate. At present, over 40 percent of the world's wheat lands are planted to semidwarf, high-yielding products of the wheat breeder's science.

Harvesting the Grain For most of the time that wheat has been cultivated, its production required tremendous amounts of hard, physical labor to prepare the soil and harvest the crop. Matured stalks of wheat were cut with hand-held stone knives (Figure 3.13), were bound into sheaves, cleaned, and the kernels collected. A family could scarcely find the time or have the energy to grow a crop sufficiently large to feed itself, much less produce grain for others, unless they were assisted by slaves or were, as frequently happened, forced to surrender what was literally the food from their own mouths. Harvesting was speeded up and made both physically easier and more efficient with the invention of the sickle. This tool seems to have made its appearance almost simultaneously throughout all of the lands where grain was grown (Figures 3.14, 3.15, and 3.16). The subsequent invention of the scythe

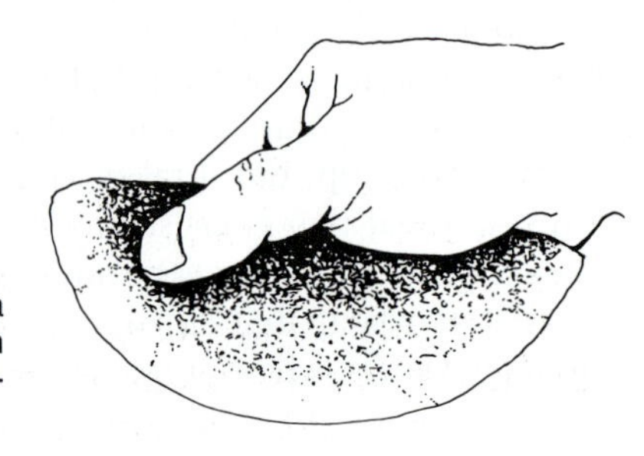

Figure 3.13 Reconstruction of a primitive stone reaping knife such as the ones found in early agricultural sites.

further enhanced harvesting and was improved in 1810 by English Quakers. They added the cradle which laid stalks neatly in piles, easier to gather and bind (Figure 3.17).

Figure 3.14 Harvesting grain in ancient China. A sickle blade was attached to a stick, easier to swing than the short-handled sickle. Taken from an impression on a clay roof tile made during the Han Dynasty, about 2200 B.P.

Figure 3.15 Harvesting grain in dynastic Egypt. Grain cut with the hand sickle was threshed by having oxen tread on the sheaves. Taken from an Egyptian tomb painting.

Figure 3.16 As late as the fourteenth century, grain was harvested with the sickle. Taken from a woodcut made in Germany in 1340.

Figure 3.17 The reaping or cradle scythe.

In 1824, Patrick Bell of Scotland invented a mower-reaper which was promptly smashed by itinerate reapers who feared for their livelihoods. Nevertheless, the idea of a mechanical reaper was born, and Obed Hussey invented a practical reaper in 1830. It had a cutting blade that consisted of teeth moving back and forth and had mechanical fingers that guided the stalks to the cutter bar. Cyrus McCormick invented his reaper in 1834 (Figure 3.18), and almost immediately sued Hussey for infringement of patent right. McCormick won the suit against Hussey's lawyers—Abraham Lincoln and his eventual Secretary of State, Edwin M. Stanton. The McCormick reaper could replace 60 people with scythes and, just as important, was an economically practical solution for reaping the large fields that were beginning to be farmed on the

Figure 3.18 A nineteenth-century reaper. Courtesy International Harvester Co.

western prairies. Massey-Ketcham reapers were on the Canadian market by 1845.

Farm machinery became available in ever-increasing numbers. Charles Newbold of New Jersey forged a cast-iron plow and moldboard, and John Lord of Chicago invented a steel plowshare. John Deere, originally from Vermont and, later, of Moline, Illinois, put these two inventions together in 1831 to make the steel moldboard plow, the first capable of breaking the thick, resistant prairie sod (see Figure 3.9). March's harvester came along in 1858, a practical threshing machine in 1876, a binder in 1878, and a combined reaper-harvester-binder in 1881. There were efficient seed drills by 1885 and, by 1890, the farmers on the prairies could purchase a complete combine incorporating reaper, binder, thresher, and bagger of the grain. Tractors, capable of pulling

Figure 3.19 Wheat harvested with a combine. Courtesy U.S. Department of Agriculture.

these combines, came into use by 1910 (Figure 3.19). The technological revolution in wheat cultivation was essentially complete.

Because of the differences in planting and maturation times, wheat harvests start in Texas in April and continue north until the final harvests of late spring wheats in Canada during late September and early October. Large companies hire combines and their expert crews for the harvesting season. A crew may live in Texas until the start of the spring harvest and then never stop working until the crew reaches Edmonton, Saskatoon, or Winnipeg in October.

It is impossible here to discuss the social and economic history of wheat since the beginning of its cultivation. Certainly wheat cultivation was a major factor in the development of civilizations. Nevertheless, grain cultivation was a mixed blessing for large numbers of people. The system of slave owning was one consequence of settled agricultural communities, in which some members of the community did not till the soil, but served in other capacities that the community believed were of sufficient importance that it assumed the responsibility for feeding these people. Greeks and Romans, and, before them, the Chinese and peoples of the Middle East and of North Africa raided other communities for strong males and females who could work in the fields.

The feudal system in Europe did nothing for the slaves and peasants except to brutalize them as well as those who ate the wheat and barley obtained from slave labor. One can get an inkling of this in Edwin Markham's poem "The Man with a Hoe." The Renaissance was profitable for the nobility and for the developing merchant class; peasants still worked for other people's benefit. The invention of the scythe in the middle of the eighteenth century certainly reduced physical labor, but the peasant and slave were required to work faster. In addition, since fewer people could cut a field of grain, many peasants were summarily dismissed, doomed to wander, with their families, from place to place in search of work. Thomas Hardy wrote of this in several of his books.

With the notable exception of black and American Indian slaves in the North American colonies, the concept of a peasant class did not exist in North America. The freemen of most of the United States and Canada took this name proudly, for their toil redounded to their own advantage. Even indentured servants knew that they, too, could eventually clear their debt and become freemen. Word of this spread throughout Europe, and by the beginning of the eighteenth century, the first of many waves of immigrants from northern and western Europe began to arrive in North America. A large wave of people broke on these shores in the 1820–1850 period with some moving to the maize-growing country of the Midwestern United States and the agricultural lands of Quebec and Ontario. The hegemony of North American wheat, however, was the lure for many immigrants who occupied the open lands of Manitoba to the Alberta mountains and the U.S. prairies west of the Mississippi River.

ADDITIONAL READINGS

Hawkes, J. "The Birth of Wheat and Maize Farming." *UNESCO Courier* 20 (1967): 6–25.

Inglett, G. E. (ed.). *Wheat: Production and Utilization.* Westport, Conn.: Avi Books, 1974.

Kahn, E. J., Jr. "The Staffs of Life. III. Fiat Panis." *The New Yorker* (December 17, 1984), pp. 57–106.

Kent, N. L. *Technology of Cereals with Special Reference to Wheat.* New York: Pergamon Press, 1966.

Needham, J. *Science and Civilization in China.* Vol. 6, *Biology and Biological Technology.* Part II, *Agriculture,* by F. Bray. Cambridge, England: Cambridge University Press, 1984.

Morgan, D. *Merchants of Grain.* New York: Viking Press, 1979.

Peterson, R. F. *Wheat.* New York: Wiley Interscience, 1965.

Trager, J. *Amber Waves of Grain.* New York: Arthur Fields Books, 1973.

Wailes, R. *The English Windmill.* London: Routledge and Kegan Paul, 1954.

Bread

For the Lord, your God, is bringing you into a good land, a land of brooks of water, of fountains and springs flowing forth in valleys and hills, a land of wheat and barley . . . a land in which you will eat bread without scarcity.

Deuteronomy 8:7–9

The technology that allows you to pick up a loaf of bread for your morning toast or luncheon sandwich is complex. The production of sheaves of golden wheat and the collecting of the brown kernels from the heads is part of the story of agriculture. Conversion of grain into flour is part of the story of agribusiness, and the conversion of flour into bread is essentially the story of civilization.

Wheat, barley, and millet have been cultivated for our bread for thousands of years, but except for dynastic Egypt, these cereal grasses were grown by the people who ate them; grain was not a cash crop.

MILLING

The grain or kernel of wheat is, botanically, a one-seeded fruit enclosed in husks (the chaff) with the seed coats (the bran) firmly bound to the seed itself (Figure 3.20). The seed consists of the embryo (the germ), which is the young wheat plant, and the starchy endosperm, which is the food supply for the growth of the wheat seedling. White flour is the ground-up endosperm. The bran is indigestible by humans, and the germ with its vitamins, oil, and minerals makes flour heavy and reduces

Figure 3.20 Anatomy of a grain of wheat.

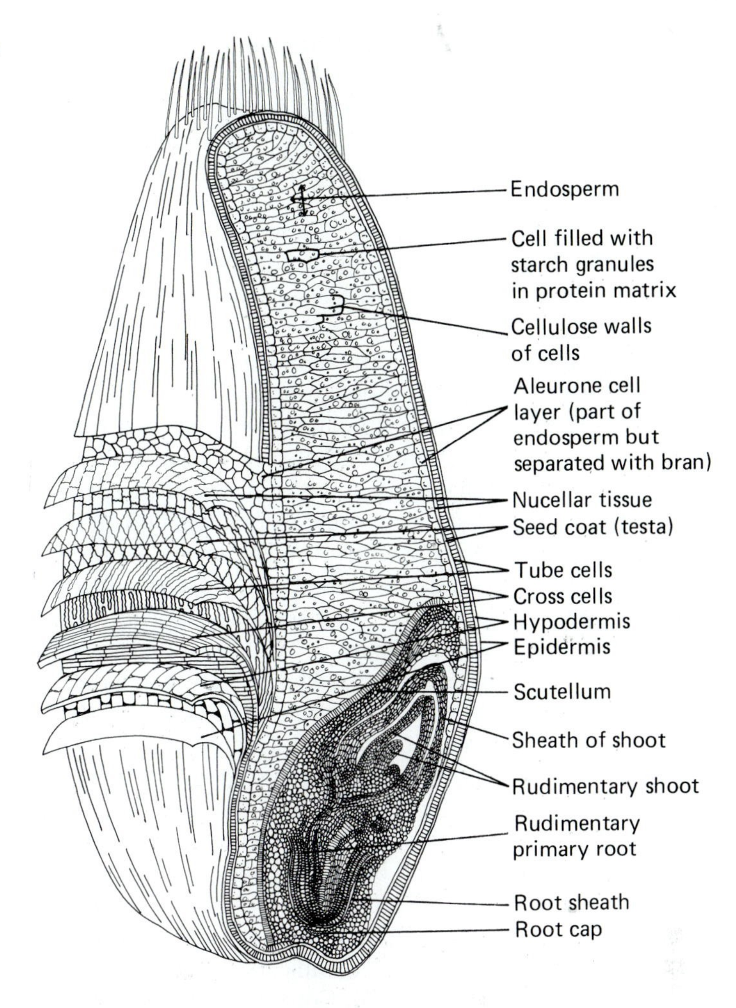

its ability to rise. To make fine white flour, both bran and germ must be removed and the endosperm cells pulverized. This conversion of wheat into flour has occupied Man's attention for well over 4000 years.

People learned that the kernels could be freed of their husk by passing a head of wheat quickly through a flame, the process of parching. For centuries the whole grain was eaten, sometimes raw and sometimes softened by boiling or merely soaking for several days. If grain was pounded with a rock, a coarse mixture of whole and broken kernels was obtained that would cook more quickly and was more digestible. Several thousand years ago, humans learned that if whole grains were rubbed on a stone, at least some of the indigestible bran was removed, giving a better product. This is the *bulgar* of the Middle East or the groats of Europe. When roasted and then mixed with water, oil, and sometimes honey, a nutritious and tasty gruel or porridge could be made. This was the famous Roman *pulz*. It could be eaten hot or cold, or placed on hot stones or the ashes of a fire to produce a product that

Figure 3.21 The saddle stone used for grinding grain into coarse flour. Taken from an Egyptian tomb painting.

Figure 3.22 The mortar and pestle, still used in tribal societies.

Figure 3.23 The handmill, a small, double-stoned grinder.

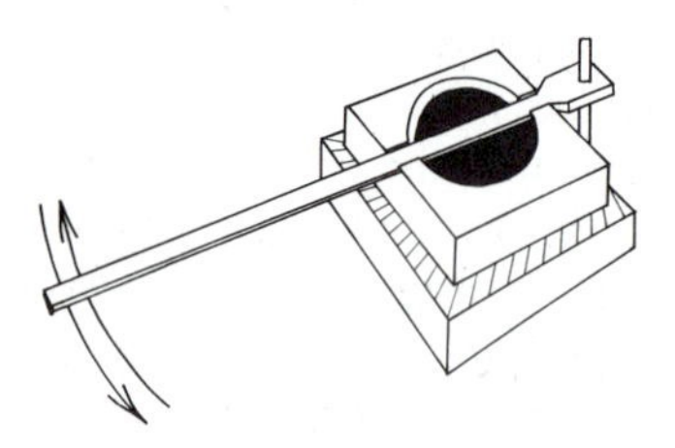

Figure 3.24 The rocker mill, invented in Greece, was used in Europe until the late Middle Ages.

had increased storage life. The first flat bread must have been hard to chew and even harder to digest. Flatbread has been excavated in homes of the Swiss Lake Dwellers of 8000 B.P. Alternatively, groats plus dried fruit could be soaked in water or milk for several days until the mixture fermented and softened, forming the *frumenty* eaten in European farm homes.

Coarse milling of grain was, apparently, independently invented by all of the cultures that ate any of the edible grasses. The Mayans and Aztecs of the New World who ground maize into meal, the Orientals who ground millet for noodles, and the peoples of the Middle East and the Mediterranean basin who grew wheat and barley, each developed a saddle-shaped grinding stone (Figure 3.21), on which grain was placed to be broken and powdered by a round stone. The resulting flour was coarse and full of stone chips. It contained all of the bran and most of the germ, but it was finer than groats and could be mixed with water to a paste that could be baked into a flat cake. The mortar was, again independently, used by several cultures (Figure 3.22). Around 2800 B.P. it was discovered that flour would be finer and the tedious grinding reduced if one placed the grain between flat stones and turned the upper stone (Figure 3.23). Small millstones were so vital a part of each household that Mosaic Law stated that "No man shall take the nether or the upper millstone to pledge, for he taketh a man's life to pledge." A small rocker mill (Figure 3.24) was an obvious improvement that allowed one person to grind grain. The Greeks invented milling machines in which the upper stone, instead of being flat, had carved grooves which prevented clogging. Larger mills were built to be operated by slaves (Figure 3.25), and a bit later, by domestic animals. About 2200 B.P., those marvelous Roman engineers discovered a way to harness the power of falling water to turn their mills and, with subsequent modification, the water mill was used for over 2000 years (Figure 3.26). Countries with limited water power relied on the ox or the horse until the twelfth century, when someone in Holland realized that the windmills used to raise water could also turn millstones (Figure 3.27).

Still, the flour was dark with bran flakes. People knew that bran was not nutritious, was a laxative, and spoiled flour. Medieval Britons said, "Bread having moche branne fylleth the bealye with winde and excrement," and people wanted bran removed from their bread. Coarsely woven cloth was used to bolt or sieve the flour. This removed most of the germ and the larger particles of bran. As cloth technology developed, bolting fabrics became finer, and the bread became whiter and, correspondingly, more expensive; the wealthy ate bread that was, at best, a light brown in color. The poor ate coarse bread similar to our whole wheat—and they resented it.

Even the mere existence of a mill was a source of trouble. The millers could compel the soul of the free stream to be their slave. As late as 1671, Estonian peasants burned down a mill because it offended the

Figure 3.25 A large Roman mill turned by hand or with horsepower.

Figure 3.26 American grist and flour mill powered by water. Such mills were common until the end of the nineteenth century, when they were replaced by steam-powered mills and eventually by roller mills powered by electricity.

Figure 3.27 The windmill as used in Holland.

brook who told the river who complained to the sea—and the sea refused to release water causing a drought. Windmills weren't much safer. In Germany, many mills were outlawed by the nobility who feared that the wind, angered by being used, would wreak its revenge by sending hail to flatten the crops, or would become so enraged that it would become a hurricane. The Teutons, who worshipped Odin as Wind God, and the people who prayed to Jupiter or Zeus as God of the Elements, asserted that they could hear their God protesting in the thunder of the turning stones. The occasional dust explosions in flour mills were, obviously, signs of this anger. Superstition was reinforced by economic considerations. Roman law, adopted throughout Europe, provided that *"Cuius terra, eius molina"*—who owns the land, owns the mill. The feudal lords and the clergy owned both the land and the mill and could compel the peasants to have their wheat ground into flour at their mills. The miller was a servant of the noble and was paid in flour.

To survive, millers routinely short-weighed the peasants. In Chaucer's "Reeves Tale," it was said of the miller, "A theef he was for sothe of corn and mele," and it was often asserted that "the miller keeps a pile of sand behind his door." The peasant who broke a tooth on a chip of stone was kept from beating the miller because of a fear that the miller had a compact with the Devil. When, in central and eastern Europe, rye flour was made and ergotism spread throughout the community (Chapter 7), it was the miller who brought this curse upon the people.

The desire for white bread was almost overwhelming except in rye-eating lands of northern Europe and slavic countries; and even here, when wheat bread was wanted, it had to be as white as possible. Incised, grooved millstones were replaced in the fifteenth century by the French *buhr* stones, and the introduction of silk bolting cloth from China permitted the exclusion of much of the bran. Malisset in Paris invented the graduated grinding process in 1760. Here the germ was removed by the first passage through the stones set three mm apart, the bran was removed in a two mm pass, and fine flour was ground in a one mm pass. Flour was now almost as white as it is today. It was, however, expensive, and the French *sans-culottes* greatly resented the fact that the Sun King and his court were eating white bread while they, who made the flour possible, were still chewing a high bran product. One of the early acts of the postrevolutionary French government was to stipulate that *pane egalité*, the bread of equality, be white for all citizens. In England, the Tory governments of the nineteenth century were indignant to hear that the inmates of the poorhouses were demanding white bread.

The triple grinding process of Malisset was slow, and stone chips from the mills tore the delicate bolting cloths. In 1830, Müller and Sulsberger in Switzerland fitted porcelain and steel rollers together, set them in pairs at Malisset's graduated distances, and produced flour quicker. The government of Hungary subsidized construction of roller mills and, for close to 40 years, had a monopoly on high grade flour. This monopoly was broken in 1870 when the governor of Minnesota invited Hungarian technicians to build the first roller mill in Minneapolis. Minneapolis's success was short-lived. Charles Gaskill soon realized that the immense power of Niagara Falls could be harnessed to roller mills, so Buffalo, New York, became the center of milling. Steam-powered mills were developed in Italy by 1870, to grind the very hard durum wheats into seminola flour for pasta, and later to handle the hard red wheats grown on the North American prairies. Electricity entered the milling process soon after the turn of the century, and the water-turbine generators of Niagara Falls assured Buffalo's prominence as a world center of flour milling.

The very white flours desired by most people required the product not have the light yellow hue caused by natural pigments in the endosperm cells. Over the years, many agents were employed to bleach out these *xanthophylls*. Alum, ammonium carbonate, and even chalk were added to the point where the whitening agent actually became an adulterant. Tobias Smollet, an English author, had his character Matthew Bramble say,

> The bread I eat in London is a deleterious paste, mixed up with chalk, alum and bone ashes, insipid to the taste and destructive to the constitution. The good people are not ignorant of the adulteration, but they prefer it to wholesome bread because it is whiter . . . and the miller or the baker is obliged to poison them.

Chemical bleaching agents were introduced in 1903. There were many reactions against finely milled, bleached flour. In the middle of the nineteenth century natural food advocates were recommending a return to Christian morality and whole wheat flour. The most vocal of these food fundamentalists was Sylvester Graham, who went so far as to assert that the depauperate white bread of the 1840s was an evil that led to drunkenness and sin (type not specified). Graham correctly observed that removing the germ resulted in flour that lacked "nutriemental value." A bowel fanatic, he also recognized that white bread did not provide the "bulk that promotes good intestinal cleanouts." He invented Graham crackers as a healthy and tasty substitute for white bread, but since one can't make a sandwich with them, their popularity as a substitute for a slice of bread was short-lived.

After World War I, when North America assumed the responsibility of feeding a world in which other dietary sources of minerals and oils were in short supply, it was realized that Graham had a point. It wasn't until the 1930s, when vitamins were discovered, that it was recognized that white bread had nutritional drawbacks. The virtue of bran is questionable; a moderately well-rounded diet will provide all the bulk cellulose that the body requires, but people demanded white bread without the vitamin-rich wheat germ. In a series of conferences involving nutritionists, biochemists, botanists, government, and the milling and baking industries, it was decided that vitamins and minerals removed with wheat germ should be replaced. Enriched breads were developed when synthetic B vitamins and minerals were added and milk solids and high vitamin yeasts were used. New milling techniques allow 70 percent of the natural vitamins to be retained in the flour. Today, virtually all bread flours are enriched during milling or baking, and wheat germ is sold separately as a food supplement. This apparently topsy-turvy way of handling the problem is really quite logical, for, when wheat germ is retained, the flour cannot be stored for long periods of time because wheat germ oils will turn rancid.

THE MIRACULOUS LEAVENING

In very general terms, raised or leavened bread is produced when a dough of flour and water is acted upon by carbon dioxide gas, which forms gas bubbles in the dough. Unless these bubbles are stabilized, raised dough would collapse like a poorly made soufflé. The dough constituent that confers elasticity and stability to the bubbles is a protein called glutenin, which constitutes about 9 percent of wheat flour. To produce bread with small, even holes, the gas is punched out of the dough and a second rising occurs before baking. There are a number of leavening agents. Air itself is a leavening agent for beaten products such as angelfood cakes, in which the batter is stabilized by the proteins of egg white. Baking powder is a combination of baking soda (sodium bicarbonate) and an organic acid that reacts with water to form carbon dioxide:

$$NaHCO_3 + \text{tartaric acid} \rightarrow \text{sodium tartrate} + H_2 + CO_2$$

If baking soda is used alone, the same chemical reaction occurs except that the acid is supplied in the dough mixture as buttermilk, fruit juices, or molasses.

For most raised dough, the leavening phenomenon is a biological process involving living organisms, primarily the fungi called yeasts (*Saccharomyces cerevisiae*). The production of carbon dioxide is a metabolic process in which yeasts and all other living cells convert carbohydrates into chemical energy that the cell uses for its growth and for the synthesis of all cellular constituents (Chapter 1). The overall reaction can be summarized as follows:

$$\text{starch} \rightarrow \text{glucose (6-carbon sugar)} \rightarrow 2 \text{ pyruvic acid} \rightarrow \\ 2\ CO_2 + 2 \text{ ethyl alcohol}$$

A small amount of sugar added to bread flour starts the fermentation process, enzymes in the flour and from the yeasts convert starch to sugar, and production of carbon dioxide is very rapid. Bakers add extra sugar to the dough to speed up the fermentation and to give bread a slightly sweet taste. Because the enzymatic reactions of yeast are sluggish at low temperatures, temperature control during leavening is important—as any good cook can tell you.

Early leavenings were undoubtedly natural yeasts and bacteria that fell into the flour-water mixture. Bakers soon found that a piece of risen dough contained enough leavening agent to serve as a starter for the next day's dough. I Corinthians 5:6 says, "Know ye not that a little leaven leaveneth the whole lump?" indicating that the technique is probably 3000 years old. In Egypt, another method was used to obtain leavening. Crushed wheat was soaked in white wine for three days, and the mash was added to the flour-water dough. The brewing industry became a readily available source of yeasts. Brewers supplemented their income by selling pails of beer wort or skimmed beer foam for home baking. These made excellent leavenings, and the slightly beery taste of the bread was certainly not unpleasant. When Antony van Leeuwenhoek trained his primitive microscope on beer and rising bread in 1677, he was the first person to see yeast cells, and he remarked in his letter to the Royal Society of London that the organisms multiplied rapidly in sugar solutions. This observation allowed bakers to grow yeasts in sugar-flour solutions in order to have a stable yeast which gave rapid and even rising of the dough. By the middle of the eighteenth century, it was recognized that beer yeasts were not as good for breadmaking as those selected as bread yeasts. As bacteriology developed, Charles Fleishmann founded the yeast-making industry in the United States in 1869. Yeasts were grown in sugar solutions, the excess water removed, and the yeast with a starch binder was compressed into slabs that were sold to bakeries or to the home. Dried yeasts, a very stable source of active yeasts, were generally available after 1945.

THE BAKERY

Wheaten, raised breads were apparently invented in Egypt (Figure 3.28) and spread rapidly throughout the Mediterranean basin and the Middle East. Raised bread was used commonly in Greece, and unleavened barley bread was also eaten. Siphilis of Siphonos said in 2100 B.P., "Bread made from wheat, as compared with that made from barley, is more nourishing, more digestible and in every way superior," and he was nutritionally and biochemically correct. Greeks were very imaginative about their breads. They made puff cakes mixed with honey, cinnamon breads containing dried fruits, egg breads, cheese breads, breads spiced with cloves, mace, and saffron; essentially all the kinds with which we are familiar. But they didn't make rye bread, because rye was unknown in Europe until about 150. The Romans, too, had dozens

Figure 3.28 Bakery of Pharaoh Rameses. Taken from an Egyptian tomb painting.

of kinds of rolls, twists, and even raised waffles glazed with Greek clover honey mixed with white wine. In Rome, baking, like milling, was a regulated trade, but bakers, unlike the hated millers, were highly respected members of the community. Since they owned the ovens, they not only made bread but also cooked meats, pastries, and fowl for the neighborhood. This esteemed profession was regularized by the formation of guilds, trade associations that controlled the quality of the bread and also set standards for the profession. Baker's guilds continued until the nineteenth century. A boy was apprenticed about the age of eight, and by 18 was ready to become a journeyman. This was exactly what he was; he was required to travel through several countries learning different techniques and upon his return was examined to determine his fitness to become a master baker.

Most daily bread was fabricated into a round loaf, much like that given to the Roman populace. The French name for baker, *boulanger,* refers to the round loaves that came out of the oven (Figure 3.29). The quality of bread was of great concern to government. The English "Assize of Bread" in 1266 was the first set of regulations that had wide application, and equivalent laws in other countries quickly followed.

Bakers also were the makers of table service. Impervious dishes did not exist before the thirteenth century, and "plates" consisted of a thick slab of stale, partly leavened brown bread about four inches by six inches. The absorbent plate, called a trencher (Figure 3.30), gave rise to our name trencherman, one who was able to soak through a number of trenchers at a single sitting. The grease-soaked trenchers were either flung to dogs fighting under the tables or, in moments of charity, to the poor who stood around the tables waiting for them. Only when glazed dishes became generally available in the sixteenth century were trenchers completely abandoned except for their vestigial placement under broiled lamb chops or soft-boiled eggs.

The American experience with bread paralleled its development in Europe of the seventeenth and eighteenth centuries (Figure 3.31). In pioneer days, bread was primarily cornbread, with wheat bread made

Figure 3.29 Brick oven used in sixteenth-century Italy. It was called the peel oven, after the wooden spatula with which bread was moved in or out of the oven. Taken from a sixteenth-century woodcut.

from imported and expensive flour. George Washington raised some wheat, milled it, and baked it for his personal table; other wealthy planters did the same. As wheat became more available in the middle of the eighteenth century, breadmaking was done in the massive fireplaces that were also the source of light, general cooking, and home heating. The bread oven was usually of brick or stone on the side of the fireplace, and only skilled bakers produced an evenly brown loaf without a soggy center (Figure 3.32). In the cities, baking was a commercial operation almost from the beginning of the eighteenth century. The first bakery in New York opened in 1645, and professional bakers were in business throughout larger communities by 1700.

Figure 3.30 A fifteenth-century woodcut showing the use of slices of bread as trencher plates.

Figure 3.31 Wood-fired beehive oven.

The year 1850 can be taken as the watershed for baking in North America. At that time there were about 2000 bakeries in the United States employing 7000 workers who produced close to $14 million worth of baked goods. Urbanization increased in the United States as factories enlarged and as machinery began to supplant muscle power on the farm. Although home baking was still predominant in 1850, women began to form a significant proportion of the labor force, and they simply didn't have time or reserve energy to bake the family bread after a 12-16-hour factory day. By 1860, more than half the bread sold in the cities was baked outside the home. As demand increased, breadmaking began to be mechanized. The deck oven came along in 1880. It carried bread into and out of ovens on a wheeled deck that eliminated the peel and permitted almost continuous operation of the ovens. A rotary oven invented at the same time looked like a Ferris wheel turned on its side. Mixing machines, dough dividers, rounders, and proofers were in use by 1895, and continuous tunnel ovens, "a tube with flour going in at one end and bread coming out the other," were constructed by 1913 (Figure 3.33).

As people became more and more crowded into cities, retail grocery stores began to replace the bakery as the place to buy bread. Large bakeries could produce loaves faster and cheaper (albeit not better-tasting) than could the baker, and wagons, and later trucks, could place many loaves on the grocer's shelves. At the same time, the American

Figure 3.32 A nineteenth-century American peel oven.

Figure 3.33 An integrated cracker-baking factory. Courtesy Nabisco, Inc.

fetish of supercleanliness was being promoted. Wrapped bread, "untouched by human hands," was, it seemed, more sanitary. Packaging machines that enclosed bread in impervious wrappers which negated the virtue of a crust came along at the end of World War I, and the automatic slicing of bread developed in 1928. Families were asked, "Can you make better clothes than a tailor? Then why try to bake your own bread?" Brand identifications and increasing real income of the working man and woman combined to make it almost a crime to place home-baked bread on the table when company was invited for dinner. The bread companies then asked, "Aren't your children and your husband as important as your guests?" As sales of "store bread" increased, breadmaking became a fully automated process, and mass production techniques were applied to it as they were being applied to the construction of automobiles.

CLASSIC FRENCH BREAD

1 package dry yeast
1 cup warm water
1 1/2 tsp. white sugar
1 tsp. salt
3 cups all-purpose or bread flour
1/4 cup yellow corn meal
2 tsp. salad oil

1. Dissolve yeast in water, add sugar, salt, and 1 cup of the flour. Mix well until well blended.
2. Gradually add remaining flour to make a slightly stiff dough. Knead 5–10 minutes on a lightly floured board. Form into a ball, place in a well-greased bowl, and roll to coat the ball of dough. Cover with a cloth and let rise in a warm (85°F) place free of drafts for 90 minutes.

3. Punch down, remove dough from bowl, and let it rest on a lightly floured board for 10 minutes. Roll out dough into a 10-by-15-in. rectangle. Beginning at the long side, roll up tightly. Seal the seams. Place seam side down on a baking sheet sprinkled with corn meal. Slash top of loaf with several diagonal cuts. Cover and let rise for 1 hour.
4. Brush top lightly with salad oil. Bake at 375°F for 40 minutes.

BREAD IN RELIGION

Light, leavened bread could not be baked on hot slabs of rock, and the beehive oven was invented, heated by a wood fire and then brushed clean to hold the leavened dough. So important was the loaf and so mysterious was its rising that even the oven was invested with mythic importance. Bread came from a womb wherein was born the staff of life. Our expression "home and hearth" originated in Egypt, whose inhabitants despised nomadic herders as barbarians who had no hearth in which to bake their bread. As long as the Israelites were nomads during the biblical 40 years, their bread was unleavened gruel, quickly baked and quickly eaten. This is memorialized in the Passover matzos: "Seven days shall ye eat unleavened bread" (Exodus 13:6). Only when they became settled agriculturists was it again possible to leaven their bread, and they invested bread with an aura of mystery. Each household had its own oven, and only women could tend it; thus woman was both the life giver and the life sustainer. As holy food, strangers and travelers were offered God's bounty, and the Hebrews broke bread as a symbol of God's peace. Nevertheless, when offered on God's altar, the bread was unleavened, uncorrupted by the unclean (from the earth) leaven. In medieval England, from whence our word bread comes, it meant "that which is fermented."

The role of bread in Christian ritual is based on two verses in the New Testament. Matthew 26:26 and Mark 14:22 state: "Now as they were eating, Jesus took bread and blessed and broke it, and gave it to the disciples and said, 'Take, eat. This is my body.'" Luke 22:19, however, renders this differently: "And He took bread and when He had given thanks, He broke it and gave it to them saying, 'This is my body which is given for you. Do this in remembrance of me.'" Early Christians were of two minds about this discrepancy. Were they, following Matthew and Mark, to take literally the idea of "body," or were the words, as given in Luke, to be taken symbolically? Did Christ mean that in order to be saved, one must eat human flesh—a concept that has parallels in other Mideastern cultures? Or, as St. Paul suggested in I Corinthians 11:23, was the passage in Luke to be essentially a recommendation or parable of remembrance? Tertullian, Augustine, Origenes, and other early Church fathers agreed that bread represented the body of Christ but was not, in fact, the substantive body. St. Gregory of Nyassa in Asia Minor

(331–394) argued that once bread is eaten, it becomes part of the body, and thus the Eucharist is representational. St. John Chrysostom of Constantinople (347–407) disagreed, asserting that worshipers do in fact eat Christ's body transformed as the blessed bread. The matter smouldered for a thousand years until, in the Lateran Edict of 1204, Pope Innocent III declared *ex cathedra* the Dogma of Transsubstantiation. The Roman Catholic Church now said that at the Last Supper, Christ required his followers to eat of his flesh. A priest had the ability to transform bread into the fleshly body of Christ to be eaten by the faithful, thus following St. Crysostom, who said, "Why should we, the faithful, shrink from eating human flesh?" The Dogma of Transsubstantiation became a time bomb that exploded 300 years later when the role of the priest was questioned during the Reformation.

In the meantime, there were practical matters that required attention. Eucharist bread was, for the first 500 years, a large round, raised loaf of wheat and barley shared by the congregation. From about 650, each communicant was given a small wheaten loaf or roll containing some barley or other cereal grain—essentially the common bread of the people elevated by transsubstantiation. The Church decided, about 1000, to take again the Jewish injunction regarding the corruption of leaven and ruled that Eucharist bread should be unleavened. It was also agreed that while in Christ's miracle of the loaves and fishes the bread was undoubtedly a mixture of wheat, barley, and probably millet flours, the Passover bread was probably of wheat flour. St. Thomas Aquinas stated that it was almost blasphemy to assume that his Lord would have eaten anything but the best of wheat bread, and so the whitest possible flour was specified for the Host. Because even the prospective Host was almost holy, it had to be guarded carefully. Witches could steal it, or pious communicants might take some for their much loved domestic animals, feeling that their cows and sheep might profit from eating it.

The safe storage of the wafers presented several problems. In unheated, damp churches, the host could soften or even become covered with mold and have to be discarded with reverence. Much more seriously, the wafers occasionally seemed to be covered by drops of blood. The blood of Christ appearing on bread was a horrible thing to contemplate. Obviously it was caused by the Devil, and who among the residents of the community could be the Devil's agents? Why, it was the Jews, of course! In 1253, the members of the Jewish community of Beelitz, near Berlin, were burned—the Host had been bloodied. Paris in 1290, Vienna in 1298, Crakow, Posen, Breslau, and Prague in 1300, all saw the defiled Host, and human blood ran in the streets. Over 2000 Jews were killed in Brussels in 1370, and uncounted thousands died in smaller towns. As late as 1510, 38 Jews were pilloried in Berlin. Not until 1848 did Professor Christian Ehrenberg find that the "blood" was a bacterium that produced a red pigment. In spite of this knowledge, at least two pogroms in Russia in the late nineteenth century were started by rumors of bloodied hosts.

The matter of transsubstantiation, although dogma, was still being questioned. Bishop Berenger called the pope *pulpifex* (flesh eater) instead of *Pontifex* (head of the pontifical college) and was consigned to the flames. One unnamed churchman asked why the Church wished to enclose the Savior in a loaf of bread and was flayed alive for his impertinence. By the early sixteenth century, transsubstantiation became part of the larger debate leading to the Reformation. Huldreich Zwingli, a Swiss reformer, spoke of the Eucharist as a swindle and exclaimed, "If Jesus is present in the bread, we would shudder to eat of it." Martin Luther, Zwingli's mentor, was angered at this. Luther wanted not to destroy the Roman Catholic Church but to purify it. He recognized that attacks on dogma could damage the Church, and debates over parable, metaphor, and biblical interpretation had to be avoided.

For three years, the battle of words raged until the Landgrave Philip von Hessen offered his castle at Marburg—midway between Luther in Saxony and Zwingli in Zürich—as a neutral debating place where the two could defend their theses. Debate started on Friday, October 1, 1529. Its mood was set when Luther chalked on the conference table the words, "This is My Body." A week later, the debate was ended with mutual recriminations and thus was destroyed one of the last bonds that held theologians together. Zwingli was wounded in 1531 in a regional battle between Catholics and Protestant cantons in Switzerland and was captured by Catholics, who beat out his brains, had his body drawn, quartered, and then burned, and his ashes mixed with swine manure. Luther's response to all this was, "We see the judgment of God, who will not long suffer these mad and furious blasphemies."

The Swiss concept of the Eucharist as metaphor was accepted by the Church of England under Elizabeth I: "Transsubstantiation is repugnant to scripture and is the occasion of many superstitions." By 1700, a law was on the books stating that only those Englishmen who pledged allegiance to the king as head of the Church and who also declared against transsubstantiation could buy or inherit land. All others—that is—Catholics, were barred from the civil service or from living within ten miles of London. American colonists did not include the Eucharist in their religious observations, and at least some of the anti-Catholic feeling in the young United States can be traced to this schism. The French writer Montesquieu (1689–1755) was, in his youth, something of a radical. In his book *The Persian Letters,* he had Usbeck, his fictional Persian diplomat, write home about an important magician called the Pope who had persuaded the people "that bread is not bread and wine is not wine." The Chevalier LeFebre was formally indicted in Paris in 1775 for saying that he could not understand why anyone would "worship a God of dough" and was decapitated for having a heretical tongue in his head. Voltaire (1694–1778) said that Catholics profess that they eat God and not bread, Lutherans eat both bread and God, and Calvinists eat bread but not God. He was excommunicated—burning was out of fashion. Even the devout Leo Tolstoy was excommunicated from the Russian Orthodox Church in 1901 for presumably mocking the Eucharist in his novel *Resurrection.* In 1976, the Pope declared that a serious obstacle to the rejoining of Christian sects was the failure of non-Catholics to accept the Dogma of Transsubstantiation.

ADDITIONAL READINGS

Ashley, W. *The Bread of Our Forefathers.* New York: Oxford University Press, 1928.

Dunlap, F. L. *White Versus Brown Flour.* New York: Wallace and Tiernan, 1945.

Furnas, C. C., and S. M. Furnas. *Man, Bread, and Destiny.* Baltimore: Williams and Wilkins, 1937.

Jacob, H. S. *Six Thousand Years of Bread: Its Holy and Unholy History.* New York: Doubleday, 1944.

Kahn, E. J., Jr. "The Staffs of Life. III. Fiat Panis." *The New Yorker* (December 17, 1984), pp. 57–106.

MacEwan, G. *Harvest of Bread.* Saskatoon, Sask.: Prairie Books, 1969.

Matz, S. A. "Modern Baking Technology." *Scientific American* 251, No. 5 (1984): 123–134.

McCance, R. A., and E. M. Widdowson. *Breads White and Brown: Their Place in Thought and Social History.* Philadelphia: Lippincott, 1956.

Miller, B. S. *Variety Breads in the United States.* St. Paul, Minn.: American Association of Cereal Chemists, 1981.

Panscher, W. G. *Baking in America.* Evanston, Ill.: Northwestern University Press, 1956.

Pomeranz, Y., and J. A. Schellenberger. *Bread Science and Technology.* Westport, Conn.: Avi Books, 1971.

Spicer, A. (ed.). *Bread: Social, Nutritional and Agricultural Aspects of Wheaten Bread.* London: Applied Science Publications, 1975.

Storck, J., and W. D. Teague. *Flour for Man's Bread: A History of Milling.* St. Paul: University of Minnesota Press, 1952.

Wailes, R. *The English Windmill.* London: Routledge and Kegan Paul, 1954.

Rice

Is your rice bowl full?

CHINESE GREETING

Rice *(Oryza sativa)* is an annual, semiaquatic grass related to the other cereal grains and sharing many of the growth and structural characteristics of other grasses (Figure 3.34). Over thousands of years, human selection and breeding has resulted in cultivars that are adapted to a wide array of growing conditions. Cultivation extends from 53° north latitude on the Amur River of China, to central Argentina at 40° south latitude. Rice can grow in saline, alkaline, and acidic soils, in water, and on dry land and from the mountains to the hot deserts of central Asia. It's a tough plant.

As is true of wheat, statistics on rice cultivation and use are staggering. Almost 150 million hectares of land (370 million acres) are devoted primarily or exclusively to rice cultivation—roughly 11 percent of the world's arable land. The more than 370 million metric tons produced per year are the primary carbohydrate and a major protein source for 2.3 billion people, close to 60 percent of the world's population. Of this mountain of rice, China produces a third, India 10 percent, Latin America 5 percent and the United States less than 2 percent. North American production in Louisiana, California, Arkansas, Texas, and Mississippi is primarily for export, since North American consumption is less than 8 pounds (3 kilograms) per year compared with 400 pounds (160 kilograms) per year per person in the Orient. Indeed, only a few rice-consuming nations, like Thailand and the Philippines, produce surpluses. Most nations scarcely produce enough for their own population; Japan produces less than 4 percent of the world's supply and consumes close to 6 percent, making up the difference by imports and substituting wheat for rice.

DOMESTICATION AND DISPERSAL

Over 20 species in the genus *Oryza* have been found in tropical and semitropical Asia and Africa. The initial cultivar in Africa appears to have been *O. glaberrima,* while that of Asia was *O. sativa.* The genus evolved in Gondwanaland and species in the genus were distributed when this supercontinent dispersed. African rice, although still cultivated, has been largely displaced by the more vigorous and adaptable Asian species. Some dispute exists about the first domestication of rice in Asia, and it is likely that there was more than one center of

Figure 3.34 Rice *(Oryza sativa).* Courtesy of U.S. Department of Agriculture.

cultivation. One site certainly was the Thailand–Burma–South China semitropics, with carbonized grains dating from 7000 B.P. in dwellings at Non Nok Tha and Ban Chang and in Chekiang Province. Another may have been in India, although adequate documentation is lacking.

From these areas, rice dispersed throughout the Old World. It reached Japan and Korea 2300 years ago and also moved south through the Malaysian Archipelago, reaching the Philippines by the start of the present era. Alexander the Great reportedly found it in Hindustan. By the time of Strabo, the Greek geographer who died in about the year 23, it was cultivated in Babylon and Syria. The Romans appreciated it and imported African rice via Egypt and Asian rice via Persia. By the fifteenth century, rice was found to grow well in the Italian Piedmont and Lombardy; rice consumption in northern Italy rivals the pasta consumption in the south. Sir Williams Berkeley is believed to have introduced rice to the Virginia Colony in 1647, and it was grown in Louisiana by the Spanish and French by 1718. The California rice industry started from seed brought in by Spanish missionaries in 1760.

CULTIVARS AND CULTIVATION

There are two subspecies of *O. sativa, O.s. indica* and *O.s. japonica. Indica,* the long-grained rice, is a tropical plant known in China as *hsien* and grown in the valley of the Yellow River since 6500–7000 B.P. This subspecies is cultivated primarily in tropical and semitropical regions, since it needs high growing temperatures and has no requirement for a particular photoperiod to induce flowering. Because most *indica* cultivars mature in three to four months, several crops can be produced in a year. The bulk of the world's rice is obtained from cultivars of *indica. Japonica* rice (the Chinese *keng* and the Japanese *gohan*) grows better in temperate climates and is photoperiodic, requiring the shortened days of early autumn for flower induction. It is common practice in mid-China to plant winter wheat following the rice harvest and to bring in the wheat crop in May of the next year just before rice is again planted on the same land. With the exception of a few especially hardy but low-yielding cultivars, northern temperate zones are not cultivated in rice because the cold summer nights prevent flower and grain production. A third subspecies, *O.s. javanica,* flourishes only in the equatorial regions of Indonesia.

Rices are further subdivided into glutinous types, which are somewhat sticky after cooking, and nonglutinous types, which are preferred by Westerners since the grains remain separate after cooking. To some extent, the choice of sticky or dry involves the eating instrument; sticky and semisticky rice is easier to eat with chopsticks than is dry rice. The starch of sticky rice is composed of amylopectins and dextrins, while the drier rices have both amylopectins and amyloses. Over the centuries, taste preferences and cultural requirements have given rise to cultivars with all gradations of stickiness and a wonderful variety of flavors. A third type includes the aromatic rices said to have such fine odors that a single grain could perfume a whole bag of ordinary rice.

China had its own very early Green Revolution. In 1012, the Sung Dynasty Emperor Chen-Tsung learned about a rice grown in the province of Champa in Indonesia that ripened in 60 to 70 days and had good drought resistance. The ambassador was ordered to learn the details of its cultivation and to bring Champa rice to China, where the equivalent of county agents instructed the farmers in its cultivation. Within just a few years, selections of Champa rice were being grown throughout China. The grains from which the Green Revolution rices were developed in the 1960s contained Champa genes (Chapter 2).

Because of the large climatic tolerance of rice and the varying conditions of growing this crop, it is not surprising that the number of distinct cultivars is large. The International Rice Research Institute at Los Baños in the Philippines has collected 67,000 Asian cultivars, 2600 African types, and over a thousand wild strains. A duplicate set is held at the U.S. National Seed Storage Laboratory at Fort Collins, Colorado. China's seed bank has 40,000 cultivars. These are maintained under conditions that preserve viability to form the genetic pool for further breeding of rice to increase yields and resistance to diseases and insects. Successful resistance to several viruses has already been inserted in the dwarf strains that now constitute over half the rice grown. It is, however, still common practice for Oriental farmers to grow five or six cultivars simultaneously to ensure that at least one will produce a satisfactory crop.

Almost 80 percent of the world's rice is grown by the paddy method, in which the plants are in standing water until ready for harvesting. Paddy methods are more efficient and dependable than upland or rain-fed cultivation techniques, which depend on rainfall or irrigation for their water supply. Paddy methods can yield up to 8000 kilograms per hectare compared with 1600 kilograms per hectare for upland rice, although some of the new strains of upland rices (IRRI 13146) show promise of doubling the yield.

For centuries the construction of rice paddies has required grueling human labor. Using simple tools—digging sticks, crude wooden spades, and woven carrying baskets—many generations have spent a significant portion of their lives building, maintaining, and repairing rice paddies. On Luzon Island in the Philippines, in China, Japan, and Indonesia, people carved out paddies and paddy terraces across mountain slopes so steep that climbing ropes are used to reach them (Figure 3.35). Terracing was practiced in China at least 4000 years ago. For each retaining wall, tons of earth and stone had to be carried up from river beds and huge volumes of fertile soil were redistributed. These labors were collective efforts that required social ties and interlocking networks of relationships on which the entire life of a village depended. To provide the necessary water, people dug uncounted miles of irrigation

Figure 3.35 Rice terraces in Banaue, the Philippines. The terraces were carved into the sides of Mayaoyao and Carballo mountains over 2000 years ago by the Ifugaos tribe of Luzon Island. Photo courtesy of the Philippines Embassy.

ditches over uncountable centuries, frequently with their bare hands. The Grand Canal of China is one outstanding example of the labor needed to provide the water required to grow this staff of life. In swampy areas of south China, reclamation of land started at least 3000 years ago by methods similar to those used centuries later to reclaim the sea lands of Holland.

For paddy cultivation, rice is sown in small, highly fertile plots; the seedlings are allowed to grow for several weeks prior to being transplanted into the cultivated and flooded paddy. This is labor-intensive (Figure 3.36). Broadcast seedings into paddies, as practiced in the United States and a few parts of the Orient, usually result in lower yields because the individual plants do not form side shoots (tillers) as readily. In general, however, the small size of fields and the narrow terraces on mountainsides do not lend themselves to machine planting, cultivation, or harvesting. Upland or rain-fed rice cultivation is basically similar to that used for the other cereal grains.

Rice is a "heavy feeder," requiring large amounts of fertilizer, particularly nitrogen, for optimum growth and yields. In poorer countries, where funds for commercial fertilizer are limited, composts and sewage (night soil) are generally used. Chinese and other Oriental rice growers recognized centuries ago that when the standing water has a

Figure 3.36 Cultivation of rice in Japan. Courtesy Japan National Tourist Organization.

blueish tint or has a floating mat of the water fern *Azolla,* the rice develops better. The blue color is due to free-living, nitrogen-fixing blue-green algae; related blue-green algae grow in association with the *Azolla.* Weeding is done by hand or with herbicides, and attention must be paid to the many insect pests of the crop.

Because of limited machinery and small fields, paddy rice is usually harvested with sickles after the paddies have been drained. In industrialized countries with larger fields, rice is harvested with combines, as is wheat. Hand flailing and winnowing is still widely practiced, although simple machinery is now more available. The hulls (chaff) of rice are removed and the resulting brown rice has 12 percent protein. Few people in the rice-eating countries eat brown rice; the grains are usually polished to remove the brown seed-coat layers. Unfortunately, this reduces the protein content to 7 percent and also removes most of the B vitamins. People who depend on a rice diet, inadequately amended with vegetables, can develop the nutritional disease beriberi (Chapter 4).

RISOTTO ALLA MILANESE

(for four)

4 tbsp. butter
1 tbsp. finely chopped onion
2 cups arborio or Uncle Ben's rice
1/2 cup dry white wine
4 cups chicken broth or water
1 tsp. salt
1/2 tsp. white pepper
large pinch of saffron
4 tbsp. grated Parmesan cheese
1 tsp. chopped parsley

1. Melt butter in large saucepan and sauté onion until translucent. Add rice and cook 3–4 minutes until opaque but not browned.

2. Add wine and chicken broth, salt and pepper, and saffron. Bring to boil and then reduce heat. Simmer very gently, stirring to prevent rice from sticking to the pan. Cook until rice is tender but not mushy and all liquid is absorbed. If necessary, add more water or chicken broth during the cooking.
3. Add cheese and mix in thoroughly with a fork just before serving. Sprinkle with parsley.

ADDITIONAL READINGS

Chandler, R. F., Jr. *An Adventure in Applied Science: A History of the International Rice Research Institute.* Manila, Philippines: International Rice Research Institute, 1982.

Chang, T. T. "The Origin, Evolution, Cultivation, Dissemination and Diversification of Asian and African Rices." *Euphytica* 25 (1976):425–441.

Grist, D. H. *Rice.* 4th Ed. London: Longmans, 1953.

Hanks, L. M. *Rice and Man.* New York: Aldine-Atherton, 1972.

Higham, C. F. W. "Prehistoric Rice Cultivation in Southeast Asia." *Scientific American* 250, no. 4 (1984):138–146.

Keightley, D. N. (ed.). *The Origin of Chinese Civilization.* Berkeley: University of California Press, 1983.

Luh, B. S. (ed.). *Rice: Production and Utilization.* Westport, Conn.: Avi Books, 1980.

Moorman, F. R., and N. van Breeman. *Rice, Soil, Water, Land.* Los Baños, Philippines: International Rice Research Institute, 1978.

Ochi, Y. *Rice in Asia.* Tokyo: University of Tokyo Press, 1975.

Swaminathan, M. S. "Rice." *Scientific American* 250 (1984):81–93.

Tsunoda, S., and N. Takahashi (eds.). *Biology of Rice.* New York: Elsevier, 1984.

Maize

The corn is as high
As a elephant's eye.

RODGERS AND HAMMERSTEIN
Oklahoma!

All too often a name is applied that causes confusion but becomes so deeply rooted that it just can't be changed. *Corn* is just such a word. The word *corn* is applied to the leading cereal crop grown in a particular area. Wheat, barley, oats, or millet are corn to the European, and rice is the corn of the Orient. Even buckwheat, which isn't a cereal grass, is called corn in the Soviet Union. Ruth gleaned corn in Boas's barley

Figure 3.37 An early illustration of Indian maize. Taken from a woodcut in *Herbarz* by Mattioli, published in 1562.

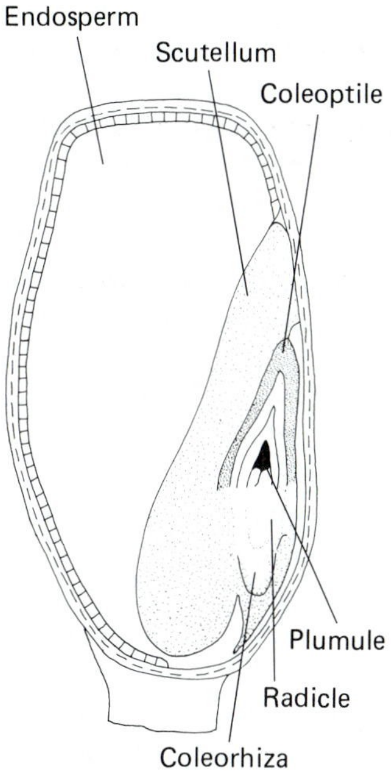

fields, and Joseph organized a granary system in Egypt to store the (wheat) corn for the seven lean years. It was, therefore, quite natural for Europeans who first saw the maize plant to call it corn and, as Indian corn, it entered the English language (Figure 3.37). It is more accurate for us to use the Indian name *maize,* but any attempt to change our habits would probably be futile. The Latin name, *Zea mays,* was coined by Linnaeus. *Zea* means "the cause of life," and *mays* is a rendering of the Haitian *mahiz,* first heard among Europeans by Columbus.

The ear of corn, botanically a female flower cluster enclosed in leaflike husks, bears several hundred fruitlets, botanically known as kernels, on a rigid cob. The silks are the stigmas and styles through which the pollen tube grows to fertilize the individual ovules on the cob-receptacle. The male flowers, the tassel, are borne at the top of the plant. Corn is both self-fertile and cross-fertile, accepting pollen from itself or other corn plants. The pollen is wind-borne but is usually heavy enough simply to fall from the anthers down to the silks, and selfing occurs almost as frequently as does cross-pollination. The plant is botanically unique; no equivalent of the ear of corn exists elsewhere in the plant world.

Tremendous amounts of corn are grown throughout the world. Over a hundred million hectares are planted yearly, with close to half this amount planted in North America. The Soviet Union, China, and South America account for most of the remainder of the crop. Worldwide, over a hundred 56-pound bushels are produced per hectare, although in the corn-producing states, 250 bushels per hectare are not unusual.

Five major types are grown. Two of these, flint and dent, are field types, grown primarily for animal food, for cornmeal, grits, and for industrial use. In dent corn, the endosperm—the starch-storing portion of the kernel—has a core of soft starch that shrinks on drying to give a dented or wrinkled appearance to the kernel (Figure 3.38). Flint corn endosperm lacks this soft starch, and the dry kernels are smooth and rounded. Sweet corn is bred for high moisture and sugar content. Flour corn has a soft endosperm starch that, upon drying, tends to be floury and is grown in Central and South America for tortillas and tamales. Popcorn is now a specialty crop whose use increased with the development of drive-in movies and TV.

Endosperm
Scutellum
Coleoptile
Plumule
Radicle
Coleorhiza

Figure 3.38 Diagram of the maize kernel. The endosperm consists of starch-filled cells; the scutellum cells contain starch, protein, and some oil. The coleoptile is a sheath protecting the first leaf, and the radicle is the embryonic root.

The United States produces over 8 billion bushels and exports over 2 billion bushels of corn per year. That retained for domestic use has become so much a part of our everyday lives that we are usually unaware we are using a corn product. Most of our pork and beef are from corn-fed animals. We use corn oil directly in cooking and on salads, but consume even more as hydrogenated fats and margarines. Corn starch thickens gravies, and even more is used as adhesives, as sizings for cloth and paper, and as bulkers for other powders. A minor industry is centered in Washington, Missouri, where a specially bred strain of corn is fabricated into corncob pipes. Most beer (Chapter 8) is brewed with some corn, and a considerable quantity is used to make bourbon and Canadian whiskey (Chapter 8). Stalks and leaves are converted to silage for winter feeding of cattle and are used industrially as a source of cellulose for rayon and other synthetic fibers. Corn extracted with water is used as the nutrient for the production of antibiotics and to grow those fungi and bacteria that produce the world's supply of citric, butyric, and other acids. Children in the Midwest grew up on pancakes and waffles slathered with corn syrup sometimes adulterated with synthetic maple flavoring or a trace of the "real thing." Corn sugar syrups are having a large negative impact on the cane and beet sugar industries (Chapter 9).

ORIGIN OF CORN

Of the more than 300,000 named species of flowering plants, Man has adapted for his use fewer than 400. With the sole exception of corn, the wild plant(s) from which our cultivars were derived is known. But corn, a basic food plant, is incapable of reproducing itself except under cultivation. If an ear falls to the ground, the kernels will germinate, but the seedlings will twist about themselves so inextricably that most will die. Failure to identify the wild parents of corn is a dilemma to botanists, and a considerable effort has been made to solve the riddle of its botanical origin. This aura of mystery has been a spur to continued research and is a source of considerable debate, for it is a rule of thumb in research that the less known about a topic, the more contentious the debators.

Figure 3.39 Mature fruiting head of *Tripsacum mexicana,* one of the presumed ancestors of corn.

Research interest has focused on several wild plants. One, a wild grass called *Tripsacum* (Figure 3.39), looks nothing like maize but can be crossed with maize to yield a hybrid. It must, therefore, have chromosomes in common with maize. Another plant, teosinte *(Zea mexicana),* looks like maize and has been placed in the same genus (Figure 3.40). It, too, will hybridize with maize, but the ear, characteristic of maize, is not formed. Teosinte and *Tripsacum* do not hybridize readily, and it is unlikely that they are the only parents of maize. A third wild plant may be involved, possibly pod corn (Figure 3.41). When pod corn is crossed with the Mexican popcorn (a true maize), a plant with many maizelike characteristics results. Attempts to cross the pod corn X popcorn hybrid with teosinte and *Tripsacum* to see if this will result in a plant like modern maize have been made. This is not merely a sterile academic exercise. Genes present in these wild plants can provide the hereditary potential needed for improving the yields, conferring disease resistance, and increasing the adaptability of corn to various cultural conditions.

Figure 3.40 Mature fruiting head of teosinte, one of the presumed ancestors of corn.

Figure 3.41 Mature fruiting head of guarany pod corn, one of the presumed ancestors of corn.

INDIANS AND CORN

Whatever the parentage and genetic background of maize, we know that it has been cultivated for a very long time and has been a part of culture for many individual civilizations (Figure 3.42). Pollen of maize has been discovered in archeological sites dated to 6000 B.P. Cobs the size of pencil erasers but bearing a few typical maize kernels were unearthed in Tehuacán, Mexico, and carbon-dated to 7000 B.P. Modern maize, virtually indistinguishable from that grown today, has been dated with certainty to 4000 B.P. Most investigators agree that modern corn originated in the highlands of central Mexico (Figure 3.43) and had spread south by 3500–3000 B.P. through Colombia, Peru, and Chile, as well as north into Arizona and New Mexico. By 2400 B.P. corn was a staple crop of all agricultural peoples of the New World. The population explosions in Central and South America that started about this time are believed to be due to advances in corn selection and agricultural improvements. Hunting tribes of the Great Plains traded buffalo skins for parched corn to sustain them in the winter.

Corn was so important for survival that it became a central focus of religions of Indians of the Americas—just as wheat is still part of the religions of the Europeans. The Aztec Corn God, Centoetl, was anthropomorphized as a spirit with green and gold feathers. The plumed serpent, God of lightning and rain, was said by the Mayans to have mated with Mother Earth to give Ghanan, the corn plant, who sustained the people. Pueblo peoples believe that corn was dropped to their ancestors by a giant bird sent by a Great Spirit. Even the name *maize* has religious connotations. When, on November 5, 1492, sailors of Columbus's fleet went into the interior of Haiti, they saw fields of corn. They returned to the Santa Maria with "a sort of grain they [the natives] call mahiz which tasted well whether bak'd, dry'd and made into a flour." *Mahiz* means "our mother," and the crop was planted, tended, and harvested with reverence.

The cosmology and attitudes of the Hopi and other Pueblo peoples is, like that of the Aztecs, Mayans, Incas, Zapotecs, and other civilizations of the Southwest and Central Americas, replete with legends and ceremonies of the corn (Figure 3.44). Hopi children are told, "When you put the seeds in the ground, you give them to the Earth. She is their mother. She nourishes them so they grow strong and bring forth seeds of their own. Let us make an offering to the Earth Mother so She will feel pleased. Let us feed Her where we plant our corn so She will be strong and bring forth strong plants." Ceremonies accompany all stages of cultivation; prayers and dances for rain, dances to ward off disease, and paeans of thanksgiving are part of the rhythm of life of these corn-planting cultures. Women, closer to Earth Mother than males, guard the growing plants, urging Earth Mother to give birth to healthy plants. Intercourse was contraindicated during the time of pollination. A newborn child is irrevocably united with corn. The umbilical cords of

Figure 3.42 Peruvian Maize Gods. Taken from a vase dated from 500.

Mayan babies were cut over an ear of corn, and the blood-soaked grains were planted to sustain the children until they were old enough to plant a crop. Thus, the children would, for the rest of their lives, eat their blood-brother, the corn.

The Indians used corn in many ways. Young green corn was boiled and eaten as we eat sweet corn or was mixed with boiled beans to produce succotash. Mature corn, scraped from the cobs, was the staple for the winter season. Whole grains were boiled in water containing wood ashes that yielded lye to make hominy. Dried corn was ground in stone mortars or pounded in hollow trees (Figure 3.45) to produce a meal from which a variety of boiled, baked, or fried products could be made. The Incas first bred popcorn and it, too, spread throughout the Americas as a staple of the diet.

WHITE MAN'S CORN

The history of corn's exploitation by the white man can be followed as an exercise in the history of the United States, although it is convenient to abbreviate and simplify it. When the Spanish conquered the Indians of Central and South America, they realized that corn was good food for their slaves and encouraged its cultivation. But they were greedy for the wealth of their colonies and allowed the elaborate irrigation systems of the Indians to fall into disrepair. At the time of Cortez, there were 25 million people in Mexico; by 1605 there were fewer than 1 million, and many died from starvation. English and French explorers of North America found corn cultivated by the settled tribes from Georgia to Nova Scotia. One ship's scribe wrote that "the grain is about the bigness of our ordinary English Pease . . . but is of divers colours, some white, some red and some blew. All of these yield a very sweet flavoure and, being used according to its kind, it maketh a very good bread." French voyageurs took avidly to cornmeal mush and cornbread, eventually using it to the exclusion of wheat flour, which rises but tends to harden when wetted on portages.

Figure 3.43 Glyph of a Mayan Corn God sowing seed. Taken from a wall painting dated from 500 and copied from the *Madrid Codex*.

Captain John Smith of the Virginia Colony realized that the self-sufficiency of the colony depended on raising a staple crop; he required all members of the company to grow corn. Since bread made with cornmeal does not rise, the colonists preferred to grow tobacco as a cash crop, selling it to England for wheat flour. A few years later and 700 miles north, the ill-equipped and mostly city-bred Pilgrims landed at Plymouth on December 21, 1620. Their stores depleted, they were sustained only by gifts of corn and unearthed caches of corn. Even with this, half of them died, and by spring the survivors were desperate for food. Squanto, an Indian who spoke some English, taught them to plant several grains of corn in a hole together with a dead alewife fish to supply fertilizer. In the autumn of 1621, to praise their God for

Figure 3.44 Hopi Indian in ceremonial dress for a corn festival. Taken from a nineteenth-century painting.

sustaining them and to thank the Indians, a feast was held which became the U.S. Thanksgiving holiday. A few years later, before 1650, several Salem goodwives were executed as witches after being accused of "hexing" the corn crop.

By 1660, the colonies were tied to a corn economy. Thomas Ash, Clerk of His Majesty's ship *Richmond,* wrote in 1682, "Their provision . . . is chiefly Indian corn . . . of which they make a wholesome Bread and good biskit which gives a strong, sound and nourishing Diet. At Carolina, they have lately invented a way of making with it a good sound Beer; but it is strong and heady."

CORN AND AMERICAN HISTORY

Increased population, rising prices for corn as a result of European wars, and growing demand for meat, lard, and cheese, all made it apparent that additional agricultural land must be secured. The British colonies were hemmed in between the ocean and the first range of mountains to the west, and, except for the wilderness areas of northern New England (too cold) and the southern swamps (too wet), additional land just didn't exist. The reports of hunters and trappers about the wonderful land just over the mountains whetted appetites. This poorly explored territory was not English but was owned by France. The land between the mountains and the "giant river of the west" had been explored by Joliet, Champlain, Cartier, Brulé, Nicolet, and others in the name of French kings.

Settlement to the west was also limited by the presence of only a few difficult trails from the east. The Mohawk Valley route to the Great Lakes teemed with hostile Indians. The Delaware Gap route through Pennsylvania and the Cumberland Gap trail from Maryland to the Ohio River were safer, but well patrolled by the French. Some Virginians were beginning to organize armed parties to invade Kaintuck country, and the Virginia House of Burgesses commissioned Colonel George Washington in 1753 to warn the French not to attempt to bring

Figure 3.45 Pounding corn into meal, a technique adopted by early New England colonists from the Indians.

French-Canadians into the Ohio Valley. The French commander told Washington that this was French territory and that they would fight, if necessary, to retain the Ohio River and would incite their Indian allies to kill any English settlers who attempted to cross the mountain passes. Provoked by this—and other mutual insults—the two sides began the French and Indian War. British victory allowed England to take possession not only of the Ohio Valley but of all lands east of the Mississippi except for New Orleans. With supply lines stretched, and for other reasons, France ceded Louisiana to Spain.

The end of the Indian troubles unleashed a flood of immigrants. Indentured servants, having almost given up hope of finding land to secure their promised advantages, were among the first to leave the settled eastern areas. Kentucky and Tennessee were settled in 1773–1775. Virginia claimed this land and offered each settler 400 acres at $0.25 per acre on the condition that a house be built and a corn crop planted within a year. By 1789, there were 20,000 people in western Virginia and corn covered the land. It was here that bourbon whiskey was created (Chapter 8).

Land north of the Ohio River was opened up by the military exploits of George Rogers Clark and his rangers, who crossed the Ohio River and attacked the French fort at Kaskaskia, now in Illinois, on July 4, 1778. The following year, Clark took Vincennes, now in Indiana, and the Ohio Valley was open to settlement. People poured in from Virginia, Maryland, and southern New England. The Continental Congress offered land at $6 per acre—payable in corn—and by 1800 there were 45,000 settlers in Ohio and fewer, but growing numbers, in Illinois and Indiana. To subjugate the Indians, settlers and their military allies burned the standing and stored corn and easily killed or relocated the starving people. The topsoil of this land was amazingly fertile. It was a deep rich brown, extending down 5 or more feet, soil such as the settlers had never seen. Few trees had to be felled, drainage was excellent, and the climate was ideal for raising corn: 30, 40, even 50 bushels per acre was common. Pigs, cattle, horses, and chickens fattened at rates that amazed the farmers. Settlers wrote that they had discovered the biblical land of milk and honey.

Heaven, was, however, flawed. The abundant harvest could not readily be moved east to market. Mountain passes were inadequate for any conveyance bigger than a horse carrying kegs of whiskey or sacks of cornmeal. One of the first acts of the United States government was to sign an agreement with Spain to open the port of New Orleans to corn products and preserved meats shipped down the Ohio and Illinois rivers and on to New Orleans via the Mississippi. The treaty provided that the shipper could "export them hence without paying any other duty than a fair price for the hire of the stores." When France again took possession of Louisiana and proved less amiable than did the Spanish, it was obvious that this port had to be secured. Jefferson's Louisiana Purchase of 1803 was a notable act of intelligence and was, as contemporary records show, occasioned largely by demands of corn farmers.

Corn was moved to New Orleans by flatboats floating down the river. Flatboatmen were a breed apart, and legends of their brawling, drinking, and wenching caused Ohio farm boys hoeing corn to dream of escape as rivermen. Since it was impossible to move the rafts back up to Ohio, they were sold for scrap lumber and the proceeds spent on "a good time" before the boatmen started the 1500-mile walk back home. The era of the flatboatmen was short-lived; there were sufficient goods shipped to New Orleans to warrant introduction of Fulton's steamboat to the Mississippi by 1811. This ushered in a new era of American romanticism exemplified in the writings of Mark Twain. New Orleans was still sin city, its customers being the paddleboat crews who were more affluent and hence more sinful than flatboatmen. Bawdyhouse music became, by 1880, jazz, and it moved up the river to Kansas City.

Even the Mississippi wasn't fast enough to handle the freight from the corn fields. As early as 1802, Ohio insisted that as part of the negotiations leading to statehood, Congress had to agree to use money collected from the sale of public lands to build roads from the Ohio Valley to eastern markets. It was, the farmers asserted, unfair and expensive to have to ship to New Orleans and then reship all the way to the East Coast. Since road construction through the mountains was expensive, a system of canals was started. The Hudson–Mohawk canal was built in 1817, the Erie canal by 1825, and the Ohio canal by 1830. Roads were built soon thereafter and so were the railroads. The Albany–Buffalo lines, the Philadelphia–Cincinnati lines, and the Baltimore and Ohio lines completed a road–water–rail system that permanently linked the Midwest to the East. People moved in, land prices soared, and the good, rich soil supported the growth of more and more corn.

Chicago, the "place of the wild onion," was chartered as a town in 1833. Precariously situated in the mud on the southwestern shore of Lake Michigan, it served as a trading post and fort. It was, however, a natural center for railroad builders, whose lines of steel radiated out to the south and moved west. With generous subsidies and grants of land from the federal government, the rapid building program was dictated by Chicago's location. It was almost a fringe benefit that these lines traversed the best corn-growing country in Ohio, Indiana, Illinois, and Iowa and also extended to Kansas City, Dodge City, and other towns where cattle drives terminated. A few enterprising families—Swift, Wilson, Armour, Cudahy—built cattle pens to accept range cattle and farm hogs, fed them on corn, butchered them, and shipped the meat east to the established centers of population. By 1850, the slaughterhouses advertised themselves whenever the wind blew, and Chicago isn't called the "windy city" without good reason. Destroyed in the big fire of 1871, the Chicago Stockyard was rebuilt on the south side of the city where, in Carl Sandburg's words, it was "hog butcher to the world," and in the boast of one of the meatpackers, it "used everything but the squeal." Leather, glue, hog bristles for shaving brushes, fertilizer, soap, and myriad other products spawned by the corn–cattle–hog–interaction were produced in the city. As the natural center for the corn trade, a corn-products industry followed the stockyards to make Chicago wealthy albeit smelly. The Chicago Board of Trade recognized the city's debt by placing a gilded statue of Ceres on the top of its building where for a time she dominated the city. Cattle feed lots were built in many states (Figure 3.46).

PELLAGRA

Although Europeans recognized the superior virtues of corn for animal feeding, they displayed relatively little interest in trying to grow it themselves. By 1520–1525, Spanish explorers brought corn back home, where it was grown in limited amounts in Andalusia. People didn't like its smell, and, since Christ's bread was wheat, corn was fed to the animals. Venice bought corn from Spain for sale to North Africa and established it as a crop on Crete. Corn was being grown as cattle feed in the Danube Valley by 1630. By 1590, Portuguese Jesuits introduced it to China, where it was known as *fan mai* (Western barbarian wheat). Portuguese traders brought it to west Africa by 1680 as food for slaves. In northern Italy growing conditions were adequate for corn and the use of *polenta*—cornmeal—as a dietary staple developed around Milano. Only after World War II did corn become important in Europe as a major cattle and hog food. With the development of strains well adapted to European conditions, corn is now an important crop in

Figure 3.46 Cattle feedlot in Texas with a capacity of 70,000 head. Courtesy U.S. Department of Agriculture.

France and the Soviet Union. China now grows over a billion bushels per year.

In northern Italy and in a few isolated parts of Spain, cornmeal became the exclusive cereal food of the desperately poor. In 1730, Dr. Casel noticed that those on a corn diet became weakened after a time, and Italian physicians named the condition *pellagra*—the disease of rough skin. Although the cause was unknown, it was obvious that corn was involved, and advising people without other food not to eat corn was useless. Pellagra was characterized not only by a roughened skin. There were disturbances in digestion, diarrhea, sore mouths, nervousness bordering on paranoia, and a characteristic butterfly-shaped rash over the nose and cheeks.

Corn cultivation almost died out in Europe by 1800, but its use as the primary food among the black slaves of the American South was on the rise. After the American Civil War, the numbers of poor, both black and white, increased in the South, and the diet of the poor was fatback, cornbread, and molasses. In 1914, alarmed (finally!) by the debilitation caused by pellagra and the increasing number of deaths, the U.S. Public Health Service placed Dr. Joseph Goldberger in charge of pellagra research. Goldberger found that pellagra was caused by the absence of something from corn that could be replaced by meat, milk, and fresh vegetables. Dried yeast also contained the antipellagra factor and was distributed to the poor. Only during the 1930s when vitamins were discovered was it determined that the antipellagra factor was one of the water-soluble B vitamins, the one called niacin (Chapter 4).

HYBRID CORN

From the time of the Aztecs, ears of corn were saved from the previous harvest and the kernels carefully put aside for the next year's crop. Over the centuries, kernels were selected for plumpness, desirable taste, degree of starchiness, and other characteristics. By 1850, farmers vied with one another at state fairs to exhibit large, full ears with even rows of grains and slim cobs. By 1870 there was, throughout the corn country, general consensus as to what constituted good corn. Jacob Leaming's 'Lancaster Sure-crop,' James Reid's 'Illinois Yellow Dent,' and Krug's 'Famous' were preferred. Yet farmers could never be sure that all their corn would be true to type, and while yields were satisfactory, they had plateaued by 1890–1910 in spite of the introduction of better fertilizers and machinery. Agricultural experts agreed that little or no improvement in corn was possible. Yet as early as 1716, Cotton Mather of the Plymouth Colony published observations on natural crosses of corn. James Logan, governor of Pennsylvania, experimentally crossed corn types in 1735, and a few curious people in North America and Europe were defying their ministers and obscenely dusting pollen on plants to see what kind of offspring would result

(Chapter 1). Charles Darwin, one of the most inquisitive, studied the crossing of many plants, including corn, and concluded that the offspring showed increased vigor—due, he reported, not to the mere act of crossing but to the interaction of whatever might be the physical basis of heredity.

William Beal, a student of the famous Harvard botanist Asa Gray, moved to Michigan's corn country and decided to study corn. In 1877 he planted alternate rows of flint and dent corn, detassled the dent so it could not self-pollinate, and found that its progeny formed some large kernels. He reported that yields of his "mule corn" were greater than those of either parent. His experiments were confirmed, but none could explain the results. In 1900 the genetic work of Gregor Mendel was rediscovered after 30 years of burial in an obscure Austrian periodical, and the impact of this new science, named genetics, permitted reorientation of plant breeding. Mendel's work provided the basis for understanding the increased vigor associated with crosses. The name mule corn took on new meaning; the progeny of crosses between corn types was, like the mule, more vigorous because it had—using the new terminology—genetic potential of two parents that complemented and reinforced each other. This acquisition of complementary characteristics was termed *heterosis,* now called *hybrid vigor.*

Armed with these new ideas, George Harrison Shull continued Beal's experiments in 1908. He, too, was able to increase corn yields by 50 percent. He found that heterosis was most noticeable when the parent strains of corn were pure or inbred, having been deliberately self-fertilized for several generations. Donald F. Jones and Edward East found in 1915 that when they crossed the hybrid progeny of pure-line parents, the so-called double cross, their progeny was even more vigorous. When research started again after World War I, it was found that inbred, pure lines of the very different northern flint and southern dent types were, when crossed, capable of yielding corn that was 100 percent more bountiful than either parent (Figure 3.47). These hybrids did not look like the ideal corn of state fairs, and the idea of growing mule corn didn't appeal to farmers. Henry A. Wallace, publisher of *Wallace's Magazine* in Iowa, Secretary of Commerce in 1924, and later Secretary of Agriculture and Vice President, recognized the potential of this new-fangled corn and publicized it in his magazine and in lectures to farm groups.

The first hybrids came onto the market in 1924, and the struggling companies that bred them almost went bankrupt. Ears were slender and kernels were very hard compared with the ideal. Yet yields were bountiful and steers, hogs, and chickens ate it avidly. The young hybrid corn companies decided to promote their products actively. For several years they gave seed corn to farmers to plant next to the ones they had grown for years. Free signs were given to the farmers to indicate which hybrid they were using, and on Sunday after church, the farmers would

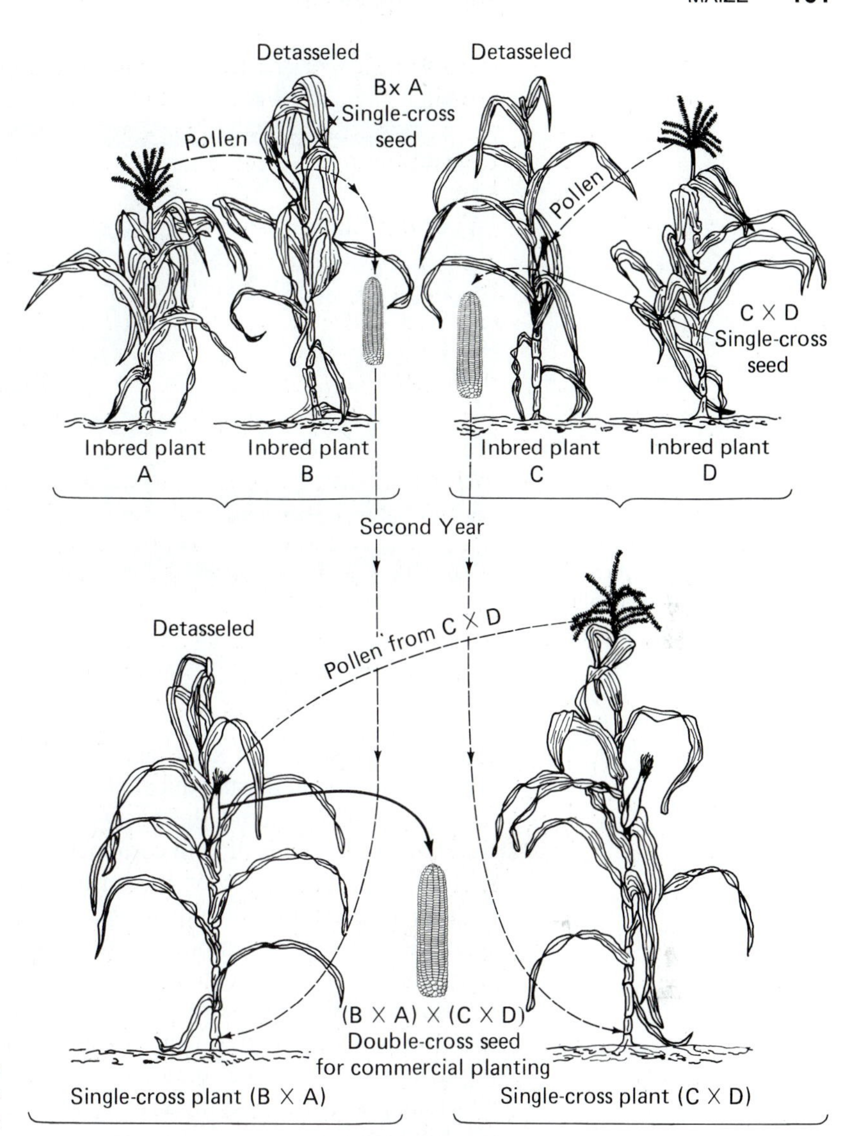

Figure 3.47 Crossbreeding for production of hybrid corn. Courtesy U.S. Department of Agriculture.

take their families for a spin in the Tin Lizzy to see how each corn crop was doing. At harvest time, seed companies had comparison viewings complete with coffee and doughnuts. In 1939–1945 the world went insane, and the U.S. Department of Agriculture coined the slogan "Food will win the war and write the peace." Corn farmers took it very seriously. Even with their sons away and farm labor impossible to hire, yields of hybrid corn were so outstanding that corn production tripled and, indeed, corn did feed a devastated world (Figure 3.48). Average yields now exceed 100 bushels per acre; the record is close to 250 bushels per acre. Although part of this increase is due to better cultural practices, a large proportion is the result of better-yielding hybrids.

Figure 3.48 Early roadside billboard advertising hybrid corn. Courtesy DeKalb Hybrid Corn Co.

HIGH LYSINE CORN

With the assistance of state agricultural experiment stations and the botanists and geneticists educated at the "aggie schools," hybrid corn seed companies entered the postwar market with hybrids tailored for different uses. Yields of field corn increased from the prewar average of 60 bushels per acre to over 100 bushels per acre. As yields climbed, it became increasingly difficult to use all that could be grown. Acreage restrictions had only temporary effects as farmers used better hybrids, more machinery, and increased fertilizer to grow even more corn on fewer acres. Although corn was now used for beef production, changes in America threatened to ruin the hog market. Pork never attained the popularity of beef. One major product of the hog was lard, used for shortenings, soaps, and glycerin. The more easily obtained vegetable oils from cottonseed, soybeans, peanuts, and corn could be converted to hydrogenated fats, soaps, and other products that cost less to produce, stored better, and lacked the objectionable odors of lard. By the late 1940s, American corn grain elevators were bursting at their seams, and the bottom dropped out of the corn market; it sometimes cost more to grow it than it brought on the open market. If the economy of the Midwest was to remain viable, additional uses had to be found for corn and some method developed to change the hog from a fat pig to an animal with more ham, bacon, and pork chops.

Professor Edwin T. Mertz of Purdue University received a research grant to study this problem. Mertz, a biochemist, decided to look at corn proteins. He confirmed early reports that the concentrations of several of the amino acid building blocks for protein synthesis were low.

These amino acids, particularly lysine and tryptophan, were essential amino acids, ones that could not be synthesized in the bodies of hogs. Mertz also knew that protein synthesis is an outstanding example of the law of the limiting factor, propounded in the nineteenth century by Justus von Liebig, one of the pioneers in agricultural chemistry. This fundamental law in all of biology can best be understood by analogy. Suppose that one wishes to bake cakes—chocolate for example:

1. Melt 4 oz. dark chocolate with 1/2 cup milk and 1/4 cup sugar in a double boiler over hot water until thick and smooth. Cool and reserve.
2. Sift 2 cups flour with 1 1/4 tsp. soda and 1 tsp. salt.
3. Cream 4 tbsp. butter and 4 tbsp. vegetable shortening with 1 1/4 cups sugar. Blend in 3 eggs, one at a time. Beat for 1 minute. Add the cooled chocolate mixture.
4. Measure out 1 cup of milk and add small volumes of the milk alternating with small volumes of the dry flour mixture to the creamed shortening–egg mixture, beginning and ending with the flour.
5. Blend thoroughly after each addition, using the low speed on an electric beater.
6. Pour into two well-greased and lightly floured 9-in. round layer pans and bake in a 350° F oven for 30–35 minutes.
7. Cool, remove from pans. Spread raspberry jam between the layers. Frost with dark chocolate frosting.

Now, if one has all the flour, sugar, milk, chocolate, butter, and shortening needed for five cakes and has only three eggs, only one cake can be baked. So it is for protein synthesis. If one amino acid is in low dietary concentration, it is the limiting factor for protein synthesis. Mertz concluded that corn protein could not be the sole supply of amino acids to make a hog meatier unless the lysine and tryptophan contents were raised. If this could be done, the accumulated corn surpluses would be used up, and the economy of the farms in the Midwest could be enhanced.

Together with Drs. Ricardo Bressani and Oliver Nelson, Mertz embarked on the long and difficult task of genetic engineering. They started cross-breeding experiments with corn types called "floury" and "opaque," given these names because of endosperm characteristics. When genes of opaque were introduced into standard varieties, lysine content of corn protein increased by 70 percent. They tested the new corn on rats, since, many years before, Thomas Osborne had found that rats fed only corn were malnourished and that symptoms of inadequate protein formation were corrected by additions of lysine and tryptophan. The new corns allowed rats to develop normally. The complete research report was published in 1964, and small amounts of high lysine seed corn began to be grown for pigs (Figure 3.49). It was a few years before

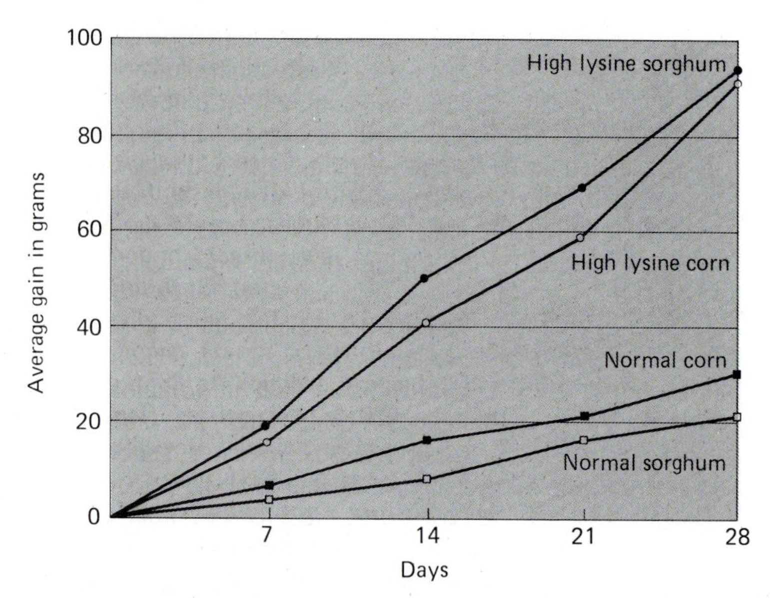

Figure 3.49 Growth rates of pigs fed normal or high lysine corn or sorghum. Taken from data obtained by the College of Agriculture, Purdue University.

a medical researcher got the idea that if the new corn was good for hogs, it might be good for people. The Indians of Colombia were outstanding examples of people whose nutrition was corn-based and whose children showed symptoms of the protein-deficiency disease kwashiorkor. In 1967–1968, Dr. Alberto G. Pradilla found that 6-year-old children whose development was that of 2-year-olds could recover rapidly on a high lysine corn diet; stunted bodies grew, and dull minds sharpened.

People are very much creatures of habit, and few of us willingly change our eating habits. The first types of the new corn were soft, almost floury. They didn't mill well to make the meal used by the Colombians. The ears were unduly susceptible to rots, and because each kernel was small, yields per acre were reduced. A corn geneticist on the staff of the Rockefeller Foundation's agriculture unit in Colombia transferred the high lysine genes into cultivars acceptable to tropical peoples. Disease resistance, resistance to the stresses imposed upon plants growing in the tropics, and a host of other factors are being considered in the continuing genetic research designed to improve corn for people and domestic animals.

INDUSTRIAL UTILIZATION OF CORN

Until the early part of this century, corn was used almost exclusively as human and animal food, with little attention to its industrial potential. Most of the industrial products that have been developed utilize both flint and dent corns, with dent being the more widely used. The most obvious products of corn are the oils contained in the germ or embryo and the starchy endosperm. Corn oil, in addition to being a highly unsaturated food and cooking oil, is used as a semidrying oil for many

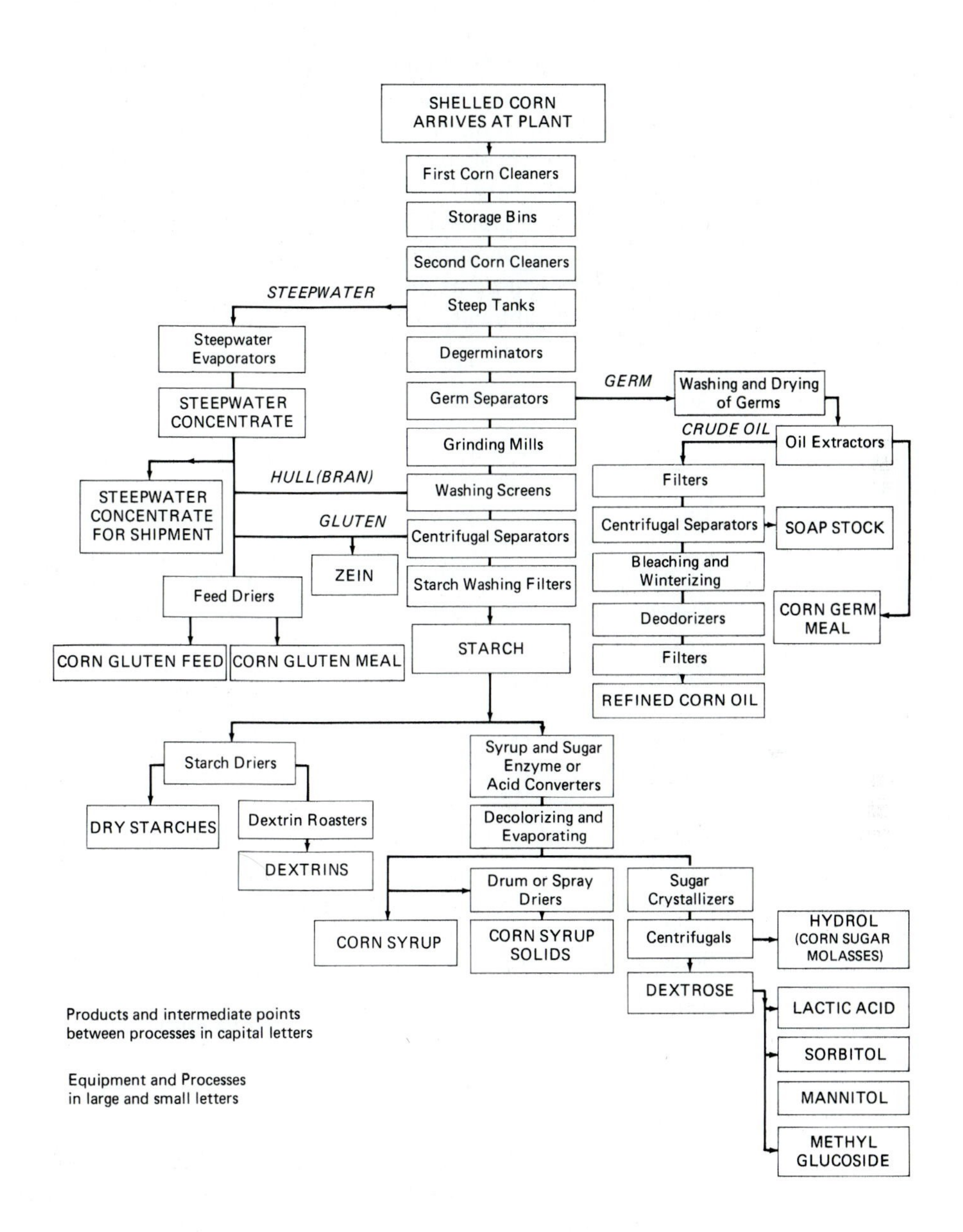

Figure 3.50 The corn refining process. Courtesy Corn Refiners Association, Inc.

purposes. The kernels are steeped in weak acid and then ground; the starch is extracted for food use and as a starting material for the production of a wide array of products. Acted upon by enzymes and other reactants, corn starch can be converted into a variety of sugars and allied compounds (Figure 3.50, p. 165), including the liquid sugars that are economically competitive with cane and beet sugars.

The carbohydrates of corn, both sugars and starch, can be fermented by yeasts and bacteria into pure (200-proof) ethyl alcohol. Worldwide, 450 million bushels of corn per year are used to produce alcohol that can at least partly substitute for gasoline in the automobile engine. Some gasolines, called *gasohols,* contain up to 10 percent ethyl alcohol and sell at competitive prices. The residues of alcoholic fermentation contain proteins used as dietary supplements for milk cattle, hogs, and other domestic animals.

ADDITIONAL READINGS

Bauman, L. F., et al. (eds.). *High Quality Protein Maize.* Stroudsburg, Pa.: Dowden, Hutchinson & Ross, 1975.

Billard, J. B. (ed.). *The World of the North American Indian.* Washington, D.C.: National Geographic Society, 1974.

Carpenter, K. J. (ed.). *Pellagra.* New York: Van Nostrand Reinhold, 1981.

Flannery, D. V. "The Origins of Agriculture." *Annual Reviews of Anthropology* 2 (1973):271–310.

Galinet, W. C. "Botany and the Origin of Maize." In E. Halfiger (ed.). *Maize,* pp. 6–12. Basel, Switzerland: Ciba-Geigi, 1979.

Giles, D. *Singing Valleys: The Story of Corn.* New York: Random House, 1940.

Iltis, H. H. "From Teosinte to Maize: The Catastrophic Sexual Transformation." *Science* 222 (1983):886–894.

Inglett, G. E. (ed.). *Corn: Culture, Processing, Products.* Westport, Conn.: Avi Books, 1970.

Nelson, O. E. "Genetic Modifications of Protein Quality in Plants." *Advances in Agronomy* 21 (1969):171–194.

Sprague, G. F. (ed.). *Corn and Corn Improvement.* Madison, Wis.: American Society of Agronomy, 1977.

Wallace, H. A., and W. L. Brown. *Corn and Its Early Fathers.* East Lansing: Michigan State University Press, 1956.

Waters, F. *Book of the Hopi.* New York: Viking Press, 1963.

Other Grains and Seeds

The greatest service which can be rendered any country is to add a useful plant to its culture.

THOMAS JEFFERSON

Six other cereal grains and several seeds have played and continue to play seminal roles in human food history. Although they are, in terms of amounts produced and people fed, not as important today as the big three—wheat, rice, and maize—literally millions of people use them.

BARLEY

Barley is the most ancient of cereal grains. Grains dated to 17,000 B.P. have been found in the Nile Valley, where present climatic conditions cannot support the plant and where it was probably a wild species gathered rather than cultivated. The cultivated barley is native to southwest Asia and has been found in 10,000-year-old archeological sites on the flood plains of the Euphrates River in modern Syria. It was grown very early in Mesopotamia (Figure 3.51), the Abyssinian Basin, Ethiopia, and as far west as Tibet, from where it entered China about 5500 B.P. By 5000 B.P., it was a staple crop of the settled world. Hard and nourishing biscuits were made from barley; it was used as a gruel or was boiled with meat and vegetables to thicken and bulk out soups and stews. The Egyptians considered barley a gift from Isis, and the peoples of the Testaments grew more barley than wheat because the plant is much more resistant to drought. The Greeks considered it a sacred grain to be used during the great festival at Eleusis that honored the Gods who controlled agriculture. Although widely used by the Romans, barley was considered inferior to wheat. During Roman rule, it was the major cereal grain in Britain.

Figure 3.51 Six-rowed ear of barley. Taken from a coin struck about 2500 B.P.

Tolerant of inferior, dry soils and quite cold-hardy, barley is now grown in all but very humid, hot regions; different cultivars are selected for various climatic conditions. The cultivated barley is *Hordeum vulgare* (= *H. sativum*). The species is *polyphyletic,* with several wild species contributing to its genetic constitution. The original cultivated plants were derived from *H. spontaneum,* a species with two rows of grains on a stalk. The more common type today has six rows of grain; its ancestor appears to be *H. distichon* or *H. deficiens*. Spring-cultivated varieties are hardier than winter-planted cultivars and can be grown further north, since they can mature in 60 to 80 days. A four-rowed subspecies is even hardier but is more difficult to grow. The two-rowed types are ancestral, although four-rowed and six-rowed plants can easily be dated to 9000 B.P. in the Middle East.

Barley today has only limited use as a food crop. It is an ingredient in gruels, and pearl barley is used in a rich winter soup in central Europe. Pearling involves rubbing the grain against abrasive disks to remove the hulls, much in the way rice is polished. Barley flour is a component of the Indian flat bread *chapatis*. Barley has, for centuries, been the premier grain in the production of malt from which beer is made (Chapter 8). Records from 2500 B.P. in Egypt, the Middle East, and the Mediterranean all show that barley was cultivated for malt (Chapter 8). The plant reached North America in 1602, to be used for beer malt and for human consumption. It reached Nova Scotia by 1606.

BARLEY AND MUSHROOM SOUP

(for four)

2 lb. beef chuck cut in small pieces
1 sliced carrot
1 stalk celery with leaves
1 sliced onion
1/4 cup sliced parsnip
1/2 tsp. chopped parsley
8 dried mushrooms, soaked
3 tbsp. washed pearl barley
1 large potato cut in 1-in. cubes
salt & pepper to taste

1. Bring beef to boil in 2 qts. water, skim foam, cover, and simmer gently for an hour.
2. Add carrot, celery, onion, parsnip, and parsley. Chop mushrooms coarsely and add together with soaking liquid.
3. Simmer for 1 hour and add the barley and potato cubes. Simmer until meat and barley are tender. Add additional water if necessary during the simmering.
4. Skim any fat from surface, season with salt and pepper to taste, and serve very hot.

OATS

In Samuel Johnson's *Dictionary of the English Language* of 1755, he defined *oats* as food for people in Scotland and food for horses in England; someone suggested that this is why England had such good horses and Scotland had such good men. In Scotland, people still eat oats as breakfast porridge and as oatcakes. The grain is one of the main ingredients in haggis. Many North Americans are familiar with the Quaker gentleman on the cardboard box of oatmeal.

The cultivated genus *Avena* was domesticated independently in several places about 2500 B.P. Oats are not mentioned in the Testaments. The classical Greeks knew oats only as a weed that was gathered by barbarian tribes, but the Romans, according to Virgil's *Georgics*, would eat oats if wheat or barley were scarce. People in parts of North Africa, the Middle East, and north into European Russia grow oats as a staple cereal crop for themselves and their domestic animals. Oats were introduced into North America as early as 1602, and the United States now dominates world production. Since the disappearance of the horse as a draft animal, production has fallen drastically.

Avena sativa was probably derived from a wild oat species, *A. fatua,* which has been a serious weed pest in wheat fields ever since it came to North America as a contaminant in wheat. In India, a related semidomesticated oat, *A. nuda,* is grown to feed water buffalo and sacred cows.

Oats have almost 17 percent protein, but much of the protein is not available for human needs because it is very low in several essential amino acids. Since the protein is not the glutenin type that allows wheat-flour bread to rise, leavened bread cannot be made from oat flour. Strains of *Avena sterilis* from the Mediterranean Basin have been isolated with almost double the protein level and are being cross-bred with *Avena sativa* to produce a superior grain for feeding domestic animals. The rolled oats eaten as porridge are prepared from dehulled grains—the groats mentioned frequently in Old English literature—that are then run between rollers in a steam chamber. Instant oatmeals receive an additional steaming.

OATMEAL COOKIES

1 3/4 cup flour
1 tsp. baking powder
1/2 tsp. salt
1/2 tsp. baking soda
1 tsp. cinnamon
2 cups rolled oats
1/2 cup melted shortening
1 1/4 cup sugar
2 well-beaten eggs
6 tbsp. molasses
1/2 cup chopped pecans
1 cup raisins

1. Sift flour, add baking powder, salt, baking soda, and cinnamon and sift again. Add rolled oats.
2. Combine melted shortening and sugar. Add beaten eggs and molasses. Blend well and add dry ingredients. Add pecans and raisins. Mix thoroughly.
3. Drop by tablespoonful onto a greased baking sheet. Bake in a moderate oven (325°F) for 12–15 minutes. Makes 3–4 dozen cookies to be served with milk.

RYE

Rye is more popular and is consumed in greater amounts in Europe than in North America. The rye breads available in Germany, Scandinavia, and eastern European countries span a range of textures, flavors, and colors that allows selection of types to go with just about any kind of meal. *Schwarzbrot,* the heavy, dark rye bread of central Europe, is among the most nourishing and delicious of breads. The Russians take rye bread crusts and ferment them with a special bacterial culture into a nonalcoholic *kvass* as a summer beverage; it is very much of an acquired taste. Some of the proteins in rye are glutenins so that rye flour will leaven, although not as well as wheat flour. Light rye breads, more commonly found in North America, contain wheat flour. Some rye is planted as a cover crop on bare fields to prevent erosion, and an additional amount is used to make rye whiskey.

Secale cereale is not found wild, suggesting that it is a cultivated species. It appears to be a hybrid between *S. montanum* and *S. anatolicum,* both found wild in the mountains of southwestern Asia and Asia Minor. It is likely that the wild species were weeds in barley and wheat fields at least 10,000 years ago. It has been found in the lacustrine debris of the Swiss Lake Dwellers of the Bronze Age. Rye was obtained by the Greeks from Thrace and Macedonia, but they and the Romans disdained it as human food, feeding it to their pigs and eating it only in emergencies. Since the plant flourishes in cool, dry climates, it became a dominant cereal grain when rye-contaminated wheat entered northern Europe and Russia about 4000 years ago. The plant is very hardy, growing well even when water is limited. Because it is winter-hardy, it can be planted in late fall and will mature in the spring before wheat is ready to harvest. It is grown in Norway at 67° north latitude and at 11,400 feet (3800 meters) in the Himalayas. A serious difficulty in rye cultivation is the susceptibility of the plant to attack by the ergot fungus, whose reproductive tissues cause ergotism (St. Anthony's fire—see Chapter 7). Rye entered North America via Nova Scotia in 1606 and New England in 1629.

HEARTY RYE BREAD

(for four loaves)

1 package granular yeast
1 1/2 cups warm water
1 1/2 cups medium rye flour
1 tbsp. caraway seed

Combine yeast and water in a large bowl. Stir to dissolve. Add caraway seeds and rye flour and stir to blend. Cover lightly with plastic wrap. Let stand 2 days at room temperature to sour and become sponge.

1 recipe for sour rye (above)
1 package granular yeast
1 cup warm water
1/4 cup dark molasses
2 tbsp. caraway seeds
1 tbsp. salt
3 tbsp. vegetable shortening
1 tbsp. milk
1 egg
1 cup medium rye flour
5 cups (approx.) wheat flour

1. Stir down the sponge. Dissolve yeast in water and add, while stirring, to the sponge. Add molasses and half the caraway seeds. Stir to blend.
2. Add 3 tbsp. shortening and beat to blend. Add 2 cups flour and blend with a wooden spoon. Gradually add 2 more cups flour, kneading constantly.

3. Add more flour, about 2 tbsp. at a time, until the dough attains a good, workable consistency. Turn dough out onto a lightly floured board and continue kneading for 6–8 minutes.
4. Knead and beat dough for about 10 minutes. When ready, the dough should weigh about 3 1/2 lbs.
5. Heat a mixing bowl with hot water, dry the bowl, and grease it with shortening. Shape the dough into a ball and place in the bowl. Cover lightly with a dish towel and allow to rise an hour or until doubled in bulk.
6. Divide the dough into 4 portions of equal weight. Roll each piece into a long sausage shape with your hands so that the sausages are about 15 in. long. Place on a lightly greased baking sheet or into bread molds, cover loosely with waxed paper, and allow to rise for an hour or until each loaf has doubled in bulk. Preheat oven to 375° F.
7. Using a sharp knife, make 4–5 spaced diagonal slashes on top of each loaf. Brush tops with egg beaten with milk. Sprinkle with remaining caraway seeds.
8. Bake 40 minutes or until the bread is baked through and has a nice crisp crust. All you need is sweet butter to spread on the warm bread.

TRITICALE

For centuries, people have considered the possibility of developing a cereal grain with the nutritional advantages of wheat and the temperature- and drought-hardiness of rye, but attempts to cross the two species were unsuccessful. Late in the nineteenth century it was discovered that the chromosome numbers of the two species were different so that any offspring were, like mules, sterile. There are four sets of chromosomes in most wheats and only two in rye. By modern techniques, however, it has been possible to double the number of chromosomes in rye, from 14 to 28, and to cross it with a durum wheat with 28 chromosomes. The resulting plant has been named *triticale.* The protein level in triticale is close to 16 percent, greater than in either parent; the grains are larger than those of wheat and of good nutritional quality. The plant is as cold-hardy as rye, with excellent drought resistance, but yields are usually lower than wheat; there is a tendency for the stalks to be knocked over by wind or hail (lodging), and the grains are sometimes lost before harvesting (shattering). To date, triticale is not competitive with wheat and is used mostly as animal feed. Genetic research is still in progress.

SORGHUMS

The sorghums form a complex and variable genus of grasses that, for convenience, are grouped together as *Sorghum bicolor.* In terms of quantity produced, it is the fourth largest cereal crop. The center of domestication was in the savannahs of northern and eastern Africa. There also may have been a center in India with dispersal into China by 2700 B.P., although this Asian center of domestication may simply reflect an eastward spread of the genus. In an important agricultural treatise *(Nung Shu)* of 1313, the author, Wang Shen, praised *kaoliang,* the sorghum: "The grain can be hulled and eaten and anything left over fed to livestock. It is a famine food. The tops of the stem can be made into brooms and the straw woven into traps, plaited to make fences or used to provide fuel. No part of the plant has to be thrown away."

The wild ancestors include several plants, now classified as subspecies or races, that are still found wild and are gathered by tribal groups. It is likely that domestication occurred about 5000 years ago, since the sorghums are known to have been cultivated in Egypt. Rome knew it from imports obtained from Egypt and the East Indies, and it was grown for some time in southern Italy as *sorgo melica.* In parts of Africa (the Sudan, Chad, and central regions of the continent) and in south Asia, the sorghums are much more important than wheat, although since the early part of the nineteenth century, maize has displaced sorghum in many countries.

At least superficially, the vegetative plant looks much like maize; its photosynthetic processes utilize the same, highly efficient mechanism (the C_4 photosynthetic route) that permits rapid growth in tropical regions (see Chapter 1). In contrast to maize, the flowers are perfect and the grain is borne at the top of the plant in a panicle.

There are hundreds of cultivars, most selected by the growers to meet local climatic conditions. They can be grouped into four main types.

The grain sorghums, including milo, kafir, fiterita, and hegari, are grown for their grains and are used for both human and animal food. They contain 12 percent protein, more than maize, and they have somewhat less lipid. Grain sorghum is eaten as a gruel or mush, and sorghum flour mixed with cornmeal is baked into tortillas. In Africa, sorghum is fermented into a strong beer.

The sweet sorghums have stalks with a high concentration of sugar which can be crushed to yield a sweet sap that can be boiled down into a syrup or molasses. The sugar is rarely crystallized.

The forage sorghums, sometimes classified as Sudan grasses *(Sorghum vulgare var. sudanensis),* are grown as a forage plant for silage. The foliage and stalks have over 90 percent of the nutritive value of maize for cattle, but because the young stalks contain prussic acid—a source of cyanide—only the mature plants could be harvested. This problem has almost been eliminated by crossbreeding forage sorghums with other types of sorghums.

Broom corns are grown, as the name indicates, for their brushy, thin stalks that are bound into corn brooms. They have largely been replaced by synthetic fibers and the market for broom corn has declined precipitously since 1960.

PECAN PIE

2 tsp. melted butter
1 cup light brown sugar
2 tsp. flour
2 eggs
1 tsp. vanilla extract
1/4 tsp. salt
1 cup sorghum molasses
1 unbaked pie shell
1/2 cup pecan halves

1. Mix butter, sugar, and flour. Add eggs, vanilla extract, and salt and beat into butter-flour mixture. Add sorghum molasses (or corn syrup or cane molasses) and beat lightly.
2. Pour into unbaked pie shell. Sprinkle top with pecans.
3. Bake at 375° F for 35 minutes.

MILLETS

The term *millet* includes species in at least four and possibly as many as six genera, with up to a dozen grown as grain plants in various parts of the world. Few have any importance in the Western world, although they are a staple cereal crop in parts of Asia and Africa. They are grown under the same climatic conditions as the sorghums, frequently being intercropped with legumes. In tropical regions they will mature in 3 to 4 months, even in semiarid climates where other cereals do poorly. Wetland and dry field types have been selected over centuries of cultivation. India, Pakistan, and Africa south of the Sahara are the major areas of production. Millets have been grown for centuries in China as a whole-grain cereal (Figure 3.52). In North America, the small crop is used for feeding livestock in the southern states and is one of the seeds included in birdseed. Breeding efforts have resulted in millets that grow sufficiently well in the humid southeastern states of Louisiana and Mississippi to be a useful forage crop. Many of the millets are eaten directly as grain, with some made into porridge or ground into a flour that can be baked into an unleavened bread or cracker. A surprisingly large amount is fermented into a nutritious beer.

Several millets have been cultivated for over 7000 years. Many, like the foxtail millets *(Setaria),* are of worldwide distribution, and wild ancestors have been found in Europe, temperate Asia, and eastern equatorial Africa. Italian millet *(S. italica)* is of tropical origin, probably in south India, but was dispersed early, with seeds found in the villages of the Swiss Lakes Dwellers. The Greeks called it *elumos* and the Romans, *panicum,* indicating that it was used as bread. One of the millets, *S. macrostachya,* is native to the New World and was grown by the Aztecs in the Tehaucán Valley of Mexico, but it was supplanted by maize 5000 years ago.

Pearl millet is in the genus *Pennisetum* and is almost as important as sorghum in India, where it is ground into flour *(bajra)* used to make *injara* breads. Pearl millet is the *dukhn* noted in Ezekiel's prophecies.

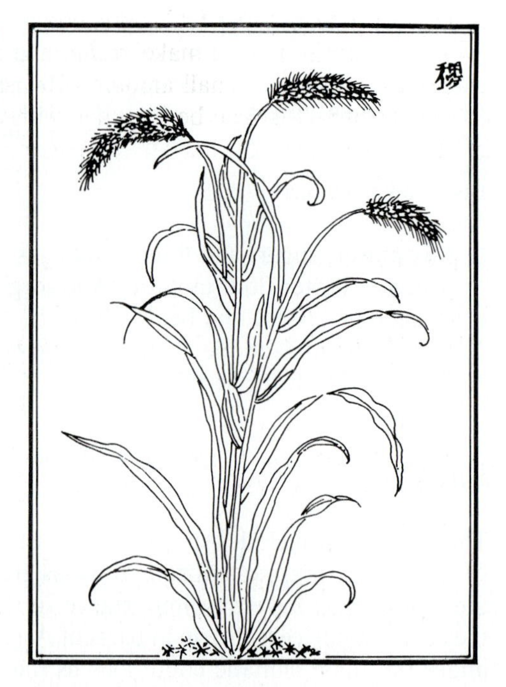

Figure 3.52 *Setaria* millet, one of the sacred grains of China. Redrawn from the *Shou Shieh Thung Khao,* an agricultural book published in 1742.

Hybrid seed, as is also true of maize, has greatly increased yields in the Soviet Union and the central parts of China, where it is believed that cultivation originated. Like the foxtail millets, the genus is found worldwide, and different species have been cultivated in Africa, India and the Americas. One, *P. setaceum,* the fountain grass, is grown as an ornamental.

The finger or ragi millets *(Eleusine)* are raised primarily in south India, the Sudan, and Uganda. Since finger millets require cooler weather and more rainfall than pearl millets, they are grown at higher altitudes. Other genera are grown in various parts of the world.

BUCKWHEAT

Buckwheat *(Fagopyrum esculentum)* is related to the weedy dock *(Rumex)* and the smartweeds *(Polygonum),* which are lawn pests. It is not a true cereal, since it is a dicot rather than a monocot grass. The plant grows well in poor soils and at high elevations and since it has a short growing season, it is suitable for northern regions where cereal grains cannot be grown. Originally cultivated in western China and still growing wild in Nepal and Siberia, it was introduced to northern Europe in the Middle Ages. The Teutons called it *heydenkorn.* It reached North America before 1625.

In North America, buckwheat flour is an ingredient in pancake flour, and the young plants are frequently used as a cover crop on bare soil. It is also grown for its flowers, whose nectar makes an excellent honey. In eastern Europe and particularly in the Soviet Union, buckwheat is a major seed grain used to make *kasha,* a groats dish. The seeds contain 11 percent protein, a small amount of which is of the glutenin type so that buckwheat flour can be leavened with yeast or baking powder; the Russian *blini* pancakes that traditionally accompany caviar are a yeast-raised bread.

KASHA VARNISHKAS

1 slightly beaten egg
1 cup granulated buckwheat
2 cups beef consommé or water
1 tsp. salt
1/4 tsp. pepper
2 tbsp. butter or margarine
2 cups cooked bow noodles

1. Stir egg into buckwheat; mix well until all the granules are coated with egg. Place mixture into a 1–2 qt. pot that has a tightly fitting cover.
2. On high heat, stir buckwheat–egg mixture until the egg has dried and the granules are very hot and mostly separate.
3. In a small saucepan, heat liquid, salt, pepper, and butter to boiling and pour over egg–buckwheat mixture in the large pot. Cover pot, reduce heat to low, and steam gently for 10 minutes or until tender.
4. Add cooked bow noodles and heat thoroughly. Fluff with a fork before serving. Makes 5 cups. A kasha pilaf similar to that eaten in China can be made by adding sautéed onions and mushrooms instead of noodles and seasoning with soy sauce instead of salt.

AMARANTH

The amaranths are an example of a food plant whose virtues have only recently been rediscovered by Western nations. It is not a cereal grain, but is a dicot (Figure 3.53). The amaranth family is a large one, with 50 genera and about 50 species in the genus *Amaranthus.* Related genera include common garden plants like the cockscomb *(Celosia),* the bloodleaf *(Iresine),* and the globe amaranth *(Gomphrena);* within the genus are found one of the tumbleweeds *(Amaranthus albus),* love-lies-bleeding *(A. caudatus),* and Joseph's coat *(A. tricolor).*

The gathering and cultivation of amaranths can be conservatively dated to at least 4000 B.P. in Central and South America. Both the Andean Indians and the Aztecs used it. Montezuma annually received tributes of thousands of pounds of amaranth seed, which were stored in special granaries. The seed was ground into flour, mixed with human blood, and baked into cakes shaped into the forms of Gods. The Spanish considered this to be a deliberate mockery of the Christian Eucharist (which was impossible, because the rituals predated the Christian era), and Cortez banished the practice upon pain of the stake. In the Andes,

Figure 3.53 Fruiting head of *Amaranthus hypochondriacus*. The plant usually grows to a height of 4–6 feet.

the plants were a regular food item, grown on islands built of logs covered with soil and floated in mountain lakes. Several amaranth species are raised in India, where they are called *rajgura* (kingseed) and *ramdana* (seed sent from heaven). There are several species of grain amaranths. The major species in the Americas, including *A. hypochondriacus,* are derived from *A. hybridus,* the red pigweed. The cultivated Asian species *(A. cruentus* and *A. caudatus)* may have moved from the New World to the Far East in the nineteenth century. The African species may have been selected from the weedy *A. polygamus,* but none of these derivations are well established.

The easily cultivated plants are annuals that are tolerant of a wide range of environmental and climatic conditions. The amaranths use the highly efficient C_4 photosynthetic pathway common to maize, sorghum, and sugar cane. Seed heads on colorful 6-foot plants may weigh up to 6 pounds and may contain a half million small, round seeds that look like miniature dinner plates. Protein levels range up to 17 percent, with unusually high amounts of essential amino acids, especially lysine. When cooked with maize, the protein combinations are basically as nutritious as those of meat. Seeds may be ground into a coarse meal or flour or heated as a popcorn that, when mixed with honey, is called *alegria* in Mexico and Central America. The leaves too are used as food, cooked much like spinach as a pot herb, but with significantly higher amounts of protein.

ADDITIONAL READINGS

Briggs, D. E. *Barley*. London: Chapman & Hall, 1978.

Bushek, W. *Rye: Production, Chemistry and Technology*. St. Paul, Minn.: American Association of Cereal Chemists, 1976.

Cole, J. N. *Amaranth: From the Past, for the Future*. Emmaus, Pa.: Rodale Press, 1982.

Doggett, H. *Sorghum*. London: Longmans, 1976.

Harlan, H. V. *One Man's Life with Barley*. New York: Exposition Press, 1957.

Janick, J., R. W. Schery, F. W. Woods, and V. W. Ruttan. *Plant Science: An Introduction to World Crops*. 3rd Ed. San Francisco: Freeman, 1981.

Müntzing, A. *Triticale: Results and Problems*. New York: Paul Parey Scientific Publishers, 1979.

National Research Council. *Amaranth: Modern Prospects for an Ancient Crop*. Washington, D.C.: National Academy of Sciences, 1984.

Norman, M. J. T., C. J. Pearson, and P. G. E. Page. *The Ecology of Tropical Food Crops*. Cambridge, England: Cambridge University Press, 1984.

Rachie, K. O., and J. V. Majmuder. *Pearl Millet*. University Park: Pennsylvania State University Press, 1980.

Schery, R. W. *Plants for Man*. 2nd Ed. Englewood Cliffs, N.J.: Prentice-Hall, 1972.

Tucker, J. B. "Amaranth: The Once and Future Crop." *Bioscience* 36 (1986):9–13.

Zillinsky, F. J. "The Development of Triticale." *Advances in Agronomy* 26 (1974):384.

Soybean and Other Legumes

How wonderful was Heaven to give Man a plant that provides both milk and meat.

CHINESE PROVERB

The legume or pea family, Leguminosae, is one of the largest families of flowering plants, with over 500 genera and several thousand species. With the exception of the grass family, legumes provide more economically useful plants than any other family. There are three subfamilies; the one forming the familiar butterfly-shaped flowers, the Papilionoideae (Latin: *papilo* = butterfly), contains genera of greatest economic importance. The fruit is a special pod—a legume—that is botanically characterized by a seam or suture on one or both sides along which the fruit splits to release the seeds. Legume seeds are large, with cotyledons that contain proteins, lipids, and starches of high nutritional and industrial value.

Many of the legumes have the ability to form root nodules (Figure 3.54) containing specific bacteria in the genus *Rhizobium* that can convert molecular nitrogen gas in the air into ammonium compounds, the primary nitrogen source for all plant cells. This allows the plants to grow in soils that are very low in nitrogen. Legume plants are correspondingly high in nitrogen compounds, which accumulate in the seeds as protein. After the seed crop is harvested, the plants can be plowed back into the ground, where their minerals, including nitrogen compounds, are released by the action of soil microorganisms and become available for plants grown the following year. The amount of nitrogen fixed by legumes is very large: close to 90 million metric tons per year, which means that a hectare (2.5 acres) of clover, alfalfa, or one of the beans will fix 500 to 600 kilograms per hectare, quite enough for the crop's own needs with enough left for a crop of another plant the following year. It would be financially and industrially impossible to produce enough chemical fertilizer to replace the activities of the legumes even in developed countries, much less in underdeveloped countries.

THE SOYBEAN

In 2221 B.P., the Chinese emperor Chin Shih Huang Ti decreed that all classical manuscripts were to be destroyed and memory of their contents

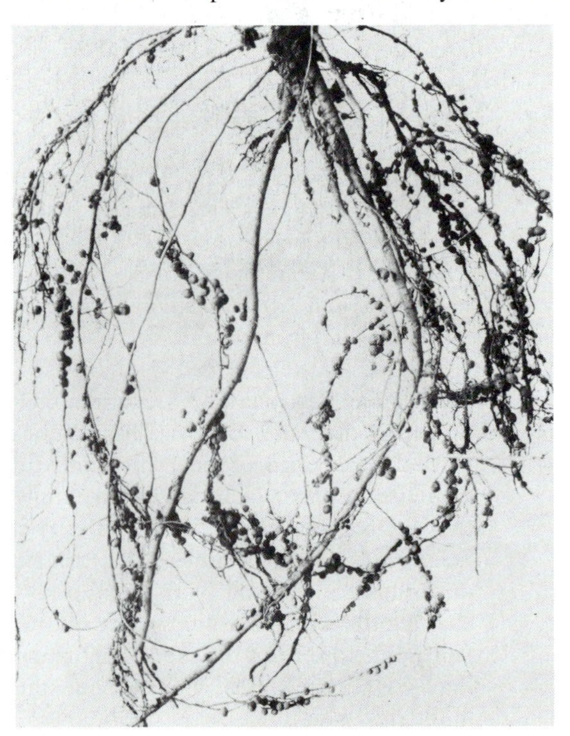

Figure 3.54 Root nodules on roots of soy. Courtesy U.S. Department of Agriculture.

expunged upon pain of decapitation. Why he so ordered is unknown, although it is thought that there had been too much attention to the classics and not enough to the changes going on in China. Among the lost works was a book of poems, the *Shih Ching,* said to have been edited by Confucius and quoted extensively in his *Analects.* In the reign of Emperor Wu-ti, about 2100 B.P., scattered fragments of the 300 poems were reassembled. Although the *Shih Ching* poems were primarily on the nature of Man, on the place of people in the scheme of things, and on the interaction of people with society at large, the book is important to botanists because plants are frequently mentioned. This source is especially valuable because, at the time of Confucius, during the Ch'u Dynasty, China had little commerce with the rest of Asia and it is assumed that plants mentioned in the *Shih Ching* could be regarded as native to the country. This is not strictly true, however. The provenance of the poems is a bit questionable, and while governments may not have much to do with other governments, people interact with people and exchanges at the local level go on. Nevertheless, these poems are an important written record of plant cultivation, although it is said that the legendary Emperor Shen Nung (3500 B.P.) "sowed the five blessed grains": wheat (*mai*), rice (*tao*), common millet (*chi*), glutinous millet (*chi shu*), and the sacred bean (*shu*). The bean, called *ta tou* or *jung shu,* was cultivated for its high protein and oil content and was, like the cereals, of fundamental agricultural importance. The Chinese also noted that when grains or cabbages were grown in fields in which shu had been grown the previous year, the crops were darker green and much healthier than they otherwise would have been, and the yields were much higher. Shu is a legume whose Latin binomial is *Glycine max.* In the West we know it as the soybean (Figure 3.55). Archeological evidence dates its cultivation in China to earlier than 5000 B.P., and it is believed to have been cultivated in Korea and then Japan long before 2000 B.P.

Glycine max is one of about ten species in the genus and is the only one cultivated for food. It was named by Linnaeus in *Species Plantarum* of 1753 as *Phaseolus max,* but since it is not a true bean, the genus was later changed to *Soja* and finally to *Glycine.* The plant is an annual, with viney types known, although most cultivars are small bushes (Figure 3.56) with small, almost inconspicuous white to purple flowers. The pods hang by short stalks and are a few inches long and about a half inch wide. There are two to four green, brown, yellow, or black seeds in each pod.

In 1690, Englebert Kaempfer, a German plant explorer in the Orient, brought seeds back to Europe, where they were grown in the Jardin des Plantes in Paris. John Ray described the soybean in *Historium Plantarum,* published in London in 1704, indicating that the soybean apparently had traveled widely in Europe in less than 15 years. It was, however, merely a garden curiosity. Benjamin Franklin brought seeds from France in 1785; they were established in several Philadelphia

Figure 3.55 Soybeans *(Glycine max)*. Courtesy U.S. Department of Agriculture.

gardens by 1804. A Dr. Mease in Philadelphia wrote a friend, "The soybean bears the climate of Pennsylvania very well. The bean ought therefore to be cultivated." There was, however, no interest in the plant since the seeds tasted strange to those accustomed to common beans. Soybeans were grown almost exclusively by home gardeners as a green manure to be planted in the spring on fallow land and plowed under in the summer to enrich the soil for other crops sown the following spring. In 1854, Commodore Matthew C. Perry started his voyage to "open

Figure 3.56 An Illinois soybean field. Courtesy U.S. Department of Agriculture.

Japan to the world," and incidentally ushered in the Meiji Restoration that revolutionized that country. The United States Patent Office's plant section—later the Plant Introduction Branch of the Department of Agriculture—charged him with the task of bringing back plants that seemed to be potentially important in agriculture. Perry collected soybeans and noted in his report their widespread esteem as food. The plants were duly grown in the Patent Office's test plots for distribution to farmers. Although the structural and biochemical details of nitrogen fixation were still to be discovered, it was recognized that *Soja* had root nodules and probably would be a good crop for improvement of maize and vegetable fields.

Soybeans as Food In North America, soybeans are most often recognized in Chinese restaurants, where bowls of soup containing pieces of soybean curd are served. But in the Orient, soy is a basic food. Consider some nutritional facts. With seed coats removed, soybeans contain close to 45 percent protein, more than any of the cereal grains (Figure 3.57). After cooking, the globulin proteins (similar to the proteins in meat) are easily digestible and contain a better balance of amino acids than do most plants. When eaten with cereal grains like rice or millet that supplement the soy's level of the sulfur amino acid methionine, the protein is nutritionally complete for humans. In cultures where animal protein is scarce and expensive, soybeans—supplemented with vegetables, other belly-filling carbohydrates, and occasional bits of meat, fish, eggs, or fowl—will meet all human food needs from the age of 2 onward. Well over 10 percent of the protein intake of the peoples of

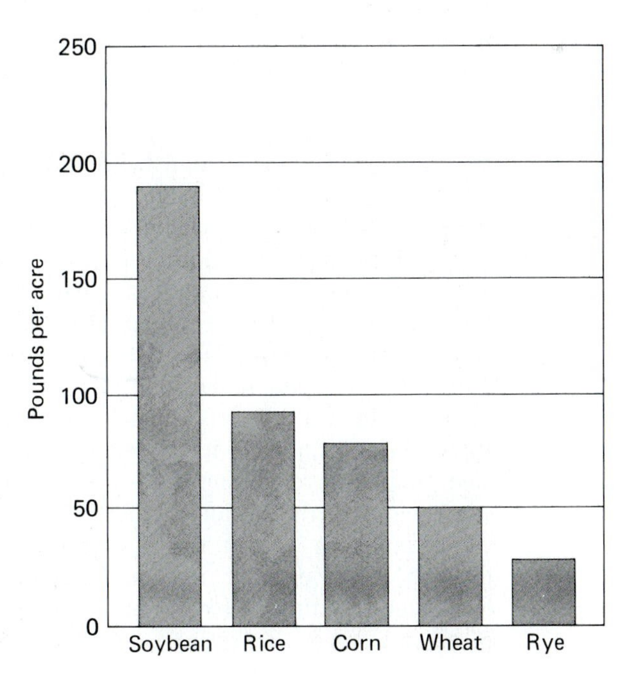

Figure 3.57 Field crops yielding large quantities of protein.

China, Korea, and Japan is derived from soybeans. The carbohydrate level exceeds 30 percent and the mineral concentration is about 5 percent, with nutritionally important amounts of calcium, phosphorus, and iron. The lipid (oil) content exceeds 20 percent of the weight of the seeds, is highly unsaturated (Chapter 4), has no cholesterol, and contains fat-soluble vitamins frequently absent from other plant foodstuffs.

In the Orient, soybeans are used in national cuisines in many different ways. Whole beans are boiled, baked, or eaten as sprouts. Equally important is the conversion into bean curd. *Tofu* (Chinese pronounce our *t* as a *d*) is made by grinding the beans into a flour and extracting the protein with hot water to make soy milk. Calcium salts are added to precipitate the protein as a curd, which is then pressed into cakes that are over 50 percent protein and 25 percent digestible lipid. When tofu is used in soups or cooked with meats, vegetables, and fish, the dish provides a full spectrum of amino acids. Soy milk is fed to young children and, like cow's milk, can be made into a high protein cheese called *sufu* in Japan; similar products are made in China (red bean cheese, or *mom yee*), Indonesia (*tempe*), and other countries. Other absolutely wonderful Oriental food products from soybeans include fermented black beans (*dow see*), brown and yellow bean sauces (*mein see*), and the hoisin sauces that contain soy, chili, garlic, and vinegar. A soup base in Japan is *miso,* a fermented product that contains both soy and wheat flours; daily per capita consumption of miso alone in Japan is over an ounce (30 grams). Soy sauce (*shoju*), too, is a familiar item. Soybeans are boiled until very soft, mixed with roasted and ground wheat, barley, millet, or rice and then inoculated with a fungus culture called *kojii.* Kojii, salt, and the soy-cereal flour are allowed to ferment slowly for almost a year. Chinese soy sauce is sweeter than the Japanese soya because of the mixture of cereals used. Synthetic soy sauce, sold in North America, is made by mixing strong acids with a paste of soy and wheat flour, heating it to boiling to change the starch into sugar, and neutralizing it with alkali. The mixture is quick-aged, bottled, and sold to an unsuspecting public. Caveat emptor!

Soybeans are a significant but usually unsuspected part of North American diets. Large amounts of soybeans are pressed to extract the oils that are used as cooking oils, in margarines, and vegetable shortenings and mayonnaise, as well as in industrial products including paints, plastics, and a long list of other manufactured goods. When the oil is purified, a compound called *lecithin* is removed and used as a thickener and stabilizer for mayonnaise, cake frostings, and ice cream. The press-cake remaining after oil removal is one of the more widely used animal feeds. Millions of metric tons of soymeal are used to fatten cattle and hogs, to put weight on chickens, and to feed cats, dogs, goldfish, and even your pet canary. Soy is part of the food chain that links plants to the Thanksgiving turkey or the cheeseburger. Soybean meal is, however, not restricted to animal feeds. Some of it is incorporated into sausages, as the label "cereal additives" indicates. It is also in breakfast foods, added to supply the amino acids, particularly lysine, that wheat, corn, or rice (puffed or unpuffed) contain in low supply. Defatted soy flour is incorporated into breads, pies, and cakes. The flour is also toasted, cooked, flavored, and colored and then extruded through thin holes to form textured vegetable protein (TVP), to be used as an additive to hamburger and to make synthetic "bacon bits" and other food supplements.

Soybean Production With the recognition of the nitrogen-fixing ability of soybeans, their high protein and oil content, and their value for animal feeding, production in North America increased dramatically. Three million bushels were produced in 1920, 70 million in 1940, 100 million in 1970, and over 125 million in 1980. In 1985, the United States produced almost as much soy as maize. Illinois is the major producing state; this Chinese plant has found a happy home in the Land of Lincoln. Of the over 80 million metric tons produced worldwide each year, 70 percent is grown in the United States. China is second, followed by Brazil, Mexico, Indonesia, and Argentina. Very little soy is grown in Japan, which has long been the major importing country, and China is now unable to grow enough to meet the dietary requirements of its billion people. Soybeans are best adapted to warm but not hot, moist but not wet climates that have at least a five-month growing season. Tropical climates are not as satisfactory, which restricts the growing area in Asian nations to north China, the northern islands of Japan, and Korea, countries whose cultures and cuisines are dependent upon soybeans. The plant is fairly tolerant of different soil types, although it does best in fairly heavy, nonacidic soils. In fact, regions where maize is grown provide almost ideal conditions.

A major barrier to increasing the areas in which soybeans can grow is the light (photoperiod) requirement for flowering. Most people are aware of the effect of photoperiodism even if they are unfamiliar with the term. Dandelions flower in the spring and chrysanthemums flower in the fall, when the daylengths are short. Most carnations and black-eyed Susans are long-day plants that form flower buds only when they have been exposed to the long days of summer. Still other plants—tomatoes, sunflowers, beans, and peas—are termed day-neutral; they will flower when they are large enough regardless of the daylength. Photoperiodism was discovered in 1920 by Drs. Garner and Allard, two U.S. Department of Agriculture plant physiologists working with tobacco. This discovery should have been recognized by a Nobel Prize because it revolutionized agriculture. It allowed agricultural scientists to determine where different crops could be grown, and it provided the basis for plant breeders to develop cultivars that, because their photoperiods could be manipulated, could then be grown where they previously would not have reproduced.

Soybeans are photoperiodic, short-day plants, requiring a photoperiod of less than about 11 hours of light per day. This means that the plant cannot initiate flowers in Illinois until early autumn. Flower development and the fertilization needed for seed initiation require almost two weeks after photoperiodic induction. The time between the initiation of the seed and final maturation for harvesting becomes critical, since early frosts would seriously reduce yields in colder parts of the temperate zone. To overcome such limitations, considerable breeding work had to be done before soybeans could become an important income crop. Long-day soybean cultivars have now been obtained that can be planted as soon as the ground warms up and will be ready for photoperiodic flower induction in midsummer. For crop maturation, early-flowering cultivars are planted in areas that normally get early frosts, while late-flowering cultivars are used for warmer climates. Types of soybeans are now available that can complete a life cycle from as short as 90 days to as long as 180 days. Soybean cultivars have now been developed to give optimum yields in any latitudinal band of 150 miles wide between the American–Canadian border and the Gulf of Mexico. No cultivars have yet been developed for tropical areas. Breeding for disease resistance and modifying cultural characteristics have resulted in an increase in yields from less than 20 bushels per acre in 1940 to over 60 bushels per acre in 1984.

Soybean production has had an impact on international relations. As populations increase, demands for adequate protein also increase; with rising—and justified—expectations for better diets, the desire for high quality protein has swelled. There are several excellent sources of plant feeds for domestic animals. Maize is still the major one, although its amino acid balance requires that for high protein conversions, maize must be amended with several amino acids. Some barley, oats, and other cereal grains are also used. An important animal feed is fish meal, a large proportion of which has been made from anchovies taken off the coast of Peru. Shifts in ocean currents and overfishing have reduced the anchovy catch, and in 1973 meat animal producers in the United States obtained an embargo on the export of soybeans to preserve domestic supplies. The international repercussions were immediate and vociferous. Although the embargo was short-lived, Japan and other countries, whose human food needs required imported soybeans, assumed that they could no longer rely upon the United States as a primary source and began a search for other areas of production.

With the accumulated knowledge about the cultural requirements of soybean and the availability of cultivars exhibiting different growth, photoperiodic, and cultural characterists, Japanese industrialists in collaboration with the Japanese government bankrolled the startup costs for large agricultural enterprises in underdeveloped countries, with the understanding that Japan would be given preference for the crop. Brazil in particular had the climate, soil, and photoperiodic characteristics. In fact, if one folds a map of the world, the soybean-growing regions of Brazil are at the same latitude as Illinois, Indiana, and Iowa, where much of the United States soybean crop is grown. Soybean production in the United States has not fallen off, since it has largely replaced fish meal for animal feeding, but the new areas of soybean production in South and Central America have had a significant impact on the export market and hence on the balance of payments of the United States.

OTHER IMPORTANT LEGUMES

Although soybeans are, in economic and nutritional terms, the most important legume crop, several other legumes have had significant impacts on civilization. These other legumes are generally lumped together as beans and peas. Another common term for this group, especially in Europe, is *pulses*.

Peanuts The peanut, also known as the groundnut, goober, or pinder and botanically as *Arachis hypogaea*, is a Brazilian contribution to the world (Figure 3.58). Brought to Europe by the Spanish and the Portuguese, it was soon planted in the African colonies of these countries as a food to sustain slave laborers. By 1550, the Portuguese missionaries had brought *lo hue sheng* to China, where it has become an important food crop. From Africa, it returned to the Americas, still as slave food, when sugar plantations were developed. Since peanuts are not photoperiodically sensitive and thrive in hot weather, they are

Figure 3.58 Peanuts *(Arachis hypogaea)*. Courtesy U.S. Department of Agriculture.

planted extensively in warmer regions of the temperate zone and in the tropics. In the United States, close to 2 million metric tons are harvested in a good year, with a third of the crop exported. Nevertheless, India and China are the major growers, with the United States production representing 10 percent of the world's crop.

Peanuts are botanically unusual in that the airborne flowers, after fertilization, are forced by the growth of the flower stalk into the ground, where the pods mature; hence the name groundnuts. The seeds are up to 25 percent protein and almost 40 percent unsaturated oil. The brown-skinned Virginia cultivars have two large seeds per pod, while the smaller, red-skinned Spanish types typically have three or more seeds per pod. In the United States, close to half the crop is ground up to accompany bread and grape jelly as a major sustenance for people under 25 years of age, or is roasted and served with beer or soft drinks. Because the oil has the highest smoke and flash point of any vegetable oil, it is preferred for french frying. Large amounts of oil are converted into margarine and solidified cooking fats. The residue after extraction of the oil is a high protein animal feed, especially fine for poultry.

Peanuts and peanut products are highly susceptible to attack by a fungus, *Aspergillus flavus,* which produces several very toxic compounds. Some of these cause kidney and liver damage, and small doses have been shown to be fatal to laboratory animals. The *aflatoxins* were discovered in the mid-1950s when extensive use of peanut meal for poultry and pigs began and many animals died of massive hemorrhages. There are several aflatoxins, identified by color and chemical structure. At least one and probably several are known to be potent carcinogens and can cause liver cancers in domestic animals and in Man. Peanut butter, roasted peanuts, or other peanut products showing any mold or having off-flavors or tastes should never be eaten.

True Beans The true beans of the genus *Phaseolus* are also important New World contributions to human nutrition. One species, *P. vulgaris,* includes dry beans (kidney, black, turtle, pea, pinto, Great Northern, and others) and the string, wax, snap, and haricot beans. The protein level of dried beans is close to 25 percent, although most do not contain much oil. Green beans have much less protein. The presence of beans throughout the Americas was reported by early explorers. Columbus wrote about "a sort of bean very different from that of Spain" and presented some to Queen Isabella—who undoubtedly would have preferred gold. Giovanni da Verrazano, for whom New York City named a bridge, wrote in 1524, "Their ordinarie food is of pulse, whereof they have a great store, of good and pleasant taste." Several types reached England from Jamaica and Barbados before 1550; their value was immediately recognized by the British. Samuel Champlain in 1605 found that beans were co-cultivated with maize, and the Pilgrims unearthed a storage pit containing beans and shelled maize which the colonists promptly boiled up and ate. Kidney beans were a prominent dish at the first Thanksgiving meal.

There are several hundred cultivars of the common bean which may be distinguished by the pattern of markings on the skins. These markings are genetically controlled and are useful in determining the precise genetics of the cultivar. By studying the markings, it is possible to evaluate the movement of bean types from one Indian group to another and to use this information in studies on cultural interchanges among groups.

There are several other nutritionally valuable species in the genus. In contrast to the other species, the scarlet runner (*P. coccineus*) is a perennial with a thickened, tuberous root eaten as a starchy vegetable in South America. Long boiling is necessary to destroy a highly poisonous cyanogenic glucoside. The tepary bean (*P. acutifolius*) is found primarily in northern Mexico. It has white, blue, and blue-black seeds and was cultivated at Tehuacán at least 5000 years B.P. Lima or butter beans (*P. lunatus*) were brought to Europe by the Spanish, who saw them being eaten by the Indians in their Andean capital of Lima, where they had been grown for at least 5000 years.

Lentils and Other Beans The lentil (*Lens culinaris*) has been cultivated since the dawn of agriculture in southwestern Asia. It spread to the Mediterranean by 5000 years B.P. Because of the seed's high protein level, lentils became the "meat" of Lent and on Fridays before the Roman Catholic prohibition against eating meat on Fridays was repealed. Almost all the lentils in North America are grown in the Palouse Hills of Idaho and Washington State.

The genus *Vigna* includes the cow- or black-eyed pea, which, when cooked with fatback and greens, is a typical southern dish. The plants originated in central Africa, traveled to India during the Sanscritic era, and reached North Carolina about 1700, when slave traders brought large numbers of east Africans to the cotton fields. Green gram (*V. aureus*) and black gram (*V. mungo*) are African and Indian in origin and are used to make porridges, including the *dhal* of India. When dhal and rice are served together, the protein value is about as high as that of meat. *V. radiata,* the mung bean, is used for bean sprouts, which are 4-day-old seedlings grown in the dark to prevent the formation of chlorophyll.

The chick-pea, cici, or garbanzo (*Cicer arietinum*) was first cultivated in western Asia, where wild relatives still grow. It was a well-known crop in the Near East by 5400 B.P. and is a staple in the Indian and Middle Eastern diet. Rich in protein (17 percent), it has recently become accepted in North America, where it is used in soup and as a salad vegetable.

Egyptians have been growing the pigeon pea or congo bean (*Cajanus cajan*) in the Nile Valley for well over 4500 years. Since it is unusually tolerant of poor soils, has good heat and drought resistance, and encounters few pests or disease, it is grown as a backup crop and a hedge against cereal crop failure in India and equatorial Africa.

Other bean crops of some nutritional importance include adzuki beans (*Adzuki angularis*); the winged pea (*Lotus tetragonolobus*); and the winged bean (*Psophocarpus tetragonolobus*), which, in addition to being a high protein seed, forms a starchy, edible tuber looking for all the world like a sweet potato, but with much more protein. In a few places, it rivals cassava or the yam.

Fava Beans Variously called the broad bean, Windsor bean, fava bean, or horse bean, *Vicia faba* had been cultivated in the Mediterranean for over 6000 years before *Phaseolus* beans reached the Old World. They were recorded in Sumerian cuneiform tablets and their God of the Fields, Enki, is described as bringing forth on the fields "abundantly its small and large beans." Fava beans were found in Fifth Dynasty Egyptian temples at Sahure dated to 4400 years B.P. They were, even then, a common food, used extensively for slaves and by construction crews, possibly including those who built the Pyramids. Fava beans, known as *tshan tou,* reached China by 2200 B.P. and entered Japan and India shortly thereafter.

Herodotus, writing about Egypt, noted that the priests considered these beans unclean and would not even look upon the bean fields, although pots of beans were placed in tombs to feed the spirits of the dead. Certainly the Israelites were familiar with fava beans; their name for the bean was *pol* (Samuel II 17:18), from which the Greek *poltos* and the Roman *puls* were derived. Hittites, Scythians, and other civilizations of the Near East grew them and called them *tares* (Ezekiel 4:9). Greek writings also discuss fava beans in some detail. Homer noted them in the *Iliad* as a good cultivated crop and was one of the few Greek writers who had anything good to say about them. Greek seers and oracles would not eat them, fearing that their prophetic visions would be impaired. Empedocles (7000 B.P.) was adamant, "Wretches, utter wretches, keep your hands from beans." Yet Diogenes (2412–2323 B.P.) wrote that "beans are the substance which contains . . . that animated matter of which our souls are particles."

The Romans were somewhat ambivalent. The major agricultural writers, Varro, Columella, and Palladius, noted the widespread cultivation of fava beans. Pliny believed that the souls of the dead were contained in the beans, an idea with which Lucian, writing a century later, concurred by noting that to eat beans and to eat the head of one's father were equal crimes. In *De Agri Cultura,* Cato the Elder observed that beans and the vetches (*Vicia spp.*) "fertilize the land," and other authors, too, recommended plowing young *Vicia* plants into the soil as a green manure before planting wheat. Yet Ceres, Roman Goddess of the Grain, is said to have refused to include fava beans in her gifts to Man because they beclouded the second sight of her priests. Cicero applauded Ceres's wisdom and recommended that the Roman citizens leave such plants to the Goths because they "corrupt the blood." When Rome was Christianized, beans were thrown at ghosts on All Saint's Eve—our Halloween. On this awesome night, women would abstain from this common dish for fear of being impregnated by the ghosts of frustrated men.

It is rather unusual for a common, highly nutritious food plant to bear the burden of such concern, and search for the basis of these fears has long been of interest. Pythagoras, a philosopher of Greece's Golden Age, is said to have been killed by the enraged people of Crotonia when he could not bring himself to cross a bean field barring his escape route. Many of his disciples were slaughtered shortly thereafter by soldiers of Dionysius who ambushed them next to a bean field. Pliny, writing several centuries later, said, "As others have reported, the souls of the dead are contained in the bean and at all events it is for that reason that beans are employed in memorial sacrifices," and that they should be cautiously used to prevent the escape of one's soul. Only when modern immunological science developed did allergists finally recognize that in genetically prone individuals with Mediterranean ancestry, beans contain a substance that causes a frequently fatal dissolution of red blood cells. Many Italians, Greeks, Israelis, and other people are susceptible to this anemia, called favism, because they lack a particular enzyme that can inactivate the toxin. Some individuals are so sensitive that they are allergic to even the faint odor of bean blossoms.

Viable for many years in storage, broad beans were a staple winter food throughout Europe. During the Middle Ages, pork boiled with broad beans was a usual meal. The Orient obtained broad beans during the Han Dynasty (about 2200 B.P.), and Japan had them shortly thereafter. Britain got the fava beans only recently; they were scarcely known before the fifteenth century. The Spanish introduced them into Colombia by 1543. The beanstalk that Jack climbed was undoubtedly a broad bean vine, and the Brothers Grimm based "The Straw, the Coal and the Bean" on a Germanic folktale of ancient lineage. By the seventeenth century, Nicholas Culpepper dubbed them "plants of Venus" and also observed that the water in which they had soaked was excellent for the "blue marks caused by blowes." Captain Bartholomew Gosnold planted them along with wheat on Elizabeth Island off the coast of Massachusetts in 1602, and the Dutch grew them in New Amsterdam (now New York) by 1640. Related species are still grown in North Africa, the Middle East, Afghanistan, and Pakistan.

The vetches belong to the same genus as the fava beans and are extensively grown as forage plants and to stabilize steep banks and road cuts. Vetch seeds are occasionally used as food, but yields tend to be scanty.

Peas Today, green peas—fresh, canned, or frozen—are as common on our dinner plates as almost any other vegetable except tomatoes. This is, however, a fairly recent development. The species, *Pisum sativum,* originated in western Asia; the name is of Sanscritic origin. Dried peas have been found in Theban tombs, in many sites in the Middle East, in the Orient, and as far south as central Africa. All available evidence suggests that peas were dried before they were used. Fresh peas were apparently not eaten in Europe until the sixteenth or possibly the seventeenth century. The practice of eating fresh peas may have originated in the French court, where, according to contemporary records, their consumption became something of a fad. The custom spread only slowly to the less exalted people of France and to the rest of Europe. In North America, half the crop is eaten fresh, but dried peas, split or whole, and canned peas are still an important dietary staple. The exception to the exclusive use of dried peas in the Orient occurred when cultivars with edible pods were developed during the Tang Dynasty (618–907) for the royal table.

Nonfood Legumes Our intimate association with legumes is not restricted to beans and peas; several tree species are very useful. The carob (*Ceratonia siliqua*), an evergreen species of the Mediterranean, produces long pods much like our honey locust tree. The pods, called St. John's bread, are occasionally eaten in North America and much more extensively in North Africa. The seeds contain an oily compound that tastes vaguely like chocolate and is used as a chocolate substitute by those who unnecessarily fear the caffeinelike alkaloid in cocoa. The famous gum arabic, used as a glaze in paintings as it was in Egypt, is derived from a North African species of *Acacia.* Herodotus recommended its inclusion in cough syrups; it is still used for this purpose.

The importance of the forage legumes as animal fodder and as restorers of soil fertility cannot be overemphasized. Alfalfa (*Medicago sativa*), the lucerne of Europe, was domesticated in the Persia of the Medes. Two and a half centuries ago the species reached Greece, where alfalfa hay was fed to chariot horses. The Spanish invaders brought it to Mexico; it spread from there to the pampas of Argentina. Prospectors brought it from Chile to California in the early nineteenth century, and winter-hardy types arrived in Minnesota when German and Scandinavian settlers brought oxen and horses to work their farms. Red, white, and crimson clovers (*Trifolium spp.*), like all the other forage legumes, are European immigrants, brought to North America for pasture improvement and as hay plants. The trefoils (*Lotus spp.*), sweet clovers (*Melilotus spp.*), bush clovers (*Lespedeza spp.*), and the vetches (*Vicia spp.*) are also imported forage plants.

ADDITIONAL READINGS

Caldwell, B. E. (ed.). *Soybeans: Improvement, Production and Uses.* Madison, Wis.: American Society of Agronomy Monograph No. 16, 1973

Dovring, F. "The Soybean." *Scientific American 230,* no. 3 (1974):14-21.

Duke, J. A. *Handbook of Legumes of Economic Importance.* New York: Plenum Press, 1981.

Gentry, H. S. "Origin of the Common Bean, *Phaseolus vulgaris.*" *Economic Botany* 23 (1969): 55–69

Hebblethwaite, P. D. *The Faba Bean* (Vicia faba *L.): A Basis for Improvement.* London: Butterworths, 1983

Hebblethwaite, P.D. (ed.). Vicia Faba: *Agronomy, Physiology, Breeding.* Amsterdam: Nijhoff/Junk, 1984.

Hesseltine, C. W. "Fungi, People and Soybeans." *Mycologia* 77 (1985): 505–525.

Kaplan, L. "Phaseolus: Diffusion and Centers of Origin." In C. L. Riley, J. C. Charles, C. W. Pennington, and R. L. Rands (eds.). *Man Across the Sea: Problems of Pre-Columbian Contacts,* pp. 416–427. Austin: University of Texas Press, 1971.

National Research Council. *The Winged Bean: A High Protein Crop for the Tropics.* Washington, D.C.: National Academy of Science, 1981.

Norman, A. G. (ed.). *Soybean Physiology, Agronomy and Utilization.* New York: Academic Press, 1978.

Purseglove, J. W. *Tropical Crops: Dicotyledons,* Vol. 1. London: Longmans, 1968.

Shurtleff, W., and A. Aoyagi. *The Book of Tofu.* New York: Ballantine Books, 1978,

Smartt, J. *Tropical Pulses.* London: Longmans, 1976.

Witcombe, J. R., and W. Erskine. *Genetic Resources and Their Exploitation: Chickpeas, Faba Beans and Lentils.* Hingham, Mass.: Kluwer, 1984.

Woodroof, J. G. *Peanuts: Production, Processing, Products.* 3rd Ed. Westport, Conn.: Avi Books, 1983.

Wolf, W. J., and J. C. Cowan. *Soybeans as a Food Source.* 2nd Ed. Boca Raton, Fla. CRC Press, 1975.

The Potato

I'm just a dacent boy just landed from Ballyfad
I want a situation; yes, I want it mighty bad
I seen employment advertised;
'Tis just the thing says I,
But the dirty spalpeen ended with:
"No Irish need apply."

AMERICAN BALLAD OF 1855

Our white potato is one of only six of the 2000 species in the genus *Solanum* that form tubers. Wild potatoes grow in the southwestern United States, Mexico, Central America, and western South America from Venezuela south through Argentina, mainly at higher altitudes. Potato's affinities to tomato, bell pepper, and eggplant are clear; the leaves, flowers, and fruit are similar. In fact, the tomato and potato were originally included by Linnaeus in the same genus, with later separation into different genera a matter of taxonomic convenience rather than necessity. The tuber is an underground storage stem formed at the ends of secondary roots which have developed from thickened underground runners, the stolons. The basic stemlike character of the tuber can be seen externally as a whorl of shoot buds—the eyes—and is obvious when the internal arrangement of the tissues is compared with that of the aerial stem.

POTATOES IN PERU

Potatoes are still the basic carbohydrate food of Indians of the Peruvian and Bolivian highlands, who live at altitudes too high and too cold to support the growth of corn. They have been cultivated since at least 2500 B.P. by peoples who selected and reselected wild plants to emphasize rapid maturation and increased tuber size, and to eliminate plants containing poisonous steroidal alkaloids. The result was a plant whose tubers were walnut-sized, with red, yellow, blue, or brown skins and flesh that could be any of these colors. The potato is basically a source of starch with very little fat and minimal protein (about 2 percent by weight). Vitamins, especially water-soluble B vitamins and vitamin C, are found just beneath the corky *periderm,* or skin

In all cultures production of primary food crops is accompanied by rituals designed to ensure a successful harvest, and the Indians of Peru were as aware of this need as were wheat, maize, or rice farmers. Their pottery includes jars in the form of potato tubers (Figure 3.59) or anthropomorphized potatoes dedicated to *Axo-mama,* the female generative spirit of the potato. Human sacrifices to Her were made, and seed potatoes were dipped in blood—preferably human, but usually animal—before planting. Planting, cultivation, and harvesting were done with the choqui-taclla, the foot plow (Figure 3.60), and plants were tended by women who conducted "female" rituals over the developing plants. During harvest, any woman who found a blood-red potato had the right to run over to the nearest man and hit him in the face with it; this is probably of deep psychosexual significance and is not recommended for anyone without Inca genes. Harvested tubers were prepared for storage by converting them to *chuños.* Tubers were spread on the ground, allowed to freeze and thaw, and then stomped with bare feet to press out the moisture, a process repeated several times until the potatoes were completely dehydrated. Fresh potatoes were also used to make *chicha* beer. Supplemented with seeds and beans and occasional meat, the potato sustained people living in one of the world's most rugged environments.

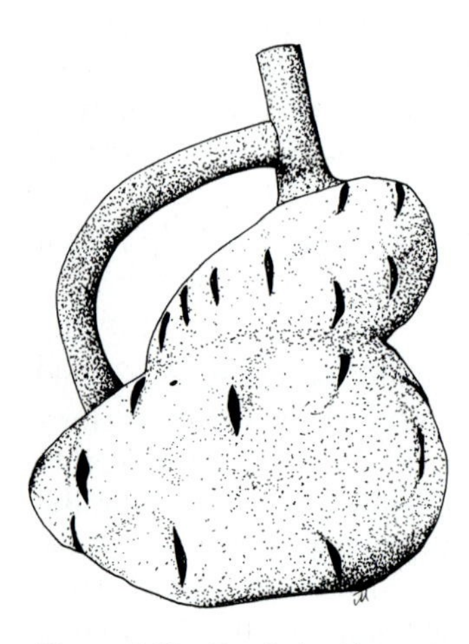

Figure 3.59 Pre-Columbian pottery jar from Peru, in the form of a potato.

By the tenth century, the highland peoples had been subjugated by Inca tribes from Lake Titicaca in Bolivia. The Incas built an imperial federation led by a priesthood and a hereditary nobility whose high civilization was based on beans, maize, and the potato. Spanish conquistadores first saw the potato in 1532, and the 1533 report of Pedro Cieze de León said: "Of provisions . . . there are two other products

Figure 3.60 Peruvian Indians planting potatoes with a foot plow. This efficient tool is still used in the Andes Mountains. Taken from a mid-sixteenth-century woodcut.

which form the principal food of these Indians. One is a kind of earth nut which, after it has been boiled is as tender as a cooked chestnut, but it has no more skin than a truffle and it grows under the earth in the same way." Gonzalo Jimenez de Quesado called the tubers "earth truffles," reported the manufacture of chuños, and noted that although the product was bitter and tough, it was life-sustaining and constituted over a third of the Indian's diet. The natives called the plant *papas,* carefully distinguishing it from the sweet potato, *batattes,* which grew in warmer climates, but the Spanish didn't make this differentiation so that our word, *potato,* is a combination of Indian words. Chuños, tubers, and seeds were sent back to Spain by 1565–1570 as curiosities. Philip II sent planting material to the pope, who distributed it to many papal legations for growing in monastery gardens. The exact source of the potatoes that reached England by 1580 is still a minor mystery. John Gerard described the plant in his *Herball of 1597* and called it the Virginia potato (Figure 3.61) because many plants from the New World were "from Virginia" and also because the English wouldn't accept any Spanish name—the memory of the Armada was still fresh in their minds. Charles d'Ecluse (Carolus Clusius) obtained plants in 1587, and in his massive *Rariorum Plantarum Historia* of 1601 published illustrations of the "Peruvian papas." Gaspar Bauhin named it *Solanum tuberosum esculentum* in 1590, a name later accepted by Linnaeus, who dropped the last word.

Figure 3.61 "Potatoes from Virginia." Taken from *The Herball* by John Gerard, published in London in 1597.

INTRODUCTION TO EUROPE

With plants available in many gardens and the knowledge that the Indians lived on them, it was natural that they would be tried by Europeans. A hospital in Seville added them to soup as early as 1573; Spanish ships carried chuños as emergency rations; James I of England ate them publicly in 1612; and in 1619 a Swiss gourmet described roasting and peeling them and the proper oil gravy to dress them with. Italians were using potatoes as cattle fodder in 1588, but most people assiduously avoided even touching them. They were not mentioned in the Bible and hence were not a food designed by God for Man. Even in the nineteenth century, John Ruskin spoke of the "scarcely innocent underground stem of one of a tribe set aside for evil" because of the obvious familial resemblance to nightshade, datura, and mandragora, all plants of witchcraft, evil, and death (Chapter 6), and Clusius earlier noted that these solanaceous plants, like others of their ilk, might contain poisons or "could be used for exciting Venus." Because of their rough and scabby skin, it was asserted that potatoes could cause leprosy or scrofula; recognizing this danger, Burgundy passed a law in 1630: "In view of the fact that the potato is a pernicious substance whose use can cause leprosy, it is hereby forbidden, under pain of a fine, to cultivate it in the territory." As late as 1710, William Salmon, an English physician, said that "the English or Irish potato increaseth thy seed and provoketh lust in both sexes." This canard was still extant in the nineteenth century. Many Italians, seeing their cattle thriving, began adding them to stews in place of turnips, and a few other brave souls were spreading the word that boiled potatoes tasted fine so long as they were well salted. Nevertheless, potatoes were not a regular part of the European diet until the first quarter of the seventeenth century.

POTATOES ARE FOR PEASANTS

The failure of the general population to take up potato eating was of concern to the upper classes and the nobility. The grand duke of Tuscany recognized that if his peasants would eat potatoes instead of bread and pasta, the more valuable wheat could be reserved for his personal profit. After all, the entire sequence of potato cultivation could be accomplished with no tools other than a spade or hoe, a crop would mature in less than four months in virtually any soil, production was tremendous (an entire family plus livestock could be fed on tubers produced on an acre of land), and its preparation for the table involved simple boiling. The potato could be the economic salvation of the landlord; efforts were made to promote its use.

Methods of weaning peasants away from expensive grain took different forms in different countries. Frederick Wilhelm, elector of Prussia, ordered potato cultivation in 1651 as "a national duty," but without much response from his otherwise dutiful subjects. His grandson, Frederick William I, passed a law stating that anyone who refused to plant and eat potatoes would, forthwith, have his nose and ears cut off, and many Germans decided that *kartoffels* were the best food available. His son, Frederick the Great, provided free seed potatoes and even distributed a recipe book, although the people generally just boiled them. Peter the Great of Russia imported potatoes from Germany, but only for the royal table. Catherine the Great obtained potatoes from Prussia and mandated their cultivation for local consumption in European Russia; Ukrainian grain was reserved for use in the colder parts of Russia and for export. The French were initiated when prisoners of war returned home from Germany at the end of the Seven Year's War with Prussia (Frederick having fed his prisoners on potatoes), bringing tubers back for planting in Normandy and the central provinces. In 1771, the Academy of Science of Besançon offered a prize for the discovery of a new food which could take the place of cereal grains in the event of famine. Antoine-Augustine Parmentier won the award for recommending that the *cartoufle* (earth-nut)—later to be dubbed *pomme de terre* (earth-apple)—was the obvious choice, citing the reasons used by the grand duke of Tuscany and his own personal experience as a prisoner of war. In contrast to standing fields of grain, potatoes could not be set afire by invading enemy troops and the tubers could be stored underground—hidden from marauding bands. Parmentier had a flair for publicity; he presented King Louis XVI with a bouquet of potato flowers on His Majesty's birthday. In return, he was given a plot of royal land on which were grown potatoes served at a dinner attended by celebrities including the good Dr. Benjamin Franklin. Recipes, including the following, were distributed:

> To make a potato pudding, first boil and blanch potatoes, and add the marrow of two bones. Add a pint of cream, ten egg yolks, five beaten egg whites, sugar, rose water, one half pound of bread, cinnamon, ginger, nutmeg, minced citron and orange, candied lemon and some sack [a light wine from the Canary Islands]. Bake in a dish with puff paste and cover with a sauce made with thick butter, sack and sugar.

The peasant boiled potatoes and ate them with salt and oil. The king told Parmentier, "France will thank you hence because you have found bread for the poor."

In 1662, the Royal Society of London met to consider the possibility of planting potatoes throughout England. John Forster, Gentleman, published these deliberations in a book entitled:

> *England's Happiness Increased or a Sure and Easy Remedy Against All Succeeding Dear Years by a Plantation of the Roots called Potatoes: Whereby (with the Addition of Wheatflower) Excellent Good and Wholesome Bread may be Made Every Year 8 or 9 Months together, for Half the Charge as Formerly; Also by the Planting of These Roots Ten Thousand Men in England*

and Wales Who Know Not How to Live, or What to Do to Get a Maintenance for their Families, may on one Acre of Ground make 30 pounds per Annum, Invented and Published for the Good of the Poorer Sort . . .

It was suggested to Charles II that a tax on potatoes be levied, the proceeds to go directly to the royal treasury: "There is no reason why his gracious Majesty should not benefit by this annual license money of 50,000 pounds, inasmuch as his loyal subjects would thus benefit by the cultivation of the potato."

Actually, there had been some cultivation of the potato in the British Isles. Thomas Hariot, estate manager of Sir Walter Raleigh, planted them in 1588 at Youghal, Ireland. They were grown in the physick (medicinal) garden in Lambeth, London, in 1656, and in Oliver Cromwell's wife's garden in 1664. Robert Morrison, botany professor at Oxford, wrote in 1680 that it was a familiar English garden plant. With royal assent, planting stock was obtained from Raleigh's estate, fixing for all times the name "Irish" potato.

THE IRISH SPUD OR LUMPER

Three areas in Britain were most in need of cheap food: Wales, Scotland, and Ireland, with the most desperate need being in Ireland. From the time of Henry VIII, when the break with the Roman Catholic Church was complete, the subjugation of the Catholic Irish became deliberate royal policy; the Irish were to be obedient to the crown, and the land was to be populated with reliable—that is, Protestant—owners. In spite of a papal bull designed to prevent Elizabeth I from ruling, the subjugation went forward; the Desmond Revolt of 1568–1582 was ruthlessly crushed. The Cromwellian upheaval of 1650 uprooted the peasants, leaving them without homes, crops, or cattle. The Penal Laws of 1690 completed the process by depriving the Irish of civil rights, ownership of a horse, the right to speak Gaelic, to get an education, to attend college, to marry a non-Catholic, or to hire a Protestant. The Irish became beggars in their own country and, each year during the seventeenth and eighteenth centuries, close to 40 percent of the people were on semistarvation diets, sustained by their religious faith and a burning desire for freedom from England's impossibly heavy yoke.

Growing conditions in Ireland were ideal for the potato—a cool, humid climate, deep and friable soil overlying chalk, and a long growing season. The prevailing moist westerlies that blew from the Atlantic checked the spread of insects carrying diseases known to reduce yields. The land tenure situation also contributed to the spread of potato cultivation. Ireland had been parcelled out to absentee landlords, all Anglicans, who used their estates for pleasure and who lived in England on profits accruing from lush grain fields, well-pastured cattle, and sheep whose mutton and wool nourished and clothed the middle and upper classes. To relieve themselves of any responsibility to the peasants, landlords granted long-term leases to middlemen, who, in turn, further leased to subcontractors and sub-subcontractors, with corresponding increases in the rents charged for the hovels occupied by the peasants. Occupying and tilling little more than an acre per family, the Irish paid their exorbitant rents, taxes, and tithes in cash (selling the pig) and in labor, including the domestic service of the women—the "Irish washerwoman" was not merely a carefree dance—and they were forever in debt to loan sharks, the *Gombeenmen* (boogymen). Prolific—to the delight of the Church; ignorant—to the delight of the middlemen; rebellious—to the delight of the red-coated military—the people were sustained by the knobbly spud or lumper.

The numbers of Irish people increased, virtually doubling between 1700 and 1800. Indeed, Europe's population, which had been stable for centuries, jumped between 1725 and 1825 from 140 million to 266 million. The technology of the industrial revolution, advances in livestock breeding, the introduction of crop rotation, and systematic manuring contributed to the availability of food, which had been a limiting factor in population growth. But demographic experts now believe that a major factor in population growth was the availability of this wonderful new food supply. By 1780 Ireland's food "is potatoes and milk for ten months of the years, potatoes and salt the remaining two."

The Irish male consumed 12 pounds of boiled potatoes a day, every day. His wife ate about eight pounds, and his children no less than five pounds. Fishers sold their catch and ate potatoes. Sheep and cattle were tended by peasants who tasted meat about three times a year, bread was over half potato starch, and drink was potato beer or *poteen,* distilled potato beer. Bloated stomachs were the least serious evidence of malnutrition; potatoes lack fat-soluble vitamins A and D, so that, even with milk in the diet, eye dysfunctions and rickets were common.

Adam Smith, in his *Wealth of Nations* of 1776, warned that "should these roots ever become . . . the favourite vegetable food of the people . . . populations would rise and rents would rise. . . ." He noted that "it is not uncommon . . . for a mother who had borne twenty children not to have two alive." Daniel Langhaus observed the same high birth and death rate in potato-eating cantons in Switzerland. Dr. Thomas Beddoes in his *Essay on Consumption* of 1800 said: "This root . . . had probably contributed to the degradation of the human species." The august *Times* of London in 1829 said that the working classes would become "miserable, turbulent drunkards" if they subsisted on potatoes. Yet there were few options for the peasant; emigration was a fearful choice, for the church warned that those who went to a pagan—that is, non-Catholic—country were in danger of losing their immortal souls. Besides, there was no money to pay for a passage to Australia (a hated "English" land), Canada (English or, what was worse, French), or America.

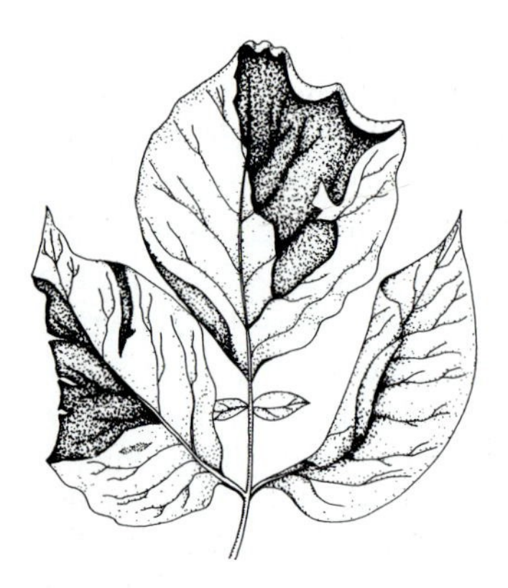

Figure 3.62 Early symptoms of potato blight on the leaves.

THE POTATO MURRAIN

Crop failure was always on people's minds. Seed potatoes were blessed and holy water sprinkled on the plants on Ascension Day. Candles were lit to Mary to intercede for good crops. In 1809 a virus disease reduced yields by 30 percent and another virus infestation in 1833 was sufficiently serious that hardier and more adventurous people worked passage to eastern Canada and to the United States. On August 23, 1845, the prestigious *Gardener's Chronicle* headlined: "A fatal malady has broken out amongst the potato crops." The potato murrain had struck the plants all across Europe and was most severe in Ireland, where the tenant farmers' patches were so close together that any murrain could sweep through the whole country. Green tops blackened and shriveled within a day (Figure 3.62), and young tubers had black patches on their still-thin skins. On September 13, 1845, the *Chronicle* stated: "We stop the press with very great regret to announce that the potato murrain has unequivocally declared itself in Ireland. Where will Ireland be, in the event of a universal potato rot?" The answer was obvious—starvation. Landlords were of two minds about this. The potential loss of rent was annoying, but on the other hand, this would allow them summarily to dispossess tenants, tear down their miserable shacks, and turn potato patches into productive grazing land. A British commission suggested that things weren't too bad; even if the potatoes were rotten, the peasants could wash the starch out of them and use this to bake bread. They knew, however, that starch alone could not sustain life without vitamins, minerals, and protein.

English newspapers were filled with suggestions on how blackened, shriveled, slimy, and stinking potatoes could be eaten (by the Irish) and theories on the cause of the murrain. The respected editor of *Gardener's Chronicle* asserted that it was because wet, cool weather caused the potatoes to take up too much water. One correspondent suggested that it was due to the electricity generated in the air by the tremendous and dangerous speeds of steam locomotives. Another insisted that it was just punishment from God visited on Catholics, and some Catholics said that it was caused by not being Catholic enough. Remember that this was 1845, 20 years before Robert Koch and Louis Pasteur proved the causal connection between microorganisms and disease. Reverend M. J. Berkeley, curate of Northamptonshire, wrote in the *Chronicle* that whenever murrain struck, he was able to observe a fungus on the blighted parts of the leaves and tubers. Although others had seen the fungus, most believed that it was a consequence of decay and not the cause. Controversy raged about this matter until the fungus, named *Phytophthora infestans,* proved to be the cause, its airborne spores being carried on the lightest breezes from plant to plant (Figure 3.63).

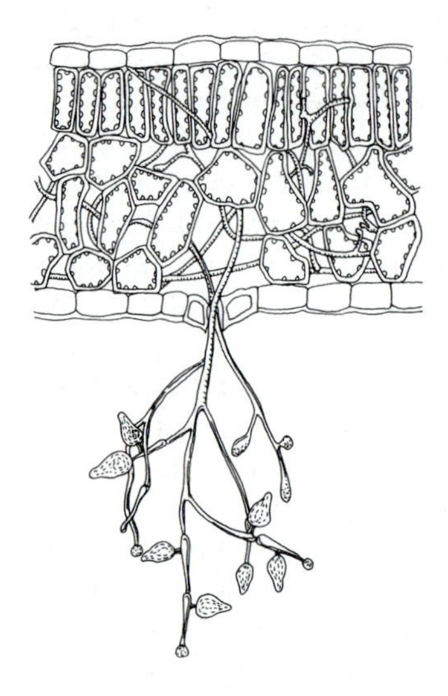

Figure 3.63 Spore-bearing hyphae of the blight fungus protruding from a stoma on the lower surface of a potato leaf.

As with most plant diseases, the severity of the blight depended on the inherent genetic resistance of the plant and on weather conditions. In Peru, it is too cold for much development of the fungus, and there is scarcely enough rain to keep potato leaves wet long enough to allow spores to germinate and invade the leaves through the stomatal pores. In 1845 the weather was just cool enough and just rainy enough, the cultivars growing in Ireland were just susceptible enough, and the fungus was just virulent enough so that about 75 percent of the total crop was a complete loss.

FAMINE

Although the summer of 1847 started well, rain, gloomy weather, and cool winds developed during July, and between July 27 and August 3 the potato blight killed close to 80 percent of the plants. Food relief was instituted on a halfhearted basis, but the people were by then so weak that only massive relief with high protein meat supplies would have stemmed the growing tide of starvation and death. Soup kitchens were reluctantly set up, but a once-a-day meal of a pint of soup made with a pound of gristle, a peck of peas, a peck of barley, and five heads of cabbage in 150 gallons of hot water or a cup of maizemeal boiled in water had little positive effect. The government congratulated itself by noting that it not only provided food but even cooked it.

The consequences were inevitable: possibly 3 million people starved to death. Corpses could not be buried before they began to rot in the streets; typhus fever and cholera became epidemic; women aborted; and child mortality approached 65 percent. Only in 1974 was it discovered that blighted potatoes contain an alkaloid causing abortion and birth defects. Those children that survived would never attain full vigor and would spend their lives with the reduced intelligence that comes from inadequate protein and alkaloid poisoning during the first years of life. With enlightened self-interest, some landlords "forgave" the rents and others hired leaky ships to take people to Canada or to the United States (Figure 3.64). The famine set into motion a wave of emigration that reduced Ireland's population from over 8 million in 1846 to 4 million in 1900. The Irish constituted 35 percent of the immigrants to the United States during this period and became 15–18 percent of the population.

The new immigrants caused a huge, almost insolvable problem for the United States. In spite of being rural, they had no farm skills, since their only real experience was spading potato patches and hoeing cabbages, and machinery was replacing scything. The immigrants had no money to move out of the port cities. The near-starvation conditions under which they had lived left them physically weak and prey to the diseases of malnutrition. They arrived in a country suspicious of Catholics and flooded a labor market unable to employ even the people who were in the cities before the Irish arrived. Ditch digging and odd jobs were the lot of the men and the women worked in factories for low wages and formed the servant class for affluent Americans. Having no

Figure 3.64 Irish emigrants leaving for New York, as depicted in *Harper's Weekly Magazine.*

knowledge of city life, they were cheated at every hand and, in self-defense, formed enclaves within New York and Boston that served not only to protect them but also to mark them as a people apart. Anxious to work, they established their base in the construction trades, where, in time, they branched out, eventually to become active in all aspects of American life up to and including the presidency. But the road was unnecessarily hard and bitter, and although these injuries are not so deep as those received from the British over a 300-year period, they still exist.

MODERN CULTIVATION

Close to 300 million metric tons of potatoes are grown throughout the world, with 80 percent of this amount produced in Europe and Russia. Both wheat and rice are produced in larger tonnages, but the land area devoted to potatoes is only 11 percent of that devoted to wheat. Based solely on total caloric value, the potato provides more carbohydrate than do other major food crops (Figure 3.65), a fact that was recognized when the rulers of Europe forced potatoes on their people in the seventeenth century. Unless very large volumes are consumed with a supply of vitamins A and D (milk, eggs, butter), a nutritionally adequate diet cannot be obtained from potatoes.

Potatoes entered North America in 1613, when English settlers brought them to Bermuda as a food for slaves. The plant reached the mainland in 1719, when a colony of Scots-Irish settled in Londonderry, New Hampshire, and grew the plant primarily as animal feed. Only a few kinds were then available, a red-skinned type, a brown-skinned

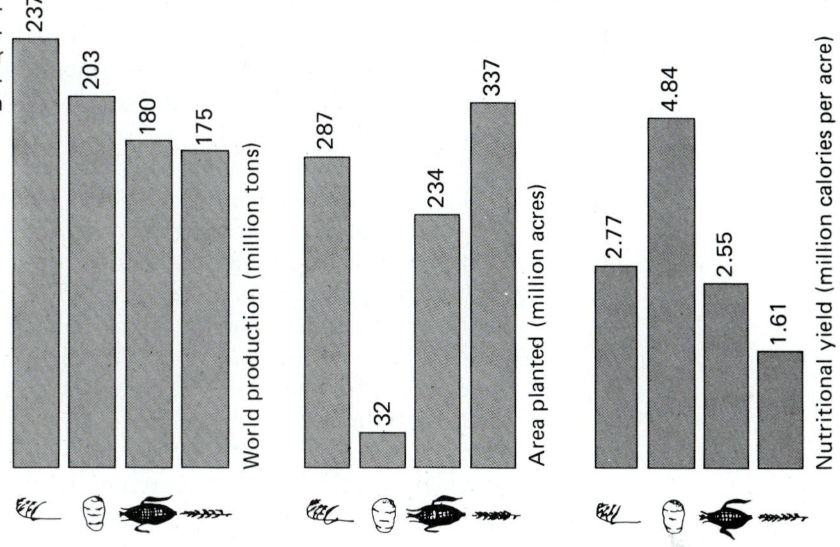

Figure 3.65 Comparison of production, areas cultivated, and nutritional yields of the four major carbohydrate crops, wheat, potato, maize, and rice. Data from 1970 harvest season.

type, and a dark-brown type, but, by 1760, there were many types with names like 'Cluster,' 'Irish Apple', and 'Cups.' Each of these types, now called cultivars (cultivated varieties), was fairly uniform, since they were propagated vegetatively from pieces of the previous year's tubers saved as "seed potatoes." The Chinese had the potato (*yang shu*) by 1650.

At least theoretically, the number of cultivars may be infinitely large, although some are genotypically so similar as to be essentially identical. To develop a new cultivar one would take seeds, plant them, and select from the mature plants to obtain phenotypes that are desirable—earliness, size, skin and flesh color, taste, or disease resistance. Prior to the Great Famine, disease resistance was not considered, for the concept of a pathogenic (disease-producing) organism was not known, but by 1880, gardeners were attempting to develop cultivars resistant to virus diseases, to wart, scab, rot, and blight. Selection for desirable growth and form was well under way by the middle of the nineteenth century. The Reverend Chauncey Goodrich of New York obtained seeds from South America in 1851, grew them in his garden, and from thousands of plants selected one named 'Garnet.' Garnet and selections from it became standard cultivars in the northeastern United States and Canada. By 1900, when the concept of sexual reproduction was established, the science of plant breeding began.

There were over 400 cultivars by 1890. The difficulty with this early breeding work was that the breeders were merely reassorting genetic characters within a relatively small genetic pool, and the statistical probability of obtaining a genotype which included strikingly new phenotypic characters such as disease resistance was low. People got the idea, in about 1910, that they should obtain wild plants whose characteristics had not been genetically selected. Nations sent plant hunters back to the mountains of Peru, Bolivia, and Mexico to collect wild potatoes for crossbreeding. The United States and other potato-growing countries established experimental potato plots to grow wild potatoes for seed to be distributed to breeders. A truly blight-resistant potato has yet to be developed, but cultivars are now available which have moderate resistance to this and other diseases.

In addition to being plagued by viruses, fungi, and bacteria, the potato is food for a host of insects, among which is the voracious Colorado potato beetle (*Leptinotarsa decemlineata*), a small insect with characteristic yellow and black stripes on its wing covers (Figure 3.66). It was first described in 1824 in Colorado, where it was fairly rare, living on leaves of wild members of the Solanaceae. As American farmers moved west before and after the Civil War, those that settled the prairies planted potatoes. The beetles, finding the potato much to its liking, began to move east. The beetle crossed the Mississippi River in 1865, reached Ohio by 1869 and Maine by 1872. In spite of rigorous quarantine against potato importation into Europe, the beetle was in England by 1875, but was quickly exterminated. Beetles were seen in Germany in 1914, but, because of the war, were not hunted down and are now munching happily away on potato leaves throughout Europe. Chemical insecticides, especially the highly poisonous chlorinated hydrocarbons, keep them from destroying the crops, but the attendant environmental pollution is becoming evident all over the world.

TROPICAL EQUIVALENTS OF THE POTATO

The warm and fairly uniform temperatures, generally high amounts of rain, and lack of photoperiodic variations that characterize the tropical regions of the world are unfavorable for the growth and maturation of the temperate-climate potato. Several high-yielding carbohydrate root crops with simple cultural requirements are the potato's dietary equivalent. Four of these are, in yields and in nutritional importance, widely grown.

Figure 3.66 Colorado potato beetle feeding on potato leaves. Courtesy U.S. Department of Agriculture.

Cassava Cassava (*Manihot esculenta*) has many common names: *yuca, manioc, sagu,* and *tapioca* all refer to the same plant. The genus is in the Euphorbiaceae, the spurge family, which also contains the poinsettia, commercial rubber trees, and the castor bean. Cassava is a perennial shrub which grows to 15 feet (5 meters) and produces large storage roots that can weigh over 10 pounds. Under favorable conditions, yields of 60 to 80 metric tons per hectare (40 tons per acre) are obtained, but yields are usually closer to 8 metric tons. Excluding local production, 100 million metric tons are produced per year throughout the world; cassava is the primary carbohydrate for a half-billion people. Although less widely consumed, the leaves contain 30 percent protein, while the roots have less than 4 percent digestible protein. Propagation is very simple. Stem segments are merely inserted into the ground, where they take root. Full maturity requires about 16 months, but by planting throughout the year, people can have a staple supply. This is important, because in contrast to the potato, cassava roots have a short storage life.

Cassava is a native of Peru and Brazil. The genus is large and its taxonomy is very complex, with many selections made by the Indians. Archeological evidence dates its cultivation, or at least its extensive use, to 3000 B.P. in Brazil, 2800 B.P. in Peru, and 2000 B.P. in tropical Mexico, the Caribbean Islands, and Colombia. Although Columbus undoubtedly saw it on his voyage to Cuba, the first descriptive report was that of Peter Martyr, who accompanied Columbus on his second voyage, in 1493: "Iucca is a roote, whereof the best and most delicious bread is made, both in the firme lande of these regions and also in Ilands." Amerigo Vespucci noted in 1497 that "their most common food is a certain root which they grind into a kind of flour of not unpalatable taste and this root is by some of them called jucha, by others chambi and by others igname." The Portuguese introduced cassava to the Antilles and to west Africa in 1558, and the Spanish brought it to the Philippines. By the seventeenth century it was taken to Malaysia, from where it spread throughout the Old World tropics.

There are two major kinds of cassava, bitter and sweet. The bitter form, still the primary type grown in the American tropics, contains a bitter cyanogenic glucoside. When cells are cut open, enzymes in the cell sap cleave the glucoside into a sugar and the very poisonous prussic (hydrocyanic) acid. The poison is found in the skin, or periderm, and in the tissues immediately below the skin. The roots are peeled, the tissues are shredded, and the poisonous juice squeezed out. There is no information on when this technique was developed. The roots of sweet cassava are eaten after boiling, and both bitter and sweet cultivars are made into flour for unleavened bread or eaten as a gruel. In parts of Africa, boiled roots are mashed into a paste (*fufu*), for immediate consumption or for fermentation into a nutritious beer. Starch pellets from cassava are heated until gelatinous, then forced through a mesh screen to form the tapioca pearls from which the familiar pudding is made. Brazil uses this high carbohydrate root as a source of starch and sugar for the production of alcohol that, in this oil-poor country, is becoming an important automobile fuel.

In recognition of the vital importance of cassava in nutrition, efforts are under way to improve the crop. The Centro Internacional de Agricultura Tropical in Cali, Colombia, and the International Institute for Tropical Agriculture in Nigeria, both partly funded by the International Developmental Research Centre of Canada, are deeply involved in cassava breeding. Primary goals include identification of high-yielding, nonpoisonous stocks, development of disease resistance, and improvement of cultivation practices. A program has been established for the training of agricultural teachers and technicians who can instruct farmers.

Yams True yams (*Dioscorea spp.*) may have been independently domesticated in both the Old and New Worlds. Yams (*nyitti*) were noted in Sanscritic writings in India and agricultural manuals of ancient China. True yams also appear to have been seen by early Spanish explorers in South America; they were noted as a common root crop in Brazil in 1500. There are a huge number of species in different parts of the world, suggesting that the genus originated prior to the separation of Gondwanaland, which resulted in the present distribution of continents (Chapter 1). Although rarely seen in North America, true yams are the third most important tropical root crop after cassava and the sweet potato. The edible portion is a tuber—a subterranean, swollen stem—rather than a true root. The tubers can reach 100 pounds but are usually harvested when they reach 2–6 pounds. They are fairly easy to cultivate.

Sweet Potatoes *Ipomoea batatas,* the sweet potato, is a primary carbohydrate source in the tropics, exceeded in production volume only by cassava. The plant is a member of the Convolvulaceae, and the flowers look very much like those of the morning glory. Yields from this trailing vine can be up to 40 metric tons per hectare (18 tons per acre), with over 140 million metric tons produced per year. Sweet potatoes originated in the lowlands of Central America and in tropical Peru near the equator, where minimum temperatures do not fall below 24°C (75°F) and where rainfall exceeds 100 centimeters per year (40 inches). Sweet potatoes were also found in Polynesia when these Pacific islands were visited by Ferdinand Magellan, the first European to reach the Philippines, in 1521, and James Cook, who died in the Hawaiian Islands in 1779. The Maori people of New Zealand were also growing sweet potatoes before the island was first visited by Europeans, in the 1770s. This disjunction has been taken as evidence that there was intercourse between the peoples of Oceania and those of tropical western America before the fifteenth century. Thor Heyerdahl's successful voyages, in 1947, across the Pacific in vessels modeled after native ships lends some credence to this hypothesis. It is also possible that there were two distinct centers of origin. Sweet potatoes were presented to the Spanish crown by Columbus, and Iberian Roman Catholic priests introduced the plant to the Far East in the sixteenth and early seventeenth centuries. Records in China date its introduction into Yunnan by 1570, suggesting that it traveled overland from India and Burma.

The Chinese embraced the sweet potato enthusiastically. Its heavy yields, its pleasant taste, its disease and insect resistance, and its drought tolerance were greatly appreciated. Since the Chinese were skilled in root crop cultivation of the yam and in methods of cooking roots, the sweet potato quickly replaced the true yam in the diets of the south Chinese. By the time of the Yung-cheng emperor (1723–1735), officials of southeastern provinces were required to report annually on the food self-sufficiency of their districts in terms of amounts of rice and of sweet potatoes. Japan received the plant from China and, here too, production soared. Sweet potatoes are the second most important crop in Japan and can easily be purchased, piping hot, from street vendors.

There are two major subtypes of sweet potatoes. A sweet, orange-fleshed, moist type is called a yam and is grown extensively in the southern United States. The drier, mealy, yellow forms are the choice in Africa and Asia. Both have high quality carbohydrate, a high level of carotenes (the precursors of the visual pigments in the eyes), reasonable mineral and vitamin contents, little lipid, and not much protein. Candied yams nestled against slices of roast turkey at the Thanksgiving feast are a recent innovation; baked beans sweetened with maple syrup are much more in the tradition of the Pilgrims.

Taro Visitors to Hawaii invariably are confronted with a bowl containing a purplish-gray, sticky mass and are instructed to dip their fingers into it and then lick off the substance. Fascinated by this breach of Western table etiquette, most give it one try and rarely finish their portions. The dish, called *poi,* is made by steaming, mashing, and then fermenting the tuber of the taro plant (*Colocasia esculenta*). Taro, also called *eddoe* or *dasheen,* and another root crop plant, the *yautia* or *ocuma* (*Xanthosoma spp.*), are important tropical carbohydrate crops. Taro, native to Southeast Asia, moved into India and then to Oceania. The Greeks brought it to Europe and Egypt; it was grown in tropical Africa and then traveled with African slaves to the West Indies and tropical America. Yautia took the reverse route, moving from its home in tropical America to Africa, Southeast Asia, and south China, and then into India and Oceania by 1800. By 1844 slaves in the cotton fields of South Carolina were growing yautia. It never achieved the popularity of the sweet potato because of an acrid taste that can only be removed by long cooking.

ADDITIONAL READINGS

Bouwkamp, J. L. (ed.). *Sweet Potato Products: A Natural Resource for the Tropics.* Boca Raton, Fla.: CRC Press, 1985.

Correll, D. S. *The Potato and Its Wild Relatives.* Austin: Texas Research Foundation, 1962.

Dodds, K. S. "Evolution of the Cultivated Potato." *Endeavour* 25 (1966): 83–88.

Dodge, B. S. *Potatoes and People.* Boston: Little, Brown, 1970.

Edmond, J. B., and G. R. Ammerman. *Sweet Potatoes: Production. Processing, Marketing.* Westport, Conn.: Avi Books, 1971.

Gallagher, T. M. *Paddy's Lament: Prelude to Hatred. Ireland 1846–1847.* New York: Harcourt Brace Jovanovich, 1984.

Kahn, E. J., Jr. "Staffs of Life. II. Man Is What He Eats." *The New Yorker* (November 12), 1984, pp. 56–106.

Large, E. C. *The Advance of the Fungi.* New York: Holt, Rinehart and Winston, 1940.

Niederhauser, J. S. "The Blight, the Blighter and the Blighted." *Transactions of the New York Academy of Sciences. Series II* 19 (1956): 55–63.

Salaman, R. N. *The History and Social Influence of the Potato.* Cambridge, England: Cambridge University Press, 1949.

Smith, O. *Potatoes: Production, Storing, Processing.* 2nd Ed. Westport, Conn.: Avi Books, 1977.

Weber, E. J., J. C. Toro, and M. Graham (eds.). *Cassava Cultural Practices.* Ottawa: International Developmental Research Centre, 1980.

Woodham-Smith, C. *The Great Hunger: Ireland 1845–1849.* New York: Harper & Row, 1962.

Yen, D. E. *The Sweet Potato and Oceania: An Essay in Ethnobotany.* Honolulu: Bishop Museum Press, 1974.

Chapter 4

Vegetable Feast

Supermarket Botany

Mouth-watering Garden Delights to brighten summer meals
Advertisement in Burlington, Vermont, Free Press

Just as a surgeon blocks out the operating room when carving a turkey, so a botanist with a shopping list pays little attention to plant structure when pushing a cart around a supermarket. Yet the items on the shelves can form an excellent basis for an entire botanical education. Spice racks contain leaves, stems, flowers, fruits, and seeds of many species, each with an ethnobotanical history and each with a long list of physiological and chemical attributes. Beverage sections sell tea leaves, coffee and cacao seeds, and extractives from many plant parts. Toothpaste contains skeletons of diatomaceous algae that have been dead for thousands of years. Ice cream is fabricated with plant flavorings and thickened with emulsifiers from marine algae. And when you pass through the checkout counter, you pay with paper money made of wood fibers and have your groceries packed into a paper bag.

Our basic foods, either eaten directly or processed into protein by animals, are the cereal grains, members of the grass family. The cereal grains are single-seeded fruits called *caryopses,* characterized by starchy endosperm tissues that nourish the young embryo and seedling. The beer in supermarket refrigerators is made from barley and other cereal grains. Bags of cane sugar are derived from another grass, the sugar cane *(Saccharum officinarum)*; cane and sorghum can yield molasses used in baking. The liquid sugars in soft drinks, baked goods, and many other fabricated foods can also be derived from maize kernels. Even though bread has almost ceased being the North American Staff of Life, the importance of the cereal grains in human nutrition is almost as great as it was a century ago.

Seeds, too, are important parts of the human diet and are consumed in many forms. Some are easily recognized: fresh, dried, or canned seeds of legumes such as chick-peas, kidney beans, green peas, and lima beans are easily obtained (see Chapter 3). Some of our vegetable oils, cooking fats, and margarines come from legume seeds (peanut and soybean oils) and some from the seeds of cotton, sesame, and other plants. Soap, too, uses seed-derived vegetable oils. Many supermarkets now cater to the foreign-food gourmet: bean sprouts and soybean pastes, used for Japanese miso and soy sauces, are right there next to the chow mein noodles. Pine nuts enhance rice dishes, and sunflower seeds are a common and nourishing snack. So are the wide variety of nuts obtained from the fruits of many different plants. Although generally used in smaller amounts, seeds are basic seasonings in our food. Black pepper, caraway, mustard, nutmeg and mace, poppyseed and coriander are a small sampling of the culinary spices that are seeds.

Fungi, too, are well represented in the store. Many provide the distinctive flavoring for cheeses: Roquefort, bleu, and gorgonzola have bluish streaks of recognizable spores of members of the *Penicillium* genus. Fresh and dried reproductive bodies of basidiomycete fungi plus the tasty chanterelle, bolete, truffle, and other mushrooms gathered carefully from the wild are now available. Fungi play important roles in the production of fermented foods. Pickles, soy, and other Oriental sauces and, of course, beer and wine are produced with the aid of fungi. Many of the flavorings in processed foods are the products of microbial action. The citric acid in your favorite soft drink may have been made in large fermentation vats containing fungi, and a wide variety of other additives (read the fine print on the label!) are made by inexpensive fermentative processes.

SELECTION OF THE PLANTS USED FOR FOOD

As discussed in Chapter 2, we have domesticated plants that form our food, fiber, and other products so that they can be grown in desired areas and will consistently mature a crop that meets our biological, social, and economic needs. During the millennia since humans took to farming, fewer than 300 plants of the several hundred thousand species of flowering plants have been fully domesticated and widely dispersed. Of these, our food and flavoring needs have been met by fewer than 200 of these species. People initially grew wild plants and have over millennia selected from each year's crop those individuals that possessed characteristics considered to be valuable: quicker maturation, larger edible parts, better flavor, easier cooking, and so on. The plants selected were generally those that resulted either from a chance alteration in a character—a mutation—or from a cross with another plant in the field or with a wild relative growing near the field. In some cases, as with wheat, it is possible to trace back the ancestry of the modern crop plant to its wild relatives and to obtain some understanding of how it was derived, but in many other cases, this has proved to be impossible.

The selection and breeding process we can follow in plants was also done with our domestic animals. Breeds of dogs, derived from a few species of wild canine, now present an almost bewildering array of size, shape, and attributes that, if mating is not left to casual encounter, can be perpetuated. Perpetuity of uniformity is frequently easier to achieve in plants than in animals. Many plants are self-fertile. Others are cross-pollinated, but are typically grown in monocultural associations in large areas so that fertilization is with male generative nuclei, the genetic potential of which is almost identical to that in the female nuclei. Still others are replicated by vegetative propagation, which ensures genetic uniformity. We use "seed potatoes," stem sections of sugar cane, and graftings to start food plants. Our cereals and vegetables are grown from seed whose production is carefully controlled to ensure genetic and phenotypic uniformity.

Obtaining desired variations in plants is, however, less easily achieved. Deliberate or inadvertent crossbreeding has resulted in desirable new cultivars. Hybrid corn (Chapter 3) and Green Revolution cereals (Chapter 2) are outstanding examples of this, but they required sophisticated skills and many years of hard work. Occasional chance mutations have been exploited. The chance mutation of one branch of one orange tree was recognized and perpetuated by grafting buds of that branch onto other trees; the mutant produces oranges without seeds—our navel orange. The red grapefruit was a chance mutation in Texas. The thousands of cultivars of apple, pear, and other fruit trees were obtained in the same way. By selection of mutants and their propagation, we have obtained new plants. They are not new species, for they can be crossed with the parent type and differ from the parent in relatively few genetic characters; most breeds of dogs are cross-fertile—just look around the neighborhood. By selection and reselection, it is possible to obtain plants that do not look much like their ancestral wild types. A doubled and redoubled American Beauty rose looks very little like *Rosa damascena,* and a Pomeranian lapdog doesn't look or act like a wolf. Yet the Pom can mate with a wolf, and the multipetaled rose can cross with the wild species.

If we use one dictionary definition of *monster* as an organism which differs markedly from the normal, and we define "normal" as the usual or the expected, we have deliberately created monstrous plants to serve our needs. We have bred and selected for giant seeds, fruits, flowers, and vegetative organs. We have selected plants for altered or unusual metabolic pathways favoring the accumulation of chemicals we want. We have selected and perpetuated cultivars of banana, grape, sugar cane, and other plants that do not produce viable seeds.

Nowhere is this selection and perpetuation of the monstrous more apparent than in the genus *Brassica* of the mustard (Cruciferae) family. The genus contains perennial, biennial, and annual herbs with close to 100 wild species found in north temperate parts of the world. The family is easily recognized by the characteristic flower, having four petals which form a cross (hence the family name), six stamens of which two are short, and producing a special fruit, the silique. Although it is likely that the cultivated brassicas are derived from several wild species, a reasonable ancestor is the colewort *(B. oleracea)* a scrubby, perennial native to Europe and Asia (Figure 4.1). From colewort, we have selected and developed a wide array of common vegetables. One line resulted in kale (var. *gemmifera*), cauliflower, and broccoli (var. *botrytis*). A second line in the genus was selected for storage organs, giving us the kohlrabi *(B. caulorapa),* whose shortened stem enlarges into an above-ground tuberous vegetable, and the rutabaga *(B. napobrassica),* whose tuberous storage stem develops below ground. The turnip's *(B. rapa)* storage organ is also a tuber, with the true root extending down from the swollen stem. It is questionable whether all the leafy vegetable members

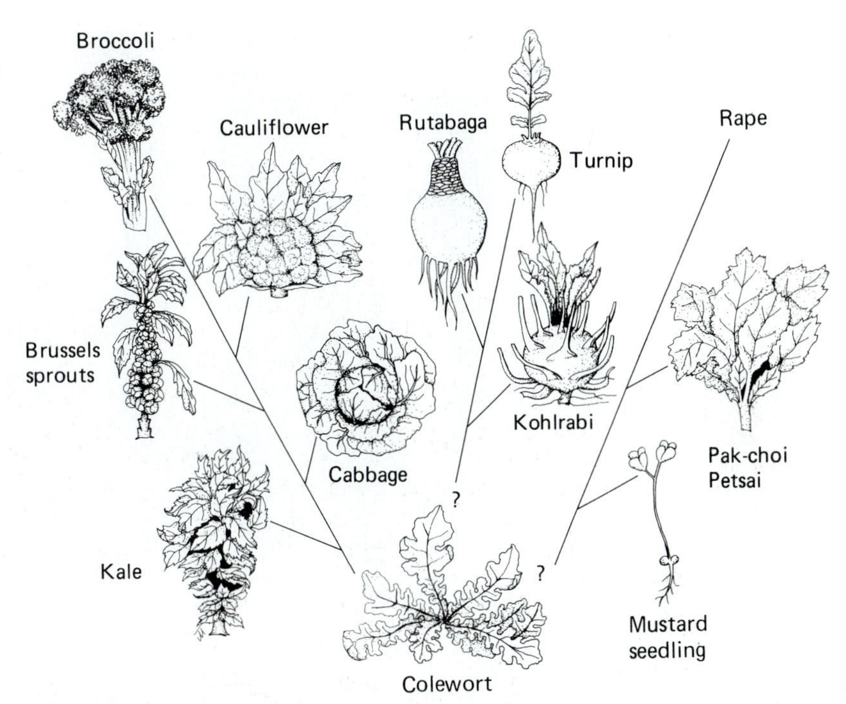

Figure 4.1 Possible evolutionary relationships among species and varieties in the genus *Brassica.*

of the genus are derived from colewort, but there is little doubt that they are all closely related. Several species of mustard are found wild and are exploited for their edible leaves *(B. juncea)* and particularly for their seeds *(B. nigra and B. alba)* used as condiments. Pak-choi *(B. chinensis)* is an important leafy vegetable in the Orient, as are the closely related petsai *(B. pekinensis)* and false pak-choi *(B. parachinensis).* Rape or colza *(B. napus)* is grown primarily for its seeds, which yield an edible oil.

WHAT YOU'RE EATING

Fruits and Vegetables When asked to contrast fruits and vegetables, botanists hem and haw, knowing that scientific definitions bear little relation to common usage of words (Figure 4.2). For botanists, *fruit* is a ripened ovary, its contained seeds, and structures which mature with the ovary, but there is no botanical definition for *vegetable.* We think of vegetables as plant parts eaten during a meal and fruits as parts eaten for dessert. Since people have no trouble with these words when they are used in ordinary speech, it is botanically useful to group plants by the organ or structure eaten.

Roots Roots anchor plants into the ground and serve to absorb and conduct water and minerals from the soil up to the stem of the plant. Many roots are also storage organs, maintaining supplies of photosynthate—sugar—transferred down via the phloem (Chapter 1)

Figure 4.2 Bountiful harvest of vegetables. Courtesy U.S. Department of Agriculture.

to cells where they can be again mobilized for upward transmission via the xylem to the above-ground portion of the plant. Thus roots may be good sources of carbohydrates, but are only poor-to-fair sources of proteins, fats, and vitamins. With few exceptions, roots used for food are taproots; fibrous roots are rarely eaten. Familiar taproots are those of carrots *(Daucus carota)* and parsnips *(Pastinaca sativa)* and celeriac *(Apium graveolens)* in the carrot (Umbelliferae) family; horseradish *(Amoracia rusticana)* and radish *(Raphanus sativus)* in the mustard (Cruciferae) family; and beets *(Beta vulgaris)* in the Chenopodiaceae. All of these plants are biennials, accumulating starch and sugar in their roots during the first year's grown and mobilizing the carbohydrates for energy needed the second year for flower and fruit production.

The sweet potato *(Ipomoea batatas)* in the morning glory (Convolulaceae) family is frequently called a yam, but the true yam *(Dioscorea spp.)* is a member of the Dioscoraceae, a monocot family allied to lilies and amaryllis (see Chapter 3). The sweet potato is a New World plant native to South America, while the yam originated in Southeast Asia. Both are more nutritious than any of the other food roots, containing up to 5 percent protein and unusually high levels of iron, calcium, and minerals. The yellow carotenoids and vitamins A and D are fat-soluble substances which can be used by people. In tropical areas, there has been considerable selection resulting in dry, mealy, and gelatinous cultivars with white-to-almost-red tissues.

In the American tropics, and now in other tropical areas where the white potato cannot grow, the major noncereal carbohydrate is manioc, or cassava *(Manihot esculenta)* of the poinsettia (Euphorbiaceae) family. Also called *yuca, sagu,* or—more familiar to North Americans—*tapioca,* this South American native is an easily cultivated root crop. Manioc roots look something like sweet potatoes, but grow to larger sizes. Cultivars called bitter manioc contain toxic concentrations of prussic acid (hydrocyanic acid) which must be squeezed from the shredded pulp before cooking. Others, called sweet manioc, are directly boiled or baked, much as we use sweet potatoes or white potatoes. Tapioca consists of dried manioc starch (Chapter 3).

A few other roots may be found in the market. Parsnip chervil *(Chaerophyllum bulbosum)* and parsley *(Petroselinum crispum)* of the Umbelliferae family are used in Europe in soups; sugar beets supply sucrose; and root beer is flavored with the roots of *Smilax spp.* Sassafras teas consist of the bark of young roots of *Sassafras albidium,* a plant in the bay leaf (Lauraceae) family.

Tubers, Corms, and Rhizomes One sure way to fail a botany examination is to classify the white potato as a root. There are three kinds of underground stems. A *rhizome* is a creeping, horizontal stem, growing parallel to the soil surface, usually a few centimeters below the ground. We eat a few rhizomes, notably the Jerusalem artichoke. Not from the Middle East but from North America and not an artichoke, the English name is a corruption of the Italian *girasole articiocco,* or edible sunflower, an accurate name because the potatolike rhizome is from a sunflower *(Helianthus tuberosus)* in the aster (Compositae) family. The stored carbohydrate is not starch, but *inulin,* a polymer of the hexose sugar fructose. Although fructose, one of the two components of sucrose, is easily metabolized, humans lack enzymes capable of hydrolyzing inulin, and the Jerusalem artichoke is a poor food source. Its starchy taste is, however, appreciated by people on carbohydrate-restricted diets. The rhizomes of arrowroot *(Maranta arundinacea)* are ground into a flour to make biscuits consumed with apparent delight by young children. Two members of the ginger (Zingiberaceae) family are important rhizome crops. True ginger *(Zingiber officinalis)* is used in Oriental cooking (Figure 4.3) and is dried and ground as a major spice in gingerbread. Turmeric is the major yellow component in curry powders and chutneys and was an important natural dyestuff for cloth.

When gardeners speak of planting crocus bulbs, they are really

Figure 4.3 Ginger *(Zingiber officinalis).* These rhizomes are used in Oriental cooking. Courtesy U.S. Department of Agriculture.

planting *corms,* the enlarged, fleshy, bulblike stem bases of plants. Few corms are seen in North American markets. Water chestnuts *(Elochoris dulcis)* in the sedge (Cyperaceae) family are rarely sold in their fresh—and infinitely more tasty—form, and the canned product is used. The famous poi of Hawaii is a paste made from the corms of the taro or dasheen *(Colocasia esculenta)* plant, (Chapter 3). Some corms, those of arum *(Peltandra virginiana),* Jack-in-the-pulpit *(Arisaema triphyllum),* and skunk cabbage *(Symplocarpus foetidus),* all in the arum (Araceae) family, are gathered by "stalkers of the wild asparagus."

Tubers are defined as thickened, subterranean stems possessing a number of undeveloped buds or eyes. The white potato is the primary example. Tubers develop at the ends of long, thin rhizomes which branch from the main stem of the plant. A close look will show that the potato tuber is distinctly stemlike. The eyes are in a spiral as are the buds of the above-ground stem. The internal arrangement of tissue systems is that of a stem and not that of a root.

Aerial Stems Few above-ground stems are used as food. Some are usually found in supermarkets. Asparagus *(Asparagus officinalis)* in the lily (Liliaceae) family originated in the Mediterranean Basin and is now grown wherever gourmets are to be found (Figure 4.4). The plant is a perennial, with the stems arising from rhizomes. The scales on the stalks are true leaves, but only when too old and tough to eat does the plant produce the fernlike cladophylls—actually stem branches—which are used decoratively. The special taste of asparagus is due to asparagine, a nitrogen-containing compound related to amino acids. Canned white asparagus is produced by mounding earth around the emerging spears to prevent the formation of chlorophyll and to etiolate them, and, as is usual with etiolated tissues, they are very tender and succulent—as well as expensive. A stop at the gourmet section will show cans of bamboo shoots (Figure 4.5). Depending on the species, this grass may have a woody stem which may grow to more than 30 m. Young shoots arising from rhizomes of several genera *(Bambusa, Phyllostachys, Guadua)* are collected, the outer scales removed, and the inner tissues used for Chinese stir-fry dishes, for boiling, and for baking. Bamboo shoots are not particularly nutritious, but they enhance the character of other plants and meats in Oriental dishes. Several stem tissues are gathered for salads, including twigs of *Portulaca,* chervil *(Anthriscus Cerefolium),* pokeweed *(Phytolacca americana),* and roselle *(Hibiscus saddariffa).* Other stems are used as herbs and spices: cinnamon bark, angostura *(Cusparia febrifuga)* bitters, and dillweed *(Anethum graveolens)* will be found in the spice section down the next aisle.

Petioles The structure connecting the leaf blade to the stem is called the *petiole* (Chapter 1). It contains vascular tissues for transmission of materials to and from the leaf, and it also serves to maintain the position

Figure 4.4 Asparagus served with hollandaise sauce. Courtesy U.S. Department of Agriculture.

of the leaf relative to the sun. Although many petioles are eaten as part of leaves, only three have been developed for food. Two of these, celery *(Apium graveolens)* in the Umbelliferae and rhubarb *(Rheum Rhabarbarum)* in the buckwheat (Polygonaceae) family, are common foods. Celery is a Mediterranean plant, but it may also have evolved independently in the Himalaya Mountains. Until the fourteenth century, it was grown for its seeds, which were used in medicine and as a food spice. Selection and breeding has been directed toward production of a long, tender, juicy petiole that has culminated in the Pascal cultivar, which dominates the market. Before Pascal, celery was bleached and etiolated to give the white-to-pale yellow-hearts of celery. The ribs and strings on

Figure 4.5 Young stem tips of bamboo are used in Oriental cooking. Not very nourishing, they add eye appeal and crispness to food. Courtesy U.S. Department of Agriculture.

the convex side of the stalk are not vascular bundles (these are small dots inside the stalk), but are bundles of strengthening cells called *collenchyma* which have thickenings of cellulose in each corner of the cell to aid in supporting the elongated petiole (Chapter 1). In celery, the base of the petiole is enlarged where it joins the short, condensed stem. When this anatomical feature is carried to extremes, the enlarged clasping petiolar bases form a bulblike structure, as seen in fennel *(Foeniculum vulgare),* a member of a closely related genus. Fennel seeds, like those of celery, cumin *(Cuminum cyminum),* coriander *(Coriandrum sativum),* dill *(Anethum graveolens),* and chervil *(Anthriscus Cerefolium)* are umbelliferous culinary herbs, but one fennel cultivar, known as Florence fennel or *finocchio,* has been selected for large petiolar bases and is eaten raw in salads or boiled.

Rhubarb originated in Asia, and one species was grown in China for its rhizome, which contains a gentle laxative. Dried rhubarb rhizomes were as valued in Renaissance Europe as were the fabled spices of the Orient. Only during the seventeenth century was the petiole found to be edible and, at the same time, the leaves shown to be poisonous. The toxic materials include oxalic acid, plus other poorly characterized compounds. The astringent taste of rhubarb sauce or rhubarb pie is due to the high acid content of the stalks, and a pinch of baking soda (sodium bicarbonate) added to the cooking water will reduce this tart taste without destroying the flavor of the rhubarb.

Although formerly restricted to the northeast part of North America, canned fiddlehead fern, young fronds of the ostrich fern *(Matteuccia pensylvanica),* can now be purchased (Figure 4.6). Tasting vaguely like asparagus, these young petioles plus immature leaves are boiled and served with butter and salt.

Apical Meristems As discussed in Chapter 1, the stem tip is composed of an apical meristem with its young leaves. The most obvious food apical meristem is the cabbage head, whose anatomy is a perfect example of an apical meristem. Brussels sprouts are undeveloped axillary buds, but their anatomy is also that of an apical meristem. Head lettuce, the most popular salad plant in North America, is *Lactuca sativa,* a member of the aster (Compositae) family. The plant is a biennial, native to Eurasia. It is depicted in Egyptian tombs dated to

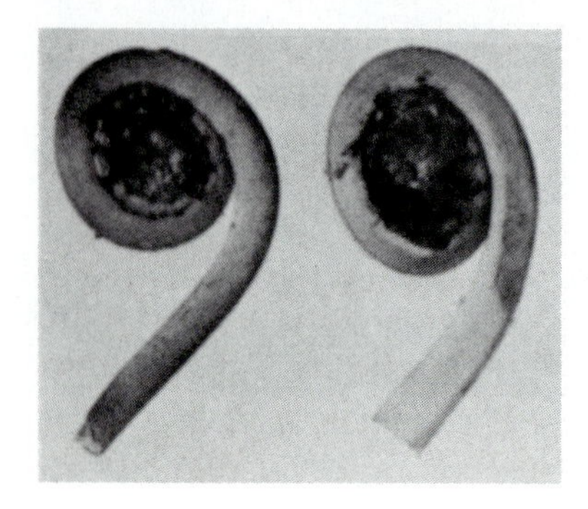

Figure 4.6 Young fiddlehead fronds of ostrich fern *(Matteuccia pensylvanica).* Courtesy U.S. Department of Agriculture.

6500 B.P., was grown in China by 2800 B.P., and was eaten in Japan by 2700 B.P., although it was not cultivated in Europe until the Middle Ages. The early cultivars were the leaf types, with crisphead cultivars selected out by the seventeenth century. It is the heading types which are apical meristems surrounded by enlarged, overlapping leaves.

Leaves Many leaves are used as herbs, and these represent a number of botanical families. The mint (Labiatae) family is especially well represented by oregano, marjoram, basil, thyme, and the savories. Watercress, curly cress, and winter cress are all in the mustard family, as are tarragon and wormwood, used as herbs in wines and vinegars. Bay leaves *(Laurel nobilis),* the universal garnish, parsley, and several herb teas are made from leaves. Vegetables include members of the cabbage group, spinach *(Spinacia oleracea)* in the goosefoot (Chenopodiaceae) family, and the leaves of endive *(Cichorium intybus)* in the aster family, whose roots, as chicory, are used to flavor Louisiana-style coffee.

Members of the lily (Liliaceae) family are among our most common vegetables. Onions, leeks, shallots, and chives *(Allium spp.)* are all true bulbs, defined as shortened, usually underground stems with fleshy overlapping leaves (Figure 4.7). The relation of a bulb to an apical meristem is seen upon dissection; the younger leaves are higher up on the compressed stem than the older leaves. In leeks, chives, and green onions (scallions), the edible portion is the fleshy bases of the strap-shaped leaves, which form a cylinder. Garlic, cloves, and shallots consist of closely attached small bulbs or bulbils, each composed of two leaves. The thin cylindrical leaf is surrounded and protected by a surrounding, larger, and fleshier leaf, the whole surrounded by a papery epidermal

Figure 4.7 Onions, scallions, garlic, and shallots. Courtesy U.S. Department of Agriculture.

layer. The characteristic odor of *Allium* bulbs is due to an organic sulfide, allicin, which is moved in the blood stream to the lungs where it is exhaled—to the annoyance of one's neighbors—but how anyone can cook without the full array of *Allium* species is beyond comprehension.

With high concentrations of water-soluble B vitamins and minerals, leafy vegetables have high nutritional value and outstanding tastes. Some leaves are eaten in restricted parts of North America and are either ignored or made fun of in other regions. People in the southern United States esteem collard, turnip, beet (Swiss chard), and mustard greens, usually boiled with bits of bacon or fatback, and the "potlikker" drunk as a soup. Young dandelion *(Taraxacum officinale)* in the aster family is esteemed in the Northeast; nettle *(Urtica spp.),* amaranth or Chinese spinach *(Amaranthus spp.),* and seakale *(Crambe spp.)* were introduced by immigrants from eastern Europe to the prairie states and provinces.

Flowers and Their Parts Entire flowers are used as delicate flavorings; jasmine tea does contain jasmine blossoms. The unusual crystallized violets made in France are whole flowers soaked in a sugar solution which is then allowed to crystallize. Some flowers are used as spices; the clove is a flower bud, as is the caper *(Capparus spinosa)* of the aster family. The most expensive flavoring is saffron, the stamens of *Crocus sativus.* Flowers of a chrysanthemum, *Pyrethrum,* are collected in Japan and east Africa for their ovaries which contain an effective insecticide. Cauliflowers (Figure 4.8) and broccoli are the entire inflorescence of the plant, and one eats both the flowers and their fleshy peduncles or flower stalks. The globe artichoke is the floral bud of *Cynara scolymus,* a thistle in the aster family (Figure 4.9). The edible portion is the bracts surrounding the floral receptacle, plus the receptacle itself. Definitely an acquired taste, the bracts are dipped into melted butter or mayonnaise and their fleshy bases eaten. The inedible choke consists of the true floral parts and is anatomically similar to a sunflower head, but the fleshy floral receptacle is considered to be a delicacy, frequently served as artichoke "hearts."

Figure 4.8 Cauliflower *(Brassica oleracea* var. *botrytis)* is a cabbage with style. Courtesy U.S. Department of Agriculture.

Figure 4.9 Globe artichoke *(Cynara scolymus)*. Courtesy U.S. Department of Agriculture.

Fruits The simple definition of *fruit* given at the beginning of this section did not attempt to cover the structural variety of fruits that are found in the supermarket's produce section (Figure 4.10). A fruit, botanically, is usually the ovary wall and its contents. The wall is the *pericarp*, which has three cell layers—the outer *exocarp*, a middle *mesocarp*, and an inner *endocarp*. The base of the flower, the receptacle or *torus*, frequently contributes to the fruit and its contents. Seeds consist of an embryo, food storage tissues, and the seed coats, the testa. For convenience and to define the developmental processes by which fruits assume their mature forms, botanists have attempted to classify the almost infinitely broad range of fruiting structures. For our purposes, the excellent classification scheme developed by Dr. Arthur Cronquist will be used.

Figure 4.10 Bountiful harvest of fruits. Courtesy U.S. Department of Agriculture.

Figure 4.11 Berries are fruits enclosing one or more seeds. The exocarp layer forms the fruit "skin," and both mesocarp and endocarp layers develop into the edible flesh.

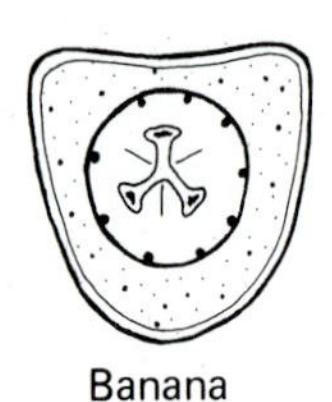

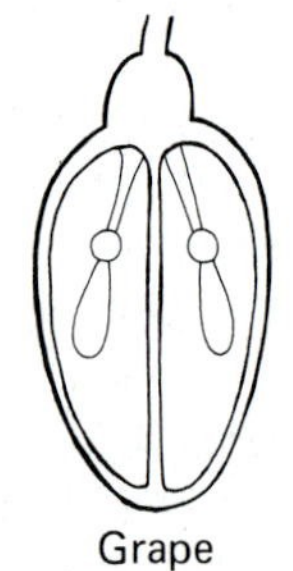

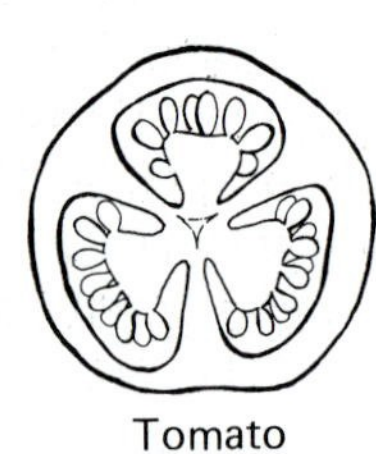

I. *Simple Fruits:* fruit derived from a single pistil or ovary
   A. Fleshy Fruits: fruits in which the pericarp cells remain alive
      1. Berries: pericarp fleshy, occasionally leathery, one or more carpels and seeds
         a. Berry: exocarp forms skin (banana, members of tomato family, date, grape) (Figure 4.11)
         b. Pepo: rind from receptacle (all members of cucurbit family)
         c. Hesperidium: rind is exocarp + mesocarp, juicy tissue is endocarp (citrus fruits)
      2. Drupe: fleshy fruit with stone; exocarp forms skin, mesocarp forms flesh, and endocarp forms woody stone surrounding seed (peach, apricots, olive, coconut) (Figure 4.12)
      3. Pome: edible flesh from enlarged receptacle, core of pericarp (apples, pears)
   B. Dry Fruits: pericarp cells dead at maturity
      1. Dehiscent fruits which open at seams at maturity
         a. Legume: fruit opens along 2 seams (peas, beans)
         b. Follicle: fruit opens along 1 seam (star anise, milkweed, kolanut)
         c. Capsule: several carpels, several seams (okra, vanilla bean)
         d. Silique: long, narrow capsule bearing seeds on margins (mustard seed)
      2. Indehiscent fruits which remain closed at maturity
         a. Acheme: one-seeded fruit, pericarp dry and free from seed (buckwheat) (Figure 4.13)
         b. Caryopsis: one-seeded fruit, pericarp dry and adhering to seed (cereal grains)
         c. Nut: like achene with woody pericarp (hazelnut, acorn)

II. *Aggregate Fruits:* fruit derived from fused ovaries and receptacles
   A. Etaerio of drupes (blackberry, raspberry)
   B. Etaerio of achenes: enlarged, fleshy receptacle bearing achenes (strawberry) (Figure 4.14)
   C. Sorosis: composite fleshy fruit composed of spike (core) + floral parts (pineapple)

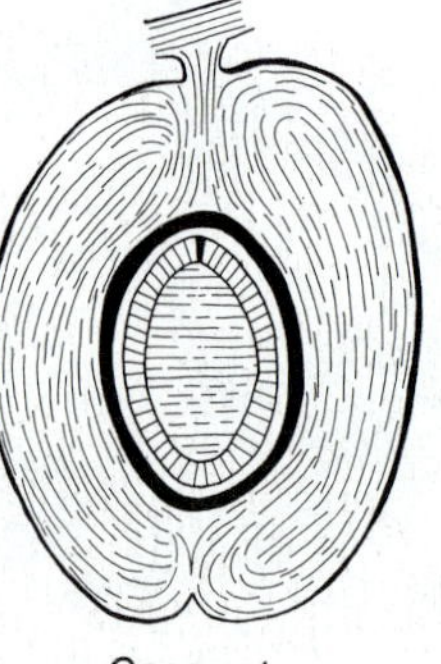

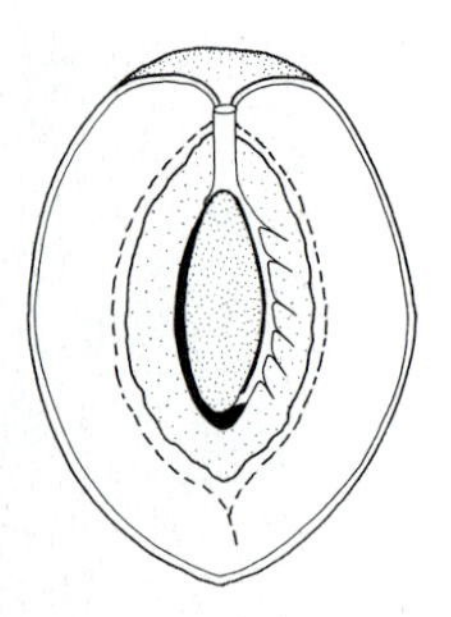

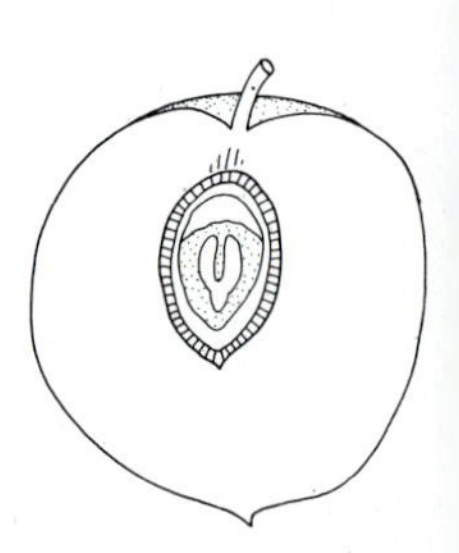

Figure 4.12 Drupes are fruits with a stone. The exocarp forms the fruit "skin" with the edible flesh developed from the mesocarp. The endocarp forms the woody case surrounding the seed. The coconut is a drupe with woody carpic tissues.

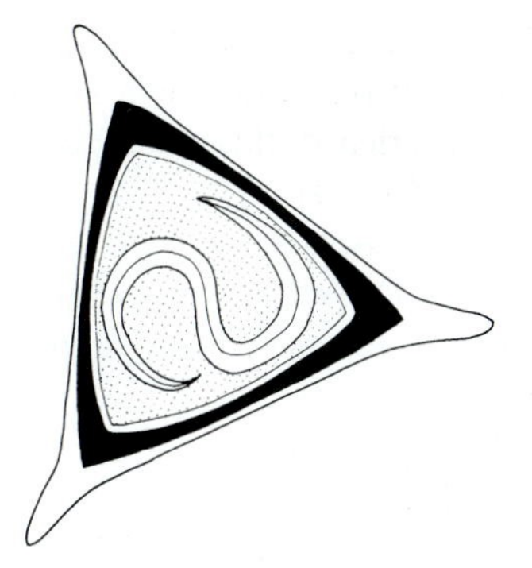

Figure 4.13 A buckwheat *(Fagopyrum esculentum)* achene. The fused carpic layers form the seed case enclosing the curved embryo surrounded by starchy endosperm.

D. Syconium: fleshy fruit composed of a swollen, hollow receptacle containing many flowers (figs)

Some fruits can be included in this classification with difficulty. The pomegranate is a berry with a leathery pericarp; the edible pulp surrounding each seed is a fleshy, juicy testa or seed coat. Rose hips, used for jam and an excellent source of vitamin C, are etaerios of achenes enclosed within a fleshy receptacle.

Considerable confusion exists among those fruits called nuts (Figure 4.15). The peanut seed is, of course, a bean. Several so-called nuts are botanical drupes. Almonds, cashews, coconuts,and macadamia nuts are the stones of the drupe, usually sold after the dry pericarps have been removed. The Brazil nut fruit is a capsule containing seeds whose testa forms the difficult-to-remove coat. Chestnuts, hazelnuts, and the Oriental litchi are true nuts, but in litchi, the delicious flesh is an aril, a special layer of cells surrounding the seed and is botanically comparable to the aril of nutmeg which we use as mace, or the edible flesh surrounding a pomegranate seed.

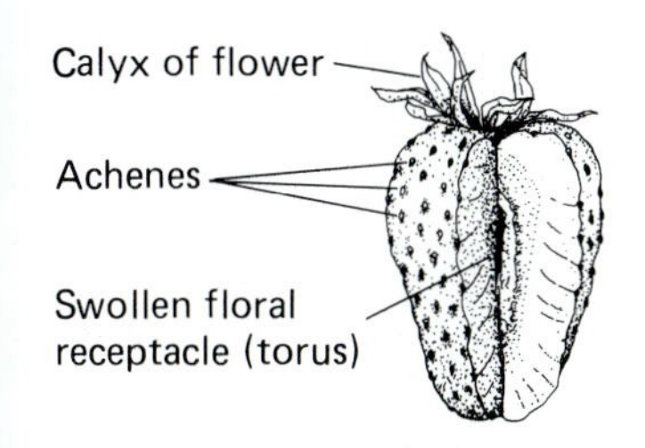

Figure 4.14 Strawberry *(Fragaria spp.)*. The edible "berry" is a false accessory fruit composed of the enlarged floral receptacle; the "seeds" are achenes.

ADDITIONAL READINGS

Bianchini, F., and F. Corbetta. *The Complete Book of Fruits and Vegetables.* New York: Crown, 1975.

Brouk, B. *Plants Consumed by Man.* New York: Academic Press, 1975.

Figure 4.15 A selection of "nuts," most of which are not true nuts. Courtesy of U.S. Department of Agriculture.

Edlin, H. L. *Plants and Man: The Story of Our Basic Foods.* Garden City, N.Y.: Natural History Press, 1969.

Forster, R., and O. Ranum (eds.). *Food and Drink in History.* Baltimore: Johns Hopkins University Press, 1979.

Harrison, S., G. B. Masefield, and M. Wallace. *The Oxford Book of Plants.* Oxford, England: Oxford University Press, 1969.

Hyams, E. *Plants in the Service of Man: 10,000 Years of Domestication.* London: J. M. Dent, 1971.

Lovelock, Y. *The Vegetable Book: An Unnatural History.* New York: St. Martin's Press, 1972.

Martin, F. W. (ed.). *CRC Handbook of Tropical Food Crops.* Boca Raton, Fla.: CRC Press, 1984.

North, C. *Plant Breeding and Genetics in Horticulture.* New York: Wiley, 1979.

Schery, R. W. *Plants for Man.* 2nd Ed. Englewood Cliffs, N.J.: Prentice-Hall, 1971.

Simmonds, N. W. *Evolution of Crop Plants.* London: Longman, 1976.

Simon, A., and R. Howe. *A Dictionary of Gastronomy.* London: T. Nelson, 1971.

Staff of the L. H. Bailey Hortorium of Cornell University. *Hortus Third: A Concise Dictionary of Plants Cultivated in the United States and Canada.* New York: Macmillan, 1976.

Sugar

Sugar and spice and everything nice,
That's what little girls are made of.

Nursery Rhyme

Although people talk about a "sweet tooth," it is small enervated buds on the tongue that perceive the five basic tastes: sweet, salty, acid, bitter, and metallic. A child wisely licks a peppermint stick, knowing that the sweet-taste buds are at the tip of the tongue. Sugars are not the only substances that activate the taste buds; they can distinguish between sweet, sour, acid, and bitter tastes. All substances chemically designated as sugars are composed of carbon atoms plus hydrogen and oxygen atoms in the two-to-one ratio in which they appear as water. This provides the generic name carbohydrate. The most abundant sugars include two monosaccharide sugars with six carbon atoms each: glucose and fructose; and the 12-carbon disaccharides composed of either two glucoses (maltose) or one glucose and one fructose (sucrose) (see Figure 4.38). Two macromolecules of glucose, with hundreds to thousands of glucoses in a long chain, are polymers (see Figure 4.39). Starch and cellulose are polysaccharides. Sugars are excellent sources of energy. Indeed, they are the primary energy sources for all living cells. The chemical energy of the mono- and disaccharides is almost immediately available, while that in starch becomes available more slowly as the long polymer is broken down into monosaccharides. If sugars are consumed in excess of daily energy requirements, the animal body converts them into storage fats. Sugars are, however, no more fattening than are any other foods eaten to excess.

Worldwide, the two most important sources of table sugar (sucrose) are the sugar cane and the sugar beet. Together they account for well over 90 percent of commercial sugar. There are, however, several other plants that are capable of providing sucrose.

HONEY

Until the eleventh century, the sweet tooth of most of the world's people was satisfied with honey, supplemented in North America by maple syrup. Honey is a bee-concentrated nectar of flowers—Aristotle called it dew distilled from the stars and rainbows—containing a mixture of 60 percent sucrose, 20 percent glucose, and 20 percent fructose plus nutritionally insignificant amounts of amino acids, minerals, and vitamins, and small amounts of ill-characterized compounds that distinguish sage or apple honey from that of thyme or clover honey. As evidenced by "a land flowing with milk and honey" from Exodus 12:8, it was highly prized and was a symbol of a peaceful, plentiful agricultural community. Bees are kept in Europe, the Americas, Africa, and Asia and the honey used directly as sweetener or fermented into sweet-tasting mead or *methgelyn* beers. Today, honey production is a moderately sized industry with 300,000 tons produced each year.

PALM SUGAR

The sap of several species of palm, including the coconut palm *(Cocos nucifera)* and the date palm *(Phoenix dactylifera),* bear large flower stalks that can be tapped to yield a thick, sweet sap with yields up to a gallon a day. This sap can be concentrated and the sugar crystallized out or remain as a syrup concentrate. In India, a wild date *(Phoenix sylvestris)* has its trunk tapped to yield a 10 percent sugar solution called *neera,* which is concentrated and crystallized into crude lump sugar *(gur).* A fermented toddy or beer can be made from the raw sap; this *tara* can be about as strong as wine, close to 14 percent alcohol by volume.

MAPLE SYRUP

No book by a Vermont resident could omit mention of maple syrup and maple sugar; one could be exiled for this sin of omission (Figure 4.16). In early spring, on a bright, sunny day after a cold night, the sun warms

Figure 4.16 Collecting sap from sugar maple tree (*Acer saccharum*). Courtesy U.S. Department of Agriculture.

the dark bark and young wood of the sugar maple tree *(Acer saccharum),* permitting enzymes to convert starches in ray cells into soluble sugars which move up the tree, providing the energy source for the eventual bursting of winter buds. The American Indian learned to subvert sap movement by gashing the trunk and collecting the clear sap, which contains up to 6 percent sucrose. By dropping hot rocks into a birchbark container of sap, the sugar was concentrated and caramelized to produce a brown syrup with close to 68 percent sugar content. Additional concentration resulted in the formation of coarse brown crystals of maple sugar. *Sisibaskwat*—"drawn from wood"—was eaten directly and used to season meats and vegetables. Some was traded to other tribes and, eventually, to settlers. Colonists improved on the Indians' methods by boring holes in the trees, inserting spouts of hollow elderberry twigs, and boiling down the sap in kettles. As a rule of thumb, it takes close to 40 gallons of sap to produce one gallon of syrup. Typical maple taste and odor develops only when the sap is concentrated by heat. The mystique of wooden sap buckets, ox-drawn sledges towing sap down a hill, and steamy sugar houses warm from rock maple fires under boiling sap caldrons has been lost. Today, electric pumps pull the sap from the trees through blue plastic tubing directly to stainless steel, oil-fired evaporating pans from which syrup flows into plastic bottles whose labels bear pictures of nineteenth-century sugaring operations.

The grade, and hence the price, is determined by color, with light amber syrups with delicate flavors commanding more interest than do darker, more caramelized syrups with more pronounced flavors. One tradition, still observed, is the sugar-on-snow party at the end of the season, when hot maple syrup is poured on fresh snow to be sopped up with raised, unglazed doughnuts; the cloyingly sweet taste is cut by bites of dill pickle. Most commercial maple syrups are compounded of cane or corn syrup to which a small volume of real maple syrup is added for flavoring, but even this token can be eliminated by adding an extract of fenugreek seeds *(Trigonella foenumgraecum),* a leguminous plant whose essential oils closely mimic the taste of maple syrup.

SORGHUM

Grasses closely related to maize, the sorghums, have been in cultivation for millennia, primarily as grain crops (see Chapter 3). The sweet sorghum, or sorgo *(Sorghum bicolor var. saccharatum),* has considerable sucrose in its juicy stems, which can be collected in the same way as the sap of sugar cane. The sugar is rarely crystallized, but is used as a light yellow-brown syrup or a dark-brown molasses. Most sorghum syrup is consumed in the Middle West and in southern states.

SUGAR CANE

The bulk of the world's sugar is obtained from two plants, the sugar beet *(Beta vulgaris)* and the sugar cane *(Saccharum officinarum).* Of the two, sugar cane accounts for about 60 percent of the total supply. *S. officinarum* is probably a cultivated hybrid of *S. spontaneum, S. robustum, S. sinense,* and *S. barbari.* All have been in cultivation for so long that we can say only that sugar cane originated somewhere in the South Pacific, possibly in New Guinea, and that it was carried to the Asian mainland by humans. No mention of cane appears in the *Veda* of 3000 B.P. but it was noted in other Indian writings of 1900 B.P. Sugar cane is a tall member of the grass family (Gramineae) capable of reaching 8 meters in height, with a stalk diameter of 10 centimeters. The stalk has a pithy center in which the sugary cell sap accumulates in cell vacuoles. At each stalk joint or node, broad leaves form. Seeds are produced in tassels and are sparingly fertile, but the plant is so genetically mixed (heterozygous) that stand uniformity can be assured only by vegetative propagation from sections of the stalk which, when planted, sprout from buds (axillary buds) at each joint. Modern cultivars, so-called noble canes, have been genetically engineered for optimum yields under varied conditions of cultivation, although a major restriction is a requirement for tropical or subtropical climates with abundant rainfall or modern irrigation. The plant matures in 12–15 months, but since new stalks, *ratoons,* develop from the rootstock, one planting may produce crops for a number of years. At harvest time, pith cells will contain close to 15 percent sucrose, all of it formed through the biochemical process of photosynthesis.

The exploitation of sugar cane started early in the history of agricultural development. By 2850 B.P. it was a staple crop in India; the Manu code provided that any traveler could take two stalks from a cane field to gain strength for the journey. Joints of cane were chewed to get the sweet sap, and, by about 2800 B.P., crude presses squeezed out the sap to be boiled down, producing a lumpy, dark-brown mass, much like dark grades of brown sugar. Small lumps were called *sakkara* (gravel) and larger lumps were called *khanda* (pebbles), from which we derived the word *candy*. Sherbets of dissolved sakkara and rose water were frozen for the rajah's table with snow carried hundreds of miles from the mountains; it was believed that this was excellent medicine to increase production of semen. China and Japan obtained cane by the third century, and the Tang Dynasty emperor, Tsai Tsung, sent agents to India to learn the secrets of making sakkara. Nearchus, general to Alexander the Great, reported that on an expedition to the Indus Valley he saw "the sweet juice of a reed which is called sakkar." Dioscorides wrote in the first century of "a sort of hard honey which is called sakkarum found in canes or reed in India, which is grainy like salt and brittle between the teeth, but of sweet taste." Sugar cane did not reach the Middle East until 550, when Persians grew it in irrigated fields. When Persia was conquered by followers of Mohammed in the early seventh century, cane was introduced into Syria, the Mediterranean islands of Sicily and Cyprus, and Egypt. Dark-brown crystalline sugar was made in Egypt about 750, and small amounts found their way into Europe, where it was a delightful but very expensive spice and medicine for coughs and stomach complaints. The crusaders saw, in the fields of Laocidia (western Syria), "certain reed-like plants . . . on account of its sweet juice we chewed against thirst," and when Jerusalem was captured and sacked in 1187, Arabian sugar was a valuable portion of the booty brought back to Europe.

By the middle of the twelfth century, there was sufficient demand for *sucra* to warrant the attempts of Venetian merchants to establish a trade center for the movement of sugar from North Africa into Europe. In about 1410, a group of merchants purchased a method for refining crude brown sugar into a sparkling white product that could be molded into cones or loaves (Figure 4.17). A Venetian monopoly was established. In 1413, Venice sent 100,000 pounds of sugar to England in exchange for British wool. Prices were so high (about $3.50 an ounce) that few people could afford it; the common people ate honey and tasted sugar only when it was prescribed to disguise bitter herb medicines (Figure 4.18).

Although cane was grown in Spain, the Moors so completely controlled its supply that Spanish sugar went first to Venice before being distributed to the rest of Europe. In 1420, Henry the Navigator, ruler of Portugal, obtained planting stock from Sicily and established a sugar plantation on the island of Madeira, on the Canary and Cape Verde islands, and on St. Thomas Island off the African coast. Columbus

Figure 4.17 Molding sugar loaves. Taken from a fifteenth-century woodcut.

brought cane to Hispaniola on his second voyage in 1493, and the first crop was harvested in 1509. The downfall of the Venetian monopoly was foreshadowed when the Portuguese established plantations in Brazil in the early sixteenth century. The conquest of tropical America was immediately followed by conversion of lush forests into sugar plantations—all of the West Indies became sugar-growing land. Impetus for expanding the sugar trade was the introduction of Arabian coffee, Chinese tea, and Aztec cocoa, all of which were more palatable with sugar. Catherine de'Medici, queen of Henri II of France, introduced Florentine pastry making and Venetian "sweetmeats" to the French court, and her contemporary, Elizabeth I of England, was so fond of sugar that her court is credited with inventing sweetened fruit pies. The addition of sugar to chocolate and the invention of marzipan, glazed fruits, and even crystallized violets hastened the development of the then crude art of dentistry.

Figure 4.18 A sugar merchant of the fifteenth century. Taken from a contemporary woodcut.

Sugar Slaves With a firm market, enterprising businessmen began to increase the volume of refined sugar produced each year. Spanish and Portuguese planters took seriously the admonition of Pope Nicholas V to expand their operations in order to "reduce to perpetual slavery the Saracens, pagans and other enemies of Christ." They grew enough sugar to eliminate the Arabian market. But in order to do so, they needed cheap labor. Sugar did not create slavery; it is sanctioned in the Testaments (Leviticus 25:44–46) and the Golden Age of Greece was built on slave labor. African slaves were brought to Brazil by the Portuguese in 1519 because the native Indians were unwilling workers and often died when put to the lash. Spain followed in a few years, and Britain imported slaves into Barbados in 1630. By the end of the seventeenth century, good Catholics were wondering aloud why Dominican, Augustinian, and particularly the Jesuit orders, dedicated to poverty, should be accumulating wealth and power from the labor of slaves. True, the blacks were baptized, thus saving their souls, but their bodies were worn out within a few years. The Jesuits were expelled from France in 1767, partly because of their use of slave labor, while the Compagnie des Indies was shipping 300 vessels packed with slaves each year to plantations in the Antilles and to Jesuit-founded plantations in Louisiana. Because of black slaves, Spanish and French sugar underpriced English sugar until almost the end of the eighteenth century. The tremendous traffic in slaves was justified according to most planters. A cost accounting in 1700 demonstrated that self-perpetuation of slave populations was not economical—infant mortality was well over 50 percent and a newborn slave had to be fed and clothed and occupied the attention of its mother until the child was 6 or 7 years old and capable of working in the cane fields. Slaves were brought to the sugar plantations until the early nineteenth century, when, in revulsion, the world demanded that the slave trade be abolished.

With the demise of the slave trade, some indigenous slavery continued in the sugar islands, but the demand for cheap labor was unabated. In an effort to obtain needed field hands, Spain, Britain, and the United States engaged in what we now realize was the greatest unforced mass migration of peoples the world has ever seen. By act of Parliament in 1837, people from India were brought as indentured laborers to the West Indies. Close to 1.5 million people immigrated to the British sugar colonies. A quarter of a million Chinese were also brought in by English planters. The Spanish in Cuba introduced Chinese workers; a long-term benefit was a cuisine that included Chinese-Cuban dishes. The Dutch imported Indians in 1800, and some Japanese and Indonesians in 1890.

Hawaiian Sugar Among the more remarkable mass-migration schemes was the one in Hawaii. Sugar cane existed on the islands for centuries; Captain James Cook ate cane in 1778 before he was killed. The first attempt to grow cane commercially was in 1803, but no market existed and the venture failed. At about the same time, a group of Calvinist missionaries left Boston to escape the "progressiveness" of the Congregational Church and settled in Hawaii to convert the heathen and build a new life. After successfully converting the rulers of the islands, they introduced heavy clothing (with drastic increases in tuberculosis and skin diseases) and Yankee concepts of land tenure. The bulk of the arable land was soon owned by misionary families. The Rev. Joseph Goodrich activated the sugar industry in 1829, and by 1840, the Rev. John Emerson, the Rev. Artemas Bishop, the Rev. Hiram Bingham, and several medical missionaries were in the sugar business. Abhoring slavery, the missionary monarchy decided to import labor (Figure 4.19), since "the native Hawaiian won't work and they can't be trusted." Chinese were imported in 1852 in large numbers, followed by Japanese,

Portuguese, Koreans, and Filipinos. Caucasians were brought in as supervisors and engineers, initiating the multicultural, polyglot society that we think of as typically Hawaiian. These successive waves of immigrations were, however, carefully orchestrated by the planter aristocracy to prevent any group from dominating the work force and demanding higher wages and benefits. Small acreages are still leased on a year-to-year basis by the Bishop estate mostly to immigrant families, while the plight of the natives continues to worsen as they are excluded from virtually all aspects of the economic and political life of what was once their land.

SUGAR BEETS

Even before sugar cane was grown commercially in Hawaii or the Philippines, a new development in sugar occurred. In 1605, a Frenchman, Oliver de Serres, wrote: "The beetroot, when boiled, yields a juice similar to syrup of sugar," but it was not until 1747 that the

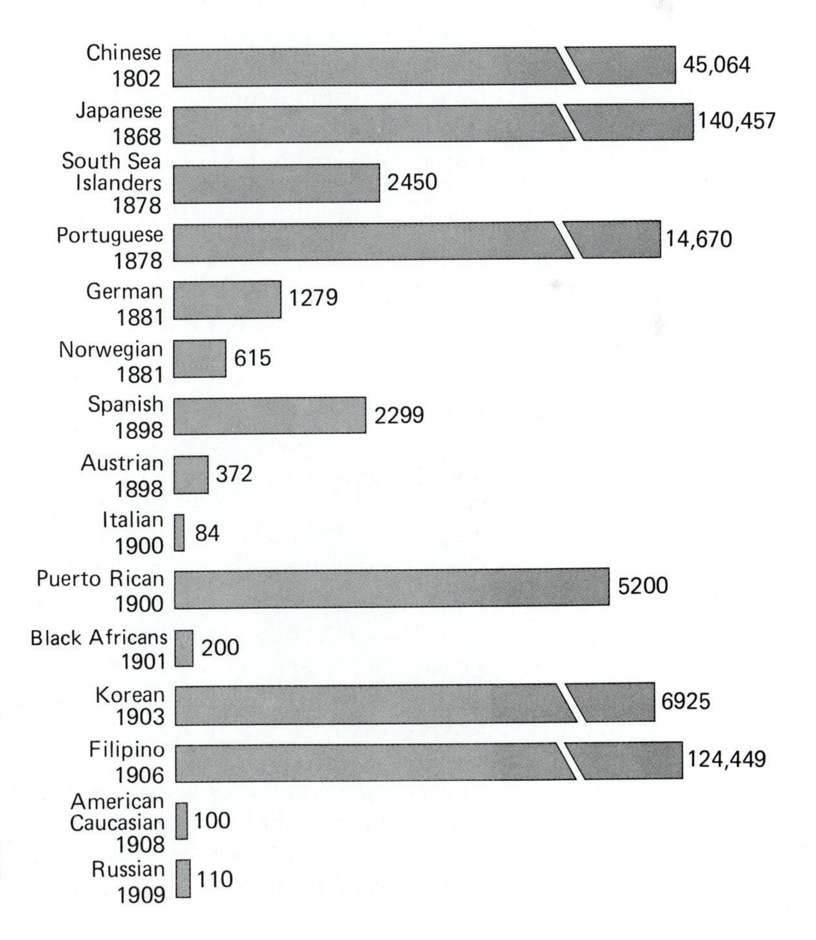

Figure 4.19 Major racial groups in Hawaii and the dates when immigration started.

German chemist Andreas Margraff proved that the beet juice was chemically identical to sugar from the cane (Figure 4.20). Margraff's student, Franz Karl Archard, planted a field of beets under the patronage of Frederick the Great and, after a successful harvest, calculated that a good grade of sugar could be obtained for less than $0.06 per pound. Authorities in France and Germany were very interested. They were paying that much and sometimes more for cane sugar produced by their economic and sometimes political rivals, the English. British sugar merchants also became interested and tried, unsuccessfully, to buy Archard off with the equivalent of $120,000. King Frederick William III supported construction of the first sugar beet factory in Silesia in 1800–1801. When news of this success reached France, Napoleon ordered beet cultivation, establishment of schools to train sugar technicians, and large-scale production of sugar from beets to break England's tight blockade. Following the battle of Waterloo and the lifting of the blockade, the infant industry collapsed not only because cheap cane sugar again became available to France but also because beet sugar was then a low-grade product with an unpleasant smell and chefs complained that it didn't "handle" the same way as cane sugar did. Germany persisted and, with its splendid agricultural schools and well-organized chemical industry, succeeded in preparing sucrose from beets with reasonable yields, low costs, and a quality only slightly below that of cane sugar. From 1870 until the start of World War II, Europe obtained a third of its sugar from Germany.

Unknown to Europeans, wild beets *(Beta maritima)* in California had, for centuries, supplied, in the words of Pedro Fages, a Spanish explorer of the 1760s, "a quantity of molasses, candy and sugar that is not unworthy." With the success of Germany and France in beet sugar production, it was only a matter of time until it was tried in the United States. Attempts were made in Massachusetts in 1838 to develop

Figure 4.20 Mature sugar beet. Courtesy Holly Sugar Corp.

economically viable enterprises, but lack of technical skills and a series of untimely mischances caused these to fail. The first successful operation was in Alvarado, California, in 1869. By 1890, extensive plantings and efficient factories were in operation in California and in the intermountain areas of Colorado, Utah, and Nevada under the financial control of the Church of Latter Day Saints. Congress aided this infant industry by authorizing a cent-a-pound bounty on all sugar produced within the continental United States and by mobilizing the resources of the Department of Agriculture to develop beets with higher sugar content and resistance to diseases.

REFINING SUGAR

The technology of refining sugar has come a long way from the primitive boiling down of sap extracted from lengths of cane. Under modern, intensive plantation practices, the mature sugar cane plants are mechanically topped, leaves removed mechanically or by setting the field on fire, and stalks harvested by machine or by the hot, dirty, and hard job of slashing with machetes (Figure 4.21). Since sugar content decreases rapidly after cutting, harvested cane is rushed to mills, where it is shredded and fed into serrated rollers capable of squeezing the juice out of 3000 tons of cane a day. The spent stalks, called *bagasse,* are used as a fuel for the mills, for paper, for cattle food, and in the manufacture of plastics. Raw juice contains, in addition to sucrose, protein, fat, and other chemical constituents of plant cells, all of which must be removed. Lime and mild heat clarify the juice by precipitating these impurities. The clear juice is passed through filters under pressure, and concentrat-

Figure 4.21 Idealized view of "Cuba, the Great Sugar Industry," from Frank Leslie's *Illustrated Newspaper of 1880.*

ed by evaporation under heat and reduced pressure to produce a thick syrup containing sugar crystals and molasses, a dark-brown mixture of caramelized sugars, and other substances. Massive centrifuges spin off the molasses (which is reworked to extract more sugar), and sucrose crystals are redissolved and recrystallized several times to produce refined sugar. Residual molasses can be used as cattle feed supplements, but much of it is used to produce rum (see Chapter 8).

Extraction of sucrose from sugar beets differs from that used for cane only in the procedure to obtain the crude juice. Cells and tissues of the beet are tough and cannot easily be broken by crushing, so that it is necessary to shred the beets. The shreds are placed in tanks of hot water, allowing the sweet sap to diffuse from the cells. Clarification, crystallization, and refining steps are essentially identical to those used in the sugar cane industry. Beet molasses and spent beetroot pulp are used as cattle fodder. Modern refining of beet sucrose has resulted in a product chemically, physically, and nutritionally indistinguishable from cane sucrose.

Table sugar is the purest bulk chemical produced, and its grading is based solely on the size of the sugar crystals, which ranges from large, rock-candy lumps to confectioners' XXXX or icing sugars that are as fine as flour. Brown sugars range from the almost white #1 grade through grade #15, a deep, brown-black product. The golden through medium browns, used in cooking, have increasing intensities of molasses flavor, but are not nutritionally superior to white sugars—they are sometimes made by adding molasses to white sugar. Liquid sugar syrups have high acceptance in baking and soft drink industries because the sugar is already in solution and also because they contain small amounts of glucose and fructose, which repress sucrose crystallization. In jelly and jam processing, boiling acid fruits with sugar also hydrolyzes some sucrose into its component monosaccharides, serving the same purpose.

SUGAR LEGISLATION

Sugar is very big business. Modern cane mills which produce raw, 96 percent pure sugar cost about $75 million, and an equivalent sugar beet factory costs over $50 million; plants that refine sugar cost about the same. The refining business is capital-intensive, requiring highly skilled workers and management in addition to the financing of large inventories of both raw and refined sugars. Current world consumption is 80 million metric tons per year with per capita use in the United States and Canada—nations with a very large sweet tooth—being close to 40 kg per person per year. With so many countries deeply committed to sugar production, sometimes to the virtual exclusion of other crops, regulation of supply and demand to stabilize prices is of concern to producers and consumers. The desire for sugar has also provided a base for direct and indirect taxation. In the United States, the first sugar duty was imposed in 1789, levied at 3 cents per pound to provide income for a young nation desperate for money to pay war debts. The duty and the tax reached a high point in 1815, when it was 12 cents per pound. Domestic supplies became available with Thomas Jefferson's Louisiana Purchase in 1803. Hawaiian sugar reached the mainland in 1835, and Philippine and Puerto Rican sugar were considered to be a domestic supply at the end of the Spanish-American War in 1898. In 1894, the United States introduced a sugar tariff to benefit and strengthen domestic suppliers. One factor in Cuba's successful struggle for freedom from Spain in 1896–1898 was the high tariff, in which a 40 percent *ad valorum* duty was imposed on Cuban sugar. This caused the planters to seek to tie their fortunes to the United States; after the revolution, Cuban sugar was accorded "preferred status."

As new strains of cane and beets were developed, as technology became more sophisticated, and as acreages devoted to sugar increased, sugar production expanded rapidly after World War I. Prices plummeted to the point that producing countries could not sell enough sugar to obtain currency for the purchase of goods and services—a situation that, in turn, affected the industrialized nations. Retrenchment during the 1920 decade resulted in a shortage of sugar. The Jones-Costigan Act of 1934 and the 1937 American Sugar Act were passed to eliminate drastic price fluctuations, to ensure a stable ratio between domestic production and the balance of payments among trading nations, and to "reward friendly foreign nations." These acts, and subsequent amendments, provided that the U.S. Secretary of Agriculture each year must determine the domestic consumption of sugar and set quotas on U.S. production and imports from foreign suppliers. Although the percentages varied from year to year, roughly half of the total quota was allotted to domestic producers, including Cuba, the Philippines, and Puerto Rico. The cost of the program was paid by a half-cent-a-pound tax on all sugar. This was so efficiently managed that between 1934 and 1974, the U.S. Treasury was enriched by $600 million above administrative costs.

With an assured market—Cuba's quota was essentially its entire output—Cuba became a one-crop economy with all its eggs in a basket made of sugar cane. Eighty percent of Cuban exports in the 1950s was raw sugar to be refined in the United States, and 40 percent of the total profit went to U.S. companies. Political stability was guaranteed by financial and military support from the United States, the country remained undeveloped since reciprocal trade agreements mandated purchase of American goods, and American capital dominated the Cuban economy. The land was cultivated to cane to the point that food was imported into this fertile tropical island. A repressive military dictatorship, national pride, and the realization that labor-intensive practices on the sugar plantations and mills served only the interests of a tiny oligarchy, resulted in a revolution in 1959. Political and economic power was transferred to leaders whose political philosophy was anathema to the United States. The United States severed relations with Cuba in January 1960 and lifted the sugar quota in July of the same year. The Cuban sugar quota, which constituted up to one-third of U.S. consumption, was allocated to other, "more friendly nations," especially to the Philippines and to the Dominican Republic, which by coincidence had been formed from Santo Domingo in 1960. Cuba, without its special status, was forced to market the product of a national monoculture on the world market, with economic consequences that had political ramifications.

The American sugar acts were quickly taken up by other countries, who passed similar legislation both to protect domestic supplies of producing nations and to ensure a reasonable proportion of the available sugar for domestic consumption. For the United States and other countries, the acts were effective during World War II, when sugar was in short supply, and they retained their effectiveness during the postwar period. Strains and cracks began to develop, however, during the 1960–1970 period as production methods became more efficient and surpluses accumulated that threatened to reduce prices to below the cost of production. In 1979, world sugar production was 90 million metric tons; it had risen only to 96 million metric tons by 1985 in spite of increases in population. Hawaii reduced its acreage by almost half and mainland production also fell significantly. This, in turn, affected other countries whose economies were closely tied to sugar as their main if not sole export crop. The American sugar acts were repealed in 1974–1975 with consequent dislocation of peasants whose livelihoods were geared to the cane fields. It is now apparent that monocrop economies are dangerous. Crop failures or overproduction in such areas cannot be counterbalanced by other crops or by a yet underdeveloped industrial sector. Dependence upon foreign capital and imported necessities binds monoculture economies to external control and possible exploitation and causes extensive human misery.

The drop in sugar utilization is directly related to the development of substitutes for pure cane or beet sugars. Blended sugars, composed primarily of sucrose but also containing variable amounts of glucose, can be sold at lower prices because they require less refining. Even greater inroads into the sweetener markets have been made by a number of liquid sweeteners. Corn syrup was developed over 75 years ago. It is a mixture of sucrose plus small, soluble fragments of starch (dextrans) which reduce the tendency for the syrup to crystallize. Invert sugar syrups are prepared by treatment of cornstarch with enzymes or acids that convert the insoluble starch into its component glucose molecules. High fructose corn syrups are made by an analogous process. These liquid sweeteners are used in place of sucrose for industrial fermentations and to increase the speed of yeast fermentations in the production of alcoholic beverages; they have almost replaced sucrose in prepared foods and baked goods. The bulk of sweeteners in soft drinks is liquid sweeteners.

Particularly in developed countries, where a social premium is placed on being slim, synthetic sweeteners have captured a significant share of the market. Saccharin is 300 times sweeter to the taste than sucrose but has a somewhat bitter aftertaste. The cyclamate sweeteners are only 30 times sweeter than sucrose but are now suspect because they cause cancer in laboratory animals. Aspartame, the most recent addition to the list of synthetic sweeteners, contains common amino acids, but its safety is still to be established. Depending on the laws of a particular country, one or more of these substances is used in diet foods and drinks and now constitutes over a quarter of the sweetener market for these products. While sugar prices in the United States remain considerably above the world price of sugar because of tariffs and subsidies, liquid sugars and synthetic sweeteners will continue to cause dislocations in the cane and beet industry.

ADDITIONAL READINGS

Aykroyd, W. R. *The Story of Sugar.* New York: Quadrangle Books, 1967.

Crane, E. *Honey: A Comprehensive Survey.* New York: Russak, 1975.

Inglett, G. E. (ed.). *Sweeteners.* Westport, Conn.: Avi Books, 1974.

Mintz, S. W. *Sweetness and Power: The Place of Sugar in Modern History.* New York: Viking, 1985.

Nearing, S., and H. Nearing. *The Maple Sugar Book.* New York: Schocken Books, 1970.

Schalit, M. *Guide to the Literature of the Sugar Industry.* New York: Elsevier, 1970.

Takaki, R. *Pau Hana: Plantation Life and Labor in Hawaii.* Honolulu: University of Hawaii Press, 1982.

Taylor, F. G. *A Saga of Sugar.* Salt Lake City, Utah: Deseret, 1944.

A Cup of Tea

Tea is naught but this:
First you make the water boil
Then infuse the tea
Then you drink it properly
That is all you need to know.

SENNO-RIKYŪ (1585)

Tea is, next to water, the most popular drink in the world. In Great Britain, a "cuppa" is akin to chicken soup in Brooklyn, a panacea for everything from chilblains to broken marriages. Linnaeus originally named this plant *Camellia japonica* because his herbarium specimen

Adapted from "The Tea Mystique," *Natural History* 84 (1975):12-29. Copyright American Museum of Natural History.

Figure 4.22 Flowering branch of tea *(Thea sinensis)*.

came from Japan, but the shrub is native to north India and China and we now classify the plant as *Thea sinensis* (Figure 4.22).

In modern Western society, the word *alkaloid* conjurs up visions of drug addiction, degradation, and death. Yet one group of alkaloids, the caffeine alkaloids, is nonaddictive and is important socially, economically, and medically (Figure 4.23). Chemically, caffeine is 1,3,7 trimethylxanthine, a compound made by plants but not animals as a branch of the same biosynthetic pathway in which nucleic acids are synthesized. It was isolated in 1820, and its chemical structure was determined by 1825. Although caffeine can be synthesized in the laboratory, isolation is cheaper than synthesis. Most supplies are obtained during the decaffeination of coffee. Caffeine affects the central nervous system since it is a mild stimulant and also serves to increase the heart rate. In some people the antisoporific action of caffeine can lead to insomnia—the "no coffee tonight or I won't sleep a wink" syndrome.

ALL THE TEA IN CHINA

During the Han Dynasty (2191 B.P.) a medical book reported that *ch'a* (Chinese for tea) could cure tumors, abscesses, and ailments of the bladder. It "lessened the desire for sleep" and "gladdens and cheers the heart." Infusions of young tea leaves in hot water gave a pleasant flavor to the flat, insipid taste of water boiled to prevent the spread of water-borne diseases. When Buddhism came to China, tea kept monks awake during all-night meditation and eased pangs of hunger during prolonged fasts. By the fourth century, reliable records on cultivation, processing, and brewing were maintained. The preferred method of processing was the tea cake technique, in which young leaves were mixed with rice, baked, the cake pounded into small pieces, and the fragments boiled with onion slices, ginger root, and orange to "render one sober or to keep one awake." Peasants had to be content with old leaves boiled in salted water. By the early Tang Dynasty (608-907), tea was an export product, taxes were assessed on it, and a three-volume, ten-part *Tea Classic* extolled the pleasure of tea drinking, giving directions for growing, storing, and brewing.

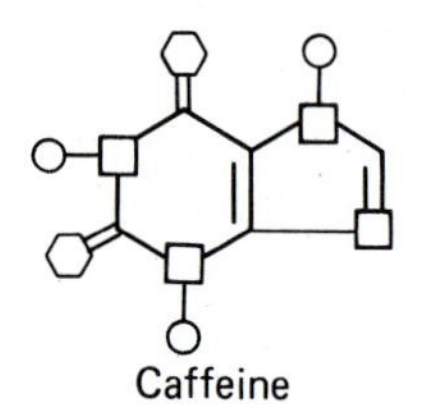

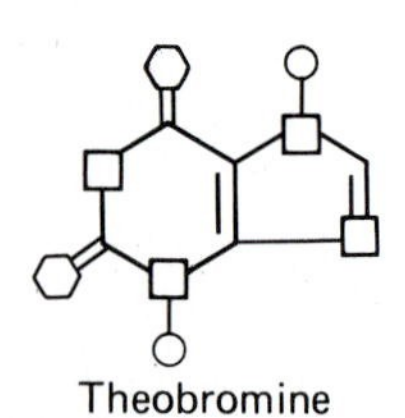

Figure 4.23 Molecular structures of caffeine, the major alkaloid in coffee, and theobromine, the major alkaloid in tea. Hydrogen atoms and carbon atoms in the rings are not shown. Carbon atoms are represented as circles, oxygen atoms as hexagons, and nitrogen atoms as squares.

One example of the aesthetic importance of tea can be seen in a letter of 730 from a Chinese gentleman to a friend:

> I am sending you some leaves of tea that come from a tree belonging to the monastery of Ou I Mountain. Take a blue urn of Ni Hung and fill it with water which has been melted from snow gathered at sunrise upon the western slopes of Sou Chan Mountain. Place the urn over a fire of maple twigs that have just been collected from among very old moss and leave it there until the water begins to laugh. Then pour it into a cup of Huen Tscha in which you have placed some leaves of this tea. Cover the cup with a bit of white silk woven at Houa Chan and wait until your room is filled with a perfume like a garden of Fouen Lo. Lift the cup to your lips, then close your eyes. You will be in Paradise.

Baked tea cake or the infusion of leaves was supplemented during the Sung Dynasty (960–1120) by a technique in which quickly dried leaves were finely powdered and whipped with hot water into a thick green froth. During this period, a branch of Chinese Buddhism, the Ch'an sect, incorporated tea into its rituals. Subscribing to the Tao ideals of simplicity with contemplation as the way to enlightenment, Ch'an Buddhists used tea to keep themselves awake during their vigils and to induce that state of calm harmony which religious contemplation requires.

THE TEA CEREMONY

For centuries, Japan was dependent upon China for protection and looked to the "Middle Kingdom" for enlightenment and culture. The Emperor Shōmu obtained tea in 729 from the Tang emperor. He invited a hundred Japanese Buddhist priests to join him in sipping this august beverage, but the Japanese did not like the bitter, almost black cloudy liquid. When Ch'an Buddhism entered Japan in the middle of the thirteenth century as Zen Buddhism, the ritual ceremony of tea came with it. The founder of the Tenryu-ji Zen Buddhist temple in Kyoto modified the ceremony to be more compatible with Japanese concepts and gave the Japanese version its name, *Cha-no-yu,* the ceremony of hot-water tea. In 1477, the Ashikaga shogun-military ruler of Japan elevated Shoku, a monk of the Shomyo-ji temple of Kyoto, to the position of first high priest of Cha-no-yu. Shoku reconstructed from oral tradition the tea ceremony of the Chinese Ch'an Buddhists in which monks successively drank tea from the same bowl while contemplating a statue or a drawing of the saint Bodhidharma, known in Japan as Daruma—patron of Zen and of tea (Figure 4.24).

The formalization of the tea ceremony had profound repercussions in Japan. Zen, initially a small sect, struck a responsive chord in the minds and hearts of the military governors of Japan and their Samurai warriors. Soldiers all, they were trained to respect simplicity, austerity, and ritual. Those that survived the bloody fighting retired to a life of contemplation instead of running for high political office. Zen offered them concepts they understood, and the symbiosis between soldier and priest resulted in the secularization of Cha-no-yu. Toshimasa, eighth Ashikaga shogun, built the famous Silver Pavilion in Kyoto and, within its walls, his tea priest built the first tea hut. The tea hut itself was copied from Chinese paintings of the Sung Dynasty. It was essentially a "rude" hut, straw-thatched, with a small door through which one entered on hands and knees. To enhance the feeling of austere harmony, the land surrounding the tea hut became a garden designed as an aid to Zen communication with and absorption into nature. The *uchi-roji* garden, a miniaturized landscape, was placed so that participants in the tea ceremony could view it through the small windows of the hut. The

Figure 4.24 Bodhidharma, Buddhist patron of the tea plant. Taken from a fifteenth-century Chinese woodblock print.

"nature as purified" roji developed into the Japanese garden as we know it today.

The hut itself had design features we now associate with Japanese life. Simplicity, elimination of luxury and ostentation, and the harmony necessary for quiet were stressed. Details of construction were codified: internal walls were of rough, gray plaster, window openings were of bamboo, beams should retain the marks of the adze, and the number of mats was specified. Although the *tokonoma* alcove was not original with the tea hut, it reached its present perfection as it came to symbolize the Zen's alcove for a statue or painting of a saint. Precious scrolls, a lovely stone on a teakwood stand, a Bonsai, or some other object of contemplation could be displayed in this alcove. It was the responsibility of the tea master to place within the tokonoma objects of such profound and quiet beauty that the observer's spirit of harmony with nature was enhanced. Although originally parts of altar decorations, flower arrangements came to be placed within the tea hut alcove and this led to *ikibana*—flower arranging (Figure 4.25). In fact, the arrangers of flowers were the tea masters. Soami was tea master and flower arranger in the early sixteenth century and developed the *rikka* style; Senno-Rikyū, his successor, developed the simple, austere, and delicate *nagiere* style; and his successors perpetuated and modified these arrangements into the schools of flower arranging that we know today.

The tea ceremony, its hut, garden, and plant arrangements embody two deeply rooted Japanese ideals. The first of these is *furyu-no-asobi,*

Figure 4.25 Flower arranging as part of the Japanese tea ceremony. Courtesy Japan National Tourist Organization.

approximately translated as an elegant amusement allowing one to lose one's self in the joy of delicate peace and tranquility. The second is the concept of *wabi,* or mellifluous simplicity in which every aspect of an act is carefully orchestrated and integrated. The tea ceremony is not simply the drinking of frothy, grass-green tea, but rather having the entire ceremony exemplify that combination of serenity, grace, and beauty which is the spirit of Zen, participating in a ritual of perfect harmony and tranquility. A family may spend huge sums on a tea hut and ancient tea utensils because, today, the tea ceremony has become overlaid with self-serving, conspicuous consumption. Although still a Zen ritual, the ceremony is also a part of the proper mode of conduct of the upper classes in Japan. Young women take courses to allow them to lead tea ceremonies, and they learn Cha-no-yu much as our children from equivalent social strata learn to ride an English saddle or serve in a receiving line (Figure 4.26).

TEA IN EUROPE AND NORTH AMERICA

Tea entered the consciousness of Europe in the middle of the fifteenth century through narratives of travelers to the mysterious East. The Portuguese, who had established themselves on Macao off China's coast, sent back a few chests of tea leaves, hoping to develop a market.

Figure 4.26 Students learning the Japanese tea ceremony. Courtesy Japan National Tourist Organization.

Little came of this venture, and traders confined themselves to silk, rhubarb for the constipated, and art curios. The Dutch tasted tea, found it pleasing, and after their conquest of Batavia, moved into the Chinese tea trade. Commercial interactions between Europeans and the tottering Ming Dynasty were poor, and the ascendancy of the suspicious Manchus in 1644 did not help; traders were confined to Canton in the south where *ch'a* was called *t'e*. The Dutch forced the Portuguese out and obtained a monopoly on the tea trade. Tea entered Russia when the Chinese Emperor presented a gift of dried leaves to Czar Mikhail Fydorovich (1596–1645), the first of the Romanov rulers. The *Court Chronicle* noted that "it is a good brew and fully satisfying when gotten accustomed to." Starting in about 1696, a regular trade in tea was established between China and Russia. The city of Hankow was the major locus of production for the Russian market. Finely powdered tea was pressed into bricks about a foot long, 8 inches wide, and an inch thick. The tea was shipped through Kiakhta, a city on China's northern frontier. Being somewhat isolated from the rest of Europe, the Russians developed a unique tea-drinking style. The samovar, glasses of tea sweetened with a spoonful of jam, and the tricky technique of sipping boiling tea through a lump of compressed sugar held between the teeth are still confined to Russia. The Russians did, however, export the use of a lemon slice.

As tea trickled down socially from European nobility to the less exalted, it came under the scrutiny of the medical profession. Dr.

Cornelius Decker, seventeenth-century publicist for the Dutch East India Company, recommended that "an obstinate fever could be cured by drinking every day forty to fifty cups of tea." French medical authorities were generally antitea, and the entire Collège de Médicine denounced "this dubious drug." Its few advocates made such extravagant claims for its medicinal value that the French gave a Gallic shrug and went right on drinking wine. Germany preferred beer, and the Latin countries were priced out of the market by the Dutch, whose hatred of Spain and Italy had an unhappy historical base.

In 1660, a London shopkeeper, Thomas Garroway, advertised the stimulating properties of tea in glowing terms: "It maketh the Body active and lusty, eliminates gripping of the Guts and pains in the Bowels." The British were, however, still coffee drinkers and were not about to pay the equivalent of $50 per pound demanded by the Dutch. Tea was obviously an article of snob appeal. Samuel Pepys tried it that year and didn't like it. The wife of Charles II, who was Portuguese, served it after her wedding, and since the nobility drank it, Pepys tried it again in 1667 and, following the example of his king, liked it. Charles II gave an absolute monopoly of importation to the British East India Company, freezing out the Dutch, who were at war with France, then an ally of England. With an assured market, the Company established a regular tea run from Canton and Amoy to England, and the dominance of coffee in Britain's social life was threatened. From 100 pounds in 1680, Britain imported a million pounds of tea by 1700. There were several hundred shops in London serving tea, but none of them catered to women. Thomas Twining built an annex to his coffeehouse, The Golden Lion, that catered solely to women, who made Mr. Twining a wealthy man. Tea was at first sipped from handleless porcelain cups as used by the Chinese (Figure 4.27). To protect ladies' fingers, the British

Figure 4.27 A family tea party in England. This portrait was made before the British added handles to the Chinese cups.

pottery firms of Spode and Wedgwood added handles to cups, and with the rediscovery in Dresden in the middle of the eighteenth century of the ancient Chinese art of porcelain making, elaborate tea services became appropriate wedding gifts. Translation of these designs into metal by Georgian silversmiths followed.

By 1780, coffeehouses had become tea houses, gin mills of the slums became tea bars, and imports rose to 14 million pounds a year. Samuel Johnson called himself "a hardened and shameless tea-drinker, who has for many years diluted his meals with only the infusion of this fascinating plant" and confessed that he would drink up to 12 cups of tea at a sitting. John Wesley, who founded Methodism, first denounced weak-willed people who substituted one evil, tea, for another, liquor, but he finally succumbed and had a special pot made for him by Josiah Wedgwood. Tea became the approved drink for the antiliquor forces, and the temperance hotels that still exist in provincial towns of England have excellent and inexpensive tea bars.

But the East India Company was in trouble. Anticipating huge sales to the American colonies, they had millions of pounds of tea in storage which the colonists wouldn't buy because of the tea tax. The colonies were drinking tea smuggled into eastern ports by the Dutch and through Canada by enterprising French-Canadians who had little enough love for the English. As textbooks say, the rest of the America tea saga is history.

Europeans developed their own tea ceremonies. Afternoon tea originated in about 1750 with Anna, wife of the seventh duke of Bedford, who complained of a "sinking feeling" in the late afternoon and decided to have a snack between the noon meal and late evening supper. George III regularly had tea and cake at court, thus sanctifying "the tea." Afternoon "low tea" was a light meal of watercress sandwiches, scones, and bread and butter with jam. The British developed precise codes of behavior governing the way a cup was to be held, proper pouring technique, and whether the milk should be added before or after the tea. One gentleman, Henry Sayville, complained bitterly about the feminization of the men who participated in these "absurd social activities" instead of rallying round the bar of the nearest pub for a pint of ale and a clay pipe. High tea became a full meal, replacing the former family supper. Mrs. Beeton, the Julia Child of the Victorian era, described a high tea as a "moveable feast which included hot dishes, fowl, hams, salad, cakes, tarts, trifles and fresh fruit."

Iced tea was invented in the late nineteenth century by Richard Blechynden, an English public relations man representing a Ceylonese tea company at a St. Louis trade fair. Arriving on a beastly hot day with a group of Singhalese waiters in full native dress, he was unable to find anyone who wanted to drink hot tea. With the genius of a good publicity agent he poured the hot tea over ice, added sugar and lemon, and created a sensation. The tea bag was the inadvertent invention of Thomas Sullivan, a New York importer and wholesaler. In 1908 he decided to send out samples in small silk bags instead of the usual tin caddy. Some unknown person dropped a bag into hot water and an entire new industry was born.

TYPES OF TEA

There are many kinds of tea. An ancient Chinese scholar said, "The essence of the enjoyment of tea lies in appreciating its color, fragrance and flavor," and these characteristics are determined by many factors. There are two major species, *Thea sinensis* from southwest China, northeast India, and Cambodia, and the less valuable *T. assamica* from Southeast Asia. The shrub can grow to about 6–8 meters, but under cultivation it is kept to about 1 meter. Ideal growing conditions include well-drained, slightly acidic soils, ample rainfall, and especially elevated altitudes. The best teas are grown in the mountains of Assam, Formosa, and Japan, with most of the North American tea grown in India. Attempts to grow tea in the Western Hemisphere have not been successful, although plantations were established in the U.S. South during the early part of the twentieth century. Parts of east and south Africa and South America show excellent potential for tea production. The fine teas of Ceylon and of India are all mountain-grown, usually at elevations of over 1500 meters.

Young bushes, started from the seed of selected plants or from rooted cuttings, are grown until they become large shrubs and are then severely pruned back to encourage formation of many twigs. At intervals of a week to ten days, the plant undergoes a flush of tip growth producing two very small, tender young leaves and an unopened leaf bud. The leaves and bud are picked by hand; as a consequence, cheap labor is necessary to keep production competitive. Experienced pluckers can gather 15–20 kilograms per day, enough for about 4–5 kilograms of tea. Older, coarse leaves bring lower prices and produce a beverage with stronger and less fragrant odors and flavors.

Independent of the cultivar grown, three major tea types can be prepared. Black tea, the most common type, is produced in a series of steps. The picked leaves are spread on racks to wither for a day in the sun. This reduces water content and initiates enzymatic processes that release aroma. From the withering racks, leaves are put through machinery which ruptures leaf cells, releasing enzymes that cause chemical oxidation of tannins and phenolic compounds. The enzymatic processes, called fermentations although they are actually oxidations, are continued in a cool, damp atmosphere until the leaves become a bright, shiny copper color, much like a new penny. Finally, the tea is dried—fired—to stop fermentation and to remove excess water. It is during the firing that tea leaves acquire the dark color we see in black tea. The withering process is not carried out for production of the green tea types which retain a slightly "grassy" taste. The rarely seen Oolong teas are produced by a very short withering process. Green and Oolong teas have delicate flavors and aromas and very light colors.

There are close to 3000 different kinds of tea, taking their names from the districts where they are grown, from the processing techniques used, from various additives, and even from the sizes and conditions of the leaves (Table 4.1). Orange pekoe, for example, now refers to a grade of black tea, although the name came from the Chinese, who had added orange blossoms to the leaves. Since most commercial teas are blends of different grades and district products, some names refer to mixtures put together for famous people—for example, Earl Grey or Queen Mary.

Table 4.1 TEAS

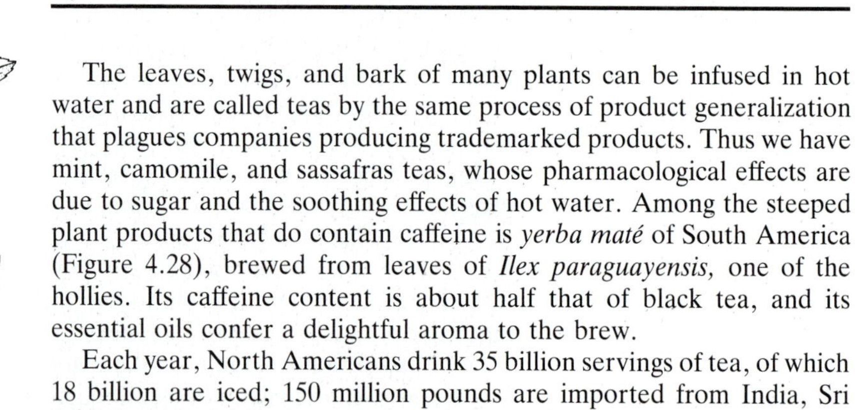

Tea grade	Type	Description	Country
Orange pekoe	Black	Thin leaves, lightly colored, pale brews	Sri Lanka, Japan China, Java
Pekoe	Black	Leaves broader, brew with more color	Sri Lanka, India, China, Java
Pekoe souchong	Black	Round leaves, pale brew	Sri Lanka, India, China
Oolong	Oolong	Medium leaves, delicate brew, pale color	Taiwan, China
Fannings	Black Oolong	Small leaves, quick brewing with good color	Sri Lanka, India
Dust	Black	Very small leaves, strong, aromatic, dark brew	India, Taiwan, Sri Lanka, Java
Imperial Gunpowder Hyson	Green Green Green	Small leaves, delicate, pale brew, faintly aromatic	Japan Taiwan China

Figure 4.28 Flowering branch of maté *(Ilex paraguayensis).*

The leaves, twigs, and bark of many plants can be infused in hot water and are called teas by the same process of product generalization that plagues companies producing trademarked products. Thus we have mint, camomile, and sassafras teas, whose pharmacological effects are due to sugar and the soothing effects of hot water. Among the steeped plant products that do contain caffeine is *yerba maté* of South America (Figure 4.28), brewed from leaves of *Ilex paraguayensis,* one of the hollies. Its caffeine content is about half that of black tea, and its essential oils confer a delightful aroma to the brew.

Each year, North Americans drink 35 billion servings of tea, of which 18 billion are iced; 150 million pounds are imported from India, Sri Lanka, east Africa, and Taiwan. Aside from green teas served in Chinese restaurants, most are black teas. The average "cuppa" contains only about half as much caffeine as a cup of coffee. Samuel Johnson notwithstanding, caffeine is neither addictive nor narcotic. Fortunes have been made from the tea trade. Our clipper ships participated in the tea runs of the nineteenth century, some bankrolled by John Jacob

Astor, who worked out a deal with the Chinese government to trade furs for tea and obtained a deferment of import duties from a complacent U.S. administration. The Great Atlantic & Pacific Tea Company (A&P) (Figure 4.29), the National Tea Company, and the Union Pacific Tea Company have all become supermarket chains. Tea financed Thomas Lipton's unsuccessful efforts to restore the crown of yachting to Britain. It is still the sales base for manufacturers of fine china—the very word reflecting the history of this most social of drinks.

ADDITIONAL READINGS

Eden, T. *Tea.* 2nd Ed. London: Longman, 1965.

Hudson, C. M. (ed.). *Black Drink.* Athens: University of Georgia Press, 1979.

Koreshoff, D. *Bonsai: Its Art, Science, History and Philosophy.* Beaverton, Ore.: ISBS/Timber Press, 1984.

Pratt, J. W. *The Tea Lovers Treasury.* New York: 101 Productions, 1982.

Quimme, P. *The Signet Book of Coffee and Tea.* New York: Signet, 1976.

Shalleck, J. *Tea.* New York: Viking Press, 1972.

Tanaka, S. *The Tea Ceremony.* Tokyo: Kodansha, 1982.

Ukers, W. H. *All About Tea.* New York: Tea and Coffee Trade Journal, 1935.

Woodward, N. H. *Teas of the World.* New York: Collier, 1980.

Figure 4.29 One of the first A&Ps. Courtesy Great Atlantic and Pacific Tea Co.

A Mug of Coffee

Recipe for good coffee:
Black as the devil, hot as hell,
Pure as an angel, sweet as love.

C. M. Talleyrand (1815)

Gemaleddin, grand mufti of Aden, a city-state on the Persian Gulf during the ninth century of the Hegira (the fifteenth century of the Christians), had occasion to travel to Persia. While in the city of Kahvah, he saw people drinking a dark-brown, aromatic beverage which did not stupify, but seemed to lessen the cares of the world. Returning to Aden, he found himself indisposed and decided to try this drink. His scribe, Schehabeddin Ben, wrote of the recovery of the grand mufti. Observing that the drink lessened the desire for sleep, Gemaleddin recommended it to the dervishes, an especially religious sect, to allow them to pass the night in prayer with greater zeal and attention. Coffee had entered the consciousness of the Arabian world.

COFFEE OF ARABIA

There are many legends of how we first learned of this precious brew. It was said that goats near a monastery in the remote city of Kahvah returned one evening much friskier than usual and that a shepherd boy found that the flock had eaten of a shrub with dark red berries (Figure 4.30). The monks mixed the berries with hot water, tasted the brew, and praised Allah for the wondrous flavor and aroma. Another story is that Mohammed was visited in a dream and was told to go to a remote spot where he knelt down next to a tree with red fruit. He ate of the fruit and was suddenly filled with energy. Since we know that Abyssinian warriors of the second century made pellets of roasted, pulverized coffee beans with grease to be carried into battle, the history of coffee is certainly older than the Moslem faith.

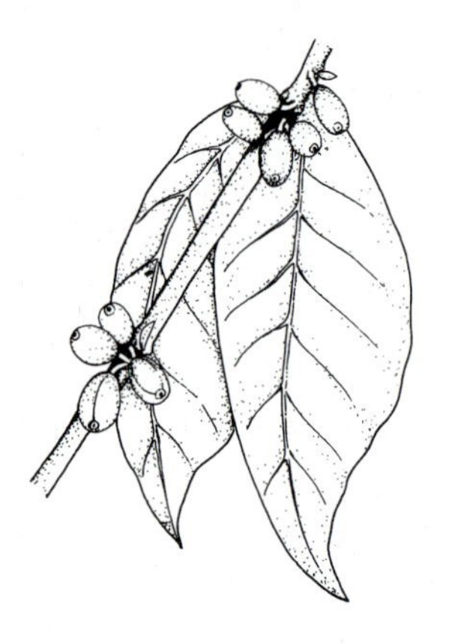

Figure 4.30 Fruiting branch of coffee *(Coffea arabica)*.

From Aden to Mecca and then to the rest of the Arab world, the coffee-drinking habit spread as religious scholars and lawyers took it up. The Imans and others in the religious hierarchy tried to restrict its use, but the pleasant taste and mild stimulation so captured the fancy of people that it moved down the social ladder to night watchmen and finally to the whole population. In about 1475 the inhabitants of Mecca began drinking coffee publicly in small *kahvah khaneh*—coffeehouses. Merchants, students, and workers gathered there in the afternoons to escape the heat and the press of trade and study. Dancing and gambling became part of the entertainment. In 1554, the opulent Café of the Gates of Salvation opened in Damascus, a model of luxury taken up by other establishments. The strict religious sects, outraged that this holy gift was secularized and thus profaned, petitioned the great Soliman to outlaw this impiety on the grounds that roasted coffee was a sort of burned wood denied the faithful. Soliman ruled that coffee was, indeed, contrary to law and the coffeehouses were shut down. This prohibition against coffee drinking was about as successful as was the prohibition against alcohol in the twentieth century; bribery, payoffs, and the equivalent of speakeasies prevailed. The question of charred wood was reconsidered, and the houses were reopened to general rejoicing, but hours were carefully regulated and the establishments were heavily taxed. Only once more were Arabian coffeehouses closed. During the reign of Mohamet IV, his grand vizier received reports that local politicians were leaking information to fellow coffee drinkers and, what was worse, were denigrating his administration. In spite of large tax losses, he closed the houses, but allowed street vendors to sell it. If found bootlegging coffee, the proprietors were bound head to tail on a donkey and flogged through the streets. When the sultan heard of this, he cancelled the edict, much to the gratification of the people.

The coffee was served thick and very black, without sugar, and sometimes with added cloves, anise, or cardamom. Turkish coffee, with the grounds left in the small cups, is identical to that prepared in the sixteenth century. Not allowed in the coffeehouses, women drank it at home, and it was also used during childbirth. If a man refused to keep his wife supplied with coffee, she had grounds for divorce.

COFFEE IN EUROPE

Coffee entered Europe in about 1615 when a Venetian brought home a supply to start a coffeehouse. It moved to France by 1640 and to England by 1650, although the English were not especially enamoured with it at first. By 1660, however, there was enough sipped in London to tax it at the rate of four pence the gallon. The revenue-generating potential of the coffeehouse was recognized throughout Europe, and various governments placed excises on the product. By 1700, there were 3000 coffeehouses in London serving as trade centers where merchants, lawyers, and civil servants could gather for discussions. Daily newspapers and desks with writing materials were available, and the customers could receive their mail there. Lloyds of London, an international underwriting and insurance firm, started in a coffeehouse in 1690. Men spent so much time in the houses that in 1750 a group of women met with the city fathers to present a petition "representing to public consideration the grand inconveniences accruing to their sex from the excessive use of this drying and enfeebling liquor." The words "drying and enfeebling" suggest that there were rumors flying about London that intemperate use of coffee might interfere with the sex lives of the men. The very existence of the houses was a political matter as well, for tavernkeepers demanded protection for their product through higher taxation on coffee, and the medical profession was raising questions about indiscriminate use of coffee without strict medical supervision. Merchants had their own coffeehouses, called "penny universities" because entrance cost a penny and one could read the *Tatler* and the *Spectator* and engage in learned discussions on politics. The rabble preferred gin. Coffee was popular in France, rivaling wine among the aristocracy. Roasted beans were sold in apothecary shops, ostensibly under medical supervision, but by 1715 there were close to 50 cafés in Paris, and over 600 in 1750.

The Arabian city of Mocha, whose name is memorialized in coffee-flavored ice cream and cake frosting, was the center of the coffee trade, extracting higher and higher prices from Europeans, who were drinking more and more coffee. It was obvious that this exceedingly valuable product could not remain a monopoly of the infidel, and attempts were made to cultivate the plant under European, and hence Christian, auspices. The Bourgermeister of Amsterdam, who was also governor of the East India Company, told the governor of Java to obtain trees—by any means possible—and establish a plantation in the hills above Batavia. This was accomplished by 1690. Louis XIV of France was given a coffee tree for his private gardens in 1714, and, charmed by the gift, loaned money to the East India Company to expand its plantations. In 1718, the Dutch planted trees in the New World in Surinam, the French followed suit in Martinique, and the English planted trees in Jamaica in 1727. Within ten years, the first plantings were made in Brazil, and by 1780 coffee trees were growing throughout much of Central America and Southeast Asia, with the bulk of Europe's supply under Dutch control. Ceylon (now Sri Lanka) became British in 1797 partly because of England's coveting of the Ceylon coffee plantations. Ravaged by disease and neglect, these plantations declined seriously, and tea again replaced coffee as the mainstay of the British home. Nevertheless, coffee was, throughout Europe, the preferred beverage of intellectuals and the aristocracy, and paeans of praise were written in its honor; J. S. Bach even composed a coffee cantata.

Today, the world thinks of coffee as a North American drink, and we consume about 40 gallons per person per year, as compared with less than eight gallons of tea. Although the first coffeehouse opened in Boston in 1670, the habit really got started in New Amsterdam because of the Dutch influence and continued when the city became New York. The nascent Americans obtained their beans from British Jamaica and, aping London, established coffeehouses in the business district. The King's Arms on Broadway near Trinity Church and several others played roles in the intellectual ferment leading to the Revolution.

BOTANY OF COFFEE

The coffee tree or shrub is a member of the Rubiaceae, a botanical family that also includes the fever-bark tree *(Cinchona),* the gardenia, and the common bedstraw *(Galium)*. There are many species in the genus, but only two, *Coffea arabica L.* and *C. canephora,* are of much economic significance. The plant grows well in tropical areas where rainfall exceeds 100 cm/year, and, because it does best at stable temperatures close to 20° C, mountain environments in South and Central America, east Africa, and Java produce superior beans. The fruit is a two-seeded berry, each seed surrounded by a sweet, yellowish, edible flesh and a red skin. A tree will flower once or twice a year, producing about 2.5 kilograms (6 pounds) of beans. A hectare of a well-managed plantation will yield close to a metric ton per year. Berries are picked by hand (Figure 4.31), subjected to a floatation process to separate defective fruit, leaves, and stems, and then depulped by machine. Residual pulp is removed by allowing the beans to ferment in tanks. The beans are thoroughly washed and dried. The flavor of the final product is the result of many interacting factors, including the variety of the plant, growing conditions, and the location of the plantations. There is general agreement that the coffees of Brazil and Colombia have the best flavors and aromas, and they sell for premium prices. Somewhat lesser grades are produced in other South and Central American countries. Almost all of these coffees are derived from isolates of the *arabica* cultivars. Most of the coffee produced in Africa (Uganda, Angola, Ivory Coast, and Zaire) is from *robusta* cultivars, discovered in 1898, but consumed only locally until the 1950s. The somewhat harsher taste of the African coffees has been exploited in making soluble coffee products, while the milder South American types are usually sold for home brewing.

The roasting process is critical in preparing beans for use. Discovered

Figure 4.31 Coffee berries are hand-picked so that only fully ripened fruit will be processed. Courtesy Pan-American Coffee Bureau.

in Turkey, roasting reduces the amount of water and brings out the inherent flavor of the beans. Lightly roasted coffee, preferred in the western parts of North America, is prepared at lower temperatures than the medium grades, which are the choice in eastern North America and most of Europe. Dark roasted beans are favored in the Middle East and southern Europe. Once roasted, beans from different plantations and regions are blended by coffee experts to provide a brew of constant flavor and aroma.

Soluble coffee is prepared from roasted, blended, and finely ground coffee. The powder is steeped in hot water, which extracts about 25 percent of the solids in the beans plus the caffeine alkaloids. This brew is either concentrated with heat or is freeze-dried. Aroma is frequently lost during processing and is restored as extracts of coffee beans. About a fifth of the coffee consumed worldwide is now soluble or freeze-dried. Decaffeinated coffee is prepared from specially blended beans that are extracted with solvents and then dried before further processing.

Prices are established through an international coffee agreement signed by most of the world's producers and consumers. Quotas are set by the agreement among the participating nations. Because producers and consumers have different goals, agreements are hammered out after negotiations that may last for months.

Probably the greatest threats to stable supplies of coffee are bad weather and disease. In 1974, close to 8 percent of Brazil's coffee trees were killed by frost, and a single frost in July 1975 damaged 70 percent of the trees. It takes three to five years to bring damaged plantations back into production, and such frosts invariably increase the cost of coffee throughout the world. Several plant diseases plague the plants, including bacterial, viral, and fungal pathogens. The most serious and dramatic pathogen is the coffee rust, *Hemileia vastatrix,* a fungus. British-owned plantations in Ceylon (Sri Lanka), which at one time accounted for a significant portion of world trade, were afflicted with the disease. It appeared in Ceylon in 1850–1860 and by 1875 had spread throughout the plantations. In 1892, coffee production in Ceylon was at an end. The Central and South American plantings were free of rust until 1970, but the problem is now receiving anxious attention from growers and consumers.

ADDITIONAL READINGS

Clifford, M. N., and K. C. Willson (eds). *Coffee: Botany, Biochemistry and Production of Beans and Beverage.* Westport, Conn.: Avi Books, 1985.

Ellis, A. *The Penny Universities: A History of the Coffee Houses.* London: Secker and Warburg, 1956.

Jacobs, H. E. *The Saga of Coffee.* London: G. Allen, 1953.

Sivetz, M., and N. W. Desrosiers. *Coffee Technology.* Westport, Conn.: Avi Books, 1979.

Starbird, E. A. "The Bonanza Bean: Coffee." *National Geographic* 159, no. 3 (1981):388–405.

Uribe, A. *Brown Gold: The Amazing Story of Coffee.* New York: Random House, 1954.

Hot Chocolate, Devil's Food Cake, and Cola

Howard Johnson's sells chocolate ice cream and 27 alternates for the real thing.

RICHARD M. KLEIN

As part of the rituals of coronation and important religious ceremonies, priests of the Aztec empire would reverently sip liquid from a golden goblet reserved solely for this libation. Legend says that the plant from which the drink was made had been a gift to humanity from the God of the air, Quetzalcoatl, who was punished by other gods for giving a thing so precious to lowly beings. Quetzalcoatl believed, however, that only through partaking of this beverage could humans communicate with Him and tell Him of their needs. *Chocolatl* was invested with sacred powers due in part to the relatively high theobromine alkaloid content. Because trees were scattered in the tropical rain forests in the Mexican highlands, they were rare and hence sufficiently precious to warrant reserving their fruit for the emperor and the priests. Westerners first saw the *cacahual,* the pod-shaped fruits, when Columbus was given several on his first voyage to the New World; he included them in the gifts laid at the feet of Ferdinand and Isabella when he returned to Madrid in triumph. The significance of this gift was lost on the Spanish court, for no one had any idea of what they could be used for, and, eventually, when the enthusiasm for what Columbus had discovered died, the pods were simply discarded.

CHOCOLATE IN MEXICO

Only when Cortez reached Mexico in the early sixteenth century did the Europeans learn the delicious secret contained in the pods. The Aztecs believed that a pale-skinned god would come to their land in a vessel with white wings, so when the Spanish fleet hove into sight in 1519, the ships' crews and bearded chief god clothed in a shiny skin were greeted with great joy. Coming when they did, just at the time of the coronation of Montezuma II, was a most auspicious concordance. Cortez was led to the royal palace at Temistitian, and with all the pomp the priests could muster, the white god was presented with a golden goblet containing the ritual drink from the hands of the young emperor himself. Cortez tasted a very bitter, very hot liquid compounded of dried, coarsely ground beans of the revered tree mixed with unknown spices and whipped to a foam in hot water. Not being used to it, Cortez became somewhat dizzy, but not too dizzy to note that the goblet was solid gold. According to Bernal Díaz, the scribe of the expedition, Cortez thought that Mexico must be laden with gold and, then and there, the decision was made to subjugate the land and its people and take what gold could be found; the fate of Montezuma had been sealed with his own hands.

CHOCOLATE IN EUROPE

By 1550, Spanish galleons were carrying precious metal and many plants back to Spain, among which were pods and seeds of the chocolatl plant. The court of Spain developed its own version of the drink of the gods:

> Take 100 cocoa beans, two pods of chili pepper [also a New Spain import], a handful of aniseed, six roses of Alexandria, a pod of logwood, two drachmas of cinnamon, a dozen almonds, a dozen hazelnuts and one half pound of loaf sugar. Pound these things into a paste, add one pint of boiling water and stir vigorously over a fire. Drink while still very hot.

Word of this new drink soon reached the courts of other countries of Europe. Demand for the cocoa beans from France, Portugal, Italy, and as far away as the court of the Czar reached a level high enough to warrant putting the conquered Indians, now virtually slaves, to work clearing land and starting plantations of chocolatl. Methods used to prepare the beans for making a drink was a Spanish secret for about 50 years, until an unknown Spanish monk told the method to others, thereby allowing factories to be started in France and Italy. The raw material was, however, still a monopoly of Spain. Spain advertised the virtues of chocolate throughout Europe, asserting that it was the religious drink of the Aztecs, whereupon the eastern branches of the Roman Catholic Church tried to ban its use as pagan. The wealthy commissioned chocolate sets from Meissen to match the coffee services that were "all the rage." Royal approbations and the tendency to want to keep up with the Joneses assured an interest in chocolate among the upper classes and the upper middle classes; the lower classes couldn't afford it. By the 1650s there were chocolate houses just as there were coffee and tea houses in London and Amsterdam. Although the cocoa habit spread throughout Europe, it never achieved the popularity of the other drinks. And no wonder. Public acceptance was limited not only by the price but by the very high amount of fat in the beans. This cocoa butter, a semisolid fat which constitutes up to 40 percent of the weight of the beans, was a severe problem; it made grinding difficult, became rancid, and developed an unpleasant gray color as it oxidized. Cups of hot chocolate had the annoying habit of coagulating as the liquid cooled. In spite of these drawbacks demand rose, and the Spanish developed plantations in Trinidad, the Antilles, and Venezuela, maintaining their monopoly on production. The processing business was taken over by the Dutch.

PROCESSING

With the availability of relatively inexpensive sugar, the bitter taste of chocolate could be modulated, but the hand grinding of the beans still limited production and kept prices high. By 1800, the power of the steam engine was harnessed to stones used to grind cocoa beans, and a much smoother paste could be made in a much shorter time. The steam engine's power also allowed the use of presses to squeeze out part of the cocoa butter, resulting in a more stable product with fewer annoying characteristics. In 1876 a Swiss, Daniel Peter, added milk to dark chocolate paste to produce milk chocolate, now a significant fraction of the international cocoa business. The Peter process was further modified by the Hershey Company to give a much smoother product into which nuts, raisins, and fruit could be incorporated. At about the same time, chocolate with a controlled cocoa butter concentration was developed, and cooking processes were modified to produce dark, sweet, or milk chocolates. These fondant chocolates completely replaced the less satisfctory product, lowered the price, reduced rancidity, and worldwide interest in chocolate reached new heights. From fondant, the production of a stable powdered product, cocoa, involved only a few simple manufacturing steps, steps developed by the Dutch, whose plantation of chocolate trees in Java had broken the Spanish monopoly.

The cocoa tree is a member of the Sterculiaceae, a small family that contains the cola-nut tree, whose fruit is also a source of caffeine alkaloids. Like coffee, cocoa grows best in regions of high humidity with stable temperature—conditions much like those found in a warm greenhouse—areas which include only tropical zones within 20 degrees north or south of the equator. Native to lowlands of South America east of the Andes, the trees were introduced into Africa by the Portuguese by the early seventeenth century. Roughly a dozen countries produce significant amounts of cocoa. Ghana, the Ivory Coast, and Nigeria in Africa, and Brazil, Ecuador, and Venezuela in South America produce the bulk of the world's supply. Humid, hot climates coupled with good soils and extensive cultivation are needed for production of a good crop. There are three types of cacao trees: the *criollo* type produces the best chocolate and is the primary type grown in South America; the *forastero* type is the main tree in Africa; and the *trinatario* type, apparently a hybrid between the other two, is widely grown because of the vigor of the trees. Cocoa beans develop in a pod (technically a berry) from small, odorless, pink-to-white flowers that develop directly from the trunk and main branches (Figure 4.32). Each pod is 10–12 centimeters long and contains about 40 bean-shaped seeds, each surrounded by a white, sweet, and juicy pulp (Figure 4.33). When removed from the pod, the pulp ferments and the seeds turn brown. These seeds are placed on racks in the sun for a week, during which time they lose water and begin to develop the color, odor, and flavor we associate with chocolate. Large plantations now substitute mechanical drying to save time and reduce the chances of decay. After cleaning and grading, beans are bagged for shipment to chocolate factories where they are cleaned and prepared for the critical roasting step. Heating to 140° C for several hours converts the beans into a product recognizable as chocolate.

The production of industrial chocolate involves grinding beans in steel rolling mills with sufficient heat to melt cocoa butter, to form a dark-brown, oily mass called cocoa liquor. The liquor can go through cocoa manufacturing steps or into the production stream leading to chocolate. In cocoa production, liquor is pressed to remove cocoa butter (reserved for other uses including cosmetic manufacture), leaving cocoa press cake to be crushed into a powder. The product contains up to 20 percent fat and is very bitter, still unsuitable for drinking. The bitter taste and some of the alkaloid are removed by treating the powder with alkali, the Dutch process discovered in 1829 by two Dutch firms, Droste and van Houten (Figure 4.34).

Baking and eating chocolates are manufactured from cocoa liquor by continued grinding and blending to produce an extremely smooth product with controlled amounts of cocoa butter. Emulsifiers are added to prevent separation of the chocolate and fat, and other substances may be added to impart special characteristics to milk chocolate, semisweet chocolate, and other types.

Figure 4.32 Mature cocoa pods growing from the stem of the tree. Courtesy Chocolate Manufacturers Association of the U.S.A., Inc.

Figure 4.33 Cocoa pod opened to show seeds. Courtesy U.S. Department of Agriculture.

A new era in world chocolate production was ushered in at the end of World War II when the Dutch plantations were destroyed. A few seeds, casually picked up by an African visitor to a West Indian plantation, were brought to Ghana to start the cocoa industry in Africa. With the development of plantations in Brazil, these two countries came to dominate the world market. Many plantations in Java were destroyed by a virus disease which struck in the 1930s, and the remaining plantations were decimated by monkeys, whose numbers soared after 1945 as leopard coats became fashionable. Current production is close to 900,000 metric tons, with North America the major consumer. World prices have increased severalfold since 1970 due to demand, poor growing conditions, disease, and the insistence of African producers that they obtain a fair price for their major export product.

Nutritionally, chocolate is a fine food with about 200 Calories, 3 g of protein and 16 g of fat per ounce. A cup of hot chocolate contains about the same concentration of caffeine-type alkaloids as does a cup of tea and is an effective, mild stimulant.

Figure 4.34 Droste cocoa box advertising the "Dutch Process."

OTHER CAFFEINE SOURCES

Although tea, coffee, and cocoa are the major caffeine-containing beverages, several other plants contain enough alkaloid to be economically and socially important. Most familiar to North Americans is the cola-nut tree, *Cola nitida,* a member of the cocoa family. Several other species contain caffeine, including *C. acuminata* and *C. verticillata,* but these are used locally in western Africa as a masticatory—something to chew on—and have not gained popularity in the Western world. The

Figure 4.35 Fruiting branch of cola *(Cola acuminata)*.

plant from which we get cola is a large shrub with glossy leaves and a fruit containing reddish seeds (Figure 4.35). The seeds contain about 2–3 percent caffeine plus essential oils, giving a pleasant flavor, and the extract serves, as do all other caffeine-containing products, to relieve fatigue and hunger. Dried seeds are produced on small plantations in tropical Africa in a belt extending from Ghana to Sierra Leone. The caffeine and flavorings can be extracted with hot water to make a syrup whose main economic use is in the fabrication of cola drinks.

The cola beverage of Brazil and other countries in South America is made from the fruit of the Guarana shrub *(Paullinia cupana sorbilis)*. The small, black seeds are pounded into a paste with tapioca flour and mixed with hot water. The caffeine concentration is close to 5 percent, among the highest of caffeine drinks, and there is a large quantity of tannin, giving the drink a strong taste. People in eastern Equador and parts of Peru drink a water extract of Guayusa leaves, a plant in the holly genus *(Ilex guayusa)* closely related to the Argentine maté. The leaves were originally used by the head-hunting Jivaro Indians to keep them alert for night battles, and when the Society of Jesus found out about this, they started selling the leaves as a cure-all. Although the Jesuits were expelled from Ecuador in the middle of the eighteenth century, the plantations established at the Catholic missions are still being harvested and a few new ones planted.

ADDITIONAL READINGS

Chatt, E. M. *Cocoa Cultivation, Processing, Analysis.* New York: Wiley, 1953.

Mandulay, L. S. *The Chocolate Tree.* Washington, D.C.: Pan American Union, 1966.

Minifie, B. W. *Chocolate, Cocoa and Confectionery.* 2nd Ed. Westport, Conn.: Avi Books, 1980.

Urquhart, D. H. *Cocoa.* 2nd Ed. New York: Longman, 1961.

Young, G. "Chocolate: Food of the Gods." *National Geographic* 166, no. 5 (1984):665–687.

Plants in Human Nutrition

Tell me what a man eats,
and I will tell you what he is.

ANTHELME BRILLAT-SAVARIN
French gastronome 1755–1826

In contrast to plants, which synthesize the entire organism from water, minerals, and carbon dioxide (autotrophism = self-feeding), animals must eat to live (heterotrophism = other feeding). In addition to water and minerals, people require complex organic molecules including proteins, carbohydrates, fats, and vitamins. Ultimately, the organic molecules ingested as human food are metabolized much as they are in plants, although the substances are first disassembled and then reassembled into the compounds and structures characteristic of and occasionally specific for that individual. Plant proteins are digested into their component amino acids, which are resynthesized into animal proteins; plant starch may be transformed into animal starch (glycogen).

Food serves cultural as well as nutritional needs. Feasts and fastings are linked to national and religious holidays. Bread is still broken to welcome guests, and it has considerable ritual significance in Judeo-Christianity. The Chinese politely ask if your rice bowl is filled when they want to know how you are doing. Arabians will provide hot mint tea before any weighty matters are discussed. College students are (sometimes) given their favorite foods when returning home, and food can play a central role in celebrations or at times of sorrow. Obedience to ancient religious tenets concerning food has created severe problems in modern societies whose ways of life are vastly different from those in operation at the time such food laws were written. Food may be used to influence behavior. The three-martini lunch, the state dinner for a diplomat, or a tastefully wrapped box of candy is commonly used for this purpose. Rewarding good behavior in young children or withholding food is an obvious way of influencing behavior. Less obvious, but still important manifestations include a child's refusal to eat or a person going on a chocolate binge when feeling depressed.

The need to ensure food supplies has been of paramount importance in the history of the world. Maintenance of trade routes and many of the explorations of the fifteenth through the seventeenth centuries were initiated by a desire to increase food supplies and to obtain condiments that improved the taste or the storability of foods. The expansionist and imperialist actions of many societies have been based on the rationale of obtaining land for increased food production or retaining productive lands presumed to be required by the population of a country. Food was the basis for the "winning of the West" for grain farming, and for the acquisition of Louisiana for a river route to the ocean.

Food choices and eating habits are, to a large extent, geographically and culturally determined. Because rice grows poorly in most of Europe, it never became a dietary staple there; the failure of potatoes to form tubers in tropical climates precluded their adoption by the people of these regions. Even when plant foods from other regions become available, people tend to stick to traditional diets, and there can be strong emotional reactions to unusual food items or even methods of preparation. Once a Chinese child is weaned, it rarely eats dairy products, and the Chinese find the smell of milk or cheese somewhat offensive. The British still have not made their peace with garlic, and mushrooms are avoided by people of many cultures. Few North American fruit and vegetable markets stock cassava, the tropical starch equivalent to the potato (Chapter 3), and corn on the cob is sold in Japan coated with a caramel glaze as a snack food. Not many people will eat white grubs, snakes, grasshoppers, ants, sheep eyes, or other quite edible and nutritious things. Getting children to try an unfamiliar food is an exercise that has tried the patience of uncounted parents.

Food can most easily be defined as materials that provide an organism with substances used to produce energy, form the building blocks for the synthesis of body constituents, and supply the chemicals necessary for the proper operation of enzymes and biochemical reactions. In addition to a requirement for water, which is a major component of plant foods, the human has absolute requirements for proteins, carbohydrates, fats, minerals, and vitamins. The human animal is omnivorous, capable of utilizing both plants and animals as food. By eating, literally, all parts of animals, it is possible to obtain all of the nutritional requirements; some Eskimo and Inuit peoples are entirely dependent upon animal sources. By proper selection, it is also possible to subsist on a diet composed entirely of plants, although this is more difficult than by eating a mixed diet of plant and animal foodstuffs. The reasoning behind this statement will become obvious in the evaluation of plants as protein and vitamin sources.

Nutrition includes the processes by which organisms ingest, assimilate, and utilize food. Although maintenance and proper functioning of the human body are clearly dependent upon the selection, quality, and quantity of the food that is eaten, the nutritional requirements of humans are not completely known. In part our ignorance stems from the restrictions against using humans as test organisms and in part from the fact that some essential nutrients are needed in such small amounts that they can be measured only with difficulty. It is also known that bacteria in the human gut make substances we require, but the amounts synthesized are difficult to determine. Some nutritional deficiencies do not show up in short-term studies but may take months or even years to become apparent. These considerations mandate the use of laboratory animals for nutritional evaluations, with all the cautions implicit in attempting to apply information obtained from one species to another.

The nutritional value of foods resides in five major categories of chemicals: proteins, carbohydrates, fats, minerals, and vitamins. In order to evaluate the role of plants in human nutrition, a brief résumé of the chemistry and biological roles of these substances is necessary.

PROTEINS

Proteins are macromolecules composed of 20–22 different amino acids strung together in a very long and heavy chain much as are beads on a string. If you visualize this string consisting of 20–22 different sizes or shapes of beads in some predetermined order and then twisted into a complex ball, this is a reasonable picture of a protein molecule (Figure 4.36). The combination and sequences of the different amino acids, the ratios among them, and the twisting into a specific shape are characteristic of each different kind of protein. Each amino acid is structurally unique although all have carbon, hydrogen, and oxygen atoms plus one or two nitrogen atoms connected to hydrogen atoms in a chemical organization called an *amino group*. It is through this amino group of one amino acid and the acidic group of another amino acid that amino acids are linked together in a peptide bond to form long protein chains. Several amino acids also contain sulfur atoms that bind together and are partly responsible for the three-dimensional shape of proteins. Proteins may combine with other kinds of molecules to form conjugated proteins; the hemoglobin of blood is a protein (globin) attached to an iron-containing *heme*. There are lipoproteins (fats + protein) in blood and glycoproteins (protein + sugars) that function as components of the immune system.

In the animal body, proteins play several vital roles. The amino acids of proteins can be metabolized to yield energy. Some of the proteins are structural, being major components of muscle and cell membranes; they also form parts of cell organelles. Growth depends on the synthesis of these structural proteins, and, since they break down over time and with use, the maintenance of body structure depends on the repair and the replacement of structural protein. Many other proteins have enzymatic functions. Literally thousands of enzymes, each acting to alter a different chemical, are present in every cell. Almost every chemical reaction in cells is mediated by one or more enzymes. All enzymes in all living cells are proteins and their synthesis, maintenance, and increase in number are required for growth and even survival. A third category of proteins, the fibrous proteins, form hair, skin, fingernails, and connective tissues.

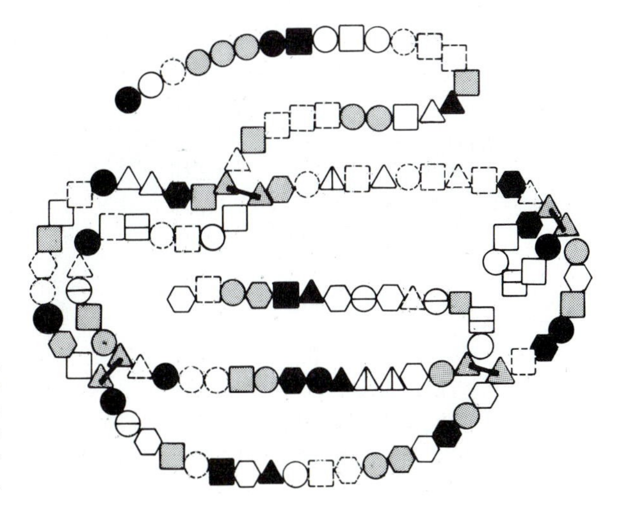

Figure 4.36 Organization of a peptide chain of amino acids into a protein. Each symbol represents a different amino acid. Bars joining triangles (the amino acid, cysteine) represent sulfhydryl bonds.

Ingested protein is digested (hydrolyzed) in the stomach and small intestine by enzymes that separate the constituent amino acids, which can then be absorbed through the stomach and intestine walls and carried throughout the body in the blood stream (Figure 4.37). The amino acids are taken up by living cells that link them together into the types of proteins required by that cell. The types, ratios, and numbers of amino acids forming each protein are coded by the DNA of the cell. Many of the amino acids can be synthesized in the human body from amino groups plus carbon-containing "skeletons" produced from sugars, but at least eight (nine or ten for babies) cannot be made and are considered "essential amino acids" that must be obtained entirely from food (Table 4.2).

A critical feature of the synthesis of protein is that all of the amino acids needed for synthesis of a specific protein must be present and available at the time of protein synthesis. If even one of the amino acids is in low supply, the amount of protein synthesis will be limited by the availability of that amino acid. Thus it is necessary to eat foods that contain a balance of amino acids in the approximate ratios in which they are found in the human body. Plant protein sources tend to have ratios of amino acids different from those of humans and other animals. This is particularly true of the essential amino acids. Compared with animal sources (see Table 4.2), the essential amino acids lysine, isoleucine, threonine, and methionine are low in plants. It is a standard nutritional procedure to rate proteins on the basis of their digestibility and on the relative abundance of essential amino acids. Milk and egg proteins have scores close to 100, but few plant proteins have scores much above 80 because of their relative deficiencies in one or more essential amino acids.

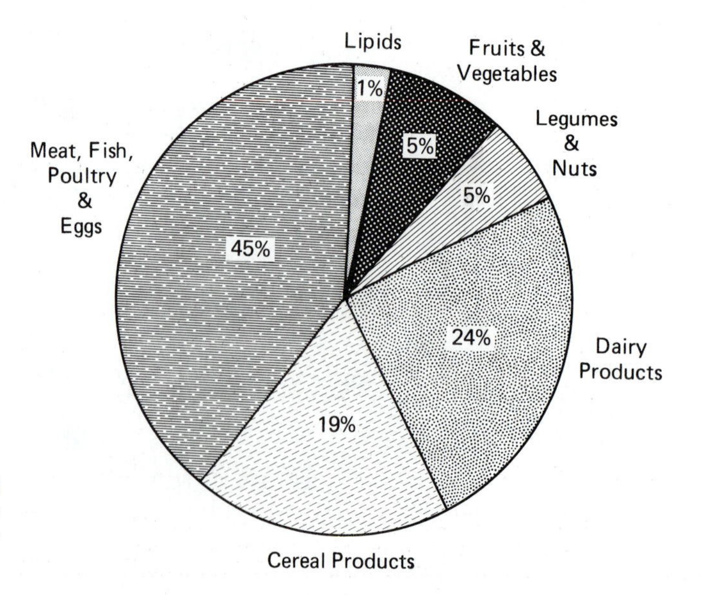

Figure 4.37 The percentages of different food groups to the proteins available in the North American diet.

Table 4.2 NUTRITIONAL REQUIREMENTS FOR ESSENTIAL AMINO AND FATTY ACIDS AND AMOUNTS PRESENT IN VARIOUS ANIMAL AND PLANT SOURCES (Per 100 grams of edible product)

	Isoleucine (mg)	Leucine (mg)	Lysine (mg)	Methionine (mg)	Phenylalanine (mg)	Threonine (mg)	Tryptophan (mg)	Valine (mg)	Linoleic acid (g)
Female, 52 kg (114 lb.)	450	620	500	350	220	300	160	650	42
Male, 70 kg (158 lb.)	800	1100	800	200	300	500	250	800	54
Whole milk	200	430	250	90	230	150	50	250	3.0
Lean beef	850	1600	1500	480	780	800	200	900	2.0
Whole egg	780	1000	860	400	900	600	180	850	1.2
Cereals									
Rice, white	300	600	250	150	350	240	95	400	0.1
Wheat, whole	870	370	370	200	590	380	140	580	0
Corn meal	350	1100	250	180	460	340	60	460	2.0
Millet	300	580	250	150	430	230	90	400	
Bread, enr. white	430	640	240	120	430	250	90	380	0.01
Pasta	350	650	180	140	470	250	100	390	0.5
Bread, enr. rye	360	560	270	130	400	270	100	440	0.05
Root vegetables									
Carrot	30	50	40	14	30	30	8	30	0.6
Cassava	50	64	67	22	49	49	19	50	0.4
Sweet potato	50	70	45	22	50	50	20	60	0.4
White potato	70	120	96	26	80	75	30	90	0.5
Legumes									
Soybeans	1900	3200	2600	500	2000	1600	500	1900	2.0
Peas, fresh	90	160	130	30	100	90	30	110	2.0
Green beans	270	460	480	60	290	250	70	300	0.8
Kidney beans	920	1700	1600	230	1200	900	220	1000	2.1
Peanuts	1000	1900	1000	340	1500	760	300	1200	2.4
Leafy vegetables									
Cabbage	50	90	50	20	50	60	20	70	0.8
Spinach	100	200	160	50	130	120	30	133	0.6
Lettuce	50	80	50	25	70	50	10	70	0.7
Fruits									
Apple	13	23	22	3	10	14	3	15	1.0
Orange	22	22	43	12	30	12	6	31	0.2
Banana	32	53	46	22	44	38	13	45	0.1
Tomato	20	30	32	7	20	25	9	24	0.5
Nuts									
Cashew	1000	1700	900	300	900	650	380	1200	2.0
Walnut	700	300	250	250	750	500	190	800	2.3

A purely vegetarian diet must be planned carefully to provide a sufficient supply of essential amino acids. This is done by eating what has been called *complementary protein mixtures* in which various plant sources are paired to ensure that there is an adequate simultaneous uptake of the required amino acids. Nutritional balance and increased protein score can also be accomplished by supplementing an all-vegetable diet with milk or other dairy products, with or without eggs.

Various cultures have, through long trial and error, devised complementary mixtures. Wheat bread and cheese (Middle East), beans and rice (Latin America), peas and rice (Caribbean), wheat cakes or gruel and legumes (India), rice and soybeans (Orient), and the familiar breakfast cereal with milk and fruit in Western cultures are examples. A mixture called Superamine was developed jointly by the Food and Agriculture Organization, World Health Organization, and UNICEF, all bodies of the United Nations, as a superior infant food. It contains wheat, chick-peas, lentils, sugar, skim milk, and a mixture of vitamins and minerals and provides all of the essential amino acids needed for the rapid growth of children. Incaparina, an all-plant mixture for the same purpose, was devised for protein-poor countries of Central and South America.

JAMAICAN RICE AND RED PEAS

(for four)

1 medium coconut
1 qt. hot water
1 cup red peas or kidney beans
1 clove garlic, minced
1 tsp. salt
½ tsp. black pepper
1 small onion, sliced
1 tsp. thyme
3 cups uncooked rice

1. Grate coconut meat, add hot water, and stir. Squeeze coconut milk out through cheesecloth or a sieve.
2. Add peas or beans to coconut milk in a saucepan. Add garlic, salt, and pepper. Cook until peas are almost tender. Add onion, thyme, and rice.
3. Add additional water and simmer until rice is done and the water is absorbed.

CARBOHYDRATES

Carbohydrates, and to a lesser extent lipids, are people's principal source of metabolic energy. Amino acids from proteins can be metabolized to provide energy, but the body preferentially uses amino acids for protein synthesis. When the level of carbohydrate is inadequate, lipids are used first—the basis for controlled weight loss—but when both carbohydrates and lipids are scarce, body proteins are hydrolyzed and the constituent amino acids are metabolized. The skin-and-bones appearance of starving people (or high fashion models) is one result of the metabolism of structural proteins.

The energy content of foodstuffs is measured as the kilocalorie, abbreviated *kcal* and often called simply *Calorie*. A Calorie is determined by burning the substance to carbon dioxide and water in a device that measures the amount of heat released in the combustion. In terms of energy content, fats are the most concentrated energy source, with protein and carbohydrate being roughly equal in available energy. One gram (1/28 ounce) of any carbohydrate or protein yields 4 kcal while a gram of fat provides 9 kcal. In cells the metabolism of carbohydrates proceeds through the respiratory process with the formation of carbon dioxide and water, which are released from the body via the lungs, and the formation of the main energy-yielding compound, adenosine triphosphate (ATP), and the heat that keeps you warm. The oxygen taken up by breathing is used in metabolic respiration. The number of kcal needed per day depends on age, sex, weight, and activity (Table 4.3); children and older people require fewer calories than active adults. Women, except when pregnant or lactating, require fewer kcal than men.

Table 4.3 ENERGY COSTS ABOVE BASAL METABOLIC RATE FOR VARIOUS ACTIVITIES*

Activity	Female	Male	Activity	Female	Male
	(kcal/hr.)			(kcal/hr.)	
Bicycling (moderate)	145	190	Playing piano		
Carpentry	130	165	(Beethoven sonata)	80	100
Playing cello	75	95	Playing piano		
Dancing	200	275	(Liszt concerto)	120	145
Dishwashing	40	50	Reading aloud	25	30
Dressing	40	50	Running	475	600
Driving automobile	50	65	Sawing wood	320	400
Eating	20	30	Sewing by hand	25	30
Ironing	50	70	Singing loudly	45	60
Jogging	485	670	Sitting quietly	25	30
Knitting	40	50	Skating	200	260
Lying in bed	5	7	Standing relaxed	25	30
Paring potatoes	30	40	Typing	50	70
Playing Ping-Pong	250	320	Walking (3 m.p.h.)	110	170
Playing piano			Walking (4 m.p.h.)	200	250
(Mendelssohn songs)	45	60	Writing	25	30

*Basal metabolic rates: female, 52 kg (114 lb.) = 47 kcal/hr; male, 70 kg (154 lb.) = 72 kcal/hr.

Over half the energy needed by a moderately active person is used simply to maintain bodily functions. This amount, close to 1500 kcal per day, is called the *basal metabolic rate (BMR)*. The BMR is used clinically to evaluate several aspects of human health. There has been considerable discussion of "empty calories," defined usually as sugars or starches unaccompanied by other nutritional necessities. The so-called junk foods are used almost entirely for metabolic energy and may be both helpful and necessary. The nutritional danger lies in the possibility that other required substances will not be eaten, but sugar has never been shown to have deleterious health effects except that it may lead to dental caries. Semirefined sources of sugars—maple or corn syrups, brown sugar, molasses, and honey—have no nutritional advantages over refined sugars, are not more easily metabolized, and do not contain metabolically significant amounts of other nutrients. Their distinctive tastes are the result of minute amounts of a number of chemicals that have no nutritional value.

Plants are excellent sources of carbohydrates and usually are the primary nutritional sources of sugars and starch. Carbohydrates include simple or monomeric sugars consisting of five or six carbon atoms combined with hydrogen and oxygen in the 2:1 atomic ratio found in water; hence the name for this class of chemicals. Carbohydrates may be found free in plants or may exist as multiples (polymers) of the simple sugars. The 5-carbon sugars are used directly for synthesis of body chemicals. The most important 6-carbon sugars are glucose, fructose, and galactose, which are present in most plant cells either free or combined with other sugars (Figure 4.38). Sucrose, the common table sugar, is a dimer of glucose and fructose; maltose, present in malted milk powder, is a dimer of two glucose monomers; lactose, or milk sugar, is a combination of galactose and glucose. These monomeric and dimeric sugars are nutritionally equivalent, although they differ in relative sweetness (Table 4.4). In plants the higher orders of simple sugar combinations are present as dextrins (polymers of a few 6-carbon sugars), starches, and cellulose. Starch is composed of hundreds of glucose units linked together in straight (amylose) or branched (amylopectin) chain arrangements (Figure 4.39). Human saliva and digestive juices contain enzymes that break up (hydrolyze) the chains into dimeric maltose or monomeric glucose molecules, which are then absorbed into the blood stream. Some of the glucose is transformed into glycogen (animal starch with the structure of plant amylose) and stored in the liver, but most of it is taken up by body cells and metabolized.

In the past few years, there has been a large increase in the manufacture of liquid sweeteners from plant starches. Some of these are glucose syrups, while others are primarily composed of fructose, which is sweeter to the taste than glucose. These products are used in processed foods, including soft drinks, bakery products, and canned goods. Artificial sweeteners such as saccharin have no nutritional value, while others consist of several plant-derived amino acids that can be metabolized.

Figure 4.38 Molecular structures of nutritionally important sugars. Hydrogen atoms and carbon atoms in the rings are not shown. Carbon atoms are represented as circles, oxygen atoms as hexagons.

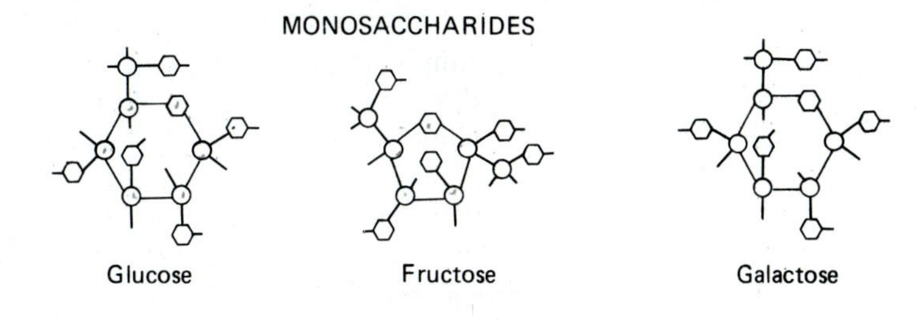

DISACCHARIDES

(glucose) (fructose)
Sucrose
(Table sugar)

(galactose) (glucose)
Lactose
(Milk sugar)

(glucose) (glucose)

Maltose
(Malt sugar)

Table 4.4 RELATIVE SWEETNESS AND ABSORPTION OF NUTRITIONAL SUGARS COMPARED TO SUCROSE

Sugar	Sweetness relative to sucrose	Rate of absorption
	Monosaccharides	
Glucose	75	100
Fructose	175	30
Galactose	30	110
	Disaccharides	
Sucrose	100	70
Lactose	15	90
Maltose	35	90
	Polysaccharides	
Starch	0	0
Dextrins	5	0
Glycogen	0	0
Cellulose	0	0

Table 4.5 NUTRITIONAL REQUIREMENTS AND AMOUNTS PRESENT IN VARIOUS PLANT SOURCES*

	Gr/avg. serving	Approx. serving	Kcal	Protein (gr)	Protein utilization coefficient	Carbohydrate (gr)	Fats (gr)	Calcium (mg)
Female, 52 kg			2100	46		200	500	800
Male, 72 kg			2700	56		250	600	800
Whole milk	240	(c)†	65	4.0	80	5	3.5	120
Lean beef	100	(4 oz.)	160	17.0	65	0	6.0	11
Whole egg	50	(1)	90	7.0	95	0	11.0	50
Cereals								
Rice, white	180	(c)	112	2.0	60	25	tr‡	10
Wheat, whole	120	(c)	330	13.0	60	71	1.6	36
Corn meal	120	(c)	360	8.0	70	75	1.2	7
Millet	130	(c)	340	10.0	55	70	3.0	30
Bread, white, enr.	25	(sl)	270	9.0	65	51	0.3	84
Pasta	130	(c)	150	5.0	60	30	0.8	16
Bread, rye, enr.	25	(sl)	240	8.0	60	52	tr‡	38
Root vegetables								
Carrot	110	(c)	40	2.0	70	10	tr	34
Cassava	150	(c)	150	1.0	60	30	0.5	33
Sweet potato	115	(c)	150	2.0	60	36	1.0	34
White potato	150	(c)	76	1.7	60	17	0.1	8
Legumes								
Soybeans	150	(c)	335	38.0	65	30	18.0	208
Peas, fresh	170	(c)	350	22.0	55	12	1.8	24
Green beans	125	(c)	32	2.0	50	6	0.1	57
Kidney beans	250	(c)	120	8.0	30	21	0.5	140
Peanuts	150	(c)	600	26.0	40	19	50.0	52
Leafy vegetables								
Cabbage	80	(c)	17	1.0	60	6	0.1	100
Spinach	180	(c)	22	3.0	55	1	0.6	80
Lettuce	135	(c)	13	1.0	55	0.3	tr	24
Fruits								
Apple	150	(1)	50	0.1	40	12	tr	6
Orange	130	(1)	80	0.6	40	9	0.1	30
Banana	100	(1)	76	0.6	55	20	0	9
Tomato	135	(1)	19	1.0	40	5	0.3	11
Nuts								
Cashews	140	(c)	560	17.0	60	30	45	37
Walnuts	125	(c)	607	20.0	60	15	60	tr

Iron (mg)	Phosphorus (mg)	Vitamin A (I.U.)	Vitamin D (I.U.)	Vitamin C (mg)	Vitamin B_1 (mg) (Thiamine)	Vitamin B_2 (mg) (Riboflavin)	Niacin (mg)	Vitamin B_6 (mg) (Pyridoxine)	Vitamin B_{12} (μg) (Cobalamin)	Folacin (μg)	Crude fiber (g)
18.0	750	4000	400	50	1.1	1.40	14.0	2.0	3.0	400	3.5
10.0	750	5000	400	50	1.8	1.60	18.0	2.0	3.0	400	3.5
0.1	95	140	50	1	0.04	0.18	0.1	3.4	0.4	8	0
2.4	10	30	0	0	0.07	0.18	4.1	0.4	1.8	3	0
2.5	90	1000	0	0	0.10	0.30	0.1	0.1	2.0	30	0
0.9	30	0	0	0	0.08	0.03	1.6	—	0		0.1
36.0	650	0	0	0	0.45	0.13	5.4	0.2	0		1.6
2.3	250	450	0	0	0.45	0.11	2.0	—	0		1.6
4.0	—	200	0	0	0.33	0.15	2.1	—	0		1.8
2.4	100	tr	0	8	0.30	0.20	2.4	0.1	0	17	0.2
1.5	60	0	0	0	1.30	0.04	1.1	—	0	0	0.4
1.6	148	0	0	0	0.41	0.16	1.3	—	0		0.4
0.8	38	5000	0	6	0.06	0.04	0.7	0.2	0	3	0.6
0.7	30	0	0	26	0.06	0.03	0.1	—	0		0.4
1.0	32	5000	0	23	0.10	0.05	0.1	—	0		0.4
0.7	62	0	0	10	0.10	0.03	1.4	—	0		0.5
6.5	400	140	0	0	1.0	0.3	2.1	—	0	—	2.0
2.0	76	600	0	25	0.4	0.2	2.6	0.16	0	8	2.0
0.8	32	500	0	17	0.1	0.1	0.5	0.08	0	6	0.8
6.7	150	30	0	4	0.5	0.3	2.3	—	0	—	2.1
1.9	412	30	0	0	0.8	0.1	16.0	—	0	—	2.4
2.0	20	3500	0	25	0.1	0.1	0.4	—	0	20	0.8
3.0	37	8000	0	60	0.1	0.2	0.6	0.3	0	29	0.6
0.5	20	200	0	7	0.1	0.1	0.2	—	0	15	0.7
0.3	14	90	0	5	0.1	0.1	0.1	0.2	0.1	0	1.0
0.4	20	150	0	40	0.1	0.1	0.2	0.1	0	45	0.1
0.5	31	200	0	11	0.1	0.1	0.7	0.5	0	27	0.1
0.6	24	700	0	26	0.1	0.1	5.0	—	0	—	0.5
3.5	500	100	0	0	0.4	0.1	2.0	—	0	—	2.0
6.0	380	300	0	0	0.2	0.1	0.7	—	0	—	2.3

*For comparison, amounts present in selected animal sources are provided.
Amounts given are 100 gm of edible product.
†(c) = cup.
‡tr = trace (nutritionally insignificant).

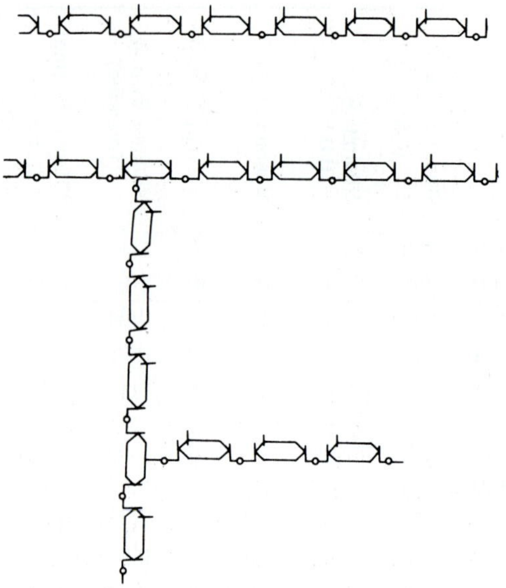

Figure 4.39 Molecular structure of the straight chain of amylose starch and of the branched chain of amylopectin starch. Both consist of individual molecules of the 6-carbon sugar glucose and both are coiled like springs (helical-shaped).

The other major carbohydrate in plants is cellulose, one of the basic substances in plant cell walls. Like starch, it is composed of large numbers of glucose units, linked together in a chemical bond different from that of starch, so that human enzymes cannot break it down into monomeric glucose. Ruminants like cattle and sheep have microorganisms in their stomachs that can hydrolyze cellulose to glucose, so that these animals can obtain sugar for their metabolism. In the human, cellulose fibers are chewed into small fragments that absorb water and swell up to increase the bulk of digested food. This "crude fiber" stimulates muscular movement of the intestines (peristalsis) and speeds food on its way. There is growing evidence that crude fiber is necessary for human health; nutritional levels for fiber have been established (Table 4.5).

Plant starch is usually present in small cell inclusions called amyloplastids (starch granules). The form of amyloplasts differs from species to species and is structurally unique in each species, so that the plant can be identified simply by examining starch grains under the microscope (Figure 4.40). Raw starch is almost indigestible because hydrolyzing enzymes work slowly on amyloplasts. Once plants or plant starch (wheat, potato, or rice flour) is cooked, the starch grains are disrupted and digestion proceeds rapidly. There is no nutritional difference among the various plant starches; those of cereal grains, potatoes, cassava, and legumes are chemically identical. As noted earlier, the starch of Jerusalem artichokes (inulin) is chemically different from other starches, and its metabolism in the human is not well understood.

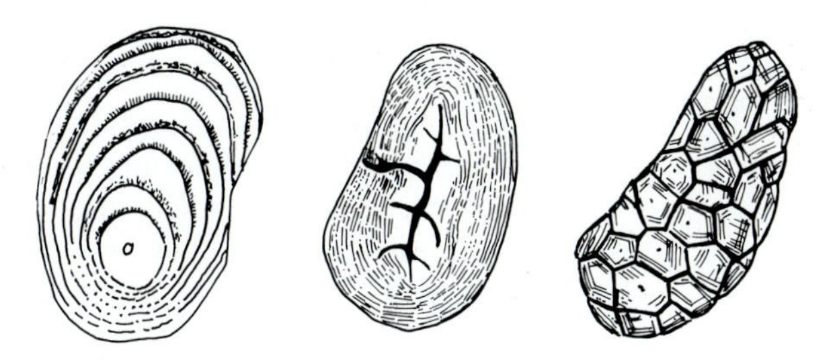

Figure 4.40 Starch grains as seen under a microscope. Left to right: potato, bean, and rice.

LIPIDS

Lipids, including those that are liquid (oils) and those that are solid (fats) at room temperatures, are essential components of all living cells. They are parts of cell membranes, internal cellular constituents, and nerve cells and can be stored as a food reserve in special body cells—the fat cells that are deplored in Western societies. In general, the fat layers beneath the skin are thicker in women than in men.

Lipids are composed of the 3-carbon compound glycerol (glycerin), to which are attached three fatty acids (Figure 4.41). These fatty acids may be all the same or may be different. Fatty acids are long, unbranched chains of carbon atoms bonded to hydrogen atoms and terminating in an organic acid unit. Lipids contain three types of fatty acids: saturated, monounsaturated, and polyunsaturated. When one

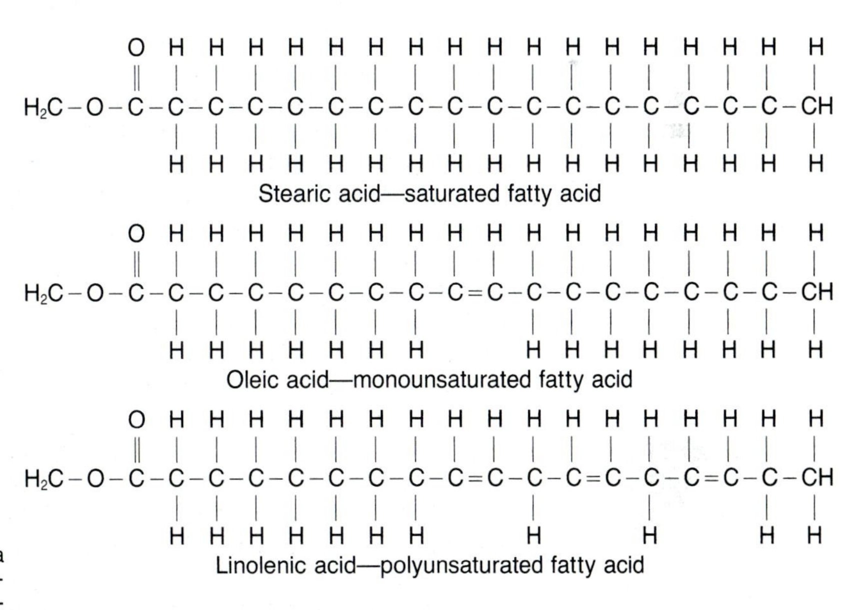

Figure 4.41 Structural formula of a lipid composed of the 3-carbon compound glycerol, bonded to three 18-carbon fatty acids differing in the degree of saturation.

atom in a pair of hydrogen atoms attached to a carbon atom is replaced by a double bond, this carbon atom is called unsaturated because it is capable of regaining the missing hydrogen atom. A fatty acid is monounsaturated when there is only one double bond and is polyunsaturated when there is more than one double bond. The greater the amount of unsaturation in a lipid, the lower is the melting point; lipids with high numbers of unsaturated carbons are usually liquid at room temperature, while those with few or no unsaturated carbons are solid at room temperature.

Oils may be converted into solidified fats by the process of hydrogenation, in which hydrogen atoms are added under pressure to the unsaturated carbon atoms. The solid cooking fats are manufactured by hydrogenation of plant oils from soybean, cottonseed, and other plants. Processed peanut butter has at least part of the peanut oil hydrogenated to prevent the separation of the oil from the ground peanuts. There is considerable medical controversy about the relative health merits of unsaturated versus saturated lipids and cholesterol as a cause of heart disease, but the dispute has not been resolved.

Some of the concern centers on the relative healthfulness of margarine versus butter. Butter has the same saturated fatty acids, though in higher proportions, as those found in the fats of meat. The health consideration, plus a price differential, is responsible for the recent shift among consumers to margarines, which have now captured two-thirds of the 'spread' market. Margarines, like butter, are emulsions of a watery phase in which a lipid phase is dispersed. The watery phase in margarine is usually nonfat milk or whey—essentially the same as the watery phase in butter—and the lipid phase is one or more of the plant-derived oils from soybean, corn, safflower, or cottonseed. Stabilizers, preservatives, and orange coloring (frequently carotene) result in a product that is at least superficially like butter. Indeed, some margarines contain butter to provide a more acceptable odor and flavor. Margarines are considerably lower in saturated fatty acids than is butter, with the soft margarines containing more unsaturated lipid than the hardened products.

At least one unsaturated fatty acid is a nutritional essential. Linoleic acid, present in virtually all plant lipids except olive oil, cannot be synthesized in the human body. This essential fatty acid is present in reasonable amounts in corn meal, some legumes, and nuts (Table 4.6). In the absence of adequate levels of linoleic acid, growth is poor and there is scaliness of the skin. Two other fatty acids, linolenic and arachidonic, appear to be beneficial but not essential to human health, suggesting that the body can synthesize part of the daily requirement. Vitamins A, D, and E are soluble in lipids and are ingested and transported in lipids. In general, plant foods very low in lipids rarely contain significant quantities of these vitamins. At best, plants are low in the fat-soluble vitamins (see Table 4.8).

Table 4.6 FATTY ACIDS IN COMMON FOODS

Product	Fatty acids as percentages of total lipid			Ratio of polyunsaturated to saturated
	Saturated	Monounsaturated	Polyunsaturated	
Butter	55	35	4	0.07
Egg yolk	32	49	12	0.40
Beef fat	48	44	3	0.06
Fish	15	25	53	3.50
Chicken	32	38	26	0.80
Chocolate	56	37	2	0.04
Margarines	26	57	13	0.50
Safflower oil	8	15	72	9.00
Corn oil	11	32	53	11.00
Soybean oil	15	20	59	3.90
Cottonseed oil	25	21	59	3.90
Peanut oil	18	47	29	1.60
Olive oil	11	76	8	0.70
Coconut oil	86	7	trace	—

METABOLIC INTEGRATION

The metabolism of proteins, carbohydrates, and lipids in all living cells in highly integrated and interdependent. As seen in Figure 4.42, amino acids and fatty acids can be converted into sugars that can be metabolized to yield energy, and sugars can be converted into amino acids for protein synthesis or into fatty acids and glycerol for storage as concentrated sources of metabolic energy. Thus excess intake of protein or of carbohydrate can lead to lipid deposition, while dietary restrictions can lead to weight reduction and loss of body fat. The presence in the diet of adequate but not excessive carbohydrate will allow the amino acids to be used preferentially for protein synthesis rather than for energy metabolism, an important consideration in those cultures where protein is scarce and consequently expensive. In the human, the interconversions do not include the essential fatty acids or the essential amino acids.

MINERALS

Proteins, carbohydrates, lipids, and the vitamins are organic molecules, composed of compounds of carbon, hydrogen, and oxygen atoms, with nitrogen and sulfur in proteins. Together with water, these organic molecules comprise over 95 percent of total body weight. The minerals make up the remaining 5 percent but are no less essential. Over 30

Table 4.7 MINERAL ELEMENTS KNOWN AND PRESUMED TO BE REQUIRED FOR HUMAN HEALTH

Classification	Elements	Percent of body weight
Macronutrient elements essential for human nutrition(> 0.005% body weight or 50 p.p.m.)	Calcium	1.5–2.2
	Phosphorus	0.8–1.2
	Potassium	0.35
	Sulfur	0.25
	Sodium	0.15
	Chlorine	0.15
	Magnesium	0.05
Micronutrient elements essential for human nutrition (< 0.005% body weight)	Iron	0.004
	Zinc	0.002
	Selenium	0.0003
	Manganese	0.0002
	Copper	0.00015
	Iodine	0.00004
	Molybdenum	
	Cobalt	
	Chromium	
	Fluorine	
Elements for which essentiality has not yet been established, although there is evidence of their participation in certain biological reactions	Vanadium	
	Barium	
	Arsenic	
	Bromine	
	Strontium	
	Cadmium	
	Nickel	
Elements found in the body but for which no metabolic role has been elucidated	Gold	
	Silver	
	Aluminum	
	Tin	
	Bismuth	
	Gallium	
	Lead	

mineral elements have been isolated from animal tissues, although only 17 have been shown to be required. Another 7 are known to be involved in some metabolic processes in laboratory studies, but their essentiality for humans has not been established (Table 4.7). The rest are probably present as a result of ingestion of plant or animal foods that also acquired them by chance. In elevated concentrations, members of the third and fourth groups of elements can be toxic or fatal; they are acquired by plants as a result of water or air pollution or may be present in soils and taken up by the plants and animals we eat. In fact, imbalances among the essential minerals can also have undesired consequences. Excess levels of sodium relative to potassium or the reverse can seriously affect metabolism.

Essential minerals are conveniently separated into two classes, depending on the amounts needed to maintain maximum health (Table 4.7). The macronutrient minerals are required in amounts exceeding

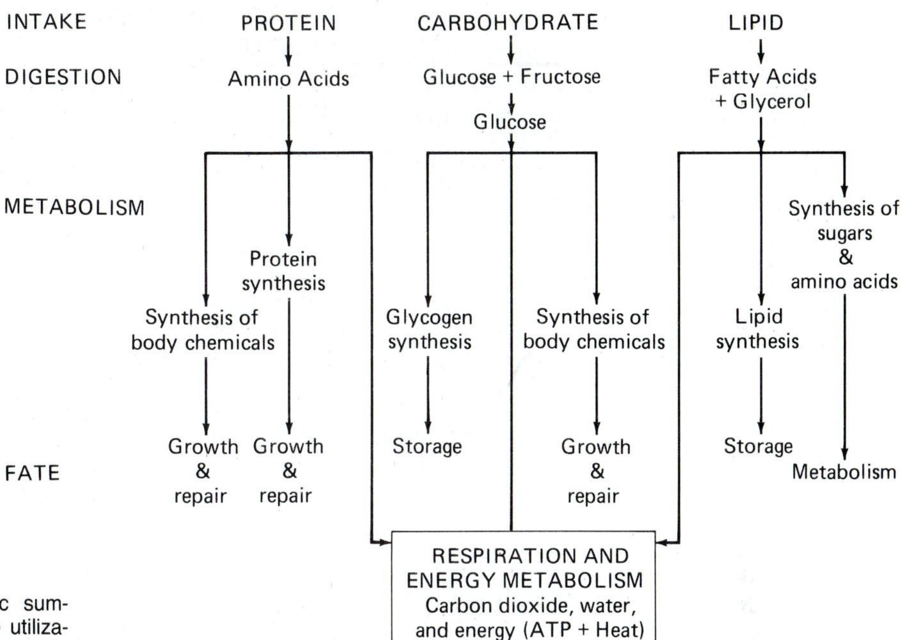

Figure 4.42 Diagramatic summary of pathways for the utilization of dietary nutrients.

0.05 percent of body weight (for magnesium), while the micronutrient minerals range down to 1/1000 of that amount (for iodine). It is possible that some of the minerals for which essentiality has not been established may turn out to be essential, but in amounts that are at the limits of detection.

In general, minerals serve five major roles in metabolism of all living cells. Especially in animals, calcium and phosphorus are structural components of bone and teeth. The maintenance of acid–base balance in blood and body fluids is regulated by the proportions of sodium, potassium, calcium, and magnesium on the base side of the equations and by chlorine, phosphorus, and sulfur on the acid side. Water balance, too, is regulated by minerals, especially by sodium, potassium, chlorine, and bicarbonate ions derived from metabolically produced carbon dioxide. The proper functioning of many enzymes depends on the presence of minerals either as parts (coenzymes or metal-enzyme complexes) of enzymes or as activators or regulators of enzyme action. Sodium, potassium, and calcium are required for the transmission of nerve impulses, the electric signals for nerve and brain function.

A mineral may play several roles in body function. Calcium, for example, is essential for all five metabolic activities and is needed for the activity of cell membranes. It is not surprising that it is required in the highest amount of all minerals, with 99 percent being deposited in bone. Many people, particularly lactating women and the elderly—again, especially women—develop fragile bones (osteoporosis) as a result of the removal of calcium from the bone structures; the recommended daily intake levels of calcium for the elderly may be 50 percent higher than for the young adult. While dairy products are the most common source of calcium, bread (frequently made with milk), whole wheat, legumes, and cabbage are excellent sources of dietary calcium. Spinach, rhubarb, and other plants contain oxalic acid, which interferes with calcium absorption. Because these foods contain other nutritional substances, however, their use need not be discontinued.

Although iron is required in low amounts, iron deficiencies are sufficiently common to suggest that iron, like calcium, is frequently absorbed in inadequate amounts. As a part of the red hemoglobin of blood, iron is needed for red blood cell formation. Inadequate iron leads to simple anemia, a condition in which each red blood cell contains less than normal levels of oxygen-carrying hemoglobin. This results in an oxygen deficiency at the cellular level, poor energy metabolism, and consequent feelings of tiredness. Iron is also a part of several enzymes and respiratory compounds (cytochromes) in muscle and body cells. Although iron is recycled from worn-out red blood cells, some is excreted in feces and must be replaced daily. There is a small but significant loss of iron during menstruation; blood loss through wounds also must be compensated for by additional iron intake. Meat and eggs are common sources of iron. Although whole grain, legumes, and nuts may provide adequate amounts, plant sources are generally low in available iron (See Table 4.5). Most commercially available wheat flours and breads are fortified with iron. Iron incorporation into body cells is facilitated by vitamin C, and there is some interference of absorption by crude fiber (bran) of cereal grains and the tannins of tea.

Most of the body's phosphorus is combined with calcium in bone, indicating that during growth, high phosphorus intake is crucial. This element is also required for energy metabolism, both as an intermediate in respiratory reactions and as a constituent of the energy-yielding compound ATP (adenosine triphosphate). It is also part of the nucleic acids that exert genetic control over protein synthesis. A few structural components in nerve cells also contain phosphorus. Plants are excellent phosphorus sources. Because cereal grains and seeds (legumes and nuts) can easily provide adequate nutritional levels of phosphorus, deficiencies are rarely seen except during starvation.

Of the other essential macro- and micronutrients, few are in low supply in the normal diet. Plant cells contain more potassium than animal cells, but moderate use of table salt (sodium chloride) and plant foods will satisfy the requirements for both these elements that are needed for acid–base balance and for nerve impulse transmission. The sulfur required by humans is present in protein, primarily in the form of the amino acid cysteine and the essential amino acid methionine. Plant cells contain sulfur amino acids, and substantial amounts are present in plants as the sulfate ion ($-SO_4^{-2}$). Magnesium, required for bone formation and as a cofactor for several enzymes, is present in adequate nutritional amounts in plants. Iodine is a part of the thyroid hormone thyroxin. Enlarged thyroid glands (endemic goiter) in adults and reduced mental capacity (cretinism) in children are consequences of inadequate iodine uptake before and just after birth. In regions near the ocean, where iodine is present in sea water, in seaweeds, and in food plants grown in soil receiving sea spray, iodine is in sufficient supply in plants, but in inland areas, plant foods are inadequate as an iodine source. In such localities, iodine is best provided in iodized salt, although small amounts are found in whole wheat, peanuts, and spinach.

VITAMINS

The word *vitamin* refers to a number of structurally varied organic chemicals, usually of small size, that play a variety of vital metabolic roles, some of which are known only partially. Most of the B vitamins function as cofactors for enzymes, frequently in conjunction with mineral cofactors; deficiencies of these substances lead to metabolic failures that can be manifested by a range of clinical symptoms (Table 4.8). For convenience, and with some nutritional significance as well, they are divided into those that are fat-soluble and must be ingested in lipid-containing foods and those that are water-soluble.

Early evidence of a need for special nutritional substances was provided in the eighteenth century when it was discovered that sailors on long voyages came down with scurvy (a vitamin C deficiency disease), which could be prevented by citrus fruits. The term *limey* as slang for British sailors had its origin in the inclusion of lime juice in the rations. In the early nineteenth century it was observed that Japanese sailors came down with beriberi (a vitamin B_1 deficiency disease) when forced to subsist solely on white rice without vegetables. The recognition that the absence of specific dietary substances resulted in a number of severe and frequently fatal nutritional diseases prompted searches for the preventing substances. In animal studies dating from the early twentieth century, the missing dietary components were identified; in recent years, many of them have been synthesized.

With one exception, vitamin D, vitamins are not synthesized by the human body, although microorganisms in the human gut provide at least part of the daily requirements. Microorganism syntheses allow a person to survive without deficiency symptoms for long periods of time even when vitamin intake is below required levels. While popping vitamin pills is a well-developed compulsion in wealthy industrial nations, this practice is unnecessary for healthy adults who eat a balanced diet. For pregnant women, babies, young children, and the elderly, vitamin supplements are frequently recommended. The fat-

Table 4.8 VITAMINS NEEDED IN HUMAN NUTRITION AND MAJOR SOURCES

Vitamin	Some deficiency symptoms	Important sources
	Fat-soluble	
Vitamin A (retinol)	Dry, brittle epithelia of skin, respiratory system, and urogenital tract; night blindness and malformed rods	Green and yellow vegetables and fruit, dairy products, egg yolk, fish-liver oil
Vitamin D (calciferol)	Rickets or osteomalacia (very low blood calcium level, soft bones, distorted skeleton, poor muscular development)	Egg yolk, milk, fish oils
Vitamin E (tocopherol)	Male sterility in rats (and perhaps other animals); muscular dystrophy in some animals; abnormal red blood cells in infants; death of rat and chicken embryos	Widely distributed in both plant and animal food—including meat, egg yolk, green vegetables, seed oils
Vitamin K (phylloquinone, etc.)	Slow blood clotting and hemorrhage	Green vegetables
	Water-soluble	
Thiamine (B_1)	Beriberi (muscle atrophy, paralysis, mental confusion, congestive heart failure)	Whole grain cereals, yeast, nuts, liver, pork
Riboflavin (B_2)	Vascularization of the cornea, conjunctivitis, and disturbances of vision; sores on the lips and tongue; disorders of liver and nerves in experimental animals	Milk, cheese, eggs, yeast, liver, wheat germ, leafy vegetables
Pyridoxine (B_6)	Convulsions, dermatitis, impairment of antibody synthesis	Whole grains, fresh meat, eggs, liver, fresh vegetables
Pantothenic acid	Impairment of adrenal cortex function, numbness and pain in toes and feet, impairment of antibody synthesis	Present in almost all foods, especially fresh vegetables and meat, whole grains, eggs
Biotin	Clinical symptoms in humans are extremely rare, but can be produced by great excess of raw egg white in diet; symptoms are dermatitis, conjunctivitis	Present in many foods, including liver, yeast, fresh vegetables
Nicotinamide	Pellagra (dermatitis, diarrhea, irritability, abdominal pain, numbness, mental disturbance)	Meat, yeast, whole wheat
Folic acid	Anemia, impairment of antibody synthesis, stunted growth in young animals	Leafy vegetables, liver
Cobalamin (B_{12})	Pernicious anemia	Liver and other meats
Ascorbic acid (C)	Scurvy (bleeding gums, loose teeth, anemia, painful and swollen joints, delayed healing of wounds, emaciation)	Citrus fruits, tomatoes

soluble vitamins can be stored for relatively long periods of time in body cells, but excess amounts of vitamin C and B vitamins are eliminated quickly.

Fat-Soluble Vitamins Vitamin A includes several related compounds produced only by plants. Animals obtain their vitamin A from the plants they have eaten, especially from plants (including marine algae) that contain large amounts of the orange pigment carotene. The carotene, present in all green, photosynthetic cells and in large amounts in carrots and sweet potatoes, is converted enzymatically in animal cells into the chemical form in which the animal can utilize it. The major function of vitamin A is as light-capturing pigments in the eye, the retinal pigments rhodopsin and retinal. In laboratory animals vitamin A is also required for bone growth, although this role has not been fully demonstrated in the human. Vitamin A deficiency is manifested in poor night vision, dryness of the tissues of the eyelids, and dry, scaly skin. Young children with prolonged lack of vitamin A may become blind. Although it is difficult to retain excess amounts of carotene or vitamin A with a normal diet, use of supplements can lead to vitamin A toxicities, with increases in pressure on the brain, excessive fatigue, and other symptoms as possible results. In rare instances, the skin and eyeballs appear yellow.

Orange, yellow, and dark-green vegetables are excellent sources of carotenes, the precursors of vitamin A (see Table 4.5). Yellow corn and millet are important cereal sources, but some yellow plants (oranges, wax beans, apples) contain a derivative of carotene (xanthophyll) which cannot be used for vitamin A synthesis.

Rickets, a disease of children in which bones do not calcify, is the major symptom of vitamin D deficiency. With the photochemical synthesis of vitamin D and its use to fortify milk, rickets has almost disappeared in the industrial world. It can still be seen, however, in adults who were children in the early decades of this century, and it is not uncommon in poor countries. One major metabolic role of vitamin D is to facilitate the retention and transport of calcium to bone. Vitamin D is a sterol, chemically related to sex hormones. Healthy individuals can satisfy adult requirements for the vitamin by normal exposure of the face and hands to sunlight, which converts a precursor of vitamin D into the active compound. Adult requirements are in fact, small (see Table 4.5). The pernicious practice of sunbathing has little effect on vitamin D formation in adults, but it puts the person at risk for skin cancer, sunburn, and senile elastosis (the skin wrinkling seen more frequently in women than in men of the same age). Excessive storage of vitamin D (hypervitaminosis D) is potentially dangerous, leading to abnormalities in calcium metabolism and kidney damage. The active wavelengths of sunlight are in the ultraviolet range and are not transmitted through window glass. Children can usually obtain enough vitamin D in their milk supply, but the elderly, at risk for bone resorption, may require vitamin supplements. No plant sources of vitamin D are known.

Working with laboratory rats, investigators in the 1920s found that a fat-soluble factor, named vitamin E, was necessary for reproduction. Renamed *tocopherol* (Greek *tokos* = "childbirth"; *pherin* = "to bear"), it is a yellow oil that can now be synthesized. Its role in the human is poorly understood, although it is now suggested that vitamin E prevents destructive oxidation of cell constituents. While a range of deficiency symptoms have been reported in laboratory animals, none have been found in humans, although megavitamin supplementation of diets can lead to toxicity. Liver, eggs, and kidney are the best animal sources, and many vegetable sources of vitamin E are known. Vegetable oils from the embryos of cereal grains (wheat germ), cottonseed, and soybean oils and solid fats made from these oils are high in vitamin E.

When chicks were raised on highly purified diets, they developed abnormalities that could be cured by supplementing the diet with whole grain cereals. The substance, first named vitamin K, was later identified as several related quinones, substances that are found in relatively large amounts in leaves and in the products of bacterial metabolism. Because the bacteria in the human gut make this vitamin, deficiency symptoms have rarely been observed in healthy people, although individuals whose blood-clotting time is very long may have a vitamin K deficiency (there are other causes of poor clotting). Vitamin K is in adequate supply in vegetables, including spinach, cabbage, and cauliflower.

Water-Soluble Vitamins Vitamin C is a small molecule chemically related to the sugars. It is formally known as ascorbic acid, reflecting one of its primary functions—the prevention of scurvy, a metabolic disease involving capillary fragility. It also plays a role in protein and lipid metabolism, in the utilization of iron, and in keeping body chemicals from oxidizing. Megadosing with natural or synthetic vitamin C has both advocates and detractors. The evidence linking vitamin C to better athletic performance or reduction in the number and severity of common colds is scientifically inadequate at this time. There is no evidence for and considerable evidence against its having any role in preventing cancer, heart disease, or other ailments. Vitamin C is not found in nutritionally significant amounts in animal foods and must be obtained from citrus and other fruits and from green leafy vegetables and root crops (see Table 4.5). Vitamin C is unstable, being destroyed by sunlight and heat. Fruit juices should be kept in the refrigerator. Canned vegetables have lower vitamin C levels than fresh produce.

The B vitamins include a substantial number of chemically and biologically unrelated compounds, linked primarily by their water solubility and their roles as cofactors in enzyme action. Several are cofactors in the cellular respiratory pathway from glucose to carbon dioxide, water, and ATP energy formation that is common to both plants and animals. Others are involved in nerve signal transmission, in nucleic acid synthesis, and in other vital processes. Since the clinical symptoms of vitamin B deficiencies involve a wide variety of body functions, impairment of energy metabolism is not the only biochemical consequence of deficiency; other metabolic processes must also involve these vitamins, but not all are understood.

The first to be discovered was vitamin B_1, chemically called thiamin.

Severe clinical deficiency is seen as beriberi, a malfunctioning of the nervous system, while less severe deficiencies are noted as irritability, emaciation, loss of appetite, and weakness in legs and arms. Thiamin is found in small amounts in most plant foods and in high concentrations in wheat germ and in yeast (see Table 4.5). Breads and pasta are frequently enriched with thiamin. Although heat-stable, its solubility leads to its extraction in cooking water, which is frequently discarded.

Vitamin B_2, riboflavin, is a bright yellow-orange compound stable in air and to heat but inactivated by light. It is part of respiratory cofactors. Deficiency symptoms include skin lesions, inflamations of the lining of the eyes, and general body weakness. Riboflavin is one of the chemicals added to flour, breads, and pasta, although dairy products and meats usually account for over half the daily intake of this vitamin. Plants are adequate sources (see Table 4.5).

Pellagra, a serious and frequently fatal nutritional disease caused by a lack of the B-complex vitamin niacin (nicotinic acid), is now seen primarily in food-poor areas where access to animal products, including fish, is limited (Chapter 3). Whole grain cereals and peanuts are excellent sources (see Table 4.5), coffee contains a reasonable amount, and brewer's yeast (and hence beer) is also a fine source. The body can synthesize at least part of its niacin requirement from the amino acid tryptophan, although this essential amino acid is rarely in high concentrations in plant proteins and is preferentially used for protein synthesis.

Three closely related chemicals are considered to be interchangeable as vitamin B_6; the general name is pyridoxine. For the most part, the B_6 vitamins play their major role as enzyme cofactors: there are over 50 pyridoxine-requiring enzymes involved in cellular respiration and amino acid metabolism. In young children, a lack of the B_6 complex leads to anemia, skin lesions, and reduced antibody formation. Adults may show dermatitis with red, rough, and scaly skin and nervous irritability. Both animals and plant foods contain pyridoxines, with whole grain cereals being excellent sources.

Vitamin B_{12} is a red, cobalt-containing compound. Its absence causes severe intestinal disorders. Dietary imbalances and infestations with tapeworms may both cause deficiencies. In people in their 50s and 60s, pernicious anemia is not uncommon and can be cured with large intakes of B_{12}. Because plant foods do not contain B_{12}, vegans must rely on bacterial synthesis in the gut (effective up to about age 45–50), vitamin supplements, or large intakes of brewer's yeast, a moderately good source of the vitamin.

Folacin (folic acid) is a B-complex vitamin whose absence leads to anemia in women, swelling of the tongue, and disturbances in intestinal function. Proper absorption of folacin depends on an adequate supply of vitamin B_{12}; lack of vitamin C allows folacin to be oxidatively destroyed. High alcohol use also interferes with folacin utilization; chronic alcoholics are notable examples of folacin deficiency. In addition to liver, folacin is found in adequate amounts in leafy vegetables such as lettuce and cabbage and in the legumes (see Table 4.5).

There are several other B vitamins whose roles in human nutrition are not well known since most of the research has been done with laboratory animals. Based on what is known about vitamin nutrition of the human body, pantothenic acid and biotin both are important in enzyme-regulated reactions and are found in satisfactory amounts in plant food sources.

ADDITIONAL READINGS

Altschul, A., and H. L. Wilcke (eds.). *New Protein Foods.* Vol. 5, *Seed Storage Proteins.* New York: Academic Press, 1985.

Crispeels, M. J., and D. Sadava. *Plants, Food, and People.* San Francisco: Freeman, 1977.

Gerard, I. D. *The Story of Food.* Westport, Conn.: Avi Books, 1974.

Hansen, R. G., B. W. Wyse, and A. W. Sorenson. *Nutritional Quality Index of Foods.* Westport, Conn.: Avi Books, 1979.

Lee, F. A. *Basic Food Chemistry.* Westport, Conn.: Avi Books, 1983.

Lowenberg, M. E., E. N. Todhunter, E. D. Wilson, J.R. Savage, and J. L. Lubauski. *Food and People.* New York: Wiley, 1979.

Lusk, G. *Elements of the Science of Nutrition.* 4th Ed. New York: Academic Press, 1984.

Nasset, E. S. *Nutrition Handbook.* 3rd Ed. New York: Harper & Row, 1982.

Pennington, J. A. T., and A. N. Church. *Food Values of Portions Commonly Used.* 14th Ed. New York: Harper & Row, 1984.

Pomeranz, Y. *Functional Properties of Food Components.* New York: Academic Press, 1984.

Pyke, M. *Man and Food.* New York: McGraw-Hill, 1970.

Reichcigl, M. (ed.). *CRC Handbook of Nutritional Value of Processed Foods.* Vol. 1, *Food for Human Use.* Boca Raton, Fla.: CRC Press, 1982.

Robinson, R. K. *The Vanishing Harvest: A Study of Food and Its Conservation.* Oxford, England: Oxford University Press, 1983.

Rose, A. H. *Fermented Food.* New York: Academic Press, 1981.

Weiss, T. J. *Food Oils and Their Uses.* Westport, Conn.: Avi Books, 1970.

Wilson, E. D., K. H. Fisher, and P. A. Garcia. *Principles of Nutrition.* 4th Ed. New York: Wiley, 1979.

Yamaguchi, M. *World Vegetables: Principles, Production and Nutritive Values.* Westport, Conn.: Avi Books, 1982.

Spices and Savory Herbs

Are you going to Scarborough fair?
Parsley, sage, rosemary and thyme.

English Folksong

Authorities on spices and herbs suggest that the desire for these substances in the Middle Ages stemmed from the wish to control the development of strong odors and the taste of meats that had been sitting around too long. It is reasonable to believe that they were also employed just as we use them, to make food more appetizing and less boring. Consider a stew made without bay leaf, pepper, an herbal bouquet, a dash of mace, and another dash of ground cloves—dull and uninteresting. Pickles made only with salt are merely salted cucumbers, and steamed spinach without a sprinkle of nutmeg has turned millions of children against this succulent green vegetable. North American supermarkets display well over 50 spices and herbs, but we use few of them because of lack of imagination. We also have the North American ignorance or disdain for the wonderful cooking of non-European cultures.

There is considerable nomenclatural confusion about spices and herbs. The word *herb* is itself misleading. Botanically, it refers to a plant that is either an annual or that dies down to the ground each fall. Medically, it refers to plant material from which drugs can be extracted. Gustatorially, it means the leaf of a plant used in cooking. *Spice* is an even more obscure word; Christopher Morley defined *spice* as the plural of *spouse,* and under happy circumstances a spouse can be the spice of one's life. For our purposes, we will operationally define *spice* as a plant part such as a seed, stem, or root. The two terms are not mutually exclusive; the confusion can be left to lexicographers and ignored by botanists and chefs.

The physiological responses to spices and seasonings are complex. Basically, we can taste only a limited array of chemical flavors: acidity or sourness, sweetness, saltiness, and bitterness. These sensations are not specific; the malic acid of apples, the citric acid of lemons, or the sharp taste of vinegar are indistinguishable to the tongue. Similarly, the saltiness of table salt or that of other inorganic salts cannot be told apart. One exception to this is the sharpness or burning associated with peppers, whether the black pepper (from amides of vanillylamine), the mustards, or the capsaicins of chili peppers. Menthols of mint leaves produce a cool feeling in the mouth. Most of our appreciation of spices and herbal seasonings in food comes from our sense of smell. The particular flavor of cinnamon is the result of a combination of tongue and mouth stimulation of one or more of the basic tastes and the simultaneous perception of the odor of this spice through the nose and mouth. The flavor components include a wide range of chemicals. Most of them have been chemically characterized by modern gas chromatographic techniques, and some have been synthesized.

Figure 4.43 Pounding horseradish root. Taken from a German woodcut published in 1493.

SEASONINGS IN EUROPE

Since our food habits are those of Europe, a look at the seasonings available in the eighth century can be instructive. The northern countries had few seasonings. Meats were dipped into pulverized mustard or into horseradish root pounded with vinegar and salt (Figure 4.43). Dill seed (Figure 4.44) and young shoots, celery seed and leaves, wintergreen and peppermint leaves, and garlic, onions, and leeks were available. Winter savory and thyme were not generally used, although the plants grew in Germany and Scandinavia. The flavoring spectrum in southern Europe was considerably broader. In addition to peppermint, the Mediterranean Basin supported the growth of many members of the mint (Labiatae) family, including basil (*Ocimum basilucum*), marjoram (*Marjorana hortensis*), oregano (*Origanum spp.*), rosemary (*Rosmarinus officinalis*), sage (*Salvia officinalis*), savory (*Satureja hortensis*), and thyme (*Thymus vulgaris*). It is interesting to note that all these go exceedingly well with tomatoes, which were introduced into Italy in the sixteenth century. The parsley (Umbelliferae) family was also well represented in southern Europe. Caraway (*Carum carvi*) was carried by Roman legions from its home in Asia Minor throughout the known world for stews, as seasoning for northern Europe's breads and sauer-

Figure 4.44 Dill *(Anethum graveolens).* Leaves and stems are dillweed; the seeds are used as spice in soups, meats, and salads. Courtesy U.S. Department of Agriculture.

krauts, and for addition to the Scandinavian aquavit and the German *kümmel* schnapps, either of which is sipped with beer while eating herring and boiled potatoes. Chervil (*Anthriscus Cerefolium*) was used as an alternate for parsley, and in England "its tender tops . . . are never wanting in our sallats." Coriander (*Coriandrum vulgare*) was used in sausages; our hot dogs still contain it. Cumin (*Cuminum cyminum*), initially from North Africa, where it is used in *cous cous,* was so valuable that it was used to pay taxes in Roman Palestine: "You tithe mint and dill and cummin and have neglected weightier matters" (Matthew 23:23). Anise (*Pimpenella anisum*) was used in baking, and Romans hung a plant in their bedchambers to prevent nightmares. Charlemagne required that the plant be grown on imperial lands as his personal condiment; its licorice flavor was used in cordials, to cover up the bad taste of medicines, and in the preparation of cookies. Its addition to cottage cheese and to the water used to boil shrimp is recommended. Fennel *(Foeniculum vulgare)* and celery *(Apium graveolens)* were grown as vegetables, and their seeds were and still are used in sausages and stuffings for fowl. Both provide a most agreeable addition to boiled vegetables.

Yet, with the exception of mustard and horseradish, these are relatively mild seasonings, and the palate-stimulating and saliva-generating flavors and odors of pepper, bay, clove, cinnamon, nutmeg and mace, ginger, and others were almost lacking from the foods of Europe. One has to say *almost,* not absolutely, because small and very precious supplies of the spices of the Orient were known as far back as the time of ancient Egypt.

In 1874 George Elbers, a German Egyptologist, discovered a long papyrus roll now known as the Elbers papyrus. It was a medical treatise in which were listed herbs and spices used to treat disease and in the medico-religious art of embalming. The body cavity of the corpse was treated with ground cinnamon, cumin, anise, and marjoram plus cloves and nutmeg. Cinnamon and cloves were soaked in sweet oil for anointing, and stick cinnamon was burned as incense to purify sick rooms. Since these plants did not grow in the Nile Valley, they were transported from Madagascar, Indonesia, and India to the shores of east Africa, through the Gulf of Aden and the Red Sea, and finally overland to Thebes. The same route was also used to carry cotton cloth from India to the land of the pharaohs (see Chapter 9). From Egypt, cargoes of spices could be carried across the Mediterranean to Greece. Herodotus reported on cinnamon sold by Arabic merchants. Besides their use in medicines, pepper, cinnamon, ginger, and other spices were added to wine and bread. Hippocrates included saffron among the effective herbal medicines and recommended black pepper in honey and vinegar for "feminine disorders." We know that in addition to the sea route, there was at least one overland route. Consumed with envy, Joseph's brothers got rid of Jacob's favorite by selling him to a caravan of Ishmaelites bearing spices to Egypt (Genesis 37:25). When the Queen of Sheba visited Solomon, she brought with her "a very great retinue and camels bearing spices" (2 Chronicles 9:1). Some were not known to the Israelites, for 2 Chronicles 9:9 states that "there were no spices such as those which the Queen of Sheba gave to King Solomon."

SPICE CARAVANS BY LAND AND SEA

By 2300 B.P., caravans of camels were carrying the vaunted spices of the Orient to the world of the Mediterranean. Some of these spice and silk routes were very ancient. One led from the port of Muziris through the Persian Gulf to Charax, then up to Aden and through the Red Sea. A land route went up the Indus Valley and along the Kabul River, climbed the mountain passes of the Hindu Kush, and entered the Middle East at Bactra. Caravans based in Bactra transported the goods to modern Iran, thence overland to Antioch or by sea to Egypt, where merchants repacked them and shipped them across the Mediterranean to Europe. A caravan would usually spend two years on a trip, and the dangers of dust storms, bandits, thirst, and starvation served to keep prices so high that a bit of cinnamon bark or a peppercorn could be worth its weight in silver. For the most part, control of the spice trade was an Arabian monopoly. Rome almost broke the Arabian monopoly when a merchant discovered that the wind systems in the Indian Ocean reversed their direction twice a year. The April–October monsoons favored the journey from Aden to India, and the October–April winds permitted the return voyage. It still took a full year for the round trip, but this was half of what the journey formerly required, it was much safer, and it allowed bigger payloads than the caravans. Prices began to drop with a corresponding increase in spice utilization. As the ultimate in conspicuous consumption, Romans piled their funeral pyres high with cinnamon, cloves, and nutmeg, annointed their bodies with fragrant oils, and added "heat" to their foods and wines with numbers and quantities of spices that modern people would find unpalatable.

The extension of Roman civilization to northern Europe and to Britain was accompanied by the introduction of spices to these lands and also contributed to the downfall of Rome. The semicivilized northern Goths tired of paying tribute and envied the foods, wines, and silks of the Romans. When Alaric appeared at the walls of Rome in 408, he demanded gold, silver, silk, and 3000 pounds of peppercorns in exchange for his promise not to sack the city. Although the tribute was paid, Alaric returned two years later and took Rome, thus ending the hegemony of the empire. Constantinople had been built in 330 to serve as a trade center for goods coming from the East. Cloves and nutmeg were brought by Indonesian sailors from the Molluca Islands to Aden, where they reached Constantinople via the Red Sea. In the early sixth century a monk, Cosmos Indicopleustes, traveled to India and Ceylon (Sri Lanka) and described the spice industries, including the cultivation of pepper and the smell of the clove trees that wafted on ocean breezes a hundred miles from the islands. Europe was interested in this report, but the descent into the Dark Ages had begun, and it became impossible for merchants to mount the expeditions necessary to exploit these resources.

Mohammed (570–632), in addition to being the prophet of Islam, author of the Koran, and founder of a legal system, was himself a spice merchant. He worked as a boy for spice merchants in Syria and became a camel driver and a caravan leader before marrying his employer, the widow Khadija. After his death, the Crescent swept from Spain to the borders of China and south into Ceylon and Java, again establishing the monopoly of spices held in pre-Roman times. Basra, situated at the head of the Persian Gulf where the Tigris and Euphrates met, became the primary trade center between East and West. Arabic apothecaries invented distillation to permit the extraction of the essences of the spices, resulting in the development of perfumes and essential oils like oil of cloves and oil of peppermint (Chapter 9). The wealth accumulated from these activities paid for missionary work, facilitated the development of arts, sciences, and medicine, and permitted the flowering of a sophisticated cultural system. It also increased the hate and envy of Christian Europe, which found itself completely dependent upon the infidel for life's necessities such as silk, pearls, jewels, and spices. The desire to wrest Jerusalem from the Moslems was not due solely to heightened religious sensibilities, but included a large component of economic self-interest on the part of the nobility and the Church. With the slow demise of feudalism in the ninth century and the development of centers of commerce, artisans organized into guilds. The merchant class, the bourgeoisie, recognized that it could make common cause with the nobility and Church if it had direct access to the goods of Cathay—a term that included both China and India.

The first Crusade was mounted in 1096 and Jerusalem was freed in 1099, with concomitant exposure of Europeans to luxury that they had not dreamed existed. Venice and Genoa were the main ports from which supplies of people and material flowed to Palestine, and these two city-states became wealthy. During the fourth Crusade (1204) Venetian merchants financed the destruction of Constantinople, thus eliminating a hated economic rival from competition in the spice and silk trade. Venetians converted Constantinople into a supply depot with financial control centered in Venice. New spice routes led to Baghdad, to Trebizond on the Black Sea, and then to Constantinople, where the goods were transshipped to Venice. In exchange, Venice bought European grain, glass, wine, and woolen cloth, and shipped them to Constantinople for transfer to India and China. Venice developed economic power never before experienced in Europe, power and wealth that matched that of the principal cities of China and India. This wealth, and the availability of leisure that accompanies wealth, culminated in a patronage of the arts and the masterpieces of Titian, Tintoretto, Veronese, and Giorgione, in the construction of Venice as we see it today, and in the magnificence of many of the art objects that constitute much of our Western heritage and contribute to our standards of beauty. Merchants in Nuremberg, Antwerp, Bruges, Paris, and London established ties with Venice through the good offices of the Vatican, and secondary centers of art and culture developed. By 1180, during the reign of Henry II, a pepperer's guild of merchants who "sold in gross" was organized in London, later to become the powerful Grocer's Company that financed the British East India Company in the sixteenth century. Apothecaries used the Arabian distilling apparatus, the alembic, to make medicinally essential oils including oil of cloves for toothaches, rubs for sore muscles, and unguents (see Chapter 9).

In the middle of the thirteenth century, Nicolo and Maffeo Polo, businessmen of Venice, traveled east to establish connections with the great Khan, Mongolian emperor of the Yüan Dynasty of China. In 1269, Nicolo's son Marco accompanied his father and uncle on a second trip, and 26 years later, three ragged men returned to Venice carrying pearls, diamonds, sapphires, and emeralds in the seams of their clothing. While Marco, then a middle-aged man, was a prisoner of war in Genoa, he wrote his *Travels,* which whetted Europe's interest in things Chinese. Marco described the mouth-watering flavor of ginger-spiced pork; Peking duck rubbed with nutmeg, star anise, and honey; and other foods and drinks unknown to Europeans. He provided detailed descriptions of pepper, nutmeg, clove, and cardamom plants and told almost unbelievable stories of marble palaces, paper money (Chapter 9), and a thick, black liquid that burned with a hot blue flame. Venice was encouraged to bring more and more of these wonders to the wealthy of Europe, and merchants happily obliged.

The safety of the caravans was, however, a matter of concern. Constantinople had fallen to the Turks in 1453, and Moslems were reasserting control over the caravan routes through Persia and the fabled cities of Damascus and Samarkand. Each caravan was harassed and tribute exacted. This increased costs, and although the costs were passed on to the customers, demand was falling off. Nevertheless, Europe kept Venice rolling in wealth—and this was translated into economic, social, political, and religious power. The doges of Venice could and did dispute the pope, and their arrogance and greed was unfavorably noted throughout Europe.

AROUND THE CAPE OF GOOD HOPE

Under the leadership of Prince Henry the Navigator, Portugal established a naval college at Sagres to train ship's officers. Henry believed that the economic future of his small country depended on the exploration and development of new sea routes for trade and commerce. Navigators, astronomers, and geographers staffed the school; new and better charts were prepared; and more efficient and accurate instrumentation was developed. Working their way slowly down the western coast of Africa, Portuguese ships found the Madeira and Cape Verde islands. In African Guinea, they found the aromatic spicy seeds of *Amonum melegueta,* the Melegueta pepper—a plant related to ginger but one that could partly replace true pepper and could be sold cheaper than the black pepper controlled by Venice. In 1488, Dias circumnavigated the Cape of Good Hope, demonstrating that the Indian Ocean could be reached by sea without need to cross the Red Sea, which was controlled by the Arabs. Stimulated by the reports that the Genoese admiral Columbus had found a western route to India, King Manuel I ordered Vasco da Gama to find a southern route to the Orient. Da Gama charted the eastern coast of Africa, and in 1498 reached Mozambique before sailing northeast across the Arabian Sea to Calicut, on the west coast of India. Waiting almost six months for the winds to shift, da Gama retraced his outward journey and in spite of an outbreak of scurvy, reached Lisbon in August of 1499 with precious jewels, spices, and an agreement for gold, silver, and coral. Halfway around the world, Columbus's crews discovered allspice in the West Indies, red (*Capsicum*) peppers, and tobacco.

Anxious to secure all trade rights, the Portuguese mounted another voyage to Calicut and in 1505 sent out 16 ships under the command of Don Francisco de Almeida. Capturing and staffing naval supply bases at Quiloa and Mombasa, the fleet subdued Ceylon, Malacca, and other spice ports, established factories for the processing and packing of spices, and—through agreements with local rulers—assured the control of all the seaport-based spice trade of Asia with the exception of China. The Portuguese East India Company monopolized the pepper trade and drove up prices to the point that many spices were reserved solely for the very wealthy. Ferdinand Magellan, trained in Prince Henry's naval college, hired himself out to Charles V of Spain to find another route to the Spice Islands, since Spain realized that Columbus had merely come upon a New World devoid of cinnamon and black pepper. Although Magellan was killed in the Philippines, one ship of his fleet completed the circumnavigation of the world in 1522. Cloves, nutmegs, mace, and cinnamon carried by the survivor more than paid the Spanish crown for the cost of the trip. Captain Sebastian del Cano, master of the surviving ship *Victoria,* was rewarded with a handsome pension and a coat of arms that included two crossed cinnamon sticks, three nutmegs, and a dozen cloves.

EAST INDIA COMPANIES

With the success of the Portuguese and the Spanish as a goad, Britain, Holland, and France formed East India companies to find new sources for spices and new routes. Sir Martin Frobisher reached Labrador while searching for a northwest passage to India. Hudson and Cabot for England, Champlain for France, and the founders of New Amsterdam were actively probing the northeast coast of America for inland waterways, while Sir Francis Drake visited the Spice Islands on his voyage around the world. Spain was exploiting Mexico, Peru, and

Central America. Portugal established bases in Brazil, and red pepper, allspice, and vanilla were being packed into the galleons sailing back to the Iberian peninsula. Venice, being too conservative to mount expeditions beyond the Mediterranean, was reduced to the status of a tourist's museum. By the beginning of the seventeenth century, Britain and Holland were—by guile and force of arms—undercutting Portugal in the Orient. By 1630 the Dutch had seized Malacca and secured control of most spice production in the east. England bullied her way into control of India and established tenuous trade relationships with the Ch'ing Dynasty through Canton and Singapore, driving the Portuguese from Macao and other offshore islands. Britain and Holland vied economically and militarily for hegemony over the lands of the Indian Ocean. In 1824, after more than a century of sometimes bitter fighting, the two countries signed a treaty in Vienna under which Holland received the Malay archipelago except for North Borneo, and England controlled India, Ceylon, Singapore, Hong Kong, and much of the coastal areas as far west as Siam. These islands were, as it turned out, valuable not only for spices but also for rubber, tea, and other plantation crops.

During the fourteenth to sixteenth centuries, spices were collected from wild trees scattered throughout forested areas. Pepper vines were cultivated, ginger planting was a cottage industry, and poppy seeds were imported from modern Pakistan. The Europeans recognized that yields could be improved and production efficiency increased if spice trees were grown in plantations. Cinnamon was introduced into the Seychelles by France, which was considered theft and piracy by the British and Dutch, who imposed the death penalty for anyone caught smuggling spice-planting stock. Even today, authorities of Zanzibar are empowered to kill anyone caught smuggling cloves into Kenya, but such restrictive measures have little meaning; indeed, North America is supplied with many Oriental spices now being grown on plantations in Central and South America.

Figure 4.45 Pepper *(Piper nigrum)*. Taken from *Historia Generalis Plantarum* by Jacques d'Alechampes, published in 1587.

PEPPERS

Although it is not feasible to discuss all or even the majority of spices and herbs available to us, certain of them require some discussion. Among these, black pepper is vital (Figure 4.45); in terms of volume consumed, it is our most important spice. Pepper is apparently native to the Malabar coast (also the home of cardamom, bananas, and sugar cane); cultivation moved eastward to the Malay peninsula, Indonesia, India, and China. Indeed, pepper has been cultivated throughout Asia and Micronesia for untold centuries. Our name for the spice is derived from the Sanskrit *pippali.* Botanically, the plant is *Piper nigrum* in the Piperaceae, a small family with few other economically important plants except the peperomias grown as house plants. Although it is now grown in South America, the bulk of the world's supply comes from India, Indonesia, and the Malagasy Republic (Madagascar). Grown from cuttings to ensure genetic uniformity, the fruits are harvested just before they ripen, and they darken into the typical black peppercorn as they dry. The trendy green peppercorns are picked before they begin to turn color. When the entire fruit is ground, black pepper results, but if the outer hull is removed, white pepper is produced. White pepper is less pungent than black. The sharp taste of pepper is due to the presence of a resin, an essential oil, and an alkaloid.

The fruits of two species of the genus *Capsicum* in the tomato family are almost as important a seasoning as is true pepper. The genus contains the familiar bell peppers, which are not hot at all, but the fruits of *Capsicum frutescens,* the red or cayenne pepper, can blister the insides of your mouth. The genus is native to Peru, where seeds have been unearthed from prehistoric burial sites in the Andes. Peter Martyr, who accompanied Columbus on his second voyage in 1493, brought back to Spain "peppers more pungent than those from the Caucasus," and, he added, "there are innumerable kinds . . . the variety whereof is known by their leaves and flowers." From Spain, the plants spread throughout Europe and the Near East; since they could be easily grown as far north as Denmark, they were an inexpensive substitute for black pepper. Our name *chili pepper* is Spanish, derived from the initial collection of seed in Chile. The chili con carne we associate with Mexican cooking is actually a nineteenth-century invention of the Texas *gringos;* commercial chili powder contains red pepper pods, plus the European herbs cumin, oregano, and garlic with the addition of salt, cloves, and allspice. In addition to its use in Tex-Mex dishes, scrambled eggs, tomato juice, and even soups can be spiced up with judicious additions of chili peppers either as cayenne powder or chili powder. The Chinese started to grow chilis in the sixteenth century when a few pods were brought to Canton by the Portuguese. They were gratefully accepted in Szechwan and Hunam provinces, where the people liked "hot" foods. Szechwan paste is a fiery mixture of ground chilis and garlic in oil and is a major ingredient in Kung Pao chicken, a dish named for a Pekinese official who settled in Szechwan.

KUNG PAO CHICKEN

(for two)

1 chicken breast boned, skinned, and cut into ½-in. strips
½ egg white beaten slightly
1 tsp. cornstarch
¼ tsp. salt
⅛ tsp. MSG (Accènt)
2 tbsp. Chinese brown bean
¼-½ tbsp. Szechwan paste
1 tbsp. brown sugar
½ tbsp. rice wine or dry sherry
1 tbsp. white vinegar
3 peeled and crushed garlic cloves
2 tbsp. oil
½ cup fresh, unsalted peanuts
2 whole chili peppers (more if you can stand it)

1. Combine chicken strips with egg white, cornstarch, salt, and MSG. Refrigerate for 1 hour.
2. Combine bean paste, Szechwan paste, sugar, wine, vinegar, and garlic; reserve.
3. Heat oil in wok or skillet and when very hot, add peanuts and cook until they are medium brown (about 30 seconds). Remove peanuts with slotted spoon to absorbent paper.
4. Add chicken and egg white mixture to hot oil and stir rapidly for no more than 1 minute.
5. Remove chicken, place chili peppers in oil, and heat for no more than 10–15 seconds until brown.
6. Return chicken to wok, add bean paste-Szechwan paste mixture and heat briefly. Sprinkle with roasted peanuts. Serve with steamed rice and cold beer.

Among the best-known capsicum peppers is the type used to make paprika. This fleshy pepper is native to Jamaica and reached Europe in the late sixteenth century. *Paprika* is the Hungarian term for Turkish pepper; it was assumed that it came from Asia Minor. Hungarians prepare several wonderful meat dishes—paprikashes—whose aroma and flavor are derived from liberal use of the powdered pods. Completely deseeded pods are used for the mild-flavored paprikas, while more robust cultivars are ground into hot paprikas with some of the seeds left in. Dr. Albert Szent-Györgyi, a Hungarian patriot, isolated vitamin C from paprika. He was awarded the Nobel Prize for this research, which has virtually eliminated scurvy. Spanish sweet paprika, the kind usually available in North America, is a very gentle seasoning compared with even the mild Hungarian versions. Those grown in California are Spanish cultivars.

PRAGUE VEAL GOULASH

(for four)

1½ lb. shoulder veal cut into cubes.
2 tsp. salt
pepper to taste
2-3 tbsp. Hungarian paprika
¼ cup shortening (one-half butter)
4 medium onions, sliced thinly
1 clove garlic, minced
1 cup beer, chicken broth, or water
4 oz. can tomato paste
1 tsp. caraway seed
1 tsp. dry mustard
1 cup sour cream

1. Sprinkle meat with salt and pepper and half the paprika. Heat shortening in a skillet and brown the meat on all sides. Remove meat to heavy casserole.
2. Add onions and garlic to the shortening and cook until onions are translucent, then add them to the meat.
3. Add liquid, tomato paste, caraway, mustard, and the remainder of the paprika. Bring to a boil, cover, and simmer until the meat is fork-tender.

4. Just before serving, add sour cream, but do not allow to boil. Serve with noodles or dumplings, rye bread and sweet butter, and a properly chilled bottle of white wine.

MUSTARD

Mustard is a "hot" spice, and its use as a dressing for meat dates back as far as records on cooking are available. Seeds of several species in the genus *Sinapis* (*Brassica*) are used. *S. alba* (yellow mustard seed) and *S. nigra* (black mustard seed) originated in Europe; *S. juncea* (brown mustard seed) is native to the Himalayan Mountains. Pythagoras, in about 2500 B.P., recommended it as an antidote to scorpion bites; Hippocrates prescribed it as an emetic; and Diocletian, the Roman emperor, placed a heavy tax on it, knowing that his subjects would pay without too much complaint. Greeks, Romans, and the northern Germanic tribes ate the leaves as a boiled green vegetable and ground the seeds into a paste. Matthew 13:31–32 refers to the parable of Jesus in which the kingdom of heaven is likened to a mustard seed, in that, in spite of its small size, it grows into a large plant; this demonstrates—if nothing else—that the plant was known in the Near East.

Its place on Europe's tables led Anatole France to remark that "a tale without love is like beef without mustard; an insipid dish." Powdered mustard is used in cooking and mustard seed in pickling spice, but most is consumed as prepared or salad mustard, in which the ground seed is mixed with vinegar, salt, and other spices, including turmeric to enhance the yellow color. French and German mustards utilize brown or black seeded species and may include wine or beer in the sauce. A magnificent dressing for boiled beef or smoked tongue consists of prepared mustard, horseradish, and mayonnaise—proportions adjusted to taste.

SAFFRON

Saffron, the most expensive seasoning, consists of the stamens of the saffron crocus (*Crocus sativus*), a species closely related to the garden crocus. Over 220,000 handpicked stamens are required to make a pound of saffron. The word is derived from the Arabic *zâfaran,* meaning yellow, for the stamens were used as a source of yellow dye for woolen and linen cloth. Ancient Minoan goddesses are figured with sacred snakes and the equally sacred saffron crocus flowers. Its expense and the delicate flavor imparted to foods is noted in the Song of Solomon (4:13-14). The Chinese particularly esteemed saffron, not as a flavoring, but as a cosmetic used by women. Even a few stamens impart a yellow color and a delicate flavor to the rice that complements the classic Spanish paella:

PAELLA

(for four)

2 medium onions, chopped
3 cloves garlic, chopped
3 slices bacon, chopped
3 sweet Italian sausages, sliced
2 hot Italian sausages, sliced
1 chicken cut into serving pieces
1 tbsp. salt
¼ tsp. pepper
generous pinch of saffron
paprika
olive oil
1½ cups dry rice
12 small clams, carefully washed
½ lb. peeled shrimp, parboiled 1 minute
½ lb. scallops, parboiled 1 minute
1 cup peas
sliced pimiento

1. Fry onions, garlic, bacon, and sausages until onions are golden and translucent. Remove to heavy casserole and add ½ cup water.
2. Dust chicken pieces with salt, pepper, and paprika and fry in olive oil until brown on all sides. Add to casserole. Add rice, saffron, and 2 cups of water.
3. Cover and bake at 450° F for 30 minutes until the rice is almost done.
4. Steam clams separately until they begin to open before adding them to the rice. Add shrimp, scallops, peas, and pimiento. Return to heat, uncovered, until seafood is done. Serve with a green salad, crusty bread, and a dry red wine.

CINNAMON, NUTMEG, AND MACE

Several spices are so firmly associated with baking that most people forget that they were formerly used in other ways. Cinnamon is a case in point. There are two species: *Cinnamonum zeylandica* is true cinnamon, and *C. cassia* is the spice cassia, more familiar in Europe than in North America. Both are imported from Sri Lanka, and the Seychelle Islands, where the small, bushy trees are grown in plantations. In ancient Egypt, Greece, and Rome, the bark was steeped in oil to make a fragrant unguent, in addition to being used in cooking. It was used on funeral pyres to disguise the smell of burning flesh; when Nero's wife Poppaea died in 65, a full year's supply of precious cinnamon was burned. The Portuguese controlled cinnamon until the Dutch captured Ceylon in 1656. They continued the brutal practices of enslavement of the collectors and established a price monopoly that was maintained by destroying cinnamon when prices began to fall. Their control was broken when the British took Ceylon in 1796, and by the early part of the nineteenth century, plantations were in several countries and prices dropped to roughly their present levels.

Both cinnamon and cassia are members of the Lauraceae, the same family to which the bay leaf tree (*Laurus nobilis*) and the avocado belong. The cinnamon sticks we use as swizzles in hot mulled wine are curls of the peeled bark of young twigs, the cinnamon quills of commerce. In addition to its use as cinnamon powder in baking and as cinnamon sugar on doughnuts, the addition of a small piece of stick cinnamon to a pot of spaghetti sauce should be experienced, and a dash of cinnamon powder on fish is a pleasure most North Americans have not had.

Apple desserts would not be the same without cinnamon, nutmeg, and mace; and a touch of nutmeg or mace in baked squash, sweet potatoes, or even the lowly boiled carrot transforms these into gourmet foods. Both nutmeg and mace are from the same plant and, indeed, from the same organ. The nutmeg tree, *Myristica fragrans,* is a large, evergreen member of the Myristicaceae. Because the tree is dioecious—that is, male and female flowers are borne on different plants (the term means "two houses")—plantation managers were careful to include a few male trees scattered throughout the plantings. Recently, horticulturists have found a way to graft male branches onto the female trees. The fruit looks like a small apricot with an orange-yellow pulp (Figure 4.46). Surrounding the seed is a lacy, scarlet layer called an aril. This, separated from the nutmeg seed and dried, is mace. It goes well with tomatoes, including juice and catsup, and adds an unusual flavor to fish dishes, including tuna and chowders. Nutmeg is a necessity in baking; it is the secret ingredient in Swedish meatballs and lamb stews, and no self-respecting eggnog neglects a sprinkling of nutmeg.

Figure 4.46 Nutmeg and mace *(Myristica fragrans).* One part of the fruit coat is lifted to show the aril on top of the seed (mace) and the seed itself (nutmeg).

CLOVES

The French call them *le clou,* the nails, an apt description of the dried, unopened flower buds of the clove tree (*Eugenia aromatica*) in the myrtle family (Figure 4.47). In Greece, Easter is marked by the baking of light, delicious cookies, each containing a clove to represent the nails used to fix Christ to the Cross. A native of the Spice Islands, it has been transplanted to other tropical countries where it is grown on plantations. Records from the Han Dynasty in China show that it was much appreciated, being known as the "chicken tongue" spice. It, with star anise, anise pepper, fennel seed, and cinnamon, comprise the "five spice powder" used in red-cooking roast pork and other meat dishes. Along with peppercorns, cinnamon, and nutmeg, cloves were imported into dynastic Egypt and were known in Europe by the fifth century. The Dutch maintained a rigid monopoly on cloves for more than a century, but smugglers obtained planting material and plantations were started in Malaya and on the West Indian island of Grenada. Today, much of the world's supply comes from Zanzibar. In both Africa and in the Orient, cloves are used to flavor tobacco and are chewed with betel nut. In the West, its use has generally been restricted to studding baked hams and for cakes and cookies. It adds flavor to chocolate, and, during the Middle Ages, cloves were stuck in oranges to construct the pomanders

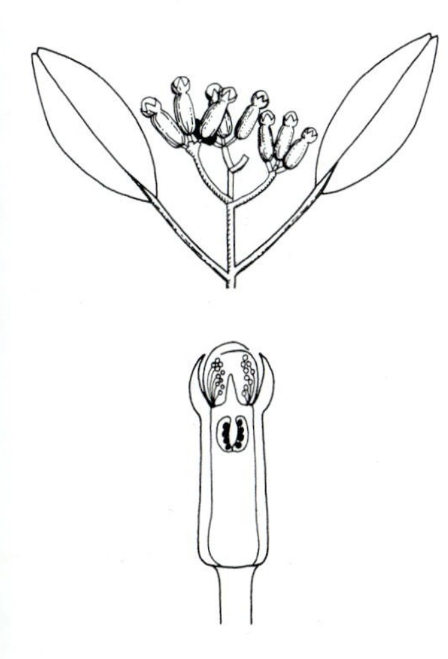

Figure 4.47 Inflorescence of the clove tree *(Eugenia aromatica)* and an enlarged view of an individual flower, the "clove."

ladies carried to avoid smelling the bodies of the great unwashed masses and, they hoped, to ward off the plague.

CHINESE PORK IN FRUIT SAUCE

(for two)

vegetable oil
½ lb. lean pork cut into 1 in. cubes
2 cloves garlic, minced
½ tbsp. brown sugar
1 tbsp. rice wine or dry sherry
⅛ tsp. ginger
½ tbsp. steak sauce
⅛ tsp. anise
⅛ tsp. ground cloves
⅛ tsp. ground cinnamon
¼ tsp. salt
1 orange, peeled and sectioned
1 tangerine, peeled and sectioned
1 pear, cored and sliced
¼ tsp. lemon juice

1. Heat small amount of oil in a wok or small pan and thoroughly brown the meat and garlic.
2. Add sugar, wine, steak sauce, spices, and salt. Add ½ cup of water and simmer gently for 30 minutes.
3. Add fruit and lemon juice and simmer 3–4 minutes.

ADDITIONAL READINGS

Andrews, J. *Peppers: The Domesticated Capsicums.* Austin: University of Texas Press, 1985.

Boxer, A., and P. Bach. *The Herb Book.* London: Octopus Books, 1980.

Clair, C. *Of Herbs and Spices.* New York: Abelard-Schuman, 1961.

Clarkson, R. E. *Herbs and Savory Seeds.* New York: Dover, 1972.

Farrell, K. T. *Spices, Condiments, and Seasonings.* Westport, Conn.: Avi Books, 1985.

Hayes, E. S. *Spices and Herbs Around the World.* Garden City, N.Y.: Doubleday, 1961.

Heath, H. B. *Source Book of Flavors.* Westport, Conn.: Avi Books, 1981.

Lopez, R., and I. Raymond. *Medieval Trade in the Mediterranean World.* New York: Columbia University Press, 1955.

Purseglove, J. W., E. G. Brown, and S. R. J. Robbins. *Spices.* 2 vols. London: Longman, 1981.

Rosengarten, F., Jr. *The Book of Spices.* Rev. ed. New York: Pyramid Books, 1973.

Root, W. *Herbs and Spices: The Pursuit of Flavor.* New York: McGraw-Hill, 1980.

Simon, J. E., L. E. Craker, and A. F. Chadwick. *Herbs: An Indexed Bibliography. 1971–1980.* New York: Archon Books, 1984.

Chapter 5

Plants in Religion

Plants of the Bible

And God said, let the earth bring forth grass. *Genesis 1:11*

The Judeo-Christian Testaments are part of the foundation of Western civilization and are quoted and misquoted frequently. We can view the Testaments as the literal Word of God, as inspired allegory, as a fairy tale, or as an historical record. For botanists, the Bible provides a record of plant use by those peoples of the Middle East who gave us our religious heritage. The Bible is not a *de novo* source of plant symbolisms; accretions from other faiths are intercalated and have become incorporated into our attitudes.

BIBLICAL SCHOLARSHIP

In order to understand these symbolisms, we should have some idea of the history of biblical literature. The Old Testament is composed of books written 2000–3000 years ago, some written shortly after the Tribes of Israel entered the now Holy Land, and others written as contemporary records of the prophets and kings. Ancillary records and surviving fragments such as the Dead Sea Scrolls validate many passages in the King James and subsequent English-language versions. These versions are based on translations from the Hebrew into Greek, the first one authorized by Ptolemy Philadelphus, patron of the library at Alexandria.

The history of the New Testament is less direct. In the first period of the new Church, the Good News (Gospels) was spread orally. The Gospel message was colored by the locale, with different plant imagery used to point up the message in Egypt, in Asia Minor, or in Greece. Written versions appeared in the second century, but not until the fourth century was the New Testament collected and collated. The language of the early New Testament was Greek, taken from oral depositions and fragmentary records in Aramaic, the common language of the peoples of the area, as well as from depositions in Greek, Syriac, Coptic, and Ethiopic. Several centuries passed before versions in Latin were prepared, the first being the Vulgate of St. Jerome in the fourth century. The first English-language version appeared at the beginning of the seventh century, and subsequent versions in English during the eighth to tenth centuries were prepared from the Vulgate. An Old English translation in the tenth century was followed by several Middle English versions; these culminated in the King James version of the seventeenth century. The American Standard version appeared at the end of the nineteenth.

This abbreviated history points up the difficulties in being sure of what plants and what plant imageries were used by Hebrews and early Christians. Many translations were inaccurate, and these errors were perpetuated in succeeding versions. Since the Gospels were bent to fit different cultures, names of plants were altered beyond our ability to be sure of the imagery of biblical peoples themselves. We have clues to these images, among the best being our knowledge of what plants were either native to the area or introduced by the waves of immigration and conquest that were the fate of these lands. Modern Israelis grow maize, tobacco, and squash, plants that did not enter Europe from the Americas until the fourteenth and fifteenth centuries and did not reach Palestine until at least 1600. To speak blithely of the apple *(Pyrus malus)* as being Eve's temptation (Figure 5.1) doesn't make botanical sense; the apple could not develop in the climate of the Holy Land. This image was used in Europe only because it was a common fruit that people were familiar with. We know that many plants that did exist in biblical times were either wiped out by human activity or are now inconspicuous parts of the flora due to conversion of natural areas into arable land. There are probably no more than 200 endemic plants in the area—plants known only from Palestine and nowhere else—and these rarely figured in biblical writings. Most plants of symbolic interest or economic importance as food or fiber either originated in the Middle East or were imported eons before the area became a cradle of Western religions. These, for the most part, are the ones mentioned in the Bible.

Figure 5.1 Adam, Eve, serpent, apple, and Tree of the Knowledge of Good and Evil. Taken from Hans Sebald Beham's woodcut *The Fall of Man.*

THE GRASSES AND GRAINS

One group of native plants, the grasses, are frequently mentioned in both Testaments, the first in Genesis 1:11, "And God said, let the earth bring forth grass." Most subsequent references to grass are allegorical: "All flesh is grass" from Isaiah and I Peter; "My heart is smitten and withered like grass" from the 102nd Psalm; or the statement in the 103rd Psalm, "as for man, his days are like grass." For nomads, the fleeting nature of pasturage involved their very survival, and thus the figurative image is perfectly understandable. Nomads of central Mongolia have the same image. References to hay (Proverbs 27:25; Isaiah 15:6) are later interpolations, since the sparse growth of natural grasses would have precluded haying. In Revelation 8:7, where "all the green grass was burnt up," it is likely that a more accurate translation would be "all the green plants"—plants used as forage—because tough native grasses will dry out but not collapse, as would broad-leaved plants.

The cereal grains, true grasses all, have figured many times in both Testaments (Chapter 3). The first of over 30 references to barley *(Hordeum vulgare)* appeared in Exodus 9:31. Barley was probably a more widely grown grain crop than wheat, because none of the other cereals can grow in as many different climatic conditions. It was planted as soon as people became agricultural; Hebrew invaders of Canaan found it in cultivation. As seen in Deuteronomy 8:8, "A land of wheat, and barley" was synonymous with fertility, and a land where "thistles grew instead of barley" (Job 31:40) was a symbol of infertility. Two barley grains made a finger's breadth, 16 made a hand's breadth, 24 a span, and 48 were the biblical cubit—about 41 centimeters. Although sprouted barley grains have for centuries been the preferred malt used in brewing, there is no statement in the Bible of this fact. Since wine was not forbidden, it is unlikely that beer was proscribed, but the absence of any references to malting or to beer is surprising (Chapter 8).

The bread of the peoples of the Bible was primarily an unleavened loaf prepared from ground barley mixed with millet, pea meal, spelt, and other seeds, but the lighter, whiter, and more stable bread prepared from wheat flour was preferred. Most references to "corn" are actually to wheat. Reuben, a successful farmer, grew wheat (Genesis 30:14), and the Psalms use "the corn" as a symbol of a fruitful harvest of grain and of souls. Matthew, Mark, and Luke mentioned cornfields in historical context. "Verily, verily, I say unto you, except a corn of wheat fall into the ground and die, it abideth alone; but if it die, it bringeth forth much fruit" (John 12:24) is a powerful image which made good sense to the hearer. Matthew used the image of a good harvest in several places (13:3–8 and 13:24–30). The sower who threw grains in stony ground and some in fertile soil and the comparison of the Kingdom of Heaven to a man who sowed good seeds in his field were immediately recognized as object lessons for the faithful. Famine was an ever-present terror, and the concept of a granary as a hedge against real or religious famines is in many verses of both Testaments.

VEGETABLES AND FRUITS

The Bible is ambiguous about the use of fresh vegetables. Most would require more water than was generally available, and many cultures in the area had little use for them. Onions and garlic were, however, used extensively. Food was dressed with spices, some grown locally, like capers *(Capparis sicula),* bay leaf *(Laurus nobilis),* saffron *(Crocus sativus),* and coriander *(Coriandrum sativum).* Cinnamon and other exotic spices arrived on the caravans from the Indies and China, and others, like cloves, came from Africa.

Many fruits are celebrated in both Testaments. The pomegranate *(Punica granatum)* was grown in Egypt from at least 2600 B.P., and it was grown in the Holy Land before the hegemony of the Jews. It was a luxury item, praised for the beauty of the plant, its flowers, and the handsome red-skinned fruit. Because of the many seeds surrounded by sweet, thirst-quenching juice, it became a symbol for abundance and especially for fruitfulness (Numbers 13:23 and 20:5). Its sexual connotations appear twice in the Song of Solomon (6:7 and 8:2), and the same symbolism appeared in Europe. The fig *(Ficus carica)* was also a sexual and fertility symbol because of its large number of seeds. Because this fruit could be dried for storage, its role as a food staple was well established (Numbers 13:23). Rabbinical scholastics suggest that the fig was the Tree of Knowledge; this suggestion has as much merit as any other theory about the plant that played this role. As a plant used in allegory and parable, figs appear many times as a symbol of peace (1 Kings 4:25), good versus evil (Jeremiah 24:1–8), literary analogy wherein "All thy strongholds shall be like fig trees" (Nahum 3:12), and parable (Luke 13:6–9; Matthew 21:18–19; Revelation 6:13).

Figure 5.2 The wondrous date palm *(Phoenix dactylifera).* Taken from Johnson's revision of Gerard's *The Herball,* published in London in 1633.

The date palm *(Phoenix dactylifera)* was used for many purposes (Figure 5.2). Fronds were used as thatch, trunks as building supports (later to be stylized as stone pillars), and the fruit could be eaten directly or made into date honey to be fermented into "strong drink." *Tamara,* a common Arabic given name for women, is translated as the palm, recalling the graceful and elegant "posture" of the tree. Leviticus 23:40 and Nehemiah 8:14 report that palm branches, as a symbol of prosperous harvests, were used to decorate the booths erected to celebrate the Feast of Tabernacles. As a symbol of triumph, it has figured in several cultures. Judas Maccabee celebrated his military successes (1 Maccabees 13:53; 2 Maccabees 10:10) with palms, and John 12:13 reported that palm branches were waved with joy upon the entry of Jesus into Jerusalem. Romans, too, used palms in symbolic fashion. When the victorious African legions returned, cohorts waved fronds of palm as they marched through triumphal arches. It is from these customs that early Christian martyrs chose the palm as a symbol of eventual triumph (Figure 5.3). On All Souls' Day, palm leaves are burned, and their smoke rising unto the sky is taken as proof of the victory of souls leaving purgatory for heaven. Legends of the miraculous properties of the plant abound in Christian and Moslem cultures but are not supported by

Figure 5.3 Cross with palm fronds and olive leaves. Taken from an embossing found in the Christian catacombs of Rome.

scriptural reference—a phenomenon common for many plant legends.

Although temperance groups believe that biblical references to wine are really to grape juice, few scholars agree with them. The grape *(Vitis vinifera)* is first mentioned in Genesis 9:20, wherein Noah planted a vineyard and got drunk on the end product. Wine and raisins were articles of commerce and trade, as evidenced by records from Egypt and Mesopotamia. The names of specific places in Palestine are linked to the vine (Figure 5.4), including Abel Kramim (Plain of the Vineyard) in Judges 11:33, Mount Carmel (Hill of the Vineyard of the Lord) in 1 Kings 18:19, and Nahal Eshcol (Brook of the Cluster) in Numbers 13:23. Many references to vines, grapes, and wine are clear statements of fact or are horticultural instructions (Isaiah). Most, however, are allegorical and symbolic. The "vine brought out of Egypt" and the "fruitful vine" referred to the Hebrews themselves. John (15:1–6) observed that Jesus referred to himself as "the true vine and my Father is the husbandman" or again, "I am the vine, ye are the branches." The Parable of the Vineyard (Luke 20:9–16) is another example. Peace and tranquility can have no clearer definition than that in 1 Kings 4:25, where every man was under his own vine, or the vision of the prophet Joel (2:24), where "the vats shall overflow with wine." "The Battle Hymn of the Republic" by Julia Ward Howe includes the powerful metaphor, "He is trampling out the vintage where the grapes of wrath

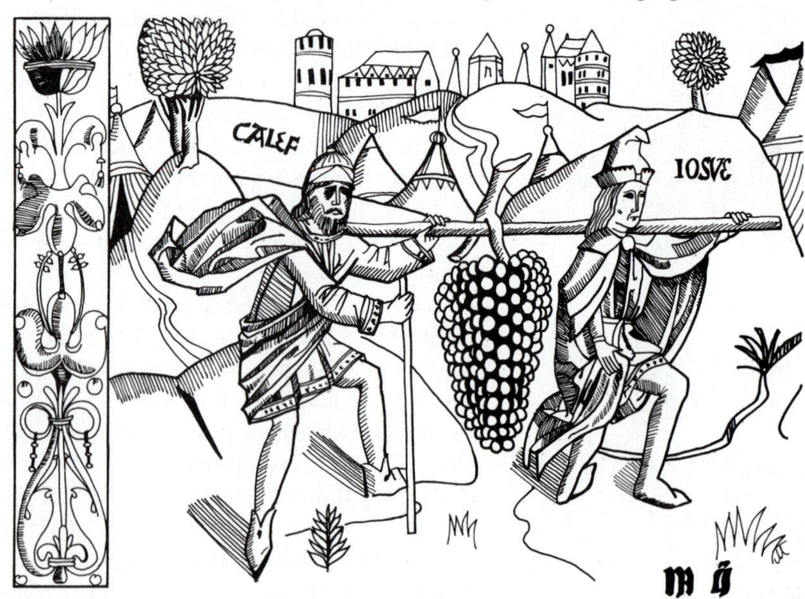

Figure 5.4 Caleb and Joshua bringing in the wine grapes. Taken from *Biblia Germanica Decinquarta,* published in Strasbourg in 1518.

are stored." Revelation 14:18–20 speaks of the "great winepress of the wrath of God," and in Lamentations 1:15 it is written, "The Lord has trodden the virgin . . . as in a winepress." The image of devastation is given in Psalms 78:47 and Isaiah 5:1–6 as the uprooting of vineyards, an act of destruction more long-lasting than virtually any other act of violence.

THE LILY

"Consider the lilies of the field, how they grow; they toil not, neither do they spin. And yet I say unto you that even Solomon in all his glory was not arrayed like one of these." This quotation from Matthew 6:28–29, repeated in Luke 12:27, is only one of many references to lilies. There are only two native species of *Lilium* that could have existed in biblical times, but neither was or is abundant. It seems unlikely that an image of such importance would be to a plant that most people would not be familiar with. This has provoked discussions as to which plant was the biblical lily. It is generally agreed that no single plant meets all the criteria imposed by the passages in which the word appears. Thus, reference in Ecclesiasticus 50:8 to "lilies by the rivers of waters" implicates the iris, of which several species are native to Palestine. The modern arabic word *Shushan,* which appears in Nehemiah, Esther, and Psalm 60, also fits well with this concept. A woman who doesn't like the given name Iris could, in all etymological accuracy, change her name to Susan.

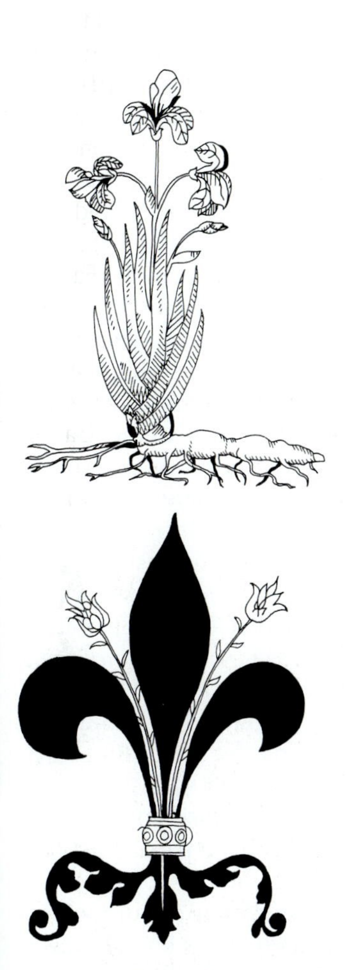

Figure 5.5 The iris. Top: drawing of lily, taken from *Commentarii* by Mattioli, published in 1579. Bottom: drawing of the fleur-de-lis, taken from a fifteenth-century engraving.

The iris (Figure 5.5) is a common spring-flowering plant found in moist habitats. Its blue and purple flowers nicely suggest the purple that early churchmen formally associated with the Passion, and the blue that clerics standardized as Mary's color. Its use in France's heraldry stems from King Clovis (d. 511), founder of the Merovingian dynasty. Clovis was pagan, but had a Christian wife, later to be St. Clotilde, who fervently wanted to convert her lord and master. Possibly after some nagging, the king agreed—so the story goes—that if he won the battle of Tolbiac against the Huns in 474, he would embrace the faith. As a symbol of his pledge, he painted the flower of the Trinity, the iris, on his battle standards and, once victorious, adopted the iris as a symbol of his reign. The crusade of 1137 under Louis VII used a somewhat stylized iris flower on its standards, and by the end of the twelfth century, the flag of France was a blue field (Mary's color) sprinkled with white iris, the fleur de Louis. This became the fleur de luce—the flower of light—and then the fleur de lis. In 1364, the number of flowers was reduced to the Trinity number. When England held part of France, the iris was included in the coat of arms of the Plantagenet kings, disappearing only in 1801, many years after France asserted its hegemony over these lands. Although the Massachusetts Bay Colony would brand an adulterous woman with an "A," the French in both the mother country

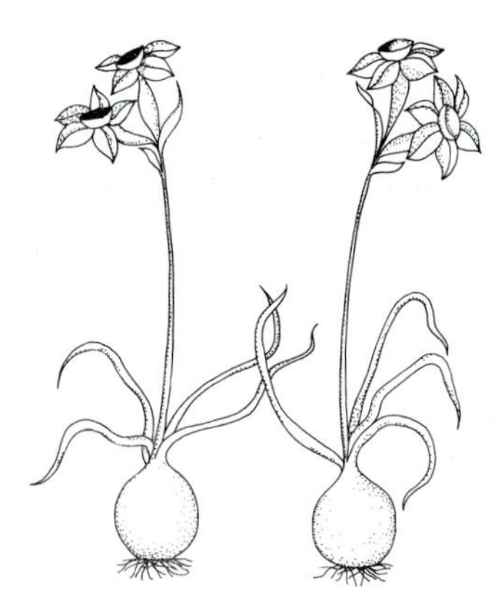

Figure 5.6 The narcissus. Taken from *Commentarii* by Mattioli, published in 1579.

and in New France branded a fleur de lis on the woman's cheek. The male partner was not marked.

In Isaiah 35:1, we read that "desert shall rejoice, and blossom as the rose," but the image is more basic than the botany. The Hebrew word used here was *chanatselet,* translated as rose, but it appears to mean bulb. Since roses don't develop from bulbs, a more adequate plant would be the narcissus (Figure 5.6), which grows in profusion on the Plains of Sharon and in fields surrounding Jerusalem. When, in Ecclesiastes 39:13, the prophet exhorted, "Hearken unto me, ye holy children, and bud forth as a rose growing by a stream of water," one may think of the oleander *(Nerium oleander).* The Lake of Galilee and brooks running into the Dead Sea and the River Jordan are lined with these poisonous but graceful and fragrant-flowered plants. The Rose of Sharon (Song of Solomon 2:1) may have been the mountain tulip *(Tulipa montana)* or even the hyacinth. There are, however, several true roses found in Palestine, but they are uncommon and none are as showy as twentieth-century roses.

The lily in Matthew and Luke was not likely to have been a lily, and either the windflower *(Anemone)* (Figure 5.7) or a crocus fits the passages. Both are spring-flowering, tending to validate our admittedly unsubstantiated impression that the teachings were made in the spring. Even today, both genera grow in uncultivated areas of the Holy Land and both have showy flowers in many different colors. The plants develop quickly after the drab days of winter and were undoubtedly appreciated, then as now, as harbingers of renewed fertility of the land. With these various interpretations of the botanical nature of the lily and the rose, one wonders where we acquired the white lily as one of the most important Christian symbols. White flowers were symbols of purity and chastity long before Christianity; white lilies were floral representations of Diana and Juno. Greeks and Romans prepared chaplets of white lilies for brides, and, when intertwined with sheaves of wheat, the benediction of a pure and fruitful married life was obvious to all. Exhibiting purity and grace, the white lily *(Lilium candidum)* was a symbol of several goddesses of the Fertile Crescent and parts of Southeast Asia, where the plant apparently originated. The lily was formally adopted by the Roman Catholic Church by the fifth century and, by the early Middle Ages, was firmly associated with the Virgin (Figure 5.8). Renaissance paintings of the Annunciation almost always include the lily, and this was formalized in 1618 when a papal edict laid down rules "as to the proper treatment of certain sacred subjects of art and the necessity of the introduction of the white lily into pictures of the Immaculate Conception." By the fifteenth century, the white lily was a traditional feature of the altar decorations in Europe. In addition to being Mary's flower, it is the floral attribute of several female saints. The device of using a common plant to facilitate the recognition of the holy in everyday life and to have symbols of saints for people who could not read can be found not only in Christianity but in all religions.

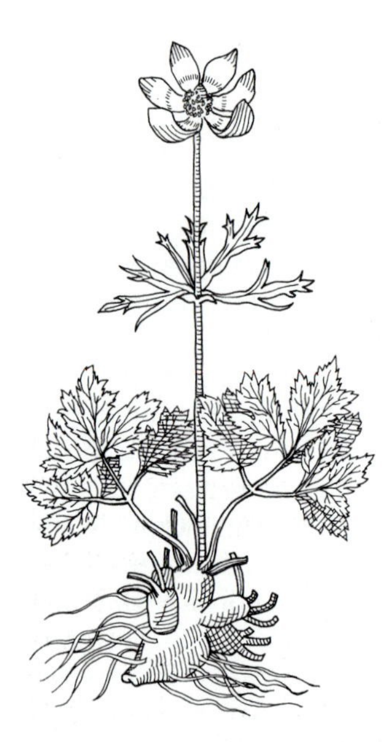

Figure 5.7 *Anemone hortensis,* a candidate for the "lilies of the field" in the Old Testament. Taken from *Rariorum Plantarum Historia* of Carolus Clusius, published in Antwerp in 1601.

In the eighteenth century, as plant explorers began bringing plants

Figure 5.8 Madonna lily, symbol of purity and motherhood. Taken from a fourteenth-century Italian woodcut.

back to Europe from the Orient, the white trumpet lily *(Lilium longiflorum)* was introduced from Japan and Korea, where it was cultivated as a garden plant and religious symbol. The plant is easier to grow than is the Madonna lily, and its perfumed flowers last many days. It can also be forced into bloom in time for Easter. In North America, the trumpet lily has almost completely replaced the Madonna lily as a potted plant. In the prurient Victorian era, it was common practice to remove the yellow stamens from lilies used as altar decoration so that the flowers would "remain ever virgin." Can you imagine the distaste (or pleasure?) with which this task was performed?

ADDITIONAL READINGS

Goor, A. "The History of the Grape Vine in the Holy Land." *Economic Botany* 20 (1966):46–64.

The Holy Bible, Revised Standard Version. New York: Collins, 1973.

Moldenke, H. N., and A. L. Moldenke. *Plants of the Bible.* Waltham, Mass.: Chronica Botanica, 1952.

Schonfield, H. J. *A History of Biblical Literature.* New York: New American Library, 1962.

Walker, W. *All the Plants of the Bible.* New York: Harper & Row, 1957.

Zohary, M. *Plants of the Bible.* Cambridge, England: Cambridge University Press, 1983.

Plants of Superstition, Myth, and Ritual

He loves me, he loves me not. . . .

Picking Daisy Petals

With a few exceptions, large urban areas are a recent development in social history. Ancient Rome, Athens, and Constantinople were within a short distance from farms and forests, and people had direct and immediate ties to the land. Plants were intimate parts of life, and it was natural for religious and secular leaders to use them for imagery, for parables, and for exhortation. So, too, did plants serve the people themselves; they were not merely for food and shelter but were intertwined with the rhythms of the seasons. A child was born "just after the corn harvest," or a man and a woman married "when the apricot trees were blossoming." When we remember that natural occurrences were mysterious at best and evil at worst, emphasis on the role of plants in human affairs was perfectly logical. Accidents and disease were caused by evil spirits or the anger of God, and one could be whole again by driving off Satan or placating the Deity with plants. The child, to be trained in the mores of the group, was presented with the myths, legends, and superstitions of the group and was provided object lessons based on attributes and properties of things that formed his or her immediate environment. These lessons to the young also served to reinforce the tenuous grasp of adults on a world that impinged so ominously upon the the people. Here again, plants—a vital ingredient for survival—were woven into the fabric of belief. We don't know where most of the legends and symbolisms of plants arose. Indeed, comparative-religion research demonstrates that cultures which could not have been in contact had similar if not identical constructs and ideas about plants. Some of these have been transmitted down to us and form part of twentieth-century rationalism.

Figure 5.9 Canterbury bells *(Campanula medium),* assigned to St. Thomas à Becket. Taken from *Plantarum sen Stirpium Icones* by Matthias Lobel, published in Antwerp in 1581.

PLANTS AND SAINTS

The association of a plant with a saint has been formally authorized since at least the sixth century. Its roots are in the pre-Christian attribution of plants to members of the pantheons of deities in all cultures. Many plants of economic importance were said to have been the visible symbol of a deity who, in kindness, sympathy, or generosity, presented the plant to Man; Athena became the patroness of Athens for giving the olive to the citizens. Honoring Christian saints with special flowers on their days served to reinforce the flow of religious faith throughout the year. The choice of a particular plant varied with the saint. St. Patrick's shamrock (actually an *Oxalis)* was a druidic mystic symbol associated with the Celtic sun wheel. The Celtic name *seanrog* (little clover) became *shamrog* and finally *shamrock.* St. Patrick (390–464) took this pagan symbol, pointed out that it really represented the Trinity, and subverted its symbolism to the new religion. Canterbury bells *(Campanula medium)* can only be associated with St. Thomas à Becket (Figure 5.9), and the attribution of spring wild flowers to St. Francis of Assisi is a concept easily impressed on even young communicants. The birthday of a saint—the nameday of many children in Latin countries—can be equated with a flower that blooms at about that time of year. The Michaelmas daisy *(Aster spp.)* for St. Michael's Day (September 29) and the Christmas rose *(Helleborus niger)* for St. Agnes's Day (January 21) are examples. For St. Peter:

> The yellow floure, called the yellow coxecombe which floureth now in the fields is a sign of St. Peter's Day whereon it is always in fine mettle in order to admonish us of the denial of Our Lord by St. Peter; that even he, the Prince of Apostles, did fall from feare and denied his Lord. So we too are fallible crethures, the more likely to a similar temptation.

Figure 5.10 Sunflower *(Helianthus annuus).* Taken from *Hortus Floridus* by Crispaen van de Passe, published in Arnheim in 1617.

At the time of the Spanish conquest of the Americas, many plants entered Europe and were given religious significance. In Peru, Pizarro

Figure 5.11 Mary's gold *(Calendula officinalis)*, one of many plants assigned to the Virgin.

saw a large yellow flower, venerated by the Incas as an image of the sun god. It was of equal interest to note that the Peruvians wore hammered gold breast plates embossed with representations of this flower. Pizarro took both seeds and breast plates back to Spain, and while the sunflower *(Helianthus annuus)* assumed only minor importance as a symbol of the Glory of God, the gold was used to finance the running battle with the heretical English (Figure 5.10). The marigold *(Calendula)* became Mary's gold (Figure 5.11) after it was imported from Mexico. A Mexican vine was asserted by sixteenth century Jesuits to be the plant that St. Francis of Assisi saw in a dream. It was named *flor de las cinco llagas* (flower of the five wounds), and by the seventeenth century each of the parts of the flower was given an association with an aspect of the Crucifixion (Figure 5.12). We now call it the passion flower *(Passiflora incarnata)*.

DOCTRINE OF SIGNATURES

Imputation of religious signifiance to plants took some strange paths, one leading to the Doctrine of Signatures. With disease a manifestation of evil and with medicine tied to the ancient Greeks, the need for curatives—"simples" that anyone could obtain—was a necessity. But unless there was some outward sign that a plant was useful and some visible indication as to its powers, how could one know which plant to use? This problem and its answer is best summed up by a quotation, given here in modern English, from Nicholas Culpepper's *English Physician* of 1680.

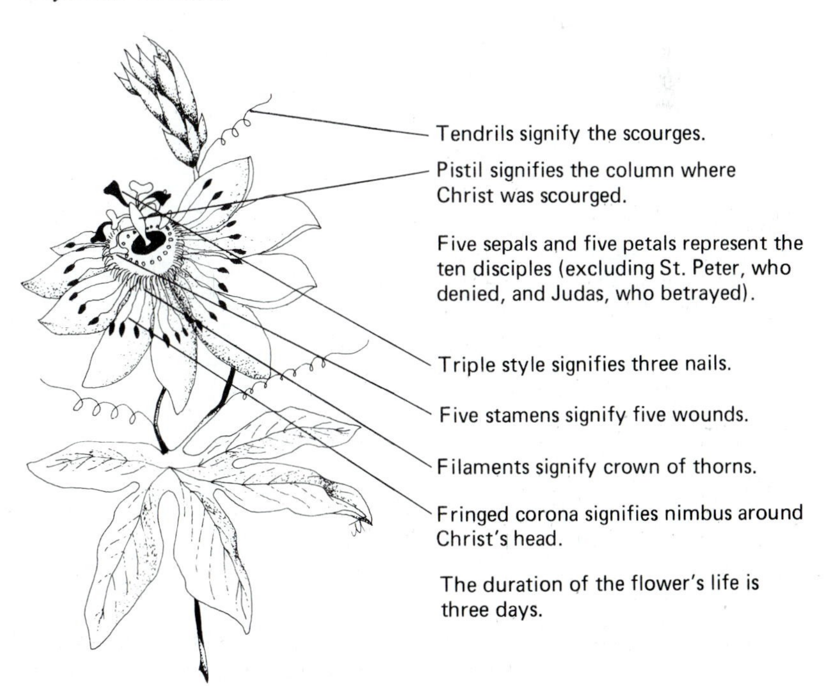

Figure 5.12 Christian symbolisms of the passion flower *(Passiflora incarnata)*.

> Though sin and Satan have plunged mankind into an ocean of infirmities, yet the mercy of God which is over all His works has made the grass to grow upon the mountains and herbs for the use of man and He has not only stamped upon these a distinct form, but also has given them particular signatures whereby men may read, even in legible characters, the use of them.

In Ecclesiasticus 38:4, it is written: "The Lord created medicine from the earth. And a sensible man will not despise them." The kidney bean looks like a kidney and would cure urinary afflictions. A walnut (Figure 5.13) looks like the human brain and should cure madness or even migraine headaches. Dutchman's-breeches *(Dicentra)* was clearly a man's plant (Figure 5.14) used to cure venereal diseases, and Dutchman's-pipe *(Aristolochia)* has recurved flowers that look like a womb (Figure 5.15)—useful in difficult births. The red sap of the bloodroot *(Sanguinaria canadensis)* was a signature that the plant was specific for blood disease. The yellow inner bark of the barberry *(Berberis)* (Figure 5.16) and the yellow spice turmeric were specific for jaundice. The spleenwort fern *(Asplenium)* and liverleaf *(Hepatica)* (Figure 5.17) were used in liver disease because their lobed fronds or leaves resembled the liver. Strap-shaped leaves were boiled into a tea to ease sores on the tongue. The list goes on and on.

Figure 5.13 Walnut as a Doctrine-of-Signatures plant.

Although the Doctrine of Signatures developed in Europe in the Middle Ages, the idea is far older. Theophrastus, an ancient botanist, reported that the root of polypody (a fern) is rough and has suckers like the tentacles of a polyp; a person who wore it as an amulet would not get rectal polyps. The American Indians, too, had an equivalent doctrine system (Figure 5.18). To eliminate worms, eat a wormlike plant part such as the tendrils of a squash; to promote lactation, use a plant with a milky sap; to control convulsions, place a gnarled piece of wood next to the patient.

BOTANICAL ASTROLOGY

Contemporary with the Doctrine of Signatures were the beliefs that plants were influenced by the heavenly bodies, that there were cause and effect relationships between stars and plants, and that these influences and relationships had direct significance for people. Philippus Aurelius Theophrastus Bombastus von Hohenheim, city physician of Basel and known throughout Europe as Paracelsus, said:

> Each plant is to be under a terrestrial star, and each star is a spiritualized plant. Each plant is under the influence of some particular star and it is this influence which draws the plant out of the earth when the seed germinates.

This confirmed the long-held belief in the importance of the phases of the moon and stars in planting and harvesting.

Figure 5.14 Dutchman's-breeches *(Dicentra cucullaria),* as a Doctrine-of-Signatures plant.

Albertus Magnus, author of *The Book of Secrets of Albertus Magnus, of Virtues of Herbs, Stones and Certain Beasts,* discussed the interrela-

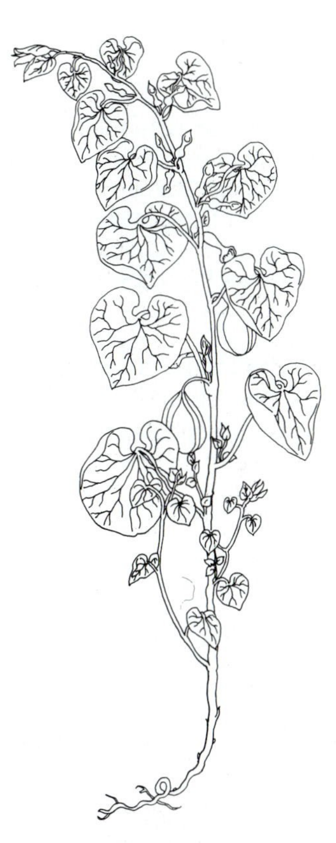

Figure 5.15 Birthwort *(Aristolochia clematitis),* as a Doctrine-of-Signatures plant.

tionships of plants and the heavens. Plantain, which Magnus called *Arnoglossus,* has roots that are "marvelous good against the pain of the head because of the sign of the Ram is supposed to be the house of the planet Mars, which is the head of the whole world." The English marigold had obvious connections with the sun "for if it be gathered, the Sun being in the sign of Leo, in August, and be wrapped in the leaf of a Laurel or Bay Tree, and a Wolf's tooth be added thereto, no man shall be able to have a word to speak of the bearer there of."

In an age when Christian physicians based their skills on the old writings of Galen, Aristotle, and Dioscorides (pagans all!), it was well known that these skills were notably ineffective. Many people utilized the Doctrine of Signatures and botanical astrology to treat injuries and diseases. Seventeenth-century England was strongly influenced by botanical astrology through the writings of Nicholas Culpepper (1616–1654). Culpepper published *A Physical Directory* to attack the physicians, calling them a "company of proud, insulting Doctors whose wits were born about five hundred years before themselves."

Given the low level of sanitation during this period, infections were common, and Culpepper devoted considerable attention to their treatment. Since, he reasoned, people were under the influence of particular stars, as were plants, the potency of medicinal plants depended upon the celestial bodies. The healer should use a particular herb under the sign of the star or planet that related to the plant; "the eyes are under the luminaries." He recommended that "the right eye of a Man and the left eye of a Woman, the sun claims dominion over: the left eye of a Man and the right eye of a Woman are privilege of the moon." Hence wormwood, a moon plant (Figure 5.19), would cure diseases of the right or left eye, and various Sun plants would cure those infections of the other eye.

Figure 5.16 Barberry *(Berberis vulgaris),* as a Doctrine-of-Signatures plant.

Figure 5.17 Liverleaf *(Hepatica triloba),* as a Doctrine-of-Signatures plant.

Figure 5.18 Snakeweed *(Polygonium bistorata),* as a Doctrine-of-Signatures plant. Taken from *De Historia Stirpium* by Leonhard Fuchs, published in Basel in 1542.

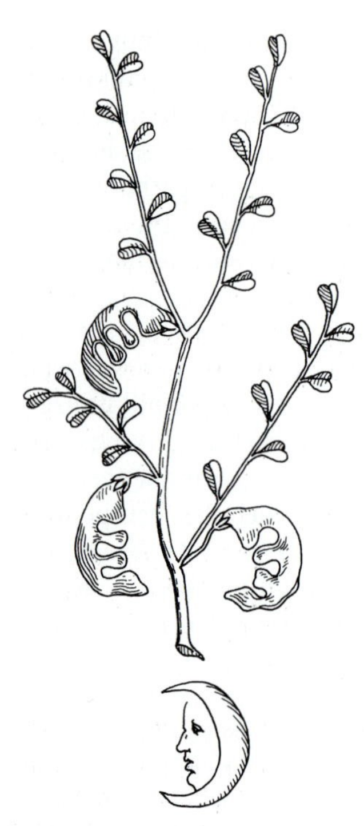

Figure 5.19 One of the "lunar" medicinal herbs. Taken from *Phytogonomica* by Giambattista Proto, published in Naples in 1591.

The physicians of the period struck back vigorously. William Cole, in his *Art of Simpling* of 1656, wrote that plants were:

> a subject as ancient as the Creation (as the Scriptures witnessed), yea, more ancient than the Sunne, or Moon, or Starres, they being created on the fourth day, whereas Plantes were on the third. Thus did God, even at first, confute the follie of these Astrologers, who goe about to maintaine that all vegetables in their growth, are enslaved to a necessary and unavoidable dependence on the influence of the Starres. Whereas Plantes were, even when the Planets were not.

While the logic of this argument rings false in today's ears, it apparently met with general agreement in the seventeenth century. Botanical astrology as a serious medical practice was essentially dead by the beginning of the eighteenth century.

PLANTS OF GOOD AND EVIL

As a corollary to plants of virtue, there were plants of evil or of contraevil. The barberry, with its thorns and fruits like drops of blood, was symbolic of the Crown of Thorns, and a chaplet over a door was a sign to the Evil One that only true Christians were within. Cloves look like nails . . . nails of the Cross . . . Christ's Passion for Mankind . . . salvation. A traditional Easter cookie in Greece has a whole clove embedded in it. Most plants associated with holidays have traditional antievil associations that predate Christianity. Evergreens, flourishing when all else is brown and sere, were symbols of enduring life, and their association with midwinter festivals was obvious. Holly *(Ilex spp.)* is a plant that was hated by witches because its berries, once white, became red with Christ's blood. The Druids used holly wood to build tiny huts as homes for woodland spirits that would protect humans, and even earlier the Romans had included it in Saturnalian festivals to ward off evil. The mistletoe, too, has significance buried in pagan belief, and its non-Christian symbolism gave title to one of the most important studies in comparative religion, *The Golden Bough* of Sir James Fraser. Christian legend says that it was originally a large tree used to make the Cross, that it shrunk to its present size, and was doomed to live off the strength of other trees. Norse legend says that the mother of the god Balder asked all living things to protect her son, but the mistletoe wasn't asked. Balder went around bragging that nothing could harm him, and evil Loki, hearing of the mistletoe's failure to promise protection, dared Balder to stand up and prove his invulnerability. Loki gave the blind Hodur an arrow made of mistletoe which killed Balder. The Goddess of Love restored Balder with a kiss, as we cheerfully restore each other in the same way.

In India, gatherers of medicinal herbs say, "Blessed be thou, plant of virtue. I pick thee with the good will of Vishnu," and in Greece they would say, "I pluck thee with good fortune, the good spirit of the gods at the lucky hour of the day that is right and suitable for all things." This parallels the sixteenth-century English incantation, "Haile be thou holie herb, growing in the grounde. All in the Mount Caluarie first wert thou found. Thou are good for manie a sore and healest manie a wounde. In the name of the sweet Jesu, I take thee from the grounde."

Not all physicians were happy about the Doctrine of Signatures, but caution was advisable; one could not offend the sensibilities of the people, nor provoke the anger of a clergy dedicated to any idea that linked sanctity and health. Gerard, author of *The Herball* of 1591, ridiculed many signature plants, but he did observe that the roots of Solomon's-seal *(Polygonatum multiflora)* will take away "any bruise gotten by falls or woman's wilfulness in stumbling upon their hasty husband's fist."

Plants of evil were rare, but included several members of the tomato family (Solanaceae) (Chapter 6) and a few other hallucinogenic and poisonous plants. Many otherwise useful or innocuous plants could be used to cast spells. St. Luke's flower, the marigold *(Calendula),* figures in one.

> Take marygold flowers, a sprig of marjarom, thyme and a little wormwood, dry them before a fire, rub to a fine powder, sift through a piece of fine lawn. Simmer with a small quantity of virgin honey in white vinegar over a slow fire. With this, anoint the stomach, lips and breast while lying down. Repeat these words three times: St. Luke, St. Luke, be kind to me. In my dreams, let me my true love see.

Up to the beginning of the eighteenth century, plants were known to possess means of communicating with people and of rectifying wrongs done to their vegetable persons. They could shriek, speak, or sing. The American Indian apologized for stripping a birch of its bark to make a canoe, and woodcutters in cultures as disparate as fourteenth-century Bavaria and seventeenth-century Japan would never enter a forest after dark for fear that the spirits of trees they had cut down would wreak a horrible revenge. Children were passed through holes in trees (Figure 5.20) to "rebirth them and make them well." One could pull petals from a daisy and know whether one was truly loved. From the beginning of the twelfth century, people spoke of the Tartarian, or Sythian, lamb, the plant-animal. Described by Sir John Mandiville in 1360, it was a large shrub bearing a full-sized lamb fixed at its navel (Figure 5.21). It fed by stretching to reach the grass surrounding the rest of the plant, but since most of the vegetable lamb plants were eaten by wolves, few people saw them. Medieval clerics argued that it grew in paradise, for the vegetable lamb was the True Lamb of God. It is likely that it was a cotton plant (Chapter 9), seen by European travelers to India. The legend continued until at least the middle of the seventeenth century, along with the goose tree, on which grew barnacles which gave birth to geese.

Figure 5.20 The birth of Adonis from a tree. Taken from a fresco by Bernardino Luini of Lombard, painted about 1500.

Tracing the history and symbolism of virtually any plant is a fascinating exercise in comparative religion, sociology, philosophy, and history. A case in point is the rose. Many of the hundreds of cultivated

Figure 5.21 The vegetable lamb described and illustrated by Claude Duret in *Histiore Admirable des Plantes et Herbes,* published in Paris in 1605.

roses are from the Orient, and hybridizations and selections over the past several hundred years have obscured the botanical and geographical origin of the roses we now grow in gardens. True roses have been cultivated for centuries and have contributed to mythology for about the same length of time. In Greek legend, Chloris, Goddess of Flowers, found a dead nymph and asked the other gods to transmute her into a special flower. Aphrodite gave beauty, the Graces bestowed brilliance, Apollo gave breezes to waft its perfumes, and Dionysus, God of Wine, gave nectar. Thus was born the rose. Cupid gave the new flower to Harpocrates, God of Silence, and this resulted in our term *sub rosa,* or "under the rose." Until the seventeeth century, a rose was hung from the ceiling of rooms in which silence and secrecy were to be observed. Ceiling paintings in council chambers still use the rose in their designs, but modern legislators apparently don't know Greek mythology.

Specific flower colors were mandated by the Council of Trent in the late sixteenth century, but the use of the red and white rose in Christian ritual predates this by centuries. Plutarch speaks of the rose "garded in modest bud" and noted that the sun is born each day from the flower bud. White as purity (remember the white rose in "Beauty and the Beast"?) and red as martyrdom were color associations before Christianity. The rose was adopted in the third century as a religious symbol to point up stages in the life of Christ.

The rose has been a heraldric device since the tenth century. The War of the Roses in 1455–1485 between the English Houses of York and Lancaster was named for the battle devices of these families, white for York and red for Lancaster (Figure 5.22). The Tudors came out on top, and, to placate the opponents, they combined the two symbols. Because of the association of roses with the Virgin, presentation of a single rose to a woman was to bestow a blessing upon her. North Americans have corrupted this practice in our penchant for extravagance. We rarely give a single flower, but often present a dozen or more. In point of fact, giving more than one rose was, during the Renaissance, a symbol of extreme and not particularly holy passion. Acceptance of a bunch of roses signaled the sealing of an unspoken bargain that the passion would be reciprocated at the earliest possible *sub rosa* moment.

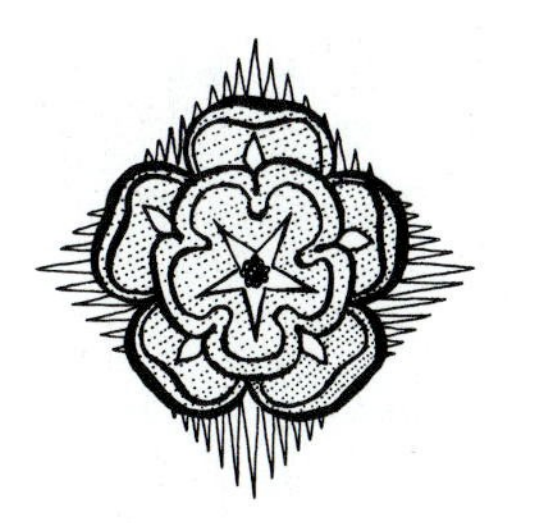

Figure 5.22 Heraldic use of the rose. Red rose of Lancaster (left), white rose of York (center), and the combined red and white rose of the Tudors. Taken from Parker's *Annales* of 1855.

ADDITIONAL READINGS

Crow, W. B. *The Occult Properties of Herbs.* New York: Weiser Press, 1970.

Eliade, M. *Patterns in Comparative Religion.* New York: New American Library (Meridian Books), 1963.

Fergeson, G. *Signs and Symbols in Christian Art.* New York: Oxford University Press, 1973.

Fraser, Sir. J. *The Golden Bough.* London: Macmillan, 1911.

Jacob, D. *A Witch's Guide to Gardening.* New York: Taplinger, 1965.

Kirk, G. S. *Myth: Its Meaning and Functions.* New York: Cambridge University Press, 1970.

Koch, R. "Flower Symbolism in the Portinen Altarpiece." *Art Bulletin* 46 (1964):70–77.

Lehner, E., and J. Lehner. *Folklore and Symbolism of Flowers, Plants and Trees.* N.Y.: Tudor, 1960.

Skinner, C. M. *Myths and Legends of Flowers, Trees, Fruits and Plants of All Ages and in All Climes.* London: privately printed, 1911.

The Tree of Life and the Rod of Aaron

Thy rod and thy staff they comfort me.

Psalms 23:4

In Genesis 2:8–9 it is written, "And the Lord God planted a garden in Eden to the eastward and there He put the man He had formed. And the Lord God caused to grow out of the ground every tree that is pleasant to the sight and good for food. And the Tree of Life in the midst of the garden and the Tree of Knowledge of Good and Evil." The Tree of Life appears only three times more in the Testaments: in Genesis 3:24, where cherubim and a flaming sword guard the way to the

Adapted from "Maypoles and Earth Mothers," *Natural History* 85(1976):4–8.

Figure 5.23 Winged bull unicorn kneeling before tree of life composed of lotus buds. Taken from an Assyrian stone relief.

Tree of Life; as an allegory in the tale of Jesse (Isaiah 11:1); and in the Apocalypse (22:1–2), where the Tree is on the banks of a river. With both Testaments replete with plant symbols, one wonders why this powerful concept was used so sparingly. To evaluate this question, we must go back to the period before the Israelites.

The Levant of the Middle East—Iran, Iraq, Syria, Palestine—was a flowing series of nation-states including Mesopotamia, Phoenicia, Chaldea, Sumeria—lands of shifting sands, peoples, and gods (Figure 5.23). They were harsh lands, where water was literally life, where obtaining food from the soil was difficult, and where natural phenomena like lightning, dust storms, disease, and death had to be explained to give meaning to life itself. All these pre-Israelite cultures had pantheons of gods, and each was venerated in forms which the people could see and feel. Rocks, solid and unchanging; water, the source of life; and trees, visible signs of renewal, growth, and fertility, were immortal objects that "lived forever." A tree, solitary and tall against a desert sky, was endowed with will, strength, and perception. It epitomized death in the dry season and resurrection when the spring rains came to the parched land. Trees were sacred symbols—visual manifestations or epiphanies—of powerful forces, and their desecration or removal was sinful. Modern women who interpose baby buggies between a shade tree and a bulldozer are repeating an act of faith which almost predates history.

Figure 5.24 Chinese version of a tree of life. Taken from a tomb of the Han Dynasty, about 2200 B.P.

Specific trees were assigned to specific deities. The sycamore *(Ficus sycomorus)* was sacred to Egypt's Ra, Hathor, and Nut; the date palm *(Phoenix dactylifera)* to Marduk of Babylonia; and oaks *(Quercus spp.)* to Zeus. Zeus's oak groves at Dodena, rustling leaves of which were the voice of the thunderer, were paralleled by those of the Celtic Druids. Hera was made manifest in willows *(Salix spp.)* that grow by springs, and who was to say that the tree was not the cause of the water? Apollo was the laurel, whose leaves contain enough prussic acid so that

Figure 5.25 "God's deer" bearing a Japanese tree of life. Taken from the *Kasuga Mandara* in Nara, drawn about 700.

His oracles could eat the God and become intoxicated with Him. As deities, trees often gave birth to humans. Yggdrasil gave birth to the first Norseman, Ask. Queen Mayha-maya gave birth to Prince Sidhartha—He who became Gautama Buddha—through strength provided by the Bo tree *(Ficus religiosa)*. The Yurucase of Bolivia know that Tiri opened a tree and brought forth the people. In east Africa, the first man was born within the baobab and fed on its breast-shaped fruits. In Persian mythology the Tree of Life stands in the sea where its seed, falling into the water, maintain the fertility of the earth. The Chinese, Japanese, Aztecs, and American Indians had such trees and developed virtually identical mythologies about them (Figures 5.24–5.26).

TREES OF LIFE

Trees of Life were the highways traveled between earth and heaven. Ra of Egypt started His daily journey through the sky from the branches of the sycamore fig (Figure 5.27). Indeed, Ra became the Sun God by climbing the Tree of Life, and His earthly counterparts, the pharaohs, reached their home in heaven by following Ra. Rulers of the Aztecs, too, moved between earth and heaven via tree ladders (Figure 5.28), as did Jacob (Genesis 27:10–14). Jack used a beanstalk. The tree-ladder symbolism is almost universal, and one doesn't walk under such a powerful symbol. Egyptians solved the riddle of why the sky doesn't fall by asserting that the universe was a huge box whose lid, forming the sky, was held up at its corners by branches of the Tree of Life. In some cultures, the Tree was upside down. The *Rig-Veda* of India says: "The branches grow towards what is low, the roots are on high and all worlds

Figure 5.26 Aztec tree of life, on which perches the parrot of wisdom. Taken from a stone relief in Mexico.

Figure 5.27 Egyptian woman worshiping the sun tree. Taken from a papyrus scroll of the Eighteenth Dynasty.

Figure 5.28 Pacal, ruler of the Aztecs, is captured in death by the cosmic tree. Taken from a bas-relief carved on his limestone sarcophagus, dated 683.

rest on it." If one sees a tree leafless against a sky, it is easy to believe that this is true (Figure 5.29).

All this was a source of embarrassment for the writers of the Testaments. Ancient Israelites invaded the lands of Canaanites and Moabites, who believed in trees of life. Their Deity, Asherah of Canaan, was symbolized by a tree stripped of branches, and Old Testament writers commented most reprovingly of such worship. The Israelites had no doubt about where they originated; they were made by Yahweh in His own image, and He did not have to show himself to His followers; His transcendence was a canon of faith. Success in agriculture, in procreation, and in battle was controlled by a single, male God with whom they had a covenant. Placation of the abode or the epiphany of a Deity was not only unnecessary but, as evidenced in the first two commandments, sinful. Yet, as Israelites married women of subjected nations, they absorbed the deities represented by trees and adopted and reshaped myths, symbols, and concepts of the conquered peoples. The Tree of Life, so basic to other Middle Eastern faiths, was just one of these symbols and, somewhat altered, it could not fail to have been included in Genesis. Nor could the symbol fail to have influenced those peoples of pagan Europe who came in contact with cultures who accepted and revered trees of life.

Implicit in Tree of Life motifs is the belief that it was at the center of the earth, that it was the *axis mundi* about which the earth turned. Virtually forgotten in our mobile Western society, this concept was central to beliefs of many cultures. As the Tree of Life was permanently wedded to the earth in the people's homeland, so do we, today, "put

Figure 5.29 Baobab *(Adansonia digitata)*, in west Africa.

Figure 5.30 Combination of Tree of Life and the Cross. Taken from a seventeenth-century Romanian house decoration.

down roots" and the wanderer is "rootless." The Cross (Figure 5.30) and the Menorah (Exodus 25) are not only trees of life; they are the center of the believer's world. Cemeteries in the Western world are planted as parks, not only for esthetic considerations, but because the plantings are symbols of the tree from which life and death spring. The annual resurrection of leaves in the spring, the changes in leaf size and color through the summer, and the fall of leaves in autumn parallels human life. The white cedar *(Thuja occidentalis)* of the Northeast and its Asiatic equivalent *(Thuja orientalis)* are both called *arbor vitae*—Tree of Life. They, like other evergreen conifers, are symbols of everlasting life.

Paradise is the Persian word for garden, a rare thing in that dry land, and our concepts of the heavenly paradise involves trees of life. Revelation 22:1–2 said that those victorious over evil "shall eat of the fruits of the Tree in the heavenly city." According to the *Mishkat al-Masabih* (a Persian commentary on the Koran), the Tooba tree in the Persian heavenly garden not only gave everlasting life to those who ate its fruit but also gave men the strength and vigor to have connection with many women, "the power of a hundred men given to a single man." When Mohammed journeyed to Heaven, he saw on God's right hand the Lote tree, made of gold and laden with jewels, with the river of Paradise flowing from its roots. Rabbinical scholastic scholars suggest that the biblical Tree of Life was forbidden Adam and Eve because it conferred immortality (Genesis 4:22), a concept that is in the belief structure of cultures all over the world. Trees, and what they stood for, could not be other than sources of awe, desire, fear, and wonder.

In Scandinavian mythology, Odin awoke the prophetess Völva and demanded that she reveal the beginning of the world. Völva spoke as follows:

> I know nine worlds, nine spheres covered by the trees of the world
> That tree set up in wisdom which grows down in the bosom of the earth
> I know there is an ash tree called Yggdrasil
> The top of the tree is bathed in watery white vapors
> And drops of dew fall from it into the valley
> It stands up green, forever, above the fountain of Urd.

Yggdrasil, sprung from the abyss, has three roots, one reaching the abode of the giants of wisdom, one touching the home of the gods of everlasting life, and one entering the cave of a dragon. One branch pierced heaven to Aasgard, and its flowers and fruits are the stars. It is on this branch that heroes enter Valhalla via a rainbow bridge, and it is on the same branch that the gods descend to earth. The first man, Ask, was born of one of the side branches. Swans, eagles, hawks, serpents, a squirrel, and other beings and symbols are appropriately placed on or near this Tree of Life.

In the Norse epic poem, the *Edda,* the awakened dragon injures the root in its cave and the tree shakes, foretelling the struggle of gods and giants. The battle itself destroys the world, as we learn from Richard Wagner's opera, *The Twilight of the Gods.* Yggdrasil is not destroyed and the Tree will again give birth to a new race of women and men who will be more nearly perfect than those presently inhabiting the earth. This legend is virtually the same for the Germanic Cosmic Tree, Irminsul, and the Finnish Tree of Life as written in the *Kalevala.*

The Persian Tree of Life stands in the lake of rain, Vouru-kasha. Its seeds were mixed with the waters and so maintained the earth's fertility. Its waters flow to the sea, but on their travel they give eternal life, promote the birth of children, give husbands to young women and horses to men. All plants are part of the Tree.

The rowan tree, our mountain ash *(Sorbus spp.),* has served as a Tree of Life in many European cultures. The British Druids believed that "roan trees and red thread put witches to their speed," and young women would weave red ribbons through its branches so that they would not be impregnated by supernatural beings. Crosses made of rowan branches, especially when held in the left hand, would frighten the Devil. No evil spirit could approach wood from this tree: Estonians used it as shepherd's crooks, Norwegian sailors kept a piece of the wood on their ships, and Finnish woodcutters placed a twig under their heads as they slept. Almost identical myths exist for other species in different cultures. The yew *(Taxus),* the oak *(Quercus),* the willow *(Salix),* the linden or basswood *(Tilia),* and others have all served as Trees of Life.

The Vedes and the Upanishads, writings of cultural groups that played important roles 2700 years B.P. in India, noted a Tree of Life, the *rukkha,* which was later adopted by both Hindu and Buddhist groups. The Tree arose from a root system in the navel center of the Supreme Being (Varuna, Buddha, and so on). Hence the Tree is a manifestation of God's existence. Illustrations and images of the Tree vary with the culture. Sometimes the trunk is built up of successive, superimposed lotus palmettes and its branches bear pearl garlands of foliage and jeweled fruits signifying the desires of humans and the procession of incessant life that extends from heaven to earth. The widespread branches extending through space illustrate the space within the mind of the believer.

In ancient China, too, trees were manifestations and the homes of unseen regulators of the fate of the believer. Different Trees of Life influenced the lives of people of different ranks. Pines served for rulers, arbor vitae for princes, the pagoda tree *(Sophora)* for high officials, the China tree *(Koelreuteria)* for scholars, and the poplar *(Populus)* for everyone else. A universal Chinese Tree of Life, *kiung,* is said to grow on K'wen Lun Mountain in central Asia, the place where the Four Great Rivers arise. Its blossoms confer immortality to those who can find the Tree.

TREES OF GOOD AND EVIL

A tree with an unusual shape or one riven by lightning acquired special powers of malignity. Those that failed to fruit could be inhabited by evil spirits. Indians believed that the banyan *(Ficus benghalensis)* could be dangerous under certain circumstances, and, since one could not know the circumstances, it was prudent to be careful; one didn't sleep under this sacred tree or carelessly knock off a branch. Before the time of Greek mythology, it was known that trees were the dwelling place of spirits, elves, or sprites who could cause suffering and had to be placated. Ornaments on Christmas trees were memorials to sacred animals and spirits that lived in that tree and were placed there to ease the insult of having their homes removed to human habitats. Witches used bits of the local Tree of Life to cast evil spells, but these could be countered by keeping branches of the Tree on one's door. A baptized man could kill a witch with such a branch or by driving a stake of the Tree through the heart of a warlock or a werewolf.

Trees could also be benevolent. A fine-looking tree or one used for food could be the home of a protective spirit. Sick children could be cured by passing them through the hole of a tree—essentially resurrecting them. Boy Scouts plant trees on Arbor Day because our ancestors planted a tree for each baby, knowing that the tree would become the home of a spirit who would, in gratitude, watch over the child. As a plant with many seeds, trees became fertility symbols. The pineapples on four-poster beds are stylized pine cones (Figure 5.31), the manifestation of several goddesses of fecundity.

Figure 5.31 Cones of the fir tree *(Picea)* were called pineapples and were fertility symbols. Taken from *Hortus Floridus* by Crispean van de Passe, published in Arnheim in 1617.

THE STAFF AND ROD

Staffs, rods, and wands are also part of this complex of symbols. In order to demonstrate that he was the envoy of a powerful Force, Moses had his brother Aaron cast his staff upon the ground, whereupon it became a serpent (Exodus 7). In the same chapter, the staff is used again, this time to turn water into blood—one of the plagues visited on Egypt. When the children of Israel thirsted, Moses was commanded to touch a rock at Horeb with his staff, and a spring gushed forth. We remember that Noah's dove brought back the branch of an olive, whose wood formed the staffs of many Mideastern deities. Later (Numbers 17:17–26), Yahweh commanded that the staffs of each of the tribes be collected before His shrine. The staff of Aaron, elder of the tribe of Levi, budded, flowered, and produced ripe almonds, testimony that priestly functions were to be assumed by the Levites. The budding staff appears in legends of St. Christopher and in the fable of the Glastonbury Thorn *(Crategus oxyacantha praecox)* in England, where it budded from the staff of Joseph of Arimathea. Charles I, in an effort to "root

Figure 5.32 Zeus with staff and bundle of lightning. Taken from a Greek amphora of 2400 B.P.

out, branch and all" the Roman Church, had the tree cut down. Representations of deity frequently involved a staff. Baal of the Canaanites carried a sceptre of cedar; Zeus a staff of oak (Figure 5.32); Anu (sky god of Sumeria) carried Kiskanu, the Tree of Life, and Dionysus, originally a Babylonian harvest God, had His staff of the Tree of Life twined about with ivy and surmounted by the pine cone of fertility. Mortal rulers use a staff, rod, or sceptre to make clear that they possess the powers of the deity; the royal umbrella of the kings of Siam was a frond of their Tree of Life, and the Christian bishop's crook is derived from the Zoroastrian priests of Persia, who carried the Tree to protect their "flocks."

Possession of a branch of God's tree conferred extraordinary powers on mortals. In Scandinavia, the ash *(Fraxinus)* and the rowan *(Sorbus aucuparia)* were sacred to Odin and Thor. The word *rowan* is derived from *rune* (to divine), and twigs, each marked with an oracular message, were selected by priests as prophetic. The ancient Judaic judicial system included having a judge choose a stick from a bundle, each stick containing the name of a litigant. Our custom of choosing long or short straws is derived from this custom. The children's game of jackstraws and the divination of tossing sticks in the Chinese *I-Ching* have similar origins. The magician's wand is a magic staff made of the Tree of Life. Modern magicians use a black rod to add the dark powers of the underworld to enhance their spells. Circe, daughter of Helios, used a magic wand to transform lustful sailors into pigs (Figure 5.33). Medea used her wand to assist Jason (and little good it did her!). The traditional broomstick of witches is a profane use of the Tree of Life. Touching with a wand or rod can either confer power—as when a notable is knighted—or bring power—as when American Indians "count coup" in battle by touching their enemies with a feathered stick.

THE CADUCEUS

Figure 5.33 Circe turning one of Odysseus's sailors into an animal. Taken from a Greek amphora.

Although the medical caduceus was not formally adopted as a symbol until the 1956 General Assembly of the World Medical Associations, it has symbolized the healer for 3000 years. Hermes, messenger of Zeus, conducted souls of the dead to Charon's boat, which transported the righteous to the Elysian fields where they were reborn as immortals or transported the unrighteous to Hell. His staff was originally a forked stick, and only later was it depicted with the forks as wings. Hermes's staff was an extension of Zeus's power, given the demi-god both for its inherent power and as a symbol of Zeus. The practice of carrying the king's staff was continued by royal envoys until the eighteenth century, and passing the baton in a relay race is similarly related. Another bearer of the Caduceus was Asklepios. He was raised by Cheiron the Centaur, who taught the boy the healing arts. Since recovery from serious illness was equated with return from the dead, Asklepios liberated people from

Figure 5.34 Asklepios, God of Healing. The sacred rod is intertwined by the serpent of renewed life.

the thrall of the underworld with a branch of the Tree of Life (Figure 5.34). The snake, depicted as coiled about one leg, was later moved to the staff as a symbol of rebirth, since snakes regularly shed their skins. The wings on the medical caduceus are a later addition (Figure 5.35).

Although we use magic staffs to search out water, their original purpose was to locate metals. In 1556, Georg Agricola, physician of Bohemia (modern Czechoslovakia), published a *De Re Metallica* in which he gave directions for using the *Virga mercuralis,* the rod of Mercury, to locate ores (Figure 5.36). As late as 1710, Jonathan Swift noted a "certain magic rod that, bending down its top, divines when o'er the soil has golden mines." Dowsing, or water witching, dates from the late sixteenth century, when the Baroness de Beausoleil popularized the art in *La Restitution de Pluton.* The rods are, of course, branches of sacred trees—oak, rowan, hazel, willow—and their users are viewed with awe, superstition, and suspicion. Does it work? It all depends upon who you ask.

Figure 5.36 Dowsing for minerals with rods made from hazel (*Corylus spp.*). Taken from *De Re Metallica* by Georg Agricola, published in 1556.

Figure 5.35 The Sigillum Chirurgorum Dresdensium, a crest used in a book on diseases published in Dresden in 1663.

THE MAYPOLE

On the first day of May, tall poles are erected in public parks and are hung with colored ribbons. Young girls in frilly dresses are crowned with flowers and dance, first clockwise and then counterclockwise, wrapping and unwrapping the ribbons about the maypole (Figure 5.37). It is a happy, innocent amusement, a symbol of the joy of spring, but few recognize that the maypole and its dance is one of the oldest and most sexual of public ceremonies involving the Tree of Life. The maypole originated in the lands of the Mesopotamians and Syrians, who all believed in an Earth Mother. Ishtar of the Assyrians became Astarte of the Chaldeans, Asherah of Canaan, Cybele, Demeter, Ceres, Diana, and many others. It was to Earth Mother that people prayed for good crops, human and animal fertility, and, through Her, intercession was asked to the male God for rain, gentle breezes, and pure, ever-flowing springs. If Earth Mother was with child, so would be the fields, the women, and the ewes; trees long dead during the dry season would burst forth with new leaves, and flowers and fresh pasturage would appear in the fields.

But gods were forgetful and had to be reminded annually of their responsibilities to the faithful, and the people simultaneously reminded of their duties to the gods. Cybele-Astarte-Demeter was symbolized by a tree, shorn of branches and carried in solemn procession to the temple, where She was erected before the awed gaze of the people. Decked with flowers, the tree was ritually worshiped with dances that included sexual orgies. Minoan Cretans painted an olive branch bright pink and planted it in a bed of grass as the imminent symbol of Isis. Old

Figure 5.37 Children dancing around a maypole in Central Park, New York. Etching of 1877.

Testament writers commented on such worship in most reproving terms. Thus the maypole was not accepted in Judeo-Christianity, and we must look to the Greek and Roman roots of our Western heritage for information.

The Hellenes probably invaded Greece from the Near East, changed the names of the gods, and made both gods and goddesses more "human," accessible, and understandable. Cybele of Phrygia became Artemis, and Demeter of Phoenicia became Maia—daughter of Atlas, mother to Hermes (courtesy of Zeus), and the month of May. We know that the May Day pole dance was retained and modified, but little else.

Roman rule was marked by changes in the names of the gods but not in their functions. Demeter became Ceres, and Cybele-Artemis became, in part, Diana. The spring ritual began in Rome on March 22, when a pine tree was cut down, debranched, decked with violets as a symbol of the male God Attis, and borne to Diana's temples. March 23, the Day of Blood, started with dancing around the erected maypole. The music of cymbals, drums, and flutes became wilder as the day progressed. Frenzied by the dance, participants lacerated themselves and dripped their blood upon Diana's epiphany. The dance culminated in self-emasculation of young men wishing to become Diana's priests. The excised organs were thrown at the tree to hasten the resurrection of the earth and its impending fertilization. The ceremonies ended when a procession, led by a human representative of the Goddess and her male consort—the King of the Woods—mated to signal the fertilization of the earth. The King of the Woods, originally the male God, also had different names in different cultures—Attis, Marduk, Virbius, Zeus, Jupiter—and was symbolized by a tree. It was usually a forest giant like the oak or the ash, one that annually lost its leaves, an indication that Earth Mother and Her conifer were immortal, but a mere male was replaceable.

With minor variations, the festival was similar throughout Europe. Young men and women went to the woods, cut down a tree, stripped off the branches, and carried it triumphantly into the village. It was decorated with flowers and other symbols of fruitfulness and set up in front of the church—to the discomfiture of the clergy. A May Queen was selected. She and the King of the Woods—called Green George, Father May, Bark King, Grass Lord, or Leaf Man—presided over the dancing. As night fell, the queen and king mated in the fields to emphasize the point of the ceremony, and this act was affirmed by the faithful. Celts of Scotland, Ireland, Normandy, and parts of Scandinavia included a fire ceremony in their May Day rituals. These *beltane* fires were lit from a flame started with a firebow whose spindle was a shaft of oak (Zeus) fitted into the slot of a board of softer wood such as willow (Hera). Tinder was dried mushrooms and puffballs. In Sweden, the *Maj Stäng* is dressed with crossarms bearing hoops of willow wrapped with flowers. Hoop rolling on the quadrangles of many colleges is derived from these garlanded May hoops except that, in the sixteenth century, the hoop stick of oak was thrown through the hoop instead of being used to propel it.

It is to England that we must look for the North American version of the maypole ceremony. Before the seventeenth century, the mating of the queen and Jack o' the Green was not pretend, and the midnight cutting of the tree was accompanied by the normal exuberance of lusty youth. Phillip Stubbes, in his *Anatomie of Abuses* of 1583, called the ritual an "act of Satan," the maypole itself "that stynking ydol," and noted that after the revels, "scarsely the thirde part of ye girls returned undefiled." In 1626, Governor William Bradford of the Plymouth Colony appointed Thomas Morton as administrator of Merrymount, a suburb of the main colony. Morton, apparently taking the community's name seriously, introduced the maypole to New England, thereby creating a scandal. Bradford noted that the celebrants invited Indian women to participate in "dancing and frisking together . . . and worse." John Milton wrote of the "dallydance" of Zephyr and Aurora "as he met her once a-Maying." The Morris dances of England were May dances, and May Queen Mary only later became Robin Hood's "Maid Marion."

By the beginning of the eighteenth century, May Day celebrations had become just celebrations and were no longer spring fertility rituals. By 1833, Tennyson could write: "You must wake and call me early, call me early Mother dear/For I'm to be Queen o the May mother, I'm to be Queen o the May." Kipling remembered the old ways in a bit of doggerell: "Oh, do not tell the priest our plight or he would call it sin/But we have been out in the woods all night conjuring summer in."

One is struck by the universality of concepts which flow around the Tree of Life. Life, death, and the perils of everyday existence must somehow be understood, for if they are not explicable, reality is without meaning. The psychologist Carl Jung seized upon this as one of the arguments to support his theory of the collective unconscious, an attempt to integrate the universal symbols human beings use to relate themselves to their environment. Whether Jung was correct in believing that homologies among symbols in all cultures represents an archetypical and universal mind, or whether the choices and beliefs were identical because the biological facts of life are identical, is an open and most intriguing question.

ADDITIONAL READINGS

Alldritt, C. *Tree Worship with Incidental Myths and Legends.* Auckland, New Zealand: Strong and Ready, 1965.

Butterworth, E. A. S. *The Tree at the Navel of the Earth.* Berlin: W. de Gruyter, 1970.

Campbell, J. *The Masks of God.* 4 vols. New York: Viking Press, 1964.

DeWaele, F. J. M. *The Magic Staff or Rod in Graeco-Italian Antiquity.* Amsterdam: 1927.

Eliade, M. *Patterns in Comparative Religion.* London: Sheed & Ward, 1958.

Hastings, J. *Encyclopedia of Religion and Ethics.* New York: Scribner, 1916.

Holmberg, U. "Der Baum des Lebens." *Annals of the Academy of Science* Fennicae B16, no.3 (1922):1–146.

Jung, C. G. *The Archetypes and the Collective Unconscious.* Princeton, N.J.; Princeton University Press, 1959.

MacMulloch, J. A. (ed.). *The Mythology of all Races.* London: Marshall Jones Company, 1928.

Porteus, A. *Forest Folklore, Mythology and Romance.* London: G. Allen, 1928.

Thiselton-Dyer, T. F. *The Folklore of Plants.* London: Chatto & Windus, 1889.

The Lotus

Om mani padme hum
Oh, Thou Jewel of the Lotus

Buddhist Mantra

Some of the plant symbolisms in religion are almost universal: the tree of life, the resurrection symbolism of new, green leaves, and the falling of leaves as a death symbol are understood by all peoples. But each religion has symbols unique to that faith, ones that aren't in other religions. The Jewish use of the *ethrog,* a citrus fruit, is not found in Christianity, and the lily has no significance in Judaism. Yet both these faiths sprang from the same roots and both drew on common sources. If we look at religions different from Judeo-Christianity, we find the universal symbolisms and, in addition, some whose roles are not in our Western experience. The lotus is such a plant. It, more than any other plant, seems to have acquired symbolic importance among many cultures that may not have had direct contact with each other. Whether this cultural difference is more apparent than real is of anthropological importance, for as we learn more about trade routes and amazing sea voyages, we realize that there may have been extensive economic and cultural intercourse among peoples who we thought were unconnected in ancient times.

The botany of the lotus is rather confusing, since many plants have been given the same common name. The botanical family to which all of these plants belong is the Nympheaceae, the water lily family, containing eight or nine genera and about 50–60 species. All are aquatic plants with a rootstock (rhizome) buried in the mud at the bottom of lakes and ponds. Leaves grow up to the surface of the water on long petioles, and the flowers, too, arise from the rootstock. Flowers are large, white, or brightly colored. In most species, the flowers open at dawn and close again at dusk. On a bright sunny day, small waves and light breezes cause the leaves and flowers to dance gracefully on the water's surface. It was thus that Linnaeus chose the family name, "flowers of the water nymphs."

Figure 5.38 Opened flower of the sacred lotus *(Nelumbo nelumbo).*

The most familiar genus is *Nuphar,* the common pond lilies, widely distributed throughout North America. A second genus, *Nymphaea,* includes the fragrant water lily *(N. odorata)* and the tuberous lily *(N. tuberosa),* whose flowers run the color spectrum from white through blue, yellow, pink, and red. There is also the blue lotus of the Nile *(N. caerulea)* and the white lotus of the Nile *(N. lotus).* One species *(N. mexicana)* is found in Central America, one is native to Europe *(N. alba),* and another is the blue lotus of India *(N. stellata).* To compound the confusion, a third genus, *Nelumbo* (Figure 5.38), contains the sacred lotus of Persia and Asia *(N. nucifera)* and the lotus of East India *(N. nelumbo).* Members of these three genera are wide-distributed throughout the world, and all are similar enough to be called lotuses. In the same way, a family's Christmas tree this year may be a Scotch pine, next year a balsam fir, and the following year a white spruce—yet they are all Christmas trees, and all have the same symbolism.

THE LOTUS IN THE NEAR EAST

Figure 5.39 Horus, God of the Rising Sun, seated on a lotus. His finger at his lips signified silence. Taken from an Egyptian wall painting.

To get a feeling for the place of the lotus in non-Judeo-Christian faith, Egypt is a good place to start. Both white and blue lotuses are found in the Nile and the tributary streams that feed the Nile on its journey from Lake Victoria. The lotus was, for the Egyptian, a symbol of the nascent life of this sacred river, a reminder of the mysterious yet bountiful fertility given yearly as spring floods covered the land with rich, black mud in which crops grew luxuriously. The Egyptian gods themselves partook of the wonder and worry of birth and death, day and night, and those other dualities that trouble humankind. Horus, the new born sun, is born daily from a lotus flower expanding at dawn under His influence (Figure 5.39), and both Horus and the day die each evening as the petals close. Osiris, father of Horus and husband of Isis, as well as king and judge of the dead, wears a crown of lotus blossoms in the closed position. When His flowers open, the entire world will be reborn. To reach the realm of Osiris, the pilgrim soul grasps a lotus as it embarks on the boat of the sun, which will carry it to the throne of judgment and eventual resurrection. Isis, in her role as Earth Mother, was manifested by young flower buds of the lotus just emerging from the waters of the deep. Atum, the primeval deity under whose influence the world took shape, is pictured as arising from a lotus bud floating on a sea without form. Heaven, Egyptian priests proclaimed, was supported by the flower stalks of the lotus and, to make this image concrete, the roofs of temples were upheld by columns whose capitals were composed of the lotus (Figure 5.40). The hippopotamus, a sacred animal living in and emerging from the equally sacred Nile, was imbued with the aura of the lotus (Figure 5.41). There is a bust of Pharaoh Tutankhamen, sculpted about 3500 B.P., in which the head of this god-king emerges from the lotus flower; pharaoh himself is a God.

Figure 5.40 Egyptian column and lotus capital.

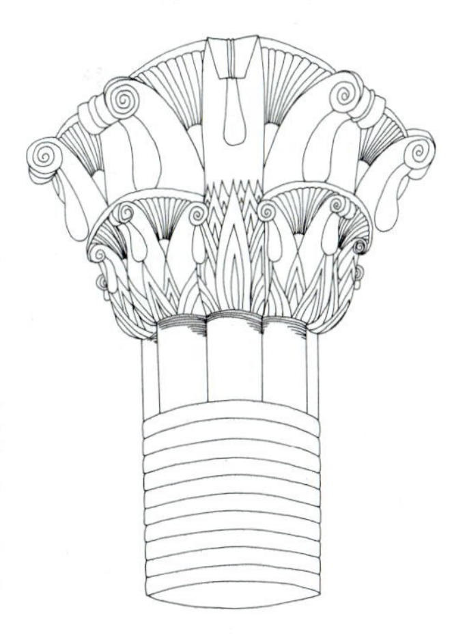

Figure 5.41 Egyptian hippopotamus with lotus design. Glazed ceramic dated from 3500 B.P.

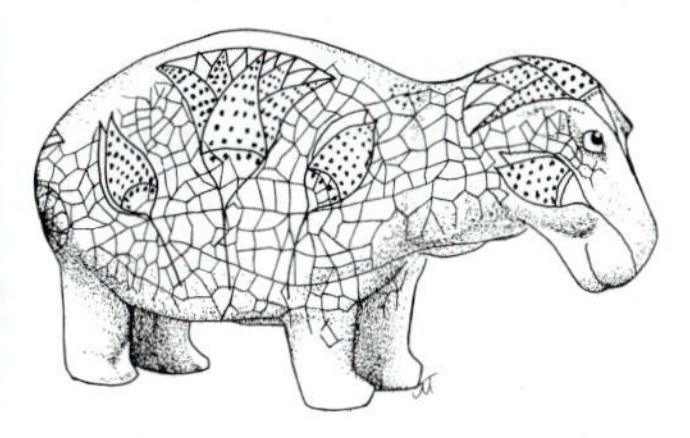

Figure 5.42 Homage to the winged globe representing the Deity, with stylized lotus as the Tree of Life. Taken from an Assyrian stone relief.

Lands in the Middle East, whose cultures are almost as old as those of Egypt, used the lotus in much the same way. In Assyria, the Moon God is figured as sitting on a disk composed of lotus petals from which wings extend (Figure 5.42). Phoenician seals depict the moon and stars arranged around a lotus flower which, in turn, is supported by wings bearing the plant to heaven. In Mesopotamia, the lotus is a symbol of conquest, the triumph of the gods of the people against those of lesser faiths. Persians venerated the opened lotus flower as a manifestation of the sun and hence of light and life. The closed lotus was a moon symbol, evidence of darkness and of death. Since the flower will again open in the morning, life has meaning because of this resurrection.

The Greek gods did not directly employ the lotus, but some of its symbolism came from Egypt. In one legend a nymph, betrayed and then deserted by Hercules (the cad!), threw herself into a pool and was drowned. The other gods had pity on her and changed her into a lotus to provide a perpetual reminder to Hercules of his philandering ways. It apparently was ineffective; Hercules continued to be the ruin of many an innocent maiden. The stylized lotus developed into the *anthemion*, a flat decoration that adorned greek temples. In the *Odyssey,* Homer noted that Odysseus was acquainted with a land in which the people lived in languorous forgetfulness—the land of the lotus eaters—and we still speak of someone who is forgetful and unmindful of the world as a lotus eater. Since no part of the lotus is narcotic, it is likely that another plant had the same common name. Most authorities suggest that it was a shrubby member of the buckthorn family, *Zizyphus,* the jujube, whose fruit can be fermented into a potent wine. Etruscan art utilized the stylized lotus in architecture, as did Greece (Figure 5.43), and traces can be found in Byzantine and Romanesque churches. The lotus appears in Western art and religion only as a pretty plant.

THE LOTUS IN INDIA

India, a land replete with interlocking cultures and religions, has employed the lotus in all of its many faiths. Prior to 4000 B.P., invaders came from the northwest, bringing with them barley—suggesting contact with peoples of the Middle East—and a mystery religion, Vedism, in which the lotus plays a prominent role. Creation legends vary. In one version, Om, the Spirit of Creation, moved on the surface of a vast sea that existed before the earth was formed. He quickened into life a wondrous golden lotus, resplendent as the sun, floating on the lifeless water. From this flower arose the Hindu Trimurti: Brahma the Creator, Vishnu the Preserver, and Siva the Destroyer (Figure 5.44). A mango tree arose from the lotus, and a seed of the mango gave birth to the daughter of the sun. She mated with a god, and their children populated the earth. In another version, the world itself is a lotus flower floating in the center of a shallow pond that rests on the back of an elephant who is standing on the shell of a turtle.

Figure 5.43 Hellenic Greek decorative frieze of the lotus.

In Hinduism the lotus is a center of vitality, partaking of the strength of the sun, and is therefore a source of tremendous power. Just as the rays of the sun can be focused to cause fire, so can the lotus focus thoughts upon universal concepts. In the Tantric tradition, there are power centers *(cacra)* in the body from which radiate understanding. Perception, bliss, sexual power, and pure thought are, accordingly, called the lotuses. In the Puranic tradition, the cosmic tree was a lotus, growing from the navel of Nārāyana *(Om)*, and both gods and people arose from this tree. Lakshmi, Goddess of love, fortune, and fecundity, rests on a lotus leaf. Even Kamadwa, Hindu equivalent of our Cupid, rides the sacred river Ganges on a lotus leaf, reinforcing the role of the lotus as a symbol of physical love and childbearing. It is said that souls of the dead sleep in lotus blossoms until admitted to a paradise containing a sacred lake covered with the flowers.

About 2500 years ago, Siddhartha Gautama was born into the Sakya clan near Nepal. At age 30, Siddhartha left his home, wife, and family to obtain enlightenment. He achieved victory over the Evil One and over himself after seven years under a Bo tree (Figure 5.45), and a new Buddha, fourth of the line, was born in the glory of the lotus, whose opened flower revealed that which he wanted to know. Having pity on Mankind, the Buddha gathered five disciples to whom he preached and whom he ordained. Soon there were many converts who were enjoined to go forth and proclaim the *dharma*. The Buddha's fame as a teacher—Padma Sambhava (preacher of the lotus)—spread throughout northeast India. For many years, He moved from place to place organizing the expanding order until, full of years, He died. His body was cremated and relics given to the holy places where, in lotus-decorated reliquaries, they are venerated.

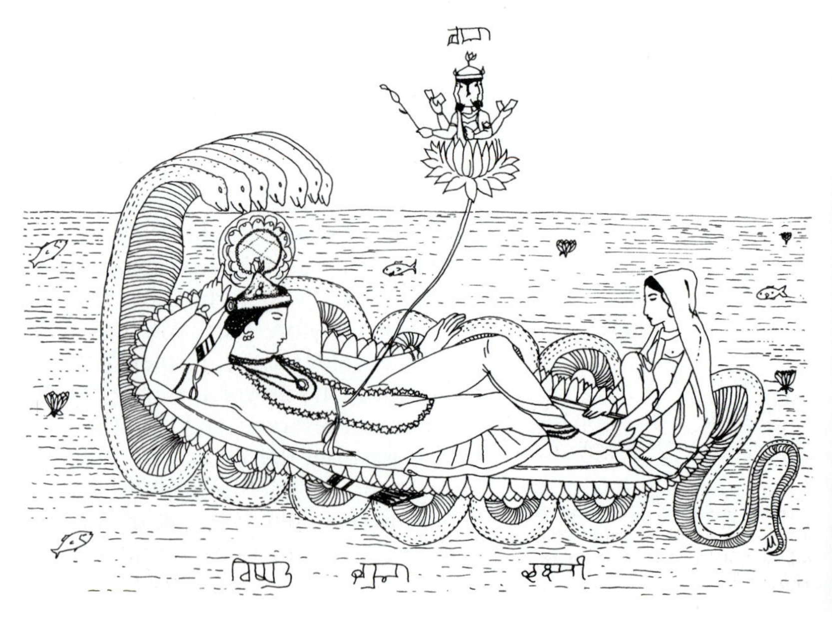

Figure 5.44 Birth of Brahma. Taken from a Hindu woodcut.

Figure 5.45 Buddha receiving enlightenment leading to Nirvana. The tree is the Bo *(Ficus religiosa)*, and the Buddha is seated in the lotus position on lotus leaves.

Buddhism, like Western faiths, showed both sectarization, in which various paths or "vehicles to enlightenment and salvation" developed, and syncretism, in which merging of philosophies and rituals occurred. Yet the place of the lotus as a symbol was little altered (Figure 5.46). The Buddha and the *Bodhisattvas* (persons who attained enlightenment but chose to remain in contact with humans) are figured on lotus pedestals just as are the Hindu Trimurti. Some bodhisattvas, such as Kwan Yin, the goddess of mercy, may carry a lotus flower and are known as *Padmapani,* the lotus bearers or, more appropriately, bearers of the good news. Maitreya, Buddha of light and living kindness, is portrayed with the fingers of His uplifted hands forming a lotus bud about to open. Amitabha, Buddha of endless light and infinite compassion, is represented by golden fish from the sea of eternity and a lotus that grew in the sea. Symbols acquired from other faiths were combined with the lotus; many of the *mandalas,* schematized representations of the cosmos (Figure 5.47), are resting—as does the Buddha—upon the lotus. In some mandalas, the lotus may be conventionalized into a wheel *(Rimbō)* with the petals representing the spokes. The wheel is the circle of birth, death, and rebirth—the rolling of time endlessly into eternity. The wheel became a rosette or was further simplified into an octagon. Mantras, words of praise to be sung or hummed, frequently incorporated the Indian word *padma* (lotus) as a symbol both of the Buddha and of the enlightenment. Yoga, a technique of bodily control to allow the unity of body *(Avalokita)* and the spirit *(Amitabha)* with the universe, uses specific body positions, including the lotus position, in which the legs are crossed so that one's feet are resting on one's thighs (Figure 5.48). This derived from the Hindu Tantric tradition, in which the lotus position signified a rebirth.

THE LOTUS IN CHINA AND JAPAN

China, Japan, and Korea used the lotus theme in some of the most beautiful art the world has ever seen. Lotus blossoms and leaves were translated, literally or after stylization, into every medium. Poetry,

Figure 5.46 Hindu decorative frieze of the lotus. Taken from a stone carving in India dated at 2250 B.P.

Figure 5.47 Mandala of Brahma resting on a lotus with the sacred swastika-in-triangle.

Figure 5.48 Korean Buddha on lotus pedestal. Taken from a painted and gilded wood carving of the twelfth century.

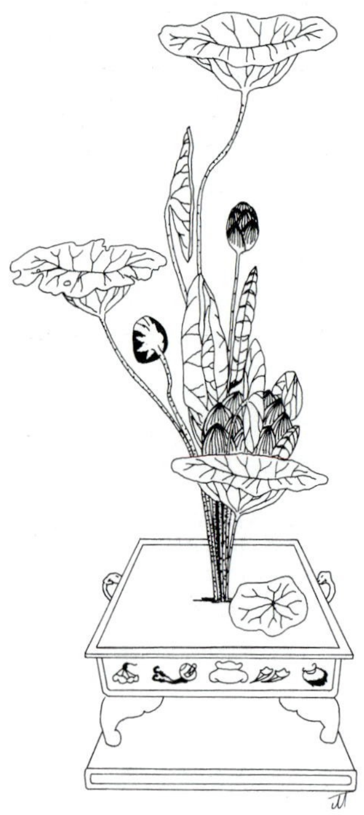

Figure 5.49 An eighteenth-century Japanese engraving of a flower arrangement using the lotus.

legend, and prose involving the miraculous lotus have passed through generations of people. In the visual arts, calligraphy, painting, sculpture, metal working, wood carving, and other techniques were used to portray the lotus. Even after Buddhism in China lost its prominence, artists continued to use the flower. No single plant, with the possible exception of the lily attributed to the Madonna, has so captured the imagination of artists.

Throughout this intense emphasis on the lotus as a symbol and epiphany of faith, the lotus was and still is a common food item. Lotus roots are grown in ponds, the rootstocks cut into rounds, dried, and used as a crunchy additive to soups. Lotus seeds are added to meat or vegetable dishes, and dried, roasted, and salted lotus seeds are eaten as we eat peanuts; there is no thought that the plant, as food, is to be equated with partaking of the Godhead.

When Ch'an Buddhism was taken to Japan and metamorphosed into Zen Buddhism, the lotus lost much of its importance, although Buddha is still depicted on the lotus throne, and the lotus position of Tantric and Yoga meditation is used in Zen. Legend tells us that once, when the Buddha was at rest, a Brahman ruler came to Him and gave Him a golden lotus flower. The Enlightened One accepted the flower and spent many hours gazing in silence at the blossom. Finally, the Buddha smiled, and this moment of complete joy was passed on to His patriarchs until the Indian patriarch Bodhidarma came to China in 520 and taught the Ch'an concepts. This intense and complete riveting of attention upon some natural object became a cornerstone of Zen. The development of flower arranging and the tea hut garden by Zen tea masters (Chapter 4) drew upon the lotus, whether it was used literally as a three-part flower arrangement (earth, man, cosmos) (Figure 5.49) or as carefully cultivated lotus blooms on a quiet temple pool.

There is only the most fragmentary evidence that the ancient civilizations of the New World were in contact with either Egypt or India, and yet Mayan frieze decorations or religious symbols depict the water lily (Figure 5.50). In Illinois, the American Indians of the Fox tribe, centered near the Fox River and the Fox Lake, used the native yellow lotus, *Nelumbo lutea,* as a symbol of purity, of the sun, death, and rebirth.

WHY THE LOTUS?

In the cultures and religions examined here, there is an obvious concordance of the symbolic significances of the lotus. Some of these are, at least superficially, fairly obvious. The opening at dawn, the round disk shape of the flowers, and the number of white or yellow-petaled species suggest the sun's course through the sky. The closed blossom can easily be related to the moon. The lotus was not the only

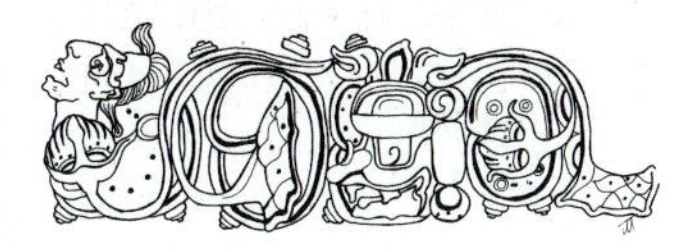

Figure 5.50 Mayan decorative frieze of the lotus. Taken from a stone carving dated at 800.

flower in which birth of souls occurred. The narcissus in Europe, the daisy in South Africa, and others have been used in an identical manner. We still tell young children that they were born in a cabbage or in a rose or another plant. In several cultures, the opened, round, yellow or golden flowers of the sunflower, the marigold, or the cone flower *(Rudbeckia)* have been used as sun symbols.

Other aspects of the varied symbolism of the lotus are, however, not so obvious, and our attempts to rationalize them may merely reflect our Western "need" to put explanatory labels and names on concepts which really elude us. Most cultures have used the triple symbols of earth-water-sky, and the lotus partakes fully of all. Rooted in the primeval mud and covered by endless water, the plant sends its leaves and, eventually, its sunlike flowers above the water into the sky. It betrays its humble, earthly origin in the mud, transcending this with leaves of bright green that are not wetted by the water upon which they float. Perhaps this characteristic, not being injured by water and rising above it, can symbolize the truly wise One who has achieved the state exemplified by the sun while still participating in and being a part of water and earth. The meanings of the lotus transcend human ability to put these concepts into words or into more representational images. The symbols, and the concepts that underlie and magnify the symbols, can be variously interpreted by individuals. Were they made concrete, they would stand for concepts that could not be put into words with which all others could agree. Just as a nation's flag means different things to different people, the lotus and other plant symbolisms mean what we, as individuals, choose them to mean or want them to mean.

ADDITIONAL READINGS

Armstrong, R. C. *Buddhism and Buddhists in Japan.* London: Society for Promoting Christian Knowledge, 1927.

Edkins, J. *Chinese Buddhism.* London: Kegan Paul, 1983.

Jung, C. G. *Man and His Symbols.* New York: Dell, 1974.

May, R. *Symbolism in Religion and Literature.* New York: Braziller, 1966.

Soothill, W. E. *The Lotus of the Wonderful Law.* New York: Clarendon Press, 1930.

Thomas, P. *Hindu Religion, Customs and Manners.* 2nd Ed. Bombay: Tara, 1910.

Peyote

A white man uses prayers out of a book; these are just words on his lips. But with us, peyote teaches us to talk from our hearts.

CHEYENNE PEYOTE LEADER

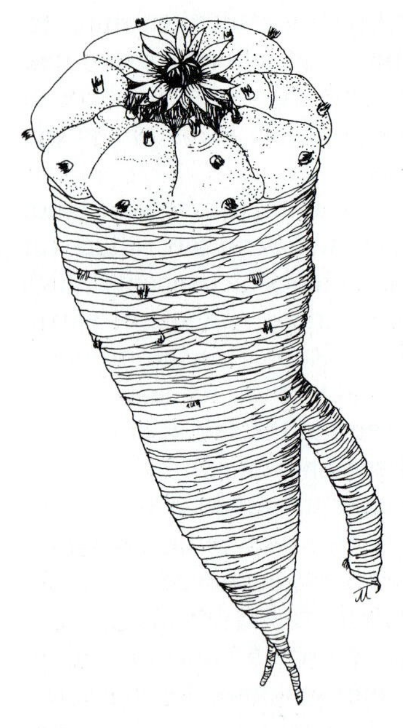

Figure 5.51 Peyote cactus *(Lophophora williamsii)*. The top is about 3 centimeters in diameter.

In the American Southwest and Great Plains, the post-Civil War nineteenth century was, for the American Indian, a period of turmoil. When the overcrowded East viewed the land beyond the Mississippi, it was considered manifest destiny to open the country for families devastated by the war. Increased immigration, a rising birthrate, railroads, and the Gatling gun made it clear that the Indians of the plains were doomed. The Indian Agency of the Great White Father was resettling hunting tribes on reservations selected to include the worst lands of the Oklahoma Territory—lands so deficient in water and good soil that they couldn't be tilled, particularly by Plains Indians, whose culture was not agricultural. The Indians could either starve or become completely dependent upon federal welfare. More insidiously, the government was supported by missionaries whose task, as they saw it, was to make the savage into a Christian content with his or her lot—for the meek shall inherit the earth. Conversion was facilitated by violating the U.S. Constitution and channeling much of the welfare through the missions. Food was distributed with covert and overt statements that the people's gods were impotent—demonstrated by the success of the Seventh Cavalry—and that their customs were savage and immoral. Not unexpectedly, the world of the American Indian fell apart and disease, sloth, and degradation became the lot of the Indians. Deprived of their culture, swindled by federal authorities, seeing their children kidnapped into mission schools that specifically denied their heritage, and buying firewater at prices inflated by the Alcohol Prohibition Act, passed by the combined efforts of the missions and the bootleggers, the people began casting about for some mechanism which would provide a modicum of dignity, hope of survival as Indians, and some spiritual sustenance.

The *deus ex machina* was a cactus botanically known as *Lophophora williamsii,* the peyote (Figure 5.51). The plant itself is an inconspicuous gray-green hemisphere with few spines, sparse white hairs, and a carrotlike tap root. When dried into *mescal* buttons, it is a hard, medium-brown disk less than a centimeter in diameter. Peyote is a modification of the Aztec *peyotl,* which means "caterpillar's cocoon" and refers to the white, wooly interior of the plant. It was in Mexico that Westerners first learned of it. In 1560, the Spanish Jesuit Bernardo Sahugan wrote, "There is another herb called peiotl; it is produced in the north country; those that eat it see visions either frightful or laughable; this intoxication lasts two to three days." Francisco Hernandez, physician to the king of Spain, noted that the Chichimecas imputed to it wonderful properties, including the ability to foresee the future, to uncover lost articles, to predict the weather, "and other things of like nature." Peyote use among the Aztecs was restricted to the priesthood, whose members used it to reach a state in which they could communicate directly with the gods. As the Cross accompanied the sword through New Spain, peyote was ruthlessly suppressed and the shamans put to torture. From 1570 until the middle of the nineteenth century, peyote was the property of *brujos* (warlocks) and *curanderos* (healing women). It was not a generally used hallucinogen but retained its mystical attributes.

CHEMISTRY AND PHARMACOLOGY OF PEYOTE

Because of the profound effects of peyote, the chemistry and pharmacology of its hallucinogenic compounds have received considerable attention. All told, there are over 50 alkaloids in peyote. Many of these are phenethylamines, of which mescaline (Figure 5.52) is by far the most active. Another psychoactive alkaloid is anhalamine, an isoquinoline. Peyote "intoxication" is manifested by a prolonged state of visual hallucinogenic responses, with a kaleidoscopic play of colors and images. Tactile hallucinations frequently accompany the visions. Vertigo, headache, nausea, profound disorientations of time and space, and a state of confusion are also commonly experienced. The visions almost always accompany the hallucinations and are characterized by bizarre and rapid changes in color, form, pattern, and complexity of normal experience. Hallucinogenic states become apparent within a half-hour after one eats the peyote buttons. The experience reaches a maximum within two hours and can persist for several hours. Visions occasionally last 24 hours and, gradually, the individual returns to the normal state. (Alkaloids are discussed in more detail in Chapter 6.)

Treatments with pure mescaline do not elicit the range, intensity, or longevity experienced with the cactus. This suggests that the other alkaloids, which singly appear to be innocuous, are probably interacting with mescaline in complex physiological and biochemical ways that are not understood. The general chemical structure of the phenethylamine and isoquinoline alkaloids is similar to some analogous compounds found normally in the human body that play roles in the activity of

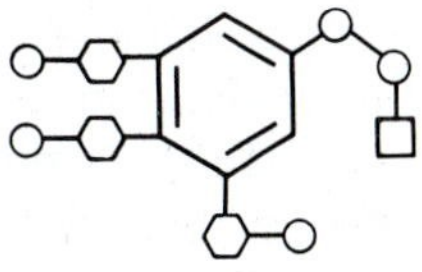

Mescaline from peyote

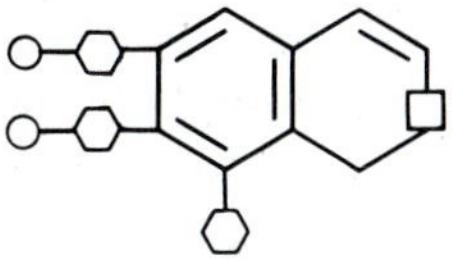

Anhalamine from peyote

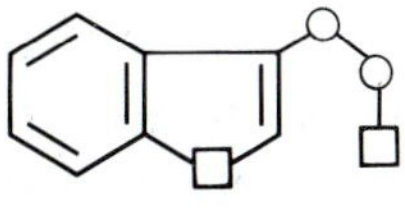

Tryptamine

Figure 5.52 Molecular structures of two major hallucinogenic alkaloid compounds in peyote. The molecular structure of tryptamine, one of the hormones in the human brain, is shown for comparison. Hydrogen atoms and carbon atoms in the rings are not shown. Oxygen atoms are represented as hexagons, carbon atoms as circles, and nitrogen atoms as squares.

nervous tissues, including brain tissues. Although this suggests that peyote-induced hallucinogenic states reflect alterations or augmentations of normal brain patterns, this has not been experimentally established.

Lophophora williamsii is not the only cactus with the capacity to induce hallucinations. *Trichocereus pachanoi,* found in north coastal Peru, is called *San Pedro* by the folk healers who use it in the diagnosis and treatment of various physical and psychological illnesses. San Pedro is usually mixed with other plants as a drink; in the induced hallucinatory state, the patient reveals to the shaman the source of the illness. There may be two peyote cacti, depending on the region in which the plant grows. *L. williamsii* is a more northerly species, reaching its limits in the high plateaus of northern Mexico. *L. diffusa* is a southerly species. The ratio of alkaloids is somewhat different, with corresponding variations in the altered states produced.

Peyote's physiological effects are due to about nine major alkaloids divided into two major categories. The first group produces a mild euphoria, accompanied by dilation of the pupils, increased blood supply to the face and body, and diminished kinesthetic responses. There is a tendency for the participant to talk a lot. This phase is relatively short-lived and is followed by lightheadedness and nausea. Only after this phase do the other peyote alkaloids begin to be effective. The participant is not sleepy but finds it more comfortable to lie down. The pupils dilate excessively, motor coordination is impaired, and reflexes are more pronounced, but the sense of touch and minor physical aches and pains are reduced. Muscular twitching, particularly of facial muscles, is pronounced. In addition, there are profound mental changes, the most dramatic being overestimation of time and an inability to focus one's attention for any length of time. Sensory hallucinations, particularly those involving vision, are noted, with the participant experiencing brilliant colors and patterns. He or she may hear the voices of ancestors, protective spirits, or gods. In some cases, unusual odors or tastes may be perceived as integral parts of hallucinogenic visions. Hallucinations are short-lived, and the participant may experience many visions during the period of maximum responses.

PEYOTE AS A CULT OBJECT

Peyote made its first appearance north of the border around 1870, when Mescalero Apaches obtained and used it ritually. It entered the Oklahoma Territory through a notable figure, Quanah Parker, son of the Comanche war chief Nokoni, and Cynthia Parker, a white woman who had been captured in 1843. Quanah attempted to bridge the gap between cultures. In 1884, at the age of 40, he became seriously ill, was treated with peyote by a Mexican curandera, and recognized that it could be a useful factor in binding his people together.

The Comanche were the right tribe at the right time. Powerful, warlike, prone to visions, and with a long history of religious ritual including self-torture, not only were they among the more mystical Indians of the Southwest, but as hunters on an agricultural-welfare reservation, they were completely demoralized. Taking cultural elements from Comanche, Kiowa, Apache, and parts of Christianity—notably (and ironically) the concept of love—Parker fashioned a series of ceremonies that were neither at variance with the beliefs of the tribes nor destructive to the cultural mix that was developing into pan-Indianism. Parker's prestige, his cultural bridge between white and red, and his superb organizational ability were focused on the development of a ritual system to provide the self-esteem that Indians desperately wanted.

Between 1880 and 1890, the peyote ritual system was restricted to Oklahoma and part of New Mexico, but between 1890 and 1910 it spread rapidly and widely. The Tonkawa, Lipan Apache, Cheyenne, Arapaho, and Navaho embraced it by 1891, the Sac, Fox, Shawnee, Mescalero, and Witchita were added by 1895, and the Algonquin, Omaha, and Pawnee by 1900. The peyote cult was introduced to the Winnebago in 1900 by John Ray. Returning to Nebraska, he converted his family, then a few friends, and eventually so many of the Winnebago and Omaha that the tribe was rent with a bitter dispute between the pan-Indian peyotists and the Christianized conservatives. Eventually, peyote reached up to Hudson Bay and Alaska. Considerable impetus was supplied the cult by the graduates of Carlisle, an Indian college whose students were effective radicals and innovators. With accretions of elements from Christianity and the inevitable cross-cultural modifications attendant upon its adoption by different tribes, tribal leaders were attracted to the cult. Church buildings were erected, and business and professional members provided respectability and a formal infrastructure. The Navaho, members of an agricultural, settled, and sophisticated group, could now make common cause with Florida Seminoles or the Potlatch culture of the Columbia River Basin. With less scholarly nitpicking than attends Judeo-Christianity, peyote practitioners held a fairly uniform set of beliefs and rituals which included the concept that God put some of His holy spirit into the peyote and gave it to the Indian. By eating this sacrament, one can absorb God's spirit. Thus, although central to the system of beliefs and practices, peyote is a means and not an end.

OPPOSITION TO THE PEYOTE CULT

As peyotism spread on reservations, opposition to it came not only from conservative tribal members but from the white establishment exemplified by the Indian Bureau and the missions. Insinuating that the peyote ceremony was a black mass and that mescal buttons inflamed ignorant savages, the missions demanded and got, in the person of William "Pussyfoot" Johnson, an Indian Bureau officer who, they believed, could crush the cult at its center in Oklahoma. Johnson was a U.S. marshal and a nondenominational fundamentalist preacher. Ostensibly, his job was to enforce laws preventing alcohol consumption by the Indians, but he stretched the law by stating that peyote was an intoxicant. His favorite trick was to raid a peyote ceremony, jail the celebrants, and release them only after they had signed a temperance pledge which he then enforced by jail sentences for breach of contract. The real objections to peyote were clearly stated by a minister, Walter C. Roe, who wrote in 1908, "One of the worst results of [peyote] use is that it creates a very strong barrier in the way of presentation of the Christian religion." Reverend Robert D. Hall, then in charge of Indian work for the YMCA, demonstrated to the psychology department of Yale University that peyote was dangerous to health, since, after eating several buttons, he vomited. What Hall didn't tell these Easterners was that vomiting is part of the ceremony; the peyotist believes that he or she is thereby ritually cleansed. A 1914 conference of missionary leaders and the Law and Order Section of the Indian Bureau resolved: "It is now well known that the increasing use of the mescal bean, or peyote, is demoralizing in the extreme. We recommend accordingly that the Federal prohibition of intoxicating liquors be extended to include this dangerous drug." This was an attempt to include peyote in the Harrison Narcotics Act of 1914. The Internal Revenue Agency ruled otherwise. Although concerned with the "ravages of peyote," the IRA, charged with enforcement of the narcotics act, published in 1916 a detailed attack on peyote as a drug and on peyotism as an immoral and anti-Christian cult: "Its followers seem to abandon Christian teachings and in their frequent nightly gatherings indulge in excesses through the midnight hours in which men and women participate. It is claimed that in the nocturnal debaucheries there is often a total abandonment of virtue, especially among the women."

Yet, as early as 1891, James Mooney, an ethnologist with the Smithsonian Institution, had published a series of reports on the culture of Indian tribes on Oklahoma reservations. Mooney was vilified when he noted that peyotists were rarely drunk although surrounded by alcoholism, that they were clean, self-aware, and emotionally stable. As an observer and participant in peyote ceremonies, he wrote of the total absence of orgy, the dignity of the ceremony, and the reverent attitudes of the celebrants.

In 1918, Congressman Carl Hayden of Arizona introduced a bill mandating total prohibition of sales to Indians of alcohol and "other intoxicants" including peyote. The hearings were held by the Committee on Indian Affairs, and it was then that the Indians finally learned how powerful were the forces arrayed against them. They realized that if the Hayden bill passed, harassment would quickly become suppression, not only of the peyote cult but of virtually every facet of non-Christian

Indian life. Testimony on both sides was marked by dramatic rhetoric, but that coming from the antipeyotists was notable for fabrications. Sexual immorality, drug addiction, and demoralization were charged, and denigration of reports of anthropologists, ethnologists, and pharmacologists duly inscribed as facts in the records of the hearings. Christianized, conservative tribal members were coached by missionaries, and the testimony of such solid citizens far outweighed that of the peyotists, many of whom could scarcely speak English or were long-haired college students. The hearings boiled down to two major points of contention. Is peyote a harmful, addictive, and intoxicating drug? Is peyotism a cult formed to foster a drug habit or is it a genuine Indian religion? For both questions, the committee found for the antipeyotists and reported HR 2614 out of committee, and the bill passed the House of Representatives. It did not get out of committee in the Senate, but the Indians knew how fragile was their victory. Clearly, protection could only come under the wings of the Constitution; the peyote cult had to become a legally recognized and established church.

THE NATIVE AMERICAN CHURCH

With the assistance of James Mooney, provisions of a charter were drawn up. It was at first decided to call the church the "First-Born Church of Christ," but this was rejected overwhelmingly, and the name "Native American Church" was chosen to emphasize intertribal solidarity. It was pan-Indian and also defined its participants as Americans. The articles of incorporation stated that it was a Christian religion, adding, "with the practice of the Peyote Sacrament as commonly understood in the several tribes of Indians." Incorporation was in the new state of Oklahoma in October 1918. For his efforts, Mooney was permanently expelled from Oklahoma by the Indian Bureau and died in 1921 without finishing what promised to be a classic study of Indian life. Opposition to the Church continued, with bills introduced (but never passed) in four Congresses. At the state level, peyote was outlawed in Kansas, Montana, the Dakotas, Arizona, Iowa, Wyoming, and New Mexico—all states with large reservations and active missions. The Indians responded by chartering the Church in these states, and by 1925 legal harassment had almost stopped.

In defense of the missionary's use of the term *intoxication* throughout the debate on the peyote cult, at least part of the problem was a failure of communication. The Aztec word *mexcalli* was given the *Agave,* or century plant, from which the term *mezcal* developed to refer to the alcoholic drink *tequila,* made from fermented and distilled *Agave* sap. In addition, there is a shrub in the American Southwest, *Sophora secundiflora,* having dark-red, beanlike fruits called mescal beans, which contain a very toxic compound causing nausea, hallucinations, and occasional death from respiratory failure. It was a custom of tequila drinkers to add crushed *Sophora* seed to their mezcal to render it more potent, and the seeds formed a basic part of the red bean dance—an oracular ceremony of the Mexican and Texas Indians. *Sophora* seeds are used by peyotists as decoration on ceremonial dress. This terminological confusion led whites to assert that the Native American Church fostered drunken orgies (the mezcal of tequila), ritual poisonings (the mescal of *Sophora* beans), and the drug habit (use of the ill-named alkaloid of peyote). Since the red bean dance also involved use of the poisonous, hallucinogenic, and possibly aphrodisiac *Datura* plant (Chapter 6), sexual debaucheries were included under the mescal rubric. This confusion was not cleared up until the 1930s, when it didn't matter any longer.

Whether the Native American Church is effective as a way of life and as a source of dignity and solace to the practitioner is difficult to determine. Its quarter of a million members know that it has reduced endemic alcoholism and has provided them with a rational set of beliefs. But if it serves to make them content to suffer the outrages that are still being imposed on them by the white man, peyote will not do what Quanah Parker had hoped that it would.

ADDITIONAL READINGS

Aberle, D. *The Peyote Religion Among the Navajo.* Chicago: Aldine, 1966.

Anderson, E. F. *Peyote: The Divine Cactus.* Tucson: University of Arizona Press, 1980.

Furst, P. T. (ed.). *Flesh of the Gods.* New York: Praeger, 1973.

Hertzberg, H. W. *The Search for an American Indian Identity.* Syracuse, N.Y.: Syracuse University Press, 1971.

LaBarre, W. *The Peyote Cult.* New Haven: Yale University Press, 1938.

Lanternarri, V. *The Religions of the Oppressed.* New York: New American Library (Signet), 1963.

Myerhoff, B. G. *The Peyote Hunt.* Ithaca, N.Y.: Cornell University Press, 1974.

Safford, W. E. "An Aztec Narcotic *(Lophophora williamsii).*" *Journal of Heredity* 6 (1915):291.

Underhill, R. M. *Red Man's Religion.* University of Chicago Press, 1965.

Chapter 6

The Drug Scene

Opium

An opiate is anything that causes dullness or inaction or soothes men or emotions

Random House Dictionary

Figure 6.1 Opium poppy *(Papaver somniferum).*

The botanical family Papaveraceae contains 25 genera and 120 species of flowering plants. Most are herbaceous annuals, although a few die back to the ground each year and form new shoots from a perennial rootstock. The family originated in Asia Minor. The genus *Papaver* contains ten species, several of which, the Iceland poppy (*P. nudicaule*) and the Oriental poppy (*P. orientalis*), are common garden plants. The California poppy (*Eschscholzia*), well known to most gardeners, is a member of another genus in the family. None of these plants produce alkaloids of medical interest. The one that does is the opium poppy (*P. somniferum*), whose specific name was chosen by Linnaeus because of the sleep-inducing properties of the gum produced in the young seed capsule of the plant (Figure 6.1).

Raw opium has been gathered by virtually the same method for ages. When the red, pink, or white flower petals fall, usually in June or July, the immature seed capsule synthesizes opium, a mixture of over 20 alkaloids plus water, sugar, several plant gums, and latex. The alkaloid content is close to 30 percent of the dry weight of opium. Within a day after the petals fall, workers go into the fields with sharp, three-bladed knives that look much like table forks with sharpened tines. They make several slits in the capsule, taking care not to puncture into the chambers where the young seeds are forming. The next day, the opium oozes out and begins to dry down (Figure 6.2). A curved spoon is used to scrape the opium from the capsule, and another set of slits causes a second flow of opium. The collected opium is pressed into cakes or balls

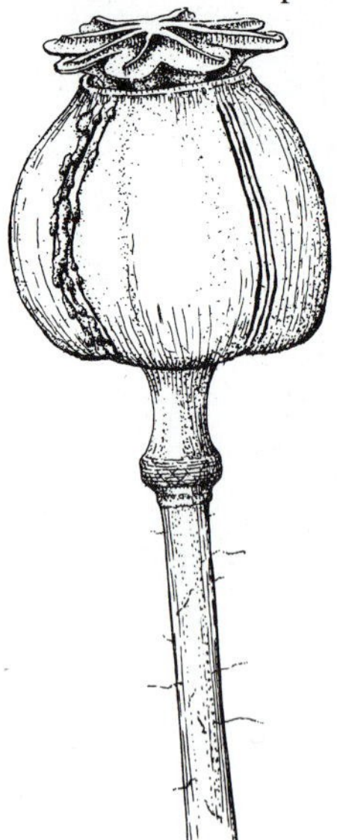

Figure 6.2 Capsule of opium flowers with raw opium gum oozing from cuts made in its side.

and allowed to dry to a rubbery-textured, yellow to brownish mass. From an acre of opium poppy (20,000 plants), an average of 10 kilograms of opium can be collected. Because of the delicacy of the operation, opium collection is entirely a hand process.

The opium poppy can be raised in regions where there is a long growing season, with extended periods of warm, sunny weather. It will also grow in colder regions, but yields are reduced. Traditionally, the opium poppy has been cultivated in Anatolia (Turkey), modern Afghanistan, India, and parts of Southeast Asia. It has been an item of commerce for centuries. Since the 1940s, however, production of opium has centered in three areas (Figure 6.3). Mexico, not previously a poppy-growing area, has become a major source of both raw opium gum and refined heroin for the North American market. Mexico now accounts for a third of the illegal supply of opium alkaloids (Table 6.1). The Golden Crescent of Iran, Afghanistan, and Pakistan is an even larger supply area; a good deal of Pakistan's production is used internally; a large increase in addiction has occurred since 1960. In 1984, Afghanistan and Iran produced close to 20 times the heroin made in Pakistan. The third major growing area is the Golden Triangle of Burma, Laos, and Thailand. Production in India is now negligible. China, a country with a history of opium addiction, no longer grows the poppy and its addiction rate is almost zero.

Table 6.1 MAJOR SOURCES OF OPIUM GUM AND OF HEROIN FOR THE ILLEGAL MARKETS IN THE UNITED STATES (1984)

Country or region	Opium gum (Metric tons)	Percent of U.S. heroin
Golden Crescent	1100	48
Golden Triangle	700	19
Mexico	1700	33

Roughly 3500 metric tons of opium gum are produced each year in these three regions. This amount can make 150 metric tons of pure

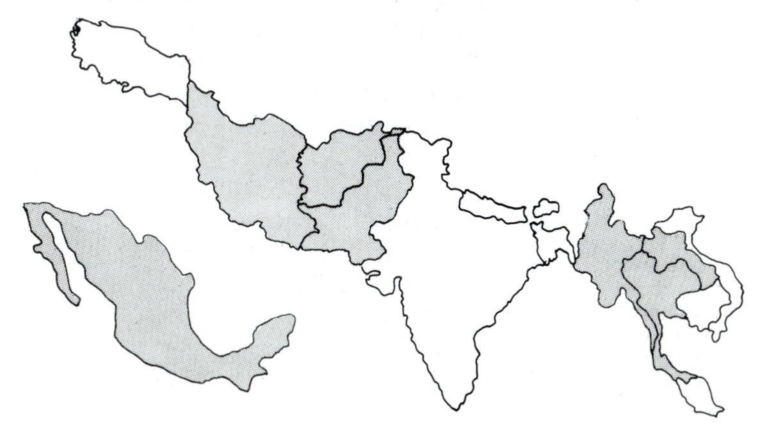

Figure 6.3 Major areas where the opium poppy is grown. The Golden Crescent of Iran, Afghanistan, and Pakistan provides close to 50 percent of the U.S. supply of heroin; the Golden Triangle of Burma, Thailand, and Laos provides almost 20 percent; and Mexico now furnishes over 30 percent. Turkey and India, formerly countries that grew poppies, are presently not involved in illegal opium production.

heroin and represents a 50 percent increase in production since 1980. A relatively small amount of opium gum, rarely exceeding 200 metric tons, is legally exported to the United States and Europe for the isolation of medical morphine and codeine, but the availability of effective synthetic pain relievers has reduced the need for medical opium. It is possible that there will be no legal market for opium within a few years. The illegal market, however, continues to grow. The number of addicts, primarily to heroin, has continued to climb both in the West—Europe and North America—and in countries like Pakistan and Thailand, whose rates of addiction were low for centuries.

ALKALOIDS

In addition to making proteins, carbohydrates, fats, and other compounds familiar to most people, plants synthesize a huge array of substances that are usually found in relatively low quantities. The general name of these substances, *secondary plant products,* gives no idea of the chemical range of the materials or of their importance. The medicinal products, the natural dyestuffs, products for the chemical industry (gums, resins, and so on), and a wide variety of everyday substances used as flavorings and essential oils (Chapter 9) are secondary plant products.

Among the secondary plant products that have long played important roles in human life are the alkaloids. Alkaloid is a purely chemical word defined as an organic chemical molecule—one containing carbon atoms—that also contains at least one atom of nitrogen. All alkaloids can be crystallized and, when dissolved in water or in alcohol, they give an alkaline reaction to the solution. The nitrogen in an alkaloid is usually found in combination with a ring of carbon atoms, the so-called heterocyclic ring (Figure 6.4). When an alkaloid is mixed with an acid, such as hydrochloric, the highly water-soluble hydrochloride is formed; this is the usual way of getting alkaloids into simple solutions. In solutions or as the crystal, they are colorless, are most soluble in alkaline solutions such as weak sodium hydroxide, and almost invariably have a bitter taste. The flavor of tonic water is due to quinine, one of the alkaloids.

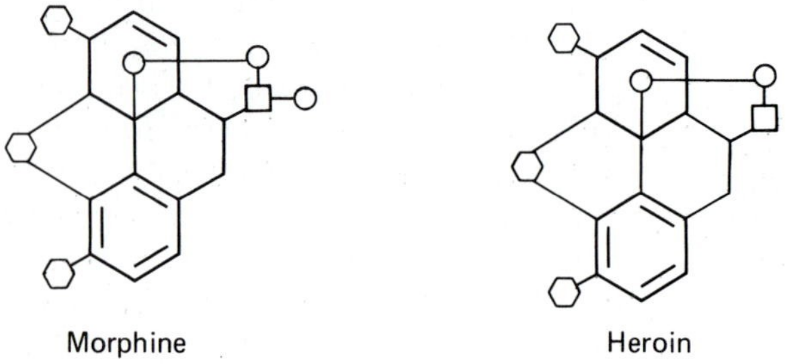

Morphine

Heroin

Figure 6.4 Molecular structures of morphine and heroin. Hydrogen atoms and carbon atoms in the rings are not shown. Carbon atoms are represented as circles, oxygen atoms as hexagons, and nitrogen atoms as squares.

Of the alkaloids found in raw opium, three are of medical importance. Close to 11 percent of opium is the single alkaloid morphine, while codeine constitutes about 2 percent and thebaine a bit less than 1 percent of the weight of the opium gum. Isolation of pure morphine and codeine is not a difficult or expensive process, and the legal price of these alkaloids has been stable for many years. Raw opium is mixed with slaked lime and then heated to remove the gums and latex, leaving an alkaloid base that constitutes about 10 percent of the original bulk of the starting opium. Further purification of morphine, codeine, and thebaine involves dissolving them in weak alkali and partitioning the alkaloids into alcohol. Water-soluble alkaloids are made by adding concentrated hydrochloric acid to form the medical morphine-hydrochloride.

In spite of a great deal of intensive research, the physiological mode of action of alkaloids in the animal body is poorly understood. Some, like the caffeine alkaloids in tea and coffee, are stimulants. Others, like the alkaloids in ergot, cause constriction of smooth muscle, and still others, like those in the opium poppy, are powerful painkillers. It is likely that all operate on or in some part of the central nervous system and that the responses reflect alterations in control over cellular function by the brain and peripheral nerve network. The physiology, pharmacology, and psychology of addiction to alkaloids are even less well understood. Certainly, not all alkaloids are addictive or even habit-forming. One *can* get along without a morning cup of coffee without experiencing withdrawal symptoms. Even for those alkaloids that are addictive, the nature of the addiction and of its consequences is a matter of intensive study; the same can be said for the physiological effects of withdrawal. There is growing evidence that the human body synthesizes substances, called *endorphins,* that are mimicked by opium alkaloids, but the biochemical and physiological interrelations between the plant alkaloids and the substances produced in the human body are still being examined.

THE WESTERN OPIUM EXPERIENCE

Raw opium has been used medically for centuries. The Swiss Lake Dwellers used opium, as evidenced by stored capsules dated to 6000 B.P. Sumerian tablets of 4500 B.P. noted that when small balls of opium were eaten or taken after being mixed with wine, the drug induced sleep and relieved pain. Homer spoke of *nepenthe,* a substance that would "lull pain and bring forgetfulness of sorrow." At the same time, Diagoras of Milos warned of addiction to *opos,* the milky sap that exuded from cut poppy capsules. Hippocrates, Theophrastus, Pliny, Dioscorides, and other ancient medical writers recommended opium for medical purposes, and it was in common use by physicians up to the present century. Thomas Sydenham, an influential British physician of the seventeenth century, said of opium, "Among the remedies which it has pleased Almighty God to give to man to relieve his sufferings, none is so universal and efficacious as opium," a statement that many physicians have echoed. Morphine, first isolated by Frederick Sertuener in 1805, quickly replaced crude opium in medical therapy. Isolation of other opium alkaloids soon followed, but it was not until 1925 that the complete chemical structure of morphine and other opium alkaloids was worked out (see Figure 6.4). Codeine is, at present, the major opium alkaloid still in common use as a local anaesthetic and as an effective cough suppressant.

In the latter half of the nineteenth century, pharmaceutical chemists began to alter the morphine molecule in order to make a compound that would be more effective than the natural alkaloid and less addictive. A German chemical company found in 1898 that simple chemical procedures yielded a compound called heroin, which was a more effective painkiller than morphine and was more effective as a cough suppressant than codeine. Up to 1917, heroin-containing cough suppressants were openly sold in North America without prescription. It is estimated that there were as many heroin addicts in New York City in 1900 as there are today, but most of them were children regularly treated for coughs, some of which were the result of tuberculosis (Figure 6.5). For reasons that still defy explanation, most of these addicts simply cured themselves.

Only after World War II did widespread morphine and heroin addiction become a social problem, reaching epidemic proportions in large cities of the United States and, to a lesser extent, in Canada. Addiction to heroin in cities of Europe developed a few years later and, today, addiction is a major problem throughout the Western world and parts of Asia and Africa. Profits are enormous. Ten kilograms of opium,

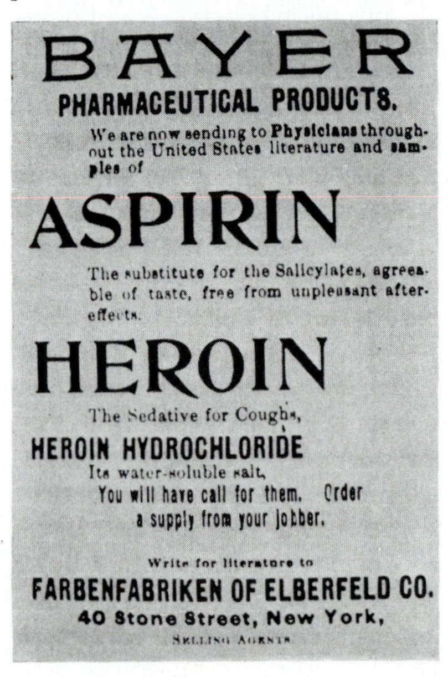

Figure 6.5 Nineteenth-century newspaper advertisement for heroin cough syrup.

which cost smugglers $25–$30 per kilogram, can be converted into one kilogram of pure heroin. This is a full year's supply for 50 addicts. The "street" price for this kilogram in New York was, in 1980, close to $70,000. When diluted with quinine, lactose, or other adulterants and packed into "bags" of two grains, the bag sells for $10. This constitutes a gross profit of $700,000 per kilogram of opium.

THE OPIUM BAN

In 1970, President Richard Nixon declared war on the international heroin traffic. He concluded that the best way to eliminate the North American supply was to prevent the growth of the opium poppy. Since Turkey was dependent upon U.S. military and economic assistance, the United States forced the government of Turkey to place a total ban on opium production. The United States offered $35 million per year to Turkey to pay the farmers not to grow the poppies and, it was hoped, to give them time to develop another crop. Seen in retrospect, these decisions were disastrous.

In the opium-growing areas, the poppy is a multipurpose crop. The most important is opium itself, accounting for 90 percent of the cash value of the plants. After the capsules mature, seeds are used as a source of oil for the peasants or are sold for use in baking; most of the poppy seeds on rolls come from the Turkish opium poppy (Figure 6.6). The oil-free meal cake is fed to cattle or ground into a flour for cookies. In early spring, young leaves are used in salads, and after the capsules mature, the stalks are used as silage. Since the peasants of Anatolia had grown poppies for centuries, they had no other cash crop. Wheat and other cereals grow poorly, the road system does not permit the development of a fresh fruit and vegetable economy, and the land receives too little rain for extensive pasturage. The impact of the ban on the peasants was immediate and profound. The *New York Times* reported that one village headman said, "We have heard that our opium becomes heroin and kills all the Americans. They are rich and that is why they use it. Their richness kills them and makes us poor." Well-to-do farmers are those who own five hectares, on which they normally planted two to the poppy. The cash income from the illegal market was about $350 for 10 kilograms. Had they planted wheat, barley, or tobacco, they would have realized about $150, assuming that they could get the crop to market. One such farmer was reported to have said, "The Americans use it and they die. We produce it and don't use it. Don't they have any brains?"

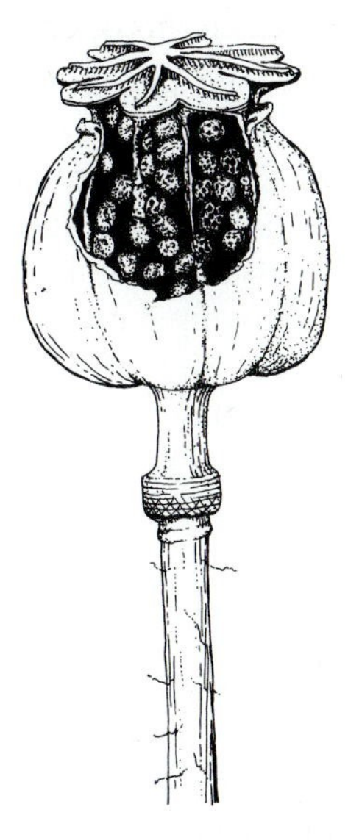

Figure 6.6 Mature capsule of opium poppy showing poppy seeds.

Keeping its part of the bargain, the Turkish government tried to eliminate the poppy plantings in 1972 and 1973. Soldiers set fire to plants, and helicopters sprayed remote poppy fields with herbicides supplied from the American stockpiles used in the Vietnam War. Afyon, a city 150 miles southwest of Ankara and the center of the poppy-growing area (*afyon* means opium in Turkish), reflected the depressed economy of the region. Debts piled up, bank loans remained unpaid, and a reasonably prosperous land was plunged into penury. During this period, thousands of young men and women, without jobs or economic resources, entered the European labor markets in Germany and France and kept their families solvent by sending home their pay. Separated from their families, lacking linguistic skills, freed from the stabilizing influences of their culture, and eating strange food, they were relegated to the dirty jobs that their affluent hosts no longer wished to do. Not unexpectedly, they caused "trouble" in Europe and were looked down upon and sometimes jailed.

Recognizing the unhappy fate of their youth and realizing that the American subsidy was both demoralizing and inadequate, the Turkish government asked the United States to reexamine the agreement. The Turkish foreign minister, Turan Gunes, said, "The sacrifices we have to bear for humanity cannot be placed on our shoulders only." The U.S. Senate voted 81 to 8 to cut off all economic aid to Turkey if it did not tighten the ban, and the U.S. State Department recalled its ambassador. The international situation became tense, and the North Atlantic Treaty Organization, of which Turkey was a member, appealed to the United States to relax its uncompromising stand. Turkey agreed to organize a poppy-growing system designed to reduce the possibility that opium would reach the illegal market. The Turks would also develop their own factories to process opium, license all growers, and would require that only unlanced, mature poppy capsules would be handled. The mature capsule yields little morphine, but contains high concentrations of codeine needed by the medical profession. The U.S. government agreed to allow a trial of this system, and the first licensing and planting was done in 1975. Nevertheless, Turkey is now almost opium-free.

The question is, of course, whether this attempt to dry up the source of opium alkaloids worked, and the answer seems to be that it did not. The United States offended and antagonized one of its allies, and after a short period of time when "street heroin" was in relatively short supply, the availability of the drug actually seemed to increase. In part, this was due to the release of stockpiles of morphine base in Turkey and heroin in France. Equally important, supplies of opium were obtained from other growing regions. Mexican heroin began to enter via Texas, California, and New Mexico; some was obtained in Iran from opium grown in Afghanistan; and a supply in the Golden Triangle of Burma, Laos, and Thailand (Figure 6.7) soon reached the North American market via Hong Kong and Japan. At this time 80 metric tons of the 700 metric tons produced each year can be made into enough heroin to saturate the illegal North American market; this amount of raw opium can be produced on about 20 square miles of suitable land. Addiction to heroin is not controllable by restricting the poppy, but must be handled

Figure 6.7 Label from Double U-O Globe heroin, manufactured in the Golden Triangle. Number 4 heroin is 100 percent pure.

by other means, including alteration of the cultural and psychological stresses that society imposes on individuals.

THE OPIUM WARS

This contemporary political and economic contretemps is far from the only one involving the opium poppy. One of the most profound of these was the Chinese Opium Wars of 1839–1842.

Opium was used in China for medical purposes for centuries, but there was only a small amount of addiction among the people. Called *chandu,* raw or processed opium was smoked in special clay pipes, and the intoxicating smoke was inhaled to induce sleep with, it is assumed, pleasant dreams. The ash remaining in the pipe was mixed with other burnable material and smoked in turn by the less affluent, but this was highly toxic, and much of the death rate attributed to opium addiction was due to smoking this *dross.* Yet, until the nineteenth century, China did not grow or process opium, and the number of addicts was very small.

Although Europeans had obtained a foothold in islands off the Chinese coast, the Ming and early Ch'ing emperors restricted commerce with the big-nosed, round-eyed barbarians who, they believed, were their intellectual and moral inferiors. China had little use for Western goods and ideas; its society was stable, and the vast country supplied all its needs. Europe, however, was extremely interested in the goods of China. Its tea, silks, spices, and porcelain commanded high prices on a market fascinated by the Orient. Although Portugal was the first country to have much trade with China, the conquest of Bengal in 1773 by Britain and the development of the world's finest merchant fleet allowed England's East India Company to obtain first a foothold in the China trade and, by 1800, to monopolize it.

After delicate and, to the English, humiliating negotiations with the Chinese government, a small island off the shores of south China was established as a base for trade. Even here, the intermediaries had to be Chinese, and the British traders were restricted in their movements. More important, the Chinese insisted that all goods be paid for in silver, since they had no desire for European products that could be used in trade. The China trade was so successful that by 1810–1815, China had acquired a good share of the silver of Europe, which, because of scarcity, was rising rapidly in price, thus reducing the profits of British merchants. Obviously, there had to be something that the Chinese wanted and, with unerring intelligence, the British decided to sell opium to the Chinese. At first, the opium was given away, and as addiction spread, prices rose accordingly. Because of its absolute control over India, the East India Company subverted the agriculture of Benares, Baher, and other areas of India to the growing of the poppy and the production of opium. Poppy cultivation was compulsory, and since the production of food crops was limited, the peoples of these Indian provinces were reduced almost to starvation.

As addiction in China rose to astronomical proportions, silver began to move out of the country and back into Europe, where its price fell and the profits of the East India Company soared. Since silver was the currency of China, taxes went unpaid and internal business was disrupted. Based on Confucian and Tao ideals, Chinese society was grounded in the philosophy of self-discipline and a hierarchical arrangement of duties to family and emperor. The addict sacrificed family, duty, and self-discipline, and the fabric of Chinese society and government began to collapse. Britain recognized this danger and forbade the legal importation of opium into England, but in spite of the pleas of a few humanitarian members of Parliament, the British government refused to order the East India Company to stop the opium trade; tax revenues from India, tea duties, and opium sales were simply too profitable.

In 1833, Emperor Tao Kwang, whose son had become an addict, ordered Commissioner Lin Tse-Hsuto to stop the opium traffic. On March 10, 1839, Lin issued a proclamation to the foreigners that asserted that it was not right to harm others for the sake of profits: "How dare you bring your country's vile opium into China, cheating and harming our people?" He demanded and obtained all opium stored in Canton and, to the horror of the British, destroyed it. The Houses of Parliament rang with the shouts of indignant members who demanded that Prime Minister Palmerston avenge this insult to the honor of Great Britain. Besides, this provided a rationale for taking over China, something that Britain had long desired.

A punitive force, assisted by the fleet of the East India Company and the merchant fleets of France and the United States, formally invaded China. The Chinese, long believing that gunpowder was for fireworks and that armies were to maintain the peace, were no match for the experienced Westerners. Britain lost 500 troops; the Chinese lost 30,000. The Treaty of Nanking, signed in August 1842, provided that all costs of this "punitive expedition" were to be borne by China, that an indemnity of $21 million was to be paid to Britain, and that the country was to be opened to Western trade. Hong Kong was converted into a British Crown Colony. France and the United States soon wrung equivalent concessions from a demoralized Chinese government. For practical purposes, China became a colonial outpost. No future Ch'ing emperor was more than a puppet; the unifying concept of "The Mandate of Heaven," upon which the dynasty was built, was lost.

Subsequent treaties extended the control of Britain. The opium tariff recognized the right of Britain to addict the population, Christian missions were imported with full protection, customs regulation was in the hands of the West, and additional indemnities were forced on the Chinese. The importation of "foreign mud" increased, and the number of addicts in China increased even more rapidly until, by 1900, it was estimated that close to 10 percent of the population of China used opium regularly. More than half of Britain's trade with China was opium from India. A few Britons protested but were told, "If we don't sell it, someone else will. Besides, Her Majesty's government needs the money for its expansion." In areas ceded to Britain, opium cultivation was started, and soon the Chinese themselves were growing opium for home consumption.

In 1906, the last of the Ch'ing emperors decided that China must give up opium, and a ten-year period was designated in which to accomplish this seemingly hopeless task. Britain reluctantly agreed to limit and then stop importation. Local production was restricted, but smuggling continued for many years. When, in 1911, the republic was organized and the Ch'ing Dynasty came to an end, the detoxification of China was continued and even accelerated, but the problem was simply too large to handle. Opium addiction continued and was enhanced during the Japanese occupation, when it became imperial policy to subjugate the Chinese peoples by supplying opium. The Mitsui family had a virtual monopoly on opium and, with the money so obtained, they used slave labor from China to dig their coal and work in their factories. Sony, Toyota, Sapporo, and Nishi-Nippon are companies whose financial base was at least partly based on opium addiction in China. They are not alone. A fair share of the wealth of Indian princes and rajahs came from converting their lands to poppy production, and some of the respected families in New England got their start by renting clipper ships to the India-China opium trade. Today, China is opium-free.

And yet . . . opium and the morphine and codeine it contains have eased more physical pain and saved more lives than the products of any other plant (Figure 6.8).

Figure 6.8 Goddess of the Night giving opium poppy to Man. Taken from a Roman cameo carved about 100.

ADDITIONAL READINGS

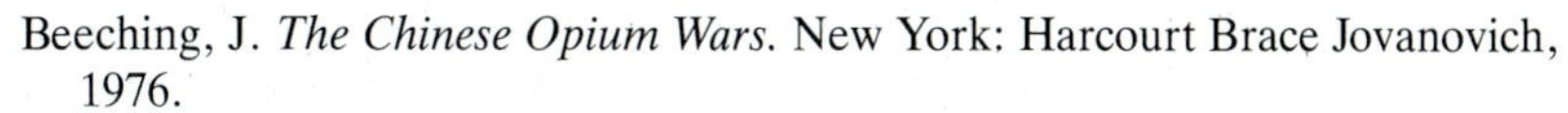

Beeching, J. *The Chinese Opium Wars.* New York: Harcourt Brace Jovanovich, 1976.

Courtwright, D. T. *Dark Paradise: Opium Addiction in America Before 1940.* Cambridge, Mass.: Harvard University Press, 1982.

Fay, P. W. *The Opium War 1840–1842.* Durham: University of North Carolina Press, 1976.

Morgan, H. W. *Drugs in America: A Social History.* Syracuse, N.Y.: Syracuse University Press, 1981.

Raffauf, R. F. *Handbook of Alkaloids and Alkaloid-Containing Plants.* New York: Wiley, 1970.

White, P. T. "The Poppy." *National Geographic* 167, no. 2 (1985):143–189.

Dama Blanca—The White Lady of Peru

I get no kick from cocaine
I'm sure that if
I took even one sniff
It would bore me terrifically, too.
But I get a kick out of you.

Anything Goes, a musical by Cole Porter, 1934

When and how people found plants that eased physical or psychic pain is unknown; possibly it followed observations that animals acted strangely after eating such plants. Because of their awesome effects, hallucinogenic plants have been invested with ritual significance; they were believed to allow the user to become one with the Godhead and to enter a reality normally screened from mundane view. In Western society, certainly since the dominance of Judeo-Christianity, all drugs except alcohol that alter consciousness are proscribed. Nevertheless, many cultures—including our own—use plant-derived substances that induce euphoria, mood alteration, changes in perception, or induce visions. One of the plants that was and still is part of its founding culture is cocaine.

Erythroxylon coca, the plant with red wood whose leaves contain cocaine, is a large shrub or small tree native to mountainous regions of Peru, Bolivia, northern Paraguay, and western Brazil. The plant is in a small family containing three genera of shrubs and small trees found entirely in the tropics of South America, Africa, and Asia; only *E. coca* produces the cocaine alkaloids. Although it flourishes at elevations of 2000–8000 feet (670–2700m), some clones can grow in the lowlands of Brazil. Depending on the location, the plant and its alkaloid-containing leaves are called *upador* (Brazil), *hayo* (Bolivia), or *khoka* (Peru). For centuries, seeds from high-yielding plants were selected and planted in clearings hacked from the bush. The Bolivian subspecies has a large leaf with a higher concentration of alkaloids than the narrow-leaved Peruvian subspecies, but since the Peruvian plant is more vigorous, total cocaine per plant is comparable. Coca leaves contain 14 alkaloids, with cocaine being highest in concentration. Mature plants may be stripped of their leaves several times a year, to yield 400 kilograms of dry leaf matter per acre per year. Since, on the average, cocaine concentration is rarely above 1 percent, an acre can provide about 4 kilograms of pure cocaine with a "street value" close to $1 million.

The effects of cocaine on the human body are well documented. Physiological responses to pure cocaine include elevated heart rate, enhancement of muscle potential, and increases in cognitive ability, the latter being the rationale Sherlock Holmes gave to Dr. Watson for Holmes's use of the drug. There is a structural relationship between part of the cocaine molecule and that of acetylcholine, a normal body chemical required for the transmission of electrical signals from one nerve cell to another. As blood flow increases, blood pressure and body temperature rise, resulting in a feeling of warmth, a reduction in fatigue, and an enhanced capacity for muscular work. As used by thrill seekers, a "mind-caressing and power-infusing jolt" lasting about an hour requires about 25 milligrams of cocaine. This is followed by an hour or so of a sense of peace and bodily ease. Reports of the drug's aphrodisiacal action are greatly exaggerated.

Cocaine has been used medically since 1884 as a local anaesthetic because it is an effective deadener of nerve endings. It has been replaced by the synthetic anaesthetic Novocaine, so that its current use in medical therapy is nil, leaving its hedonistic and conspicuous-consumption value to fill columns in newspapers. No reliable reports of true physical addiction to cocaine have appeared, although habituation is not uncommon among those who sniff or inject large amounts over a long period of time. Cocaine use is certainly not without serious dangers. It can cause fatal convulsions, respiratory failure, and cardiac collapse. Regular use can induce restlessness, impotence, liver damage, and more-or-less chronic psychological problems. Those who regularly sniff cocaine can get ulcerations of the lining of the nose.

COCAINE IN SOUTH AMERICAN CULTURES

Figure 6.9 A vase showing a Peruvian extracting lime from a calabash with a stick. The lime is added to coca leaves, which are carried in the *chuspa* bag on his left side. Taken from a moche vase.

Coca leaves have been part of the culture of South American Indians for at least 2000 years, as evidenced by the presence of dried leaves in graves, apparently placed in the corpse's mouth to provide sustenance for the dead, embarked on their long journey to the spirit world. The plant was under cultivation in Peru at least 1200 years ago, since it appears as a symbol of Incan royalty carved on tombs dated to that time. Incans believed that coca was a gift from Manco Capac, son of the Sun, who wished to bring his Father's light to wretched humans who could use it as a living manifestation of His body. As such, its use was restricted to the emperor and his priests, youths enduring the rigors of manhood rites, oral historians whose memories needed refreshing, warriors, and the sick (Figure 6.9). As far as we know, it was not used casually by the Incas.

Leaves were burned to propitiate the gods, and the priests reverently chewed leaves to enable them to enter that receptive state where they communicated directly with Inta, the Sun God, and could then return to earth—something that death does not permit—to inform the people of the Deity's desires. Divination was accomplished by scattering sacred coca leaves and foretelling the future by their position on the temple floor. For lesser prophecy, fragments were placed in bowls; water was added and then poured off, much as we do with tea leaves in a cup. A diviner might also chew a leaf and spit the juices onto his palm with two fingers pointed down; if the juice ran freely and equally down both fingers, the omens were favorable.

Although the people of the high Andes were normally under Incan rule, their homes were so remote and their land so harsh and unproductive that they were not required to restrict coca to ritual use, and it was a normal and vital part of everyday living. Moving daily from their homes at 10,000 feet (3400 meters) to their llama-grazing areas and potato fields at 15,000 feet (5000 meters) over rocky trails in an unforgiving climate, they were sustained by the juice of the coca leaf. Coca was not merely a solace; it was a physiological stimulant that allowed these people to perform hard labor under almost unbearable conditions. But they, too, knew that coca was divine. They built stone altars to the Gods and, begging strength from Apachic and Pachacamec, they threw their spent quids against these stones—which are now surmounted by wooden crosses put there by Spanish missionaries. Mother Earth was entreated to give a good potato harvest (see Chapter 3) by burning coca leaves, and piles of the leaves were allowed to blow in the wind to ensure gentle rains.

Francisco Pizarro had served under Balboa in Panama and, hearing rumors of gold to be found in the Andes, formed a company of adventurers who sailed to the west coast of Peru. In November 1532, his band of cutthroats and priests treacherously ambushed a royal welcoming party and easily won an empire of precious metals and unbaptized souls. The conquest, completed in 1533, sounded the death knell of a sophisticated culture and ushered in the degradation of close to 10 million people. This was accomplished by murdering the nobles and native priests, thus disrupting the bulwarks of culture—ritual and a legal infrastructure. To assist in this task, the Jesuits asserted that the ritual use of coca by Inca priests was a deliberate mockery of the Eucharist; they concluded that coca could be nothing else than an agent of the Devil. In fact, the bishop of Cuzco formalized the coca ban in 1551. The fifth viceroy of Cisco de Toledo published 70 ordinances concerning coca, making its cultivation, harvesting, transport, sale, or use punishable by death. Since burning at the stake was in fashion in the Spain of the Inquisition, users of the plant of evil were immediately consigned to the purifying flames. Plantations were cut down and stores of dried leaves were fired after decontamination with censer and holy water.

With demonstrable evidence that their gods were impotent or dead, the people were properly baptized, with the assurance that the wafer allowed them communication with the God of the One True Faith. Black-robed priests, in many cases, used the same altars at which priests garbed in feathered capes had stood. As with other cultures destroyed by conquerors, the people had little to sustain their wills and, not unexpectedly, coca went underground. With religious strictures on its use no longer operable, coca became a way of opposing the invaders and also an escape from the sense of shame that accompanied the usurpation of their country. No longer a ritual substance, coca was a solace for the common people, most of whom became laborers under the Spanish crown.

The Spanish badly needed this labor. Mountains of gold and silver, unexpected bonuses of precious gemstones, had to be dug from the mountains. In 1570 they imposed on all residents of New Spain a head tax that had to be paid in Spanish coin. This effectively reduced the native population to perpetual slavery. Unfortunately, the natives, on very short rations and without land or time to grow their own food, had the annoying tendency to collapse under work quotas set in Madrid or Lima. The overseers, whose very lives depended on meeting these quotas, soon became desperately short of miners and appealed to the authorities for help. The Jesuits suggested that since the hardy Indians of the upper mountains seemed capable of prodigious labor when using coca, it might be just the answer for the miners. Thus they had Philip II declare that coca was necessary for the well-being of the Indians. And it worked—the slaves were given coca leaves three or four times during the 18-hour work day; even when food rations were reduced, the amount of bullion flowing from the mines increased. Lifting the holy bans on coca presented no moral problem; the Jesuits decided that the Coca Devil would thus be forced to assist in God's work by facilitating the baptism of the Indians and providing the silver needed to pay for the forthcoming Armada, which was to subdue the English heretic and restore Roman Catholicism to Britain.

With coca again sanctioned and, indeed, its use encouraged, the value of the dry leaves increased greatly. Incan agricultural lands were graciously granted by the king of Spain for the reestablishment of coca plantations. The conversion of productive maize, bean, and squash fields further decreased the available food supply. Thus, because of starvation—as well as executions, burnings, high death rates in the mines, and diseases (including several introduced by the Spanish)—the Incan population plummeted from 10 million at the time of the conquest to less than 4 million by 1650.

The Spanish government, eager for any source of revenue, imposed ever-increasing taxes on coca leaves, over the vigorous protests of the mine operators. Nevertheless, Spanish gentlemen vied with one another to obtain rights to grow coca plants to supply the rising demands by the mine operators for the drug. Coca leaves became currency, just as tobacco did in the American colonies a century later, and bales of leaves were acceptable as payments for debts and taxes. In the words of Garcilasco de la Vega, "The greater part of the revenues of the Bishops and the Canons of the Cathedral of Cusco is derived from tithes of the coca leaves." By 1583, excise taxes on coca leaves yielded the equivalent of $500,000 to the Spanish crown (in sixteenth century value), a sum that increased to $2.5 million by 1700. Today's smuggling of purified cocaine is based on these movements of coca leaves in the sixteenth century with, in many cases, the same families being involved.

In spite of over a century of observation, the Spanish continued to be amazed at the increased capacity for muscular activity induced by sucking a quid of a few dried leaves mixed with burned sea shells, wood ashes, or lime to release the alkaloid. An Indian carrying a 70-kilogram (154-pound) load could, on level ground, travel 3 kilometers (1.9 miles) in 45 minutes and could climb from 12,000 feet (4000 meters) to 15,000 feet (5000 meters) at a rate of 2.5 kilometers per hour. The individual could do this on a diet that rarely exceeded 2000 Calories per day and would not show any signs of the debilitating mountain sickness that plagued the Spanish. In the mountains, distances were and still are measured by the number of hours that one quid (an *acullico)* would sustain a traveler. Only in the last 25 years has it been recognized that coca leaves contain other alkaloids that interact with cocaine to enhance synergistically the muscular responses without causing hallucinogenic side effects, as cocaine alone can do. It has also been shown that the Peruvian Indians, who normally will suck on about eight to ten quids a day (with a cocaine uptake of 4 milligrams per hour), are thus never exposed to large, mind-altering doses. Since cocaine is rapidly inactivated in the body, high internal concentrations are never reached.

At the beginning of the sixteenth century, Nicolas Monardes smuggled coca leaves into Spain, where they received little attention; the tobacco of the New World was much more popular. The fabulous reports of the effects of the leaves could not be duplicated because the leaves had deteriorated on the long sea voyage and the plant could not be grown in Europe. Nevertheless, a few physicians were intrigued by the plant and recommended it for everything from rheumatism (for which it probably was somewhat effective) to syphilis (for which it was not).

The nineteenth century saw the acculturation of coca in Europe. In 1814, the *Gentleman's Magazine* of London suggested that coca might replace food in times of famine or keep the ever-present poor and hungry of London happy. Mill owners were very interested, because they could foresee its value under conditions in the cotton-cloth weaving factories, where many children found even the 12-hour day difficult. The problem was that leaves were not available except from the despised Spanish (England never forgot the Armada) and so nothing was done. Besides, some of the clergy reinvoked the idea of coca as a devil's plant.

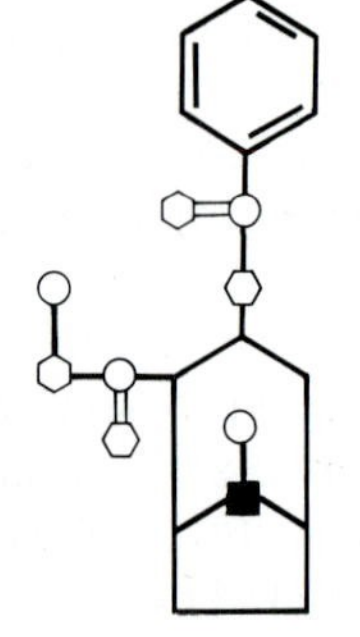

Figure 6.10 Molecular structure of cocaine. Hydrogen atoms and carbon atoms in the rings are not shown. Carbon atoms are represented as circles, oxygen atoms as hexagons, and nitrogen atoms as solid squares.

COCAINE IN EUROPE AND NORTH AMERICA

Cocaine, as a pure chemical, was isolated and chemically characterized by 1860 (Figure 6.10) and its psychopharmacological properties were first examined by Dr. Paolo Mantegazza. Testing it on himself, he reported the "deeply joyful and intensely alive" feelings it engendered and spoke of flying "through the spaces of 77,438 worlds, each more splendid than the one before." Not unexpectedly, interest in "erythroxyline" became intense. In 1884, Dr. Karl Koller of Vienna discovered that cocaine was a most effective anaesthetic for eye surgery; other physicians employed it as a local anaesthetic for inserting stitches in wounds. Somewhat earlier, medical students in Vienna and Edinburgh were chewing coca leaves, and racing cyclists were reported to complete the Tour de France with quids held firmly in their cheeks.

Without doubt, the most controversial use of cocaine was that recommended by Dr. Sigmund Freud, then a struggling neurologist hoping to find a research area that would bring him fame and allow him to get married. He was greatly impressed by the report of Dr. Theodor Aschenbrandt, a Bavarian army surgeon, who treated soldiers suffering from exhaustion and diarrhea with cocaine and achieved excellent results. Freud decided to try this new wonder drug on patients experiencing nervous exhaustion or morphine withdrawal. In a scientific report in 1884, he discounted the possibility of true addiction and praised cocaine as a most valuable treatment for his patients. He also claimed that it was an effective aphrodisiac, a finding that, as noted above, has not been confirmed. Cocaine relieved Freud's chronic depression and didn't impair his ability to work. It caused "exhilaration and a lasting euphoria." He gave cocaine to his fiancée, Martha Bernays, "to make her strong and to give her cheeks a red color." When, in addition to a series of "cocaine papers," he published a "Song of Praise" to cocaine in a staid medical journal, passions against him and this "pernicious narcotic" were—as he had hoped—aroused. His enthusiastic overreliance on cocaine resulted in the death of at least one patient; physicians following Freud's lead had comparable losses among patients who received the drug.

The eagerness with which cocaine was embraced was supported by many manufacturers. Their over-the-counter concoctions were alleged to cure just about any real or fancied disease. Dr. William Martingale, president of the Pharmaceutical Society of Great Britain, suggested in 1886 that cocaine was a better general beverage than coffee or tea. Pills, tablets, suppositories, pessaries, unguents, and nasal bougies, all containing quantities of pure cocaine that would floor an Andean Indian, appeared on the market.

Angelo Mariani, a Corsican chemist living in Paris, invented "Dr. Mariani's French Tonic," a mixture of cheap *vin ordinaire* and an extract of coca leaves. Mariana modestly advertised the wine as "a fountain of youth" that would bring vigor to aging men and women. Parisians of wealth and sensibility, including Gounod, Massenet, Poincaré, Jules Verne, Anatole France, Rodin and others of the artistic smart set, endorsed it. The wine moved to London, where H. G. Wells popularized it. The list of the patrons is long, varied, and international, and included such notables as Peter I of Serbia, Oscar II of Norway-Sweden, Albert I of Monaco, Alphonse XIII of Spain, and Thomas Edison and William McKinley of the United States. The ultimate accolade was provided by Pope Leo XIII, who presented Mariana with a gold medal for outstanding service to the Vatican and who, it was said, had his ascetic contemplations aided by the concoction. Pope Pius X was also a regular customer.

In 1886, John S. Pemberton probably modeled Coca Cola after Mariani's tonic, but he eliminated the alcohol in deference to the sensibilities of his Southern customers. This patent medicine contained caramel to provide coloring, an extract of the kola nut to supply caffeine, phosphoric acid, and Pemberton's secret mixture, which was little more than an extract of coca leaves plus sugar to disguise the bitter taste of the alkaloids. It was originally advertised as a "brain tonic . . . and a cure for all nervous affections." In 1892, Pemberton sold his formula and the rights to Asa Griggs Candler, who marketed the concoction as a beverage in his Atlanta drugstore. Adding carbonated water, he stated that "this new and popular fountain drink contains the tonic properties of the wonderful coca plant and the famous cola nut." The cocaine and the other coca alkaloids were removed from the extract of coca leaves in 1903, although an alkaloid-free extract of the leaves is still part of Coke's secret formula.

In spite of the legislation that implemented the pure food and drug acts and the later narcotics laws, cocaine intrigued post–World War I high society. The flapper and her sheik assured the success of the speakeasy, where musicians and less savory characters had established "snow" as the preeminent jolt to overcome fatigue and ennui. The combination of rotgut booze, high-purity cocaine, and the sense of gentle criminality was irresistible. In the 1930s, return of the legalization of liquor and the deepening Depression caused cocaine's popularity to wane. When, in the 1950s, cruise ships and jet travel made South America accessible to the affluent, citizens of North America were reacquainted with the white powder. It was relatively expensive, but "we're on vacation" and the effects were so fantastic that small amounts were smuggled back to New York, Toronto, and then to that center of hedonism, Los Angeles. The penchant for aping one's presumed betters has escalated the cachet of cocaine to a level still not fully realized by most of the public. U.S. government statistics indicate that 25 million Americans have tried cocaine, 4 to 5 million use it at least once a month, and about 5000 more Americans each day are at least experimenting with cocaine. Its use by the general public is increasing at about 10 percent a year.

COCAINE IN MODERN SOCIETIES

Cocaine is very big business. Its $30–$40 billion annual turnover would put it in the top ten in the list of *Fortune's* 500 companies . . . about at the same level as the Ford Motor Company (Figure 6.11). In 1984, the wholesale price for pure cocaine was $1000 an ounce (three times the

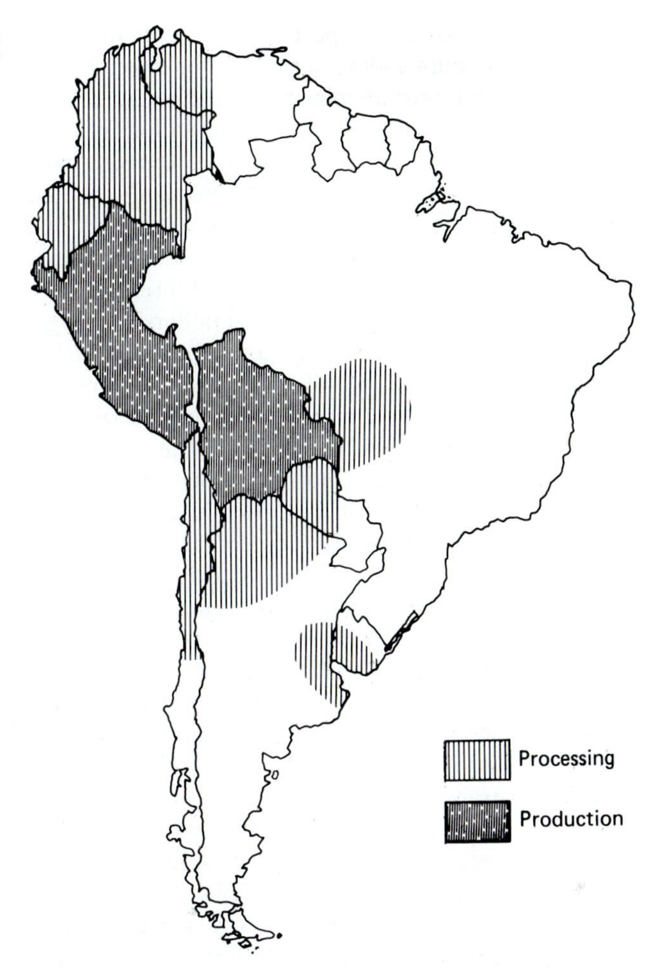

Figure 6.11 Areas of production and of processing or exporting of cocaine.

price of gold). This represents close to a 50 percent drop since 1980, a period when the consumer price index of mundane items like food and clothing increased. The reason is simple. The cultivation of the plant, the isolation of cocaine, and the smuggling operations are so profitable that cocaine production has spread from its original centers to include large areas of northern South America. Peru has three-quarters of a million acres under cultivation, Brazil has recently entered the cultivation sweepstakes, with both plantations and cocaine-purification laboratories. What is more, the industry has tripled in its original areas of Colombia and Bolivia.

That the cocaine industry is profitable can easily be calculated. A kilogram (2.2 pounds) of pure cocaine costs $8000 in Colombia or Peru and about $30,000 in New York or Los Angeles. It is immediately cut by 50 percent with milk sugar (lactose), doubling its value. As it passes through many hands, it is cut even further until a gram (1/28 ounce) of 10 percent purity will cost $300 and can jolt three to five people. Some of the greedier dealers are cutting the cocaine to less than 3 percent and substituting amphetamines, ephedrine, or quinine to magnify the physiological response. The pharmacological and psychological effects of these drug mixtures are completely unpredictable.

Independent of the physiological and any moral considerations, the cocaine traffic has caused, and will continue to cause, serious human dislocations and social problems. Whole governments in Central and South America have been subverted; many officials, if not directly part of the cocaine network, are known to have been suborned. Beneficiaries also include some banks who reap large sums by moving money around to "launder" the "narcodollars" through legitimate investments. Peasants are growing coca shrubs to the exclusion of food crops because they can quadruple their incomes; peasants can receive 600,000 pesos ($200) for 10 kilograms of dried leaves. Murder is now commonplace among rival groups. The big profits are not going to the people or the countries that produce and refine cocaine. Colombia, a major producer, receives less than a third of the money generated by its production, and Bolivia retains a smaller percentage of the profits.

Purification of cocaine is cheap and simple. Dried leaves are first soaked in a solution of potassium carbonate, water, and kerosene. The solution is evaporated to produce cocaine paste. The paste is further purified by recrystallization with laboratory equipment that costs less than $2000; a laboratory can produce 50 kilograms of paste in a day. The white, bitter-tasting powder can be compressed into thin sheets to be inserted into books or molded into any desired shape for eventual transport to centers of distribution.

In both the producing and consuming countries various national and local laws have been passed to restrict the production, transport, and sale of cocaine, but in spite of diligent enforcement, the sheer volume of processed cocaine makes any attempt to dry up the supply virtually impossible. The United States, through its Drug Enforcement Agency, has provided over $25 million to producing countries to assist local agencies in their efforts to reduce the number of plantations and laboratories, but much of this money has simply disappeared. The control of the cocaine trade must ultimately depend on the decisions of individuals not to purchase or use it.

ADDITIONAL READINGS

Ashley, R. *Cocaine: Its History, Uses and Effects.* New York: Warner Books, 1979.

Grinspoon, L., and J. B. Balaker. *Cocaine.* New York: Basic Books, 1976.

Hemming, J. *The Conquest of the Incas.* New York: Harcourt Brace Jovanovich, 1970.

Martin, R. T. "The Role of Coca in the History, Religion and Medicine of the South America Indians." *Economic Botany* 24 (1970):422–437.

Mule, S. J. *Cocaine: Chemical, Biological, Clinical, Social and Treatment Aspects.* Boca Raton, Fla.: CRC Press, 1976.

Plowman, T. "The Ethnobotany of Coca (*Erythroxylon ssp.* Erthyroxylaceae)." *Advances in Economic Botany* 1 (1984).

Schleiffer, H. *Sacred Narcotic Plants of the New World Indians.* New York: Hafner, 1973.

Stone, N., H. Fromme, and D. Kazen. *Cocaine: Seduction and Solution.* New York: Clarkson N. Potter, 1984.

The Tomato Family

Give me to drink mandragora
That I might sleep out this great gap of time.

SHAKESPEARE
Antony and Cleopatra

The tomato, chili, tobacco, bell pepper, and white potato are all New World plants in the Solanaceae family, widely distributed in both Americas long before the time of Columbus and the Spanish invaders. The introduction and adoption of the potato as a food plant occurred relatively quickly, although it did take a bit of arm twisting in Europe (Chapter 3), but it never has become an important food crop in Asia. Chili peppers have become invaluable adjuncts to the cuisines of Africa and Asia; bell peppers and pimientos (including paprika) are in use throughout the world. Eggplants, also a member of the Solanceae family, are native to the Old World; reports exist from China on their use by the fifth century and from the Mediterranean by the ninth century. In many countries, the obvious resemblance of the eggplant to the human womb has invested the fruit with magical significance (Chapter 7).

The tomato, on the other hand, has long been suspect and its ascent into culinary ubiquity took several hundred years. The beefsteak and cherry tomatoes are ultimately traceable to several weedy vines in Peru which bear red fruits smaller than a grape. Surprisingly, the Peruvian Indians did not eat tomatoes but cultivated them around their homes to ward off evil spirits. In Mexico, the fruit was called *zitomate* and was eaten raw or stewed with beans and squash. Selection resulted in plants with larger, more fleshy fruits much like our cherry tomato; it was these types that were brought back to Spain by the Conquistadores by 1550. In Europe, the tomatoes were grown as garden ornamentals and apparently were shunned as food. John Gerard asserted in 1595 that "love apples yield very little nourishment to the bodie and the same naught and corrupt." The French evidently had the same opinion; the *pomme d'amour,* so called because of its obvious resemblance to feared hallucinogenic and possibly aphrodesiac members of the family, was considered to be erotic and dangerous. French gallants pointedly presented the lady of the moment with a potted tomato plant amid much sniggering and blushing—probably the men blushed and the women sniggered.

It has been suggested that the tomato entered Italy via Morocco; but however it took place, the Italians lustily embraced the vegetable. *Pomodoro,* yellow-skinned cultivars that "gleamed like golden apples nestled in deep green leaves," were stewed into a seed-free paste to accompany polenta or pasta, or were sun-dried for future use with meat and cheese. A fine dish of spaghetti with a tomato-basil-garlic sauce, a green salad, and a bottle of Chianti, the whole consumed by candlelight, is likely to have been more effective than a courtier carrying a tomato plant to a palazzo across the piazza. By the middle of the seventeenth century, the plant was established in China, where the fruit was boiled with spices to make catsup. It spread to India and Southeast Asia, although it is still not an important part of the cuisines of these countries.

In his Garden Book for 1781, Thomas Jefferson noted that the plant grew well in his flower beds, but there is no record of his having eaten the fruit. George Washington Carver, an important agricultural scientist from the Tuskegee Institute, used to shock his early twentieth-century audiences, both white and black, by publicly eating a tomato at the end of his lecture on new food plants. Only when Italian immigrants came to North America was the culinary virtue of the tomato recognized. With over a century of prior selection and breeding, the Italian introductions included most of the familiar kinds of tomato. Today, the tomato is the second leading vegetable crop in North America, just behind the white potato. Well over 50 cultivars are commercially available, some as small as gumdrops and some the size of melons, and with skin colors that range from white through yellow to brilliant red.

The chili and bell peppers have been identified from seeds in Mexican caves dating from 9000 B.P. Hot and sweet bell peppers and pimientos are in the same genus (*Capsicum*) and have found favor throughout the world. Many ornamental garden plants are obviously tomatolike in growth habit, flower, and fruit form. The Jerusalem cherry certainly resembles a tomato, but the Chinese lantern, the butterfly flower, and the petunia are not usually thought of as being related to the tomato. And within this same economically and horticulturally important family are plants that, for untold centuries, have been involved in murders most foul and devious, in witchcraft and demonology, in military campaigns, in invitations to seduction, and in sexual orgies.

Four of the unholy members of the tomato family warrant special attention here. They are the Jimson weeds (*Datura*), belladonna, or deadly nightshades *(Atropa),* henbanes (*Hyoscyamus*), and mandrakes (*Mandragora*). The entire family is a chemist's delight; its members contain several hundred chemically distinct alkaloids. The family name itself, *Solanaceae,* is an allusion to the sedative or hallucinogenic properties of its genera. The terrible four genera, however, contain three alkaloids not found in more familiar genera. Atropine, hyoscyamine, and scopolamine are representatives of the tropane series of alkaloids, similar in structure to cocaine (Figure 6.12). Their biosynthesis is imperfectly known, but a good deal is understood about their physiological action in human beings.

TROPANE ALKALOIDS

Atropine is the most medically useful of the tropanes, serving as a stimulant of the central nervous system and a depressant of the parasympathetic nervous system. The central nervous system includes the brain and spinal cord nerves which monitor many body functions and most mental activities. As a central nervous system stimulant, atropine in moderate doses results in loss of motor coordination; at higher concentrations it causes hallucinations, delirium, stupor, and even death. The parasympathetic system, part of the autonomous network, regulates function of internal organs (heart, blood vessels, intestinal tract) over which there is little voluntary control. Atropine can increase the heart rate to 150–180 beats per minute (tachycardia) with no increase in cardiac output. It increases peristaltic movements of the intestines and interferes with the normal regulation of breathing. Atropine blocks the normal transmission of signals across synaptic junctions between nerves. Most people have experienced one parasympathetic action of atropine when an ophthalmologist puts drops in the eyes to prevent the autonomous closing of the pupil in bright light. Atropine eases the wheezing of severe hayfever victims and effectively dries up drippy noses.

Scopolamine is a depressant of the parasympathetic system, but it can depress the central nervous system. It is a mild analgesic and soporific, and an effective antidote for seasickness. Scopolamine is a truth serum which may or may not still be used depending on whose national intelligence agency is being asked.

The third tropane alkaloid, *hyoscyamine,* is similar in action to

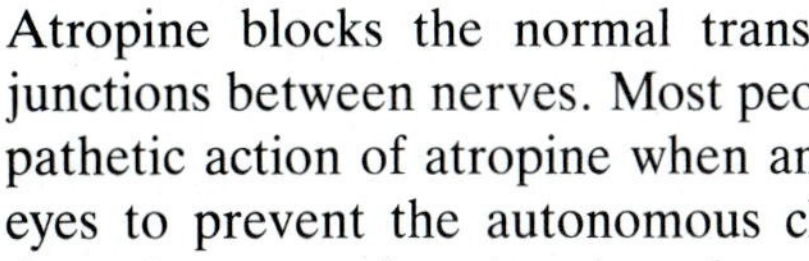

Atropine

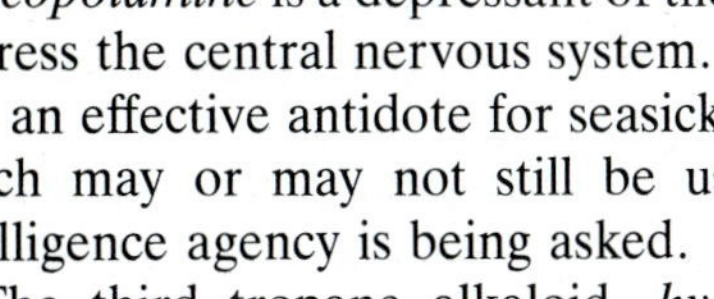

Scopolamine and Hyoscyamine

Figure 6.12 Molecular structures of atropine, scopolamine, and hyoscyamine, the major tropane alkaloids in solanaceous plants used as hallucinogens. Hydrogen atoms and carbon atoms in the rings are not shown. Carbon atoms are represented as circles, oxygen atoms as hexagons, and nitrogen atoms as squares.

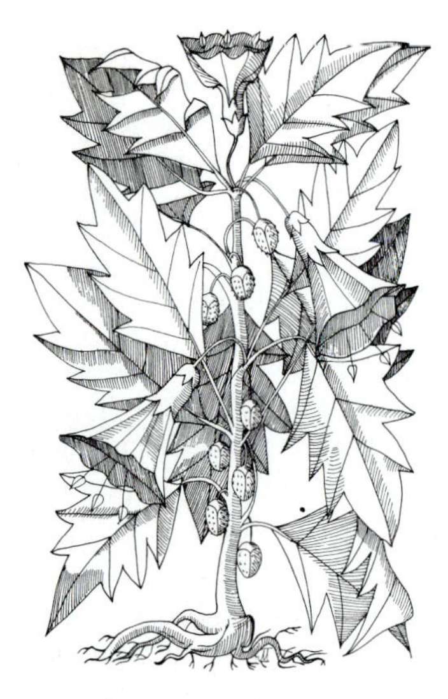

Figure 6.13 Thorn apple *(Datura spp.)*. Taken from a woodcut by Acosta, published in 1578.

atropine, but the clinical responses are sufficiently different to suggest that the receptor sites are not the same. The amounts used medically are very small; practitioners know that even slightly higher dosages, particularly when administered internally, have grave consequences and may lead to death.

DATURA

Of the four genera, we have more information on the history, anthropology, and ethnobotany of *Datura* than on the others (Figure 6.13). The genus is found worldwide, with *Datura innoxia* and *Datura metaloides* in Mexico and the American Southwest; *Datura metal* in India and Southeast Asia; *Datura sanguinea* in tropical and subtropical South America; and *Datura stramonium* in both Europe and North America. There are 50 described species; ten have been reported from Europe, and an equal number are found in Africa. Most are annuals, growing to 2 meters in good soil with a long, warm growing season. The coarse-textured leaves are usually ovate and irregularly toothed or notched. Flowers are trumpet-shaped, superficially resembling a white- or green-streaked petunia. The fruit is a spiny capsule, accounting for one common name, thorn apple. The unpleasant smell of leaves and fruit accounts for another name, stinkweed. Alkaloid content, both tropane and nontropane, may be very high in tropical climates; a goat or even a child may die after eating a few leaves or a single fruit. The genus, named by Linnaeus, is derived from the Hindu word *dhatura,* which means poison.

Datura has been used ritually in India as far back as records have been kept. Seeds were reverently offered to Siva the destroyer, and His priests ate seeds to induce hallucinogenic, oracular states. Sanscrit manuscripts recommended leaves, stems, roots, or fruit for pneumonia, mumps, heart failure, toothworm, asthma, and sexual perversion. Widows destined to die by *suttee*—burning alive on their husband's funeral pyres—were given a potion made from datura leaves to sedate them to the point that they could be led calmly to the pyre, and would not feel the fire. According to Christoval Acosta, who visited India in 1678, Indian prostitutes added ground-up seeds of datura to their patron's drinks to increase sexual excitement, and "these mundane ladies [were] adept in the use of the seed that they give in doses corresponding to as many hours as they wish their poor victim to be unconscious or transported." In the East Indies, women fed datura leaves to beetles, and then fed the resulting highly poisonous dung to faithless lovers. Datura was the original knockout drops, and thieves in India and in Europe used it for centuries. Up until at least the beginning of this century, juice of datura leaves was added to milk given young Indian girls who were to be initiated into prostitution. The drink was narcotic and, it was asserted, aphrodisiac, so that the victim was believed to have actively contributed to her own downfall. The Chinese, believing that datura was a sexual stimulant, administered it to brides on their wedding nights to calm their nerves and to make them more sexually receptive.

The European attitude toward datura parallels that in the Far East. Apollo's priests drank datura to achieve sedated, prophetic, and oracular states. The sacerdotal plant of Delphi was undoubtedly datura; the mumbling speech, trance states, and known fears of overdosage are consistent with datura intoxication. Greek physicians knew it as *nux-metal* or *neura,* a reference to its sedative action, and when extended unconsciousness was desirable, as during surgery, datura was mixed with opium. Rome followed Athens's lead in ritual and in medicine and added the drug-induced orgy in which datura mixed with wine was used to induce hallucinogenic states and to heighten sexual activity. Avicenna, a tenth-century Arabian physician, recommended *jouz-methal* not only for surgery but as an excellent treatment for anxiety. From Arabia, the medical and aphrodisiacal use of datura spread to Spain and to western Europe; northern Europe was too cold to support the growth of datura with high tropane content.

Much of the information on the use of datura in the Middle Ages and into the Renaissance has been lost, because its use became so inter-twined with witchcraft that it was suppressed. Certainly no love potion worth paying good money for was without datura, usually supplemented with extracts of other solanaceous plants and, for good measure, attar of roses, marjoram, other herbs, and a newt's tongue. Witches' sabbats, the infamous black masses that so intrigued prurient Victorians, utilized ointments and unguents containing datura. Whole leaves were inserted into the rectum or vagina, where tropanes are quickly absorbed; the broomstick developed as a symbol of this part of the ritual. Shortly before World War I, Professors Karl Wieswetter and Will-Erich Peu-kary compounded a datura ointment from an old witch recipe and rubbed it on their arms. They reported that they soon felt as if they were whirling around the laboratory. If two sober-minded German scientists thought they were flying, one can appreciate the sensations of less erudite men and women of the fourteenth century. The belief that datura was aphrodisiac was an important component in witchcraft and demonology cults. This belief flows through the drug structure of all cultures that have used datura. Our "winging it" and our "flying" are derived from the concept of hallucinogenic states.

Datura was an important ingredient in the poisonings that pervaded southern Europe from 1400 to 1700. Nobles and merchants sent members of their families to schools teaching the art of poisoning for much the same reasons that we send our children to graduate schools of business administration. In 1650, the widow Toffana was convicted of poisoning 600 people in Palermo with *aqua toffana,* an alcoholic extract of opium and datura. In her defense, Signora Toffana asserted that she was not the only professional poisoner in Palermo or, for that matter, in

Italy, and it was pointed out in her favor that the deaths caused by datura are painless. Reputable physicians of the Renaissance, still under the influence of Galen and Hippocrates, used datura as a sedative, anaesthetic, and soporific until the nineteenth century. John Gerard, in his famous *Herball* of 1596, reported that "the juice of this plant, boiled with hog's grease . . . cureth all inflamations whatsoever . . . as meiself hath found by daily practice to my grete credite and profite." Nevertheless, the potential danger of internal use was well known. One herbalist-physician reported:

> He who partakes of it is deprived of his reason; for a long time laughing or weeping or sleeping and often times talking and replying so that at times he appears to be in his right mind, but really being out of it and not knowing to whom he is speaking nor remembering what has happened.

Datura was listed in official pharmacopoeias, codexes, and dispensatories for several hundred years with specifications for extraction, formulation, and dosages. Only when the tropanes were isolated and characterized were the recommendations altered to specify the purified alkaloid. In 1905, Dr. Carl Gauss of Freiburg brought down the wrath of the Lutheran church when he used datura extracts and morphine to induce *dammerschlaf,* the twilight sleep treatment for women experiencing difficult births. The Church fathers—none of them mothers—declared that Gauss had fallen from grace and was near unto heresy because the Old Testament says (Genesis 4:16) that women were to bring forth in pain.

Datura has had great importance in puberty rites of passage. In the western Amazon basin, initiates are required to drink a datura-beer mixture presented successively by each elder of the tribe. When the boys drink as much as they can hold, they fall into trances, and additional *maikoa* is ritually inserted into the rectum until the child is comatose, remaining barely alive for several days. In the Darien and Choco regions of Central America, seeds are fed to young children. They run around wildly for a short time before falling down in a stupor, whereupon tribal elders roll the child aside and dig beneath the body to find buried treasure. The Colombian Jívaro, a notably stoic people who shrink the heads of their captured foes, punish unruly children by forcing them to eat datura seeds in the hopes that they will be visited by their ancestors who will admonish them for their social infractions; it is a rare child who does not reform after one experience. Infanticide in the Andean highlands involves smearing mother's nipples with datura juice mixed with wild boar's fat, providing a peaceful death that also ensures the baby's safe passage to the spirit world.

Datura is, as readers of Carlos Castaneda's books know, a powerful and respected hallucinogen. Mexican boys will eat datura leaves to acquire the strength of the evil sorcerer *Kieri Tewiyara* (datura person) under the watchful eyes of the *brujo,* who makes sure that they do not become permanently bewitched; the power of datura can be for good or evil, and those who use it must constantly be on guard.

The British colonists in North America were introduced to datura very early. Robert Beverly's *History and Present State of Virginia* discusses the incident which led to our common name for *Datura stramonium.* In 1676, British troops landed at the Jamestown Colony to put down a tobacco tax rebellion led by Nathaniel Bacon. Since the war office mandated inclusion of either citrus fruit or fresh vegetables in the diets of all troops to prevent scurvy, obedient regimental cooks prepared a "salide" of native plant leaves and inadvertently included datura. Beverly reported:

> The effects [were] a very pleasant Comedy, for they turn'd Fools. . . . One would blow up a feather in the air, another would dart straws at it with much fury; and another stark nakid was sitting in a corner like a monkey, grinning and makeing Mows at them. . . . a thousand such simple tricks they play'd and after eleven days return'd to themselves again not remembering anything that had pass'd.

Hence Jamestown weed, or Jimson weed (Figure 6.14). They were not the first troops to be rendered *hors de combat;* Mark Anthony's African legions made the same mistake, delaying Julius Caesar's timetable for the conquest of Egypt.

With the brutal and total repression of Indian culture by the Spanish, the no less effective repression by northern European settlers, and the less physical but equally effective repression by nineteenth-century missionaries, ritual use of datura went underground, leaving a residue of secularized use. The people seized upon datura as a vehicle to escape, if only for a little while, the pain of spiritual and physical degradation. Ritual significance was never entirely lost; the *cuarandas* and *brujos* of

Figure 6.14 Jimson weed *(Datura stramonium).* Courtesy U.S. Department of Agriculture.

the American Southwest perpetuated and guarded the ancient mysteries inherent in datura.

Figure 6.15 Belladonna *(Atropa belladonna),* the deadly nightshade. Taken from an old print.

BELLADONNA

Three sisters born of the mating of Erebus and Clotho became the Three Fates of the Greek pantheon; they spin out the thread of every man or woman's life. Atropos was physically the smallest of the sisters, but was the most feared, for it was she who snipped the thread at the appointed time and not even the major Gods could alter her decision. The deadly nightshade *(Atropa belladonna)* was given Her name by Linnaeus, who simply took one of the common names for the plant (Figure 6.15). The species name, *belladonna,* dates from the late Middle Ages. Mattioli, a famous herbalist of the late sixteenth century, reported in 1560 that Italian ladies put drops of diluted nightshade sap in their eyes to induce *mydriasis*—the deep, dark, mysterious look one gets when the pupils do not close in bright light. The pallor of Renaissance ladies as seen in portraits by Botticelli and Raphael appears, alas, to have been cosmetic, since it was brought about by rubbing their faces, arms, and shoulders with sap from the belladonna root. Like datura, deadly nightshade contains all three tropane alkaloids. The action of nightshade on the parasympathetic nervous system was known and exploited by ancient physicians; Dioscorides and Theophrastus both commented favorably on its sedative action.

Belladonna plants are usually perennial, with a thick, woody rootstock and a branched, purplish stem bearing large leaves with a dull green sheen. Inconspicuous flowers with dull purplish petals are formed in midsummer and give rise to a small berry, green when young, turning red, and finally almost black when mature. In full flower, belladonna has a pleasant scent which attracts pollinating moths in the early evening. It can reproduce by seed, by stolons arising from the roots, or even from pieces of the rootstock. The plant is common throughout Europe, southwest Asia, and parts of North Africa. It is not native to North America but was imported by early colonists for medicinal use, where it escaped cultivation and now grows in meadows, along marshy ponds, and even in tilled fields.

Greece and Rome knew it as a sedative and an hallucinogen. Bacchanalian orgies utilized new wine spiked with small amounts of the sap. It was assumed that the frenzy with which women threw themselves at the male guests was due to the aphrodisiacal qualities of the plant juices, but it must be noted that these women were paid entertainers. Jocasta, a professional poisoner of imperial Rome who specialized in the art of removing unwanted wives and mistresses, was devoted to belladonna, administering it in wine to disguise the bitter taste of the alkaloids. Arabian poisoners, as adept as their trans-Mediterranean counterparts—but usually male rather than female—knew the plant as *bu-rénguf,* the sorcerer's herb. The Norse called it *dwale,* the devil's trance plant, and believed that it was especially beloved of Satan. On Wälpurgisnacht, Satan had to go into the mountains to officiate at witches' sabbats, and the plant was transformed into a beautiful enchantress (belladonna) whose evil beauty would kill any man who saw her.

With the spread of Christianity, bacchanalian orgies became a lamented aspect of the Golden Age of Rome, but belladonna's star rose again as witchcraft and demonology captured people's imagination. One recipe guaranteed to project one into the astral sphere was an ointment of "baby's fat, juice of water parsnip, aconite (wolfsbane), cinquefoil, belladonna and soot," and another was compounded of water parsnip, sweet flag, cinquefoil, bat's blood, belladonna, and sweet oil. In his *Masque of Queens,* Ben Jonson quotes one adept as saying

> And I ha been plucking plants among,
> Hemlock, henbane and Adder's Tongue
> Nightshade, moon wort, leppard's bane.

Dryness of the throat and mouth, difficulty in swallowing, great thirst, impaired vision, and a strange rolling gait, all symptoms of tropane poisoning, are reminiscent of the symptoms of rabies, and it is no wonder that those who took these brews or rubbed on these ointments believed that they had become werewolves. A seventeenth-century coven of witches in Somerset's rolling and wild land confessed that they would anoint themselves "under their arms and in other hairy places" to be transported to the appointed place of a sabbat. They further confessed that they recommended to women who wished sexual satisfaction that before intercourse they rub their private parts with unguents containing nightshade or with fresh leaves of the plant. John Gerard said that "women with child . . . do often long and lust after things most vile and filthie" following the use of nightshade, and Francis Bacon in 1605 noted that women who used nightshade were prone to think they flew and indulged in forbidden acts. Gerard was violently opposed to the nightshade. He reported:

> It came to pass that three boies of Wisbich in the Isle of Ly did eate of the pleasant and beautiful fruit thereof; two whereof dyed in lesse than eight houres. The thirde childe had a quantitie of honey and water mixed together given him to drinke, causing him to vomit often; God blessed this meanes and the childe recovered. Banish therefore these pernitious plants out of your gardens and all places near to your houses where children or women with childe do resort.

He did suggest that nightshade had some use in the treatment of epilepsy, "but if you will follow mei councill, deale not with the same in any case."

The native North American nightshades have been placed in a separate genus, *Solanum,* from the deadly nightshades, since they are much closer taxonomically to the potato than to the poisonous members of the family. Their ripe fruits are not toxic except to very young children, but unripe fruits tend to contain fairly high concentrations of tropane alkaloids. When ripe, the fruits may be gathered and eaten raw or used in pies. There are several cultivated types that may occasionally be found on the market.

HENBANE

England calls the plant *stinking nightshade, stinking Roger,* or *soldier's wort.* In France it is *la jusquiame, potelee, morte aux poules,* or *haniban.* In German, it comes out as *schwarzes bilsen* or *dullkraut.* In Sweden, *bolmort;* in Holland, *bilsenkruit;* in Spain, *belano;* in Portugal, *folhas;* in Italy, *guisquima;* in Saudi Arabia, *bendji;* and the authors of the Gospels knew it as *sikkaron* (Figure 6.16).

Botanically, it is *Hyoscyamus niger,* the black hog killer, so named by Linnaeus because of its deserved reputation for killing swine who ate the leaves or the roots. We call it *henbane,* the chicken killer. The plant is native to Eurasia and spread throughout Europe, the Caucasus, Iran, Asia Minor, North Africa, and as far as India. In the early nineteenth century it was introduced into North America, where it has naturalized well, thriving in the western mountains at altitudes up to 7000 feet (2350 meters). Its only serious pest is the Colorado potato beetle, which seeks

Figure 6.16 An old woodcut of henbane *(Hyoscyamus).* Redrawn from *Hallucinogens and Shamanism* by Michael Horner.

out henbane in preference to potatoes.

There are several closely related species which hybridize extensively, giving rise to intermediate forms. Related species, including *H. alba,* the white hog killer, are difficult to distinguish from the black species. *H. niger* has two forms, an annual, which matures in one growing season, and a biennial. Both have coarse, hairy, alternate leaves on an erect, rarely branched stem that may grow up to 2 meters tall. In North America, the biennial is more common. During its first year of growth, the plant develops a white, fleshy, branched taproot with a rosette of leaves at the crown—something like a horseradish plant—but the leaves are distinctly tomatolike. According to the honorable Dr. Parkinson, Apothecary to His Majesty Charles II, the leaves "have a darke or eville grayish green colour and a heavie eville sopoforonous smell, fedit and rank." Others have described the odor as benumbing or heady. Merritt Lyndon Fernald, dean of American taxonomists in the early twentieth century, reported that the leaves are slimy, clammy, and fetid. He noted, as had others, that even smelling the plant can cause dizziness—a somewhat unlikely possibility. The distinctive smell is, however, somewhat of a blessing, for not even the most inquisitive child would consider nibbling on the plant.

In the second year for the biennial and at the middle of the growing season for the annual, the rootstock develops a stout, erect, and branched stem up to 6 feet (2 meters) tall, bearing flowers in a long spike. Petals of the small, tomatolike flowers are white or light green with prominant purple veins. The fruit is a rounded capsule containing many black seeds.

For medicinal use, leaves are harvested when the plants are in full bloom, since their tropane alkaloid content is highest at this time. Hyoscyamine, scopolamine, and some atropine are easily extracted from leaves; the dried leaves have been listed as an official source of tropanes in the *U.S. Pharmacopoeia* since its first edition in 1820. The leaves are extracted with alkali and the resulting solutions used to make tinctures, compound oils, electuaries (honey syrups), conserves (sugar syrups), feculars (starch pastes), or essences (in alcohol). Henbane has been used topically as a general pain killer, antispasmodic, and burn ointment, and for various unspecified aches and pains. Earaches, toothaches, and stomach aches have been treated with extracts of the plant. While henbane doesn't cure any diseases, it can make pain bearable. The relatively low tropane content compared with that of datura or belladonna made henbane the solanaceous plant of choice of thousands of physicians for hundreds of years.

Among the first records of medical uses of henbane were those found on Babylonian tablets, dating to 4250 B.P., in which chewing seeds was recommended for toothaches, a treatment also espoused by ancient Egyptians and the Aryan writers of the *Rig-Veda* of India. Hippocrates preferred the white henbane to the black because it was less likely to induce stupor: "A small dose in wine, less than will occasion delirium, will relieve the deepest depression and anxiety." Dioscorides used an early Greek name, *dioskyamos* (bean of the gods), suggesting that it may have been used ritually. His Roman contemporary, Gaius Plinus Secundus, was loath to use henbane: "Moreover, unto Hercules is ascribed henbane. Many kinds there be of it. All trouble the brain and put men beside their right wits. I hold it to be dangerous medicine and not to be used but with great heed. For this is certainly known, that if one takes of it in drink more than four leaves, it will put him beside himself."

Physicians of the Dark and Middle Ages burned henbane seeds and funneled the fumes into body orifices to expell the "tiny worms" that caused strange behavior in men and especially in women. Mixed with opium, poison hemlock, aconite, extracts of other plants, and "diverse items," henbane was an accepted anaesthetic, administered, before the crude surgery, as a cloth soaked in the mixture on which the patient sucked. John Gerard knew it as an effective analgesic; "It mitiagatheth all kindes of paine,"; his contemporary John Parkinson said that "it is good to asswage all manneres of swellings, whether of the cods or women's breasts."

All this could be expected; there were very few effective and reliable pain relievers available to the medical profession until this century. The long history of excellent responses to henbane compared to the other solanaceous plants allowed doctors to use it with a reasonable degree of confidence. Also not unexpected was the fervor with which the citizenry embraced henbane as a cheap and fairly (but not absolutely) safe material for recreational hallucinogenic uses. Here, too, the recorded history is long and detailed.

In India, a related species, *Hyoscyamus muticus,* called *kohl-bhang,* is mixed with the resins of *Cannabis* to increase the hallucinogenic effects of hashish, a practice recently revived in our western states. The famous Arabian *Thousand and One Nights* records that the smoke of burning henbane leaves was soporific: "Presently he filled a cressat with firewood on which he strewed powdered henbane and, lighting it, went around the tent with it until the smoke entered the nostrils of the guards and they fell asleep, drowned by the drug." Many innocent farm boys awoke after a night on the town to find themselves in the holds of ships sailing to foreign ports. Henbane had been included in their last beer. The poison poured into the ear of Hamlet's father by Gertrude and Claudius was probably a mixture of belladonna and henbane, since poisoning by these plants was not uncommon in Shakespeare's day. Albertus Magnus, philosopher and tutor of St. Thomas Aquinas, urged that henbane be banned from Christendom for "it hath the power to conjure demaens." Known to witches as "Insana," the plant was shunned by all God-fearing peoples. Practitioners of the black arts would mash seeds into the curdled milk of mares, pour the smelly stuff into the tanned scrotum of a bull, and sell these charms to women who wished to avoid pregnancy. Henbane seeds were featured at black masses, mixed with the sacramental wine and consumed in the false Eucharist.

In view of the concern about the plant being a favorite of the Devil, its introduction into North America is something of a puzzle. It apparently arrived in 1670, when most immigrants were religiously oriented, and it can only be speculated that either it was part of the herbal medicine of an early physician or that one or more of the members of the congregation were less (or more) than they seemed to be. Henbane never became an important hallucinogen in North America, although it has been suggested that its introduction into the western mountains was due to railroad builders who were in the habit of mixing henbane seeds with their tobacco to strengthen the smoke. And it undoubtedly worked.

MANDRAKE

Jacob, son of Isaac and brother of Esau, served Laban, son of Nahor, for seven years and was tricked into marrying Leah. She bore him four sons while Jacob labored another seven years for the hand of his beloved, but barren, Rachel. Distraught, Rachel chose in her desperation to employ the feared fertility plant of the Middle East, the awesome mandrake.

> In the days of the wheat harvest, Reuben went and found mandrake in the field and brought them to his mother Leah. Then Rachel said to Leah, "Give me, I pray, some of your son's mandrakes." But Leah said to her, "Is it a small matter that you have taken away my husband? Would you also take away my son's mandrakes from me?" When Jacob came from the field in the evening, Leah went out to meet him and said, "You must come in with me, for I have hired you with my son's mandrakes." So he lay with her that night . . . and she conceived and bore Jacob a fifth son. (Genesis 30:14)

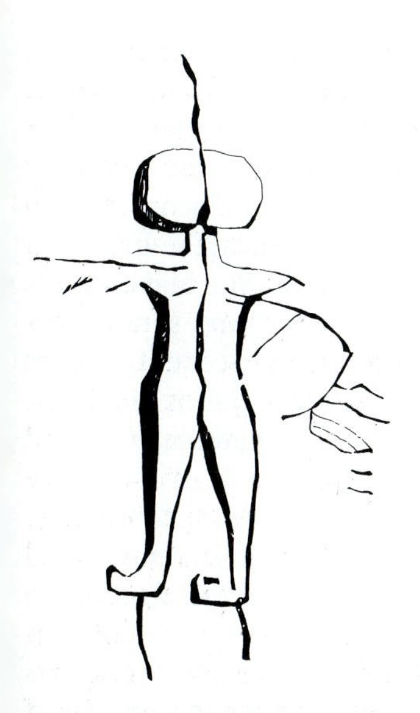

Figure 6.17 Prehistoric rock carving found in France and presumed to be of mandrake.

Apparently without the magic of mandrakes, Rachel did conceive and gave birth to Joseph.

Long before the Hebrews, the mandrake *(Mandragora officinarum)* had been associated with sexuality and sins of the flesh (Figure 6.17). The Elbers medical papyrus of 2500 B.P. listed it as *dudajm,* the fruit that excites love; Pharaoh Tutankhamen was buried with 11 mandrake roots in the sixth row of his floral collarette to ensure his potency in the next world. The Greeks named it *circeium* after Circe, who, Homer reported, lured men to her and changed them into swine—that is, into sexual pigs. They referred to Aphrodite as *Dios Mandragoritis.* Mandrake roots carved into big-hipped and -breasted fertility figurines have been unearthed at Antioch and Damascus and from tombs in Constantinople and Mersina. The Song of Solomon noted: "The mandrake gives a smell, and at our gates are all manner of pleasant fruits, new and old, which I have laid up for thee, my beloved." Pliny said that those roots which resembled penises were love charms, and Galen echoed this: "If a root falleth into the hands of a man, it will ensure women's love." By the second century, everyone knew that female elephants gathered mandrakes and presented them to male elephants. Shakespeare was very aware of the sexual power of the mandrake. Justice Shallow was described in *Henry VI* (II) as "lecherous as a monkey and the whores call him mandrake," and Falstaff cursed his page in *Henry VI* (I) exclaiming, "Thou whoreson mandrake, thou are fittest to be worn in my cap than to wait at my heels." Contemporaries of Shakespeare were equally impressed; in *Polyalbion* Michael Drayton wrote: "The power of mandrake in philtres to procure love and worn about the body to correct barrenness was unduly recognized," and again, "The fleshy mandrake was, par excellence, the love-compelling agent." In about 1513, Nicolò Machiavelli wrote a comedy, *La Mandragora,* in which mandrake root permitted a noted seducer to bed down the virgin wife of another man, and La Fontaine's fable "La Mandragora," written about 1680, was based on the same theme. In the seventeenth century, scrapings of the root were believed to be effective in curing the French disease—syphilis.

Any plant that could ensure potency and fertility must be gathered with appropriate respect and all due caution. The first written set of instructions on collecting mandrake was provided by Theophrastus, who said, "Around the mandragor one must make three circles with a sword and dig, looking to the west. Another person must dance about in a circle and pronounce a great many aphrodisiacal formulae." A more complete method was given by Flavius Josephus, scribe and historian of the Roman Jews in the first century. The virtuous man, he said, should first look for them by night, for then the leaves will shine in the dark, "but if anyone tries to pluck them, they rise in the air and fly away." They will, however, remain in the ground if a man strikes them with an iron rod. Josephus warned his readers that there was great danger attendant upon uprooting the plant, for when rudely and inconsiderately removed from the earth, they give a cry that can cause the hearer to go insane or to be driven mad. Shakespeare knew this, and told his audiences so in *Romeo and Juliet* (Act 4, Scene 3): "And shrieks like mandrakes torn from the earth"; and again in Act 8, Scene 3: "What with loathsome smells and shrieks like mandrakes torn out of the earth, that living mortals, hearing them, run mad"; and yet again in *Henry VI* (II), Act 3, Scene 2: "Would curses kill as doth the mandrake's groan." John Webster had Ferdinand in the *Duchess of Malfi* (Act 2, Scene 4) say: "I have this night digg'd up a mandrake, and I am gone mad with't." The Arabian physician Ibn Beither said that the Europeans were mistaken, for it was not the shriek that killed, rather, a root demon who lost his home would invade the digger. Ben Jonson knew of this horror, for in his *Masque of Queens* he had an old woman say: "I last night lay all alone on the ground to hear the mandrake's groan . . . and plucked him up." The hag was exceedingly brave; most people followed

Josephus's advice and tied a dog to the crown of the plant so that when the cur responded to a whistle or to a bone, it would die as the plant came screaming out of the earth (Figure 6.18). It is unlikely that Josephus invented this method; a Persian name—derived from the Assyrian—is *sag-ken,* dug by a dog.

The most elaborate technique comes down to us from the Latin herbal of Apuleius Platonicus, written in the fifth century. Apuleius was probably a Greek who lived in Rome, wrote in Latin, and collected not only the wisdom of the evening fireplace stories but also the aphorisms of philosopher-botanist-physicians. His *Herbarium* was translated into Anglo-Saxon in about the tenth century and profoundly affected the English view of the efficacy of many herbs, for echoes of Apuleius are found in the herbals of Parkinson, Gerard, and others.

> This wort is mickle and illustrious of aspect and it is beneficial. Thou shalt in this manner take it. When thou comest to it, then thou understandeth it by this, that it shineth at night altogether like a lamp. When first thou seest its head, then inscribe it instantly with iron lest it fly from thee; its virtue is so mickle and so famous that it will immediately flee from an unclean man when he cometh to it. Hence, as we before said, do thou inscribe it with the iron. But thou shall earnestly, with an ivory staff, delve the earth. And when thou seest its hands and feet, then thie thou it up, Then take the other end and tie it to a dog's neck so that the hound may be hungry. Then cast meat before him so that he may not reach it, except he jerk up the wort with him. Of this wort it is said that it hath so mickle might that what thing soever tuggeth it up, that it shall soon in the same manner be deceived. Therefore, as soon as thou seest it be jerked up and have possession of it, take it immediately in hand and twist it and wring the ooze out of its leaves into a glass ampulla.

Use of a dog as surrogate was repeated over and over again in herbals dating from the eleventh century (Figure 6.19). The *Beastiary* of Philip of Thawn, written in 1122, repeated the dog story for the benefit of French collectors. Chai Mai, an illustrious Yüan Dynasty scholar, wrote in 1285 of this plant of the masters. *Yah-puk-lu,* that "grew thousands of li to the west and could be gathered only with the aid of a dog." Ben

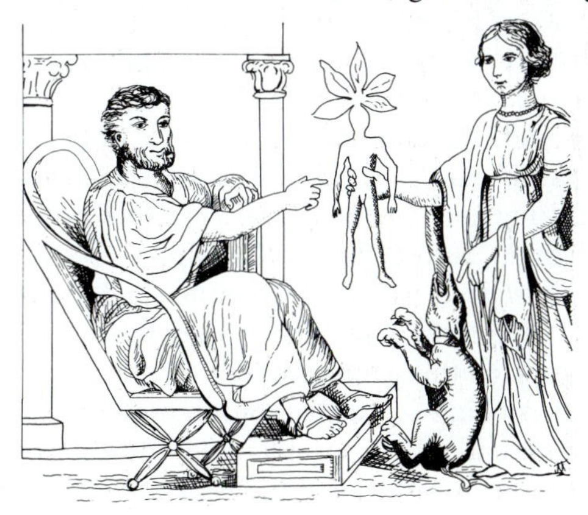

Figure 6.18 Dioscorides receiving the mandrake from a Goddess of Medicine. The dying dog represents the legend concerning the way the plant was to be dug.

Figure 6.19 Uprooting the mandrake. Taken from a sixteenth-century herbal.

Jonson used the dog story in *The Masque of Queens,* and even earlier, Archbishop Aelfric solemnly recommended it, for the dog had no soul to lose.

There were different forms of mandrake. Dioscorides spoke of black and white mandragora, but it is likely that his black form was nightshade. Pliny said that the more robust plants were male, but Galen asserted that the male plant had a single taproot, while the female plant had a forked root with the two branches "spread wide apart." By the eleventh century, woodcuts in herbarium volumes included anthropomorphic representations of male and female mandrakes with primary and secondary sexual structures (Figure 6.20). The *Grete Herball* of 1526 said:

> There be two maners, the male and the female; the female hath sharp lines. Some say that it is better for medycyne than the male, but we use both. Some say that the male hath figure of shape of a man and the female of a woman, but this is fals. For natur never gave forme or shape of mankynde to an herbe.

Figure 6.20 Male and female mandrakes. Taken from *Hortus Sanitatis* by Cuba, published in Paris in 1489.

Nevertheless, Parkinson's *Theatrum Botanicum* of 1640 still spoke of mandrakes and womandrakes.

The botany of the mandrake is fairly straightforward. The root is thick and fleshy, a taproot much like that of horseradish, but somewhat fatter; there may be a single taproot or more commonly two branches. Like horseradish or carrot, there is practically no stem, the leaves arising as a rosette directly from the root crown. The leaves may be very long, up to 30 centimeters or more, with broad blades which are hairy and have crimped or wavy margins. They tend to lie on the ground so that a single plant will cover a circular area of more than half a meter; there are few other plants that look anything like it. Flower stalks arise directly from the crown and bear solitary flowers with greenish, blue-violet, or purple petals which are unmistakably tomatolike although much larger than tomato or potato flowers. The fruit is a fleshy, many-seeded berry like the tomato and is yellow at maturity; it is sweet and edible when ripe, with an odor variously described as pleasant or mildly offensive. There are two botanical varieties, *Mandragora officinarum vernalis,* which flowers in the spring and *M. o. autumnalis,* whose flowers develop in the early fall. Both are native to the Mediterranean Basin and the Middle East and have been cultivated and naturalized throughout southern and central Europe. Both were introduced into Britain before the tenth century, possibly by the Romans, and were naturalized before William invaded in 1066.

Mandrake roots have been used medicinally for centuries; Homer reported that Petrocius treated the arrow wound for his friend Eurpylus by rubbing it "with a bitter root that took away pain." Hippocrates apparently had considerable respect for its ability to soothe: "The persons who are dull and sick and would commit suicide must take an infusion of mandrake root in the morning in a small dose, less than the amount which would cause delirium." In his *On Diseases,* II, 43, he recommended its use for malaria, and in his *On Diseases of Women,* II, 199, he said that for white or bloody flux, mandrake juice was to be mixed with sulfur, a swab dipped into the mixture and inserted into the vagina while the patient was enjoined to lie motionless and to sleep on her back. For females falling into fits (hysteria) and losing the power of speech, "a bolus . . . remedies the evil." Plato made mention in *The Republic* of its use to render insensible a sea captain who suffered from melancholia. Plato's pupil, Theophrastus, recommended dried mandrake mixed with meal as an invaluable poultice for wounds and suggested that scrapings of the root be mixed with vinegar and taken as a sleeping draught; he warned his readers that excessive use was dangerous.

Nevertheless, its soporific and analgesic properties were blessed by physicians as they blessed opium, and they usually use both to alleviate pain. Theodore Borgognoni, professor of medicine at Bologna during the early thirteenth century, developed the *spongia somnifera;* a mixture of opium and mandrake juice was compounded in vinegar, taken up in sponges, and inserted into the nostrils of patients undergoing surgery. Contemporary clerics suggested that God used the sponges to put Adam to sleep before removing one of his ribs (Genesis 2:21). Roger Bacon approved the use of mandrake juice as a cure for insomnia, and Shakespeare was aware of the sedative properties of mandrake. In *Antony and Cleopatra* (Act I, Scene 5), Cleopatra cries, "Give me to drink mandragora that I might sleep out this great gap of time my Antony is away." Iago tells Othello (Act II, Scene 8), "Not poppy nor mandragora . . . shall ever medicine thee to that sweet sleep." In *Cymbeline* (Act V, Scene 5) there is "a certain stuff, which being ta'en, would cease the present pow'r of life; but in short time all offices of nature should again do their due functions," leading one to think that it was a draught of mandrake that Friar Laurence gave Juliet. Marlow confirmed the general belief that mandrake could cause a deathlike sleep. In *The Jew of Malta* (Act V, Scene 3) Barabas says, "I drank poppy and cold mandrake juice and, being asleep belike they thought me dead and threw me over the walls." The Romans knew that this procedure was common, and there was a standing order that all men who were crucified were to be mutilated before their bodies were turned over to relatives. Is it possible that the sponge given Jesus was steeped in mandrake?

Images carved from large mandrake roots or from the counterfeit white bryony were in the form of angels (white magic) or devils (black magic) and were used appropriately. One of the charges laid against Joan of Arc was possession of a mandrake poppette; she denied it and denied even knowing what a mandrake was, but her denial was unacceptable to the court because it said that everybody knows about mandrake. Possession was punishable by death until the middle of the eighteenth century and with good reason: a well-developed root contained sufficient tropane to kill and the evil magic could "drain dry the cows grazing upon nine meadows, stop up the hens and rot the cabbages in the garden." The French knew the root as *maine de gloire* and the Germans spoke of *alraun* (Figure 6.21), both words referring to spirits inhabiting the earth. The root could be helpful to bring wealth and love, or would glow when near buried treasure. Tavern keepers kept a root on hand knowing that its presence would draw customers who would drink until their pockets were empty. A man on his way to court was enjoined, "And when thou goest to law, put Erdman (alraun) under thy right arm and thou shalt succeed whether right or wrong." Alraun protected its owner in war; Rudolph II of Austria had it emblazoned on his war shield.

But, take care! Once you have acquired a mandrake or womandrake it will never leave you: it will return like the phoenix if you throw it into a fire; it will come dripping wet back to you if you try to throw it in a river or caked with mud if you try to bury it. If you die, it will give the owners of your house no rest as it searches nightly for you in the closets and under the stairs. Beware, oh, beware!

. . . *AND OTHER HORRIBLE SOLANACEOUS PLANTS*

Figure 6.21 Fifteenth-century drawing of the *alraun* in Germany.

In addition to the terrible four, several other genera in the Solanaceae contain species that synthesize tropane and related alkaloids. Three tree species in the genus *Duboisia* had been used medically and ritually in Australia long before transported Britains settled near Botany Bay. *Duboisia hopwoodii* grows throughout western and southern Australia and, in addition to tropanes, contains about as much habit-forming nicotine as tobacco. It was smoked, chewed, and snuffed before tobacco replaced it in the eighteenth century. *D. myroporoides* and *D. leichhardtii* are small trees with a bark resembling that of the cork oak. Their flowers look like those of datura, and their fruits are similar to those of the Jerusalem cherry. The tropane content of the leaves is very high. The bush peoples actively traded leaves across the continent and kept the knowledge of the pleasure leaves to themselves. In May 1861, an exploring party spent an evening with the aborigines and smoked what the natives called *pituri* or *pedgery,* which had "a highly intoxicating effect even when chewed in small quantities." The natives mixed the dried leaves with wood ashes, rolled the mixture into a green leaf, and either chewed it like an unlit cigar or lit one end. Women and youths were forbidden its use, since tribal elders thought that the pleasure was undeserved by such members of the group. Pituri can allow a person to forget hunger and forgo misery, both facts of life in the Australian bush. One modern ethnobotanist stated that while duboisia tasted like dried manure and required considerable will power to keep on chewing, he experienced exhilaration and intoxication and lost any desire for food.

But since he also got a bellyache, he didn't repeat the experience. The Australian government proscribed the use of duboisia while maintaining a monopoly of tobacco; the natives now use tobacco and call it pituri.

In the Sibundoy Valley of the Colombian Andes there is a large tree, *Methylsticodendron amnesianum,* that the Indians calls *culebra borrachera*—the beautiful snake tree. Beautiful it may be, for the leaves contain all major tropane alkaloids in concentrations up to ten times that found in datura. The Brazilian Indians of the Amazon Basin have *Brunfelsia tastevini,* or *keya-hone,* whose alkaloid spectrum includes all three major tropanes plus coumarins. Species of *Iochroma* and *Latus* are called *arbol de los brujos*—trees of the warlocks—whose leaves and roots are used by the Indians of the Peruvian highlands for visionary rites. In Africa, the Bushmen use the juice of the fruits of *Solanum incanum* both internally and for their arrow poisons. The Zulu use *S. sodomeum*—the apple of Sodom—for the same purposes. In southern Africa, *Withania somnifera* live up to its specific epithet. It is used to prevent insomnia, treat rheumatism, ease the pain of boils, soothe colicky children, tone the uterus after birth, and combat female sterility.

With the large number of tropane and nicotine alkaloids found in solanacean plants, one wonders how innocuous the fruits really are. The alkaloid content of the edible portions of most garden solanaceous plants is negligible, although leaves of many, including the tomato, peppers, and eggplant, can contain mildly poisonous amounts, especially for children. Inhalation of the smoke from cigarettes smeared with rotten green pepper fruits is reputed to result in a mild intoxication. Potato tubers left in the sunlight will turn green and synthesize an alkaloid, solanin, which is toxic in high concentrations and has caused the death of domestic animals. Solanin is destroyed in cooking and green potatoes need not be a cause for concern, unless eaten raw.

ADDITIONAL READINGS

Emboden, W. A. *Narcotic Plants.* New York: Macmillan, 1972.

Furst, P. T. *Hallucinogens and Culture.* San Francisco: Chandlet & Sharp, 1976.

Harner, M. J. (ed.). *Hallucinogens and Shamanism.* New York: Oxford University Press, 1973.

Heiser, C. B., Jr. *Nightshades—The Paradoxical Plants.* San Francisco: Freeman, 1969.

Heiser, C. B., Jr. "The Ethnobotany of Neotropical Solanaceae." *Advances in Economic Botany* 1 (1984).

McCue, W. The History and Use of the Tomato: An Annotated Bibliography. *Annals of the Missouri Botanical Gardens* 39 (1952):289–348.

Mehra, K. L. "Ethnobotany of Old World Solanaceae." In J. R. Hawkes, R. N. Lester, and A. D. Skelding (eds.). *The Biology and Taxonomy of the Solanaceae,* pp. 161–170. New York: Academic Press, 1979.

Ott, J. *Hallucinogenic Plants of North America.* Berkeley, Calif.: Wingbow Press, 1976.

Peterson, N. "Aboriginal Uses of Australian Solanaceae." In J. R. Hawkes, R. N. Lester, and A. D. Skelding (eds.). *The Biology and Taxonomy of the Solanaceae,* pp. 171–188. New York: Academic Press, 1979.

Ricciuti, E. R. *The Devil's Garden.* New York: Walker, 1978.

Schultes, R. E. "Present Knowledge of Hallucinogenically Used Plants: A Tabular Study." *Recent Advances in Phytochemistry* 9 (1975):1–28.

Schultes, R. E. "Solanaceous Hallucinogens and Their Role in the Development of New World Cultures." In J. R. Hawkes, R. N. Lester, and A. D. Skelding (eds.). *The Biology and Chemistry of Hallucinogens.* Springfield, Ill.: Thomas, 1979.

Schultes, R. E., and A. Hofmann. *The Botany and Chemistry of Hallucinogens.* Springfield, Ill.: Thomas, 1973.

Scott, J. *The Mandrake Root.* London, 1946.

Swain, T. (ed.). *Plants in the Development of Modern Medicine.* Cambridge, Mass.: Harvard University Press, 1972.

Thompson, C. J. S. *The Mystic Mandrake.* New Hyde Park, N.Y.: University Books, 1968.

Waller, G. R., and E. K. Nowacki. *Alkaloid Biology and Metabolism in Plants.* New York: Plenum Press, 1978.

Mushroom Madness

She stretched herself up on tiptoe and peeped over the edge of the mushroom and her eyes immediately met those of a large blue caterpillar.

LEWIS CARROLL
Alice's Adventures in Wonderland

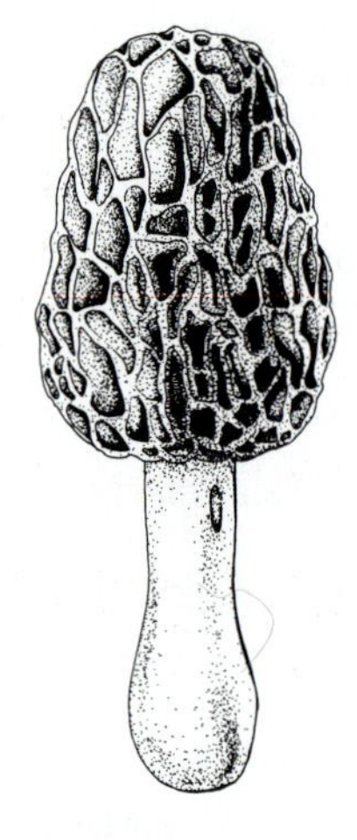

Figure 6.22 The morel *(Morchella crassipes).* Not a true mushroom—it is an Ascomycete—morels are among the most delicious of fungi.

In addition to Magna Carta, common law, Shakespeare, and overcooked vegetables, some of us living in North America inherited the Anglo-Saxon mycophobia. Many will pick at broiled mushrooms garnishing a steak, but few will eat them raw and even fewer have the courage to taste one of the four or five edible wild species which can be unequivocally identified (Figures 6.22–6.25). In Europe, there are exceptions to this blanket statement; some Scandinavians and Germans avidly search for *pfifferlings* in the spring and *steinpilz* in the summer, but the common attitude is summed up by a statement from a British farmer, "If my caows won't eat 'em, Oy ain't agon'ta."

There is a long history of mushroom distrust. Except for mention of "leprosy of a house," referring to dry rot and the notion that leavened bread is not ritually clean, neither of the Testaments speaks directly of fungi. There is a Latin saying: "If you awaken in the morning after an

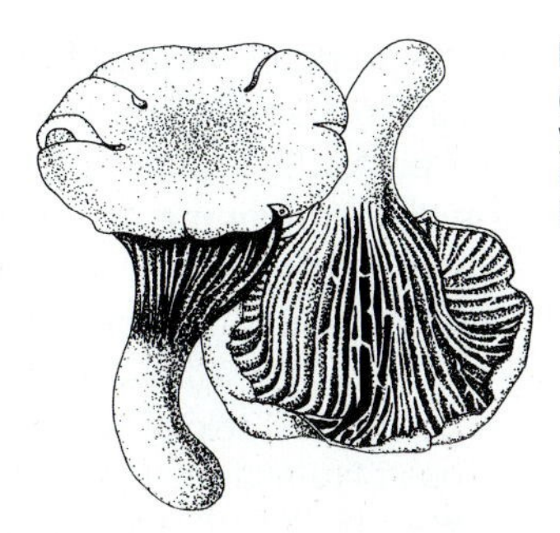

Figure 6.23 The chanterelle *(Cantharellus cibarius)*. This yellow-to-orange mushroom is readily identified and is excellent in flavor.

evening meal of wild mushrooms, you know they were good ones." Pliny asked: "What great pleasure can there be in partaking of a dish of so doubtful a character?" and Dioscorides in the first century stated: "Of fungi there is a double difference, for either they are edible or they are poisonous," and he considered that even edible sorts were hard to digest. If they are served at a banquet, Dioscorides recommended that prophylactic vomiting should be induced by swallowing hen's manure mixed with vinegar. Peter Treveris, author of a herbal, stated most forcefully:

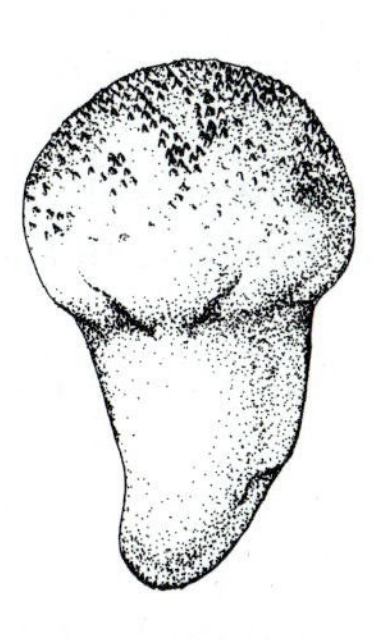

Figure 6.24 The gem-studded puffball *(Lycoperdon gemmatum)*. Puffballs are edible when their interiors are pure white and firm.

> Fungi ben mussherons. They be cold and moyst in the thyrde degre and that is shewed by theyr vyolent moysture. There be two maners of them, one maner be deedly and sleeth them that eateth of them, and be called tode stooles, and the other dooth not.

Tode is a German—a Saxon—word for death, and this concept pervades our mycophobia.

On the other hand, mycophiliac cultures esteem them highly. Egypt's pharaohs reserved them for their personal use. Wealthy Romans vied with one another to create new and different recipes, and Cicero noted that "since these elegant eaters wish to bring into high repute the products of the soil, they prepare their fungi with such high-seasoned condiments that it is impossible to conceive of anything more delicious." The favorite in Rome was the *cibus deorum*—the divine food—and in the area near Rome, this mushroom was first offered to the emperor. Julius Caesar did, however, issue an edict that those found on foreign soil could be eaten by captains of cohorts. The boletes (Figure 6.26)

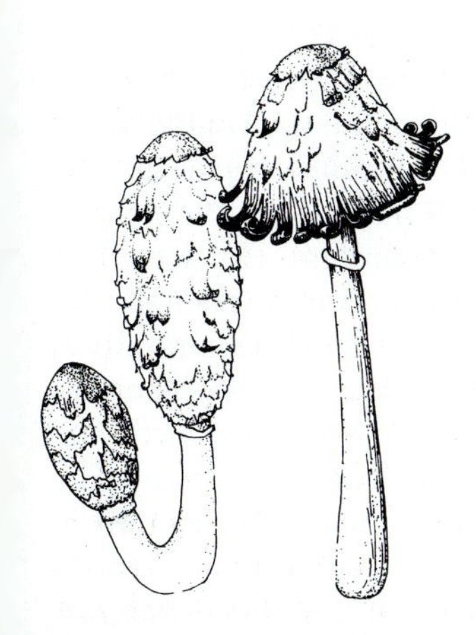

Figure 6.25 The shaggy mane *(Coprinus comatus)*. It is easily identified and very tasty when young.

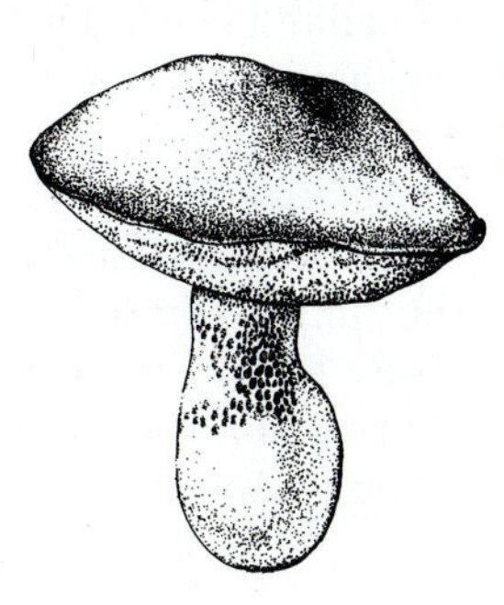

Figure 6.26 A bolete *(Boletus edulis)*, one of the prized mushrooms of the Romans.

were prepared in special cooking vessels, *boletaria,* that could not be used for less noble purposes. Emperor Tiberius (reigned 14–37) awarded one Asellius Sabinus the sum of 200,000 sesterces (about $15,000) for a recipe containing boletes, small whole birds *(beccafios),* oysters, and boned thrushes. *De Re Coquinaria,* a cookbook written by Caelius Apicus about 200, contained the following recipe:

> Boil truffles, spinkle with salt and thread them on twigs. Roast them over an open fire and then place them in a cooking pot with wine, pepper, boiled wine, oil, greens and honey. Boil, and while simmering, add flour to thicken. Arrange them around a pig while it is roasting.

Galen, however, wrote that fungi must be counted as among insipid things, for "they yield clammy juices and it is far safer to have nothing to do with them."

Medically, mushrooms are of little value. Hippocrates recommended that dried, powdered fungi be used to cauterize wounds, a practice still followed in Lapland, Sikkim, and a few other out-of-the-way places. Dried fungi were powdered, mixed with saltpeter, placed on the wounds, and set afire. Although imputed with magical properties, the fungi serve only as tinder. One fungus, a bracket type found on trees, contains laricic acid, a powerful purgative that can cause death and hence is rarely employed. In Solzhenitsyn's book *Cancer Ward,* he refers to a fungus called *chagi,* which grows on birch trees in Siberia and has a local reputation of curing cancers. Several such fungi grow on trees and one, *Polyporus hispidis,* contains a compound that Soviet scientists are evaluating as a cancer cure. Most edible mushrooms are about 90 percent water, 4 percent protein, 5 percent carbohydrate, and 0.4 percent fat—they are as nutritious as watermelon.

In North America, it is difficult to purchase any except *Agaricus bisporus,* the least tasty of the edible species. It has the sole advantage of ease of cultivation and ready recognition. Cultivation of *A. bisporus* developed in England in 1779 and was introduced into the United States about 1865 by immigrants to Pennsylvania. The mushrooms are usually grown in caves (Figure 6.27), where light will not darken the pristine whiteness we prize and where temperatures are constant. Horse manure mixed with wheat or rye straw is composted for several weeks and then placed in the bedding area where it is allowed to pasteurize naturally at the high temperatures caused by the metabolism of microorganisms. This "sweating-out" process makes the bed disease-free. Beds are inoculated with either a pure laboratory culture of the fungus or with mushroom spawn—a compressed block of the threads (mycelium) of the fungus. Small buttons begin to appear in a month, and a crop is ready to be picked in about two months. A well-cultivated bed will continue to produce for about two or three months, with yields of five kilograms of fresh weight per square meter of bed area.

Figure 6.27 A cave in which mushrooms are grown commercially. Courtesy U.S. Department of Agriculture.

THE BIOLOGY OF THE FUNGI

The familiar fungi are those whose fruiting, reproductive structures are big enough to be visible to the human eye. Most prominent are those evolutionarily advanced forms, the basidiomycetes, including the mushrooms and toadstools. The fungi are, however, among the most varied of living organisms, with well over 80,000 species divided into several hundred families. The vast majority of fungi are microscopic in size, with their vegetative structures either as single cells or as threadlike strands called *hyphae.* The fluffy, cottony masses of hyphae that reach visible size are referred to as a mycelium or, when they become twisted into ropelike masses seen under rotting bark of trees, as rhizophores. None of the fungi contain chlorophyll, and hence all are dependent on other organisms for their organic food supply. Some can obtain this food from dead plants and animals; the mycelium surrounding a dead fly on a windowpane or the fungi that cause rot in wood are examples of this *saprophytic* mode of nutrition. Other fungi live in association with algae, obtaining sugars and other compounds from the photosynthetic forms and, in return, providing water and minerals to the algae (Figure 6.28). Such associations are termed *symbiotic*—living together. Others are *parasitic,* living on or in plant or animal cells. Many diseases of plants and of animals are caused by fungi, whose activities result in large economic losses and considerable numbers of human deaths each year.

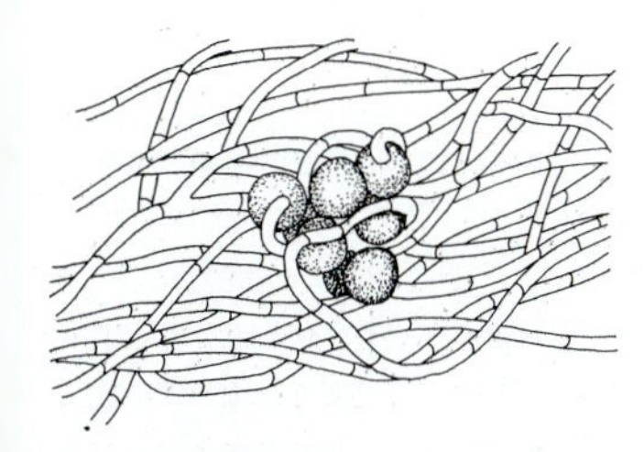

Figure 6.28 Diagram of the association of fungi and algae to form a lichen. Hyphal threads of the fungus utilize sugars formed in algal photosynthesis and provide the algae with water and minerals.

The range of habitats in which fungi are found is enormous. Some species live in the depths of the oceans, others in virtually every soil type on earth, and some can even grow in jet fuel, where they clog engines.

Leather, some plastics, paint, asphalt, soap, and other inhospitable substrates can each harbor one or more species of fungi. Fungi excrete enzymes that break down organic substrates into soluble compounds which are then taken in by the hyphal cells and converted into more fungus.

The reproductive capacities of fungi have been of scientific interest for several hundred years. Many fungi produce microscopic asexual spores, or conidia, which can be carried long distances in the air. When they land on a suitable substrate, they germinate and again form a hyphal thread. The blue, green, and black powders formed on rotting oranges are conidia of *Penicillium,* a fungus which also produces the antiobiotic penicillin (Figure 6.29). The red streaks which form on wheat leaves, giving the name wheat rust to the disease, are the spores of a pathogenic fungus, as are the black powders that form in the ears of corn as corn blight disease develops (Figure 6.30). One hyphal segment can produce millions of conidia, and the spread of plant and animal diseases "like a prairie fire" is due to this efficient asexual reproduction.

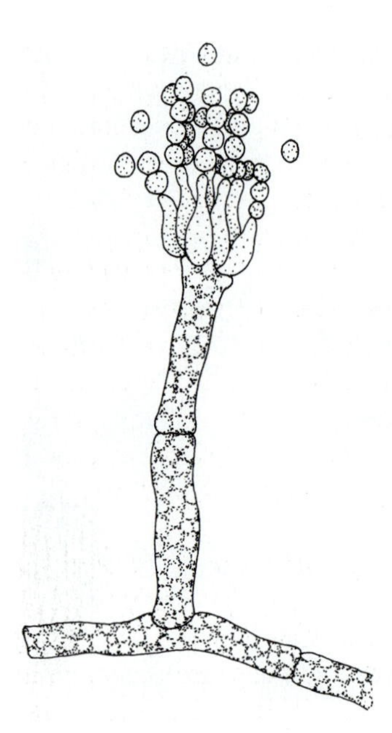

Figure 6.29 Conidial branch of *Penicillium spp.* The brushlike form of the branch was responsible for the genus name.

Fungi, like most other organisms, can also reproduce sexually—that is, by the fusion of genetically dissimilar nuclei. Many of these sexual processes are extremely complicated, involving several stages. The wheat rust fungus can only complete its life cycle with a total of five distinct kinds of spores and two different hosts—the wheat plant and the barberry. The devastating blister rust disease of white pines involves the currant *(Ribes)* plant as an alternate, obligatory host. The taxonomic separation of the fungi into classes is based on their modes of sexual reproduction. In the mushrooms, the visible entity is the reproductive structure, the terminal point in their life cycles. The mushroom itself is composed of hyphal threads closely appressed and organized into the typical form for each species. The mycelium that supplies the nutrients to the fruiting structure may ramify through soil or through trees and is rarely seen except when one digs carefully and finds rhizomorphs in the soil or in the decaying wood. Picking mushrooms does not injure the mycelium, and new fruiting structures will develop when conditions again become favorable.

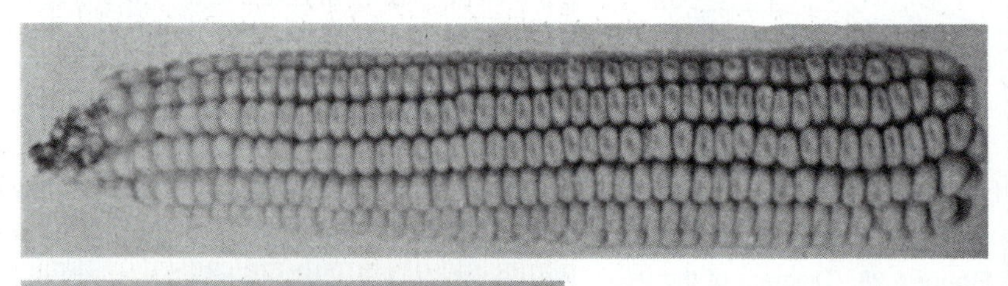

Figure 6.30 Effects of southern corn blight caused by *Helminthosporium mydis.* Courtesy U.S. Department of Agriculture.

Most emphasis has been placed on those fungi that cause the rots, blights, murrains, rusts, smuts, bunts, and so on, of plants, or the skin, lung, or systemic diseases of humans and other animals. This should not obscure the fact that the vast majority of fungi play irreplaceable roles in the ecosystems of the world. They, and the bacteria, are responsible for the return to the soil of the accumulated organic and mineral constituents that comprise plants and animals. Without this decay, the cycling of minerals and carbon-containing compounds would not occur, and the biotic world would soon come to a grinding halt. People have exploited the phenomenal capacity of fungi to synthesize compounds of use to them. The bulk of the world's antibiotics, many of our cheeses, all of our bread and fermented beverages, and compounds for the chemical industry (acetone, citric acid and so on) are produced by fungi.

Mushrooms and toadstools have been part of legend and myth in many cultures, and have inspired the work of poets, mystery writers, and other artists. Many of these stories are charming, not the least being those about the Irish elves that sit on, or in rainy weather under, giant umbrella-shaped toadstools. In England, one type of fruiting mushrooms, the puffball, is called pucksball, from Puck, the English equivalent of the Greek satyr Pan (Figure 6.31). This indirect reference to male virility led to their being considered aphrodisiacs, but this is not supported by any scientific evidence. The stinkhorn fungus *(Phallus impudicus)* is also called the devil's penis; both common names are accurate, for it has a horrible smell and its shape is anatomically correct (Figure 6.32). Sir John Mandiville wrote in 1335 about the fungus named *Hirneola auricula-judea,* stating that it could summon the Devil from hell. Mandiville declared it was first found on the tree upon which Judas hanged himself after betraying Christ. Thus, from Judas's ear to Jew's ear. Edible, but not especially tasty, it was used for several centuries as one of the Doctrine of Signature plants to cure ear inflamations (Chapter 5).

The rapidity with which the fruiting bodies can appear has been a

Figure 6.31 Puffballs *(Lycoperdon pyriforme).* Courtesy Dr. Kent H. McKnight, Agricultural Research Service, U.S. Department of Agriculture.

Figure 6.32 Stinkhorn fungus *(Phallus impudicus).*

source of superstition. Even in temperate zones, these can appear within a day after a rain, a fact that tended to emphasize their otherworldliness. Since the vegetative threads of the fungus can grow out equally in all directions from a single center, it is possible to find an almost perfect circle of mushrooms, a circle with a diameter of up to 30 meters. (Figure 6.33). People believed that these "fairy rings" were the result of the round dances of the wee people on moonlit nights. A Scottish proverb has it, "He wha tills tha fairie green, nae luck again sall hae." Young lassies wishing to improve their complexions need only rub grass dew on their faces, but if the dew came from within a fairy ring, their faces would be disfigured. Dew from within the circle could, however, bewitch a loved one.

Associations of fungi with the supernatural, with evil, rot, decay, and death pervade English literature from at least the fourteenth century (Figure 6.34). The deep woods, the gloomy, wet days following prolonged rainstorms, their rapidity of appearance, "fairy rings," the luminescence of several species as a faint glow in a dark night, and, of course, the convulsions and death resulting from eating a toadstool have reinforced our legacy of general revulsion. Alfred, Lord Tennyson, for example, wrote "As one that smells a foul-fleshed agaric," and Thomas Hardy evoked visions of "an afternoon which had a fungus smell." And remember that in order to consolidate his nefarious schemes against the lost boys in *Peter Pan,* Captain Hook "sat down on one of the enormous forest mushrooms, [for] in Never-Never Land, mushrooms grow to gigantic size."

A good deal of the aversion to mushrooms is related to the inadvertent eating of poisonous species. Not only have many people died by innocently eating them, but some have died by foul deeds most deliberate. When Agrippina, second wife to Emperor Claudius, decided to kill him so that her son, Nero, could become imperial caesar, she is said to have added the juice of *Amanita phalloides*—the death angel—to his favorite dish of cibus deorum *(A. caesarae).* Nero, when safely on the throne, was asked whether this mushroom was really the food of the gods and replied, "Yes, since it led to the deification of my

Figure 6.33 A fairy ring of mushrooms. Courtesy U.S. Department of Agriculture.

Figure 6.34 Poisonous serpents and toadstools. Taken from *Commentarii* by P. Mattioli, edition of 1560.

father." Among the poisonous species none are more deadly than *A. phalloides* and *A. virosa.* Their toxins are complex cyclopeptides which, depending on their chemical structure, will kill in 6–12 hours (the phallotoxins) or 15–20 hours (the amatoxins). If lesions in the liver are produced, mortality can run higher than 60 percent. A classical antidote was a mixture of the stomachs and brains of rabbits . . . eaten raw; this was recommended because rabbits eat *Amanita* with impunity. Even modern treatments are ineffective once the toxins begin their work on the liver and kidneys.

HALLUCINOGENIC MUSHROOMS

Our knowledge of hallucinogenic mushrooms in the Americas began with the Spanish conquest of Central America. Bernardo Sahugun wrote in his *Historia General* that "the Chichimecas possess great knowledge of plants and roots. . . . They discovered harmful mushrooms which intoxicate like wine. There are some small mushrooms which are called teonanacatl . . . only two or three need be eaten. Those that eat them see visions . . . that provoke lust and even sensuousness." The Aztec *teonanacatl* is best translated as "flesh of the gods," and contrary to the Spanish belief that they were orgiastic stimulants, the evidence is overwhelming that they were used in as holy a ritual as the Eucharist and, indeed, they served the same function as does the Body and Blood. King Philip's physician and sometime reporter of New Spain, Francisco Hernandez, piously noted that priests banned the use of the flesh of the gods by the conquered Indians because it contributed to pagan behavior, promoted idolatry, and mocked Holy Communion.

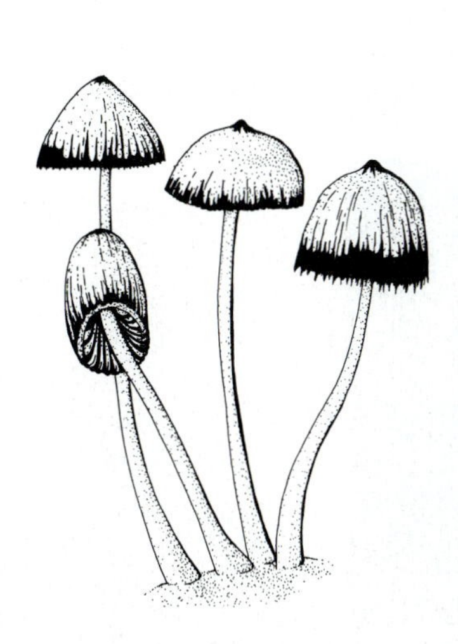

Figure 6.35 *Psilocybe spp.*

Apparently two different groups of fleshy fungi were involved in what has come down to us as mushroom cults. The teonanacatl group includes several genera, especially *Psilocybe* (Figure 6.35), of which 15 distinct species have been named; *P. mexicana* is the most important in

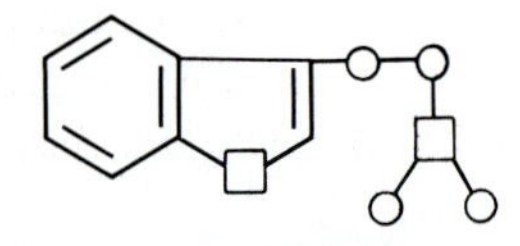

Figure 6.36 Molecular structure of psilocybine, an important alkaloid hallucinogen from mushrooms. Hydrogen atoms and carbon atoms in the rings are not shown. Carbon atoms are represented as circles, nitrogen atoms as squares.

ritual. The genera *Paneolus* (one of the Orient's laughing fungi), *Psathyrella, Conocybe,* and *Stropharia* are used as hallucinogens. All contain psilocybin (Figure 6.36) and other tryptophan derivatives, alkaloids that cause visions, muscular relaxation, hilarity, alterations in perceptions of time and space, and a feeling of total isolation from one's environment. Although they may be used by ordinary people, most of the modern rituals involve the eating of the fungus by the descendants of the Aztec priests, the male or female shamans. Under the influence of the hallucinogen, the shaman will chant the "truth" about health and disease, success or failure, and explain how to remedy the affairs of everyday life. For the Indians of Central America, the psychedelic experience awakens the forces of creation, and partaking of the flesh of the gods is a serious act, not taken lightly. It is not hedonistic but rather deeply religious.

The other hallucinogenic mushrooms of the Incas, Mayans, and Aztecs is *Amanita muscaria,* the fly mushroom (Figure 6.37), so named because water extracts of it were used by North American pioneers to kill flies. Our first record of its use came from reports of Cortez, who observed the mushroom being eaten during the coronation of Montezuma. *Amanita muscaria* contains at least five psychoactive compounds related to those found at the junctions of nerve cells in body and brain and responsible for excitation transmission along nerves. These compounds, in ratios and combinations present in the mushroom, affect the central nervous system, producing mental confusion, abnormal behavior states, visions, and, if too much is taken, convulsions and death by respiratory failure. Indians of Mexico believe that the magic mushroom came from drops of God's sperm, that it grows from the vulva of Earth Mother, that it appears above ground as a small penis that becomes more and more aroused, and that the cap (woman) is penetrated by the stipe-penis (Figure 6.38). In Guatemala, *Amanita* is identified with the thunderbolt and is believed to appear when lightning strikes the earth (Figure 6.39). The sacredness of the mushroom is attested by glyphs seen in the *Madrid Codex,* dating from the middle of the sixteenth century, in

Figure 6.37 The fly agaric *(Amanita muscaria).* Courtesy Dr. Kent H. McKnight, Agricultural Research Service, U.S. Department of Agriculture.

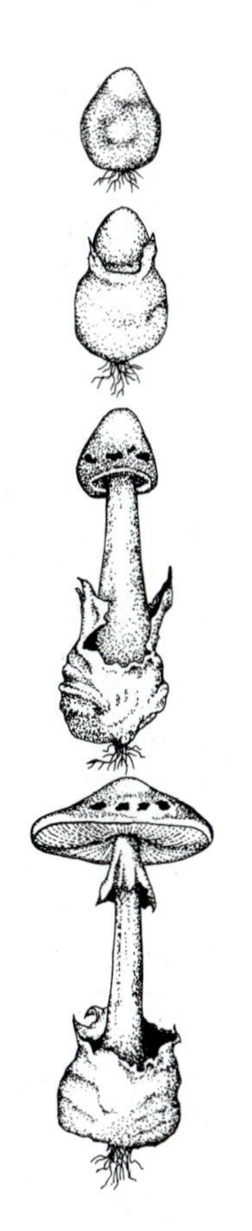

Figure 6.38 Stages in the growth of *Amanita.*

which an animal-like deity is shown presenting the mushroom to a figure seated on a throne (Figure 6.40). Even today, mushroom cults gather the fruiting structure just before dawn, at the time of Earth Mother's new moon.

Mushroom madness is not restricted to "less civilized" portions of the world. A fresco in a Roman Catholic church in Plaincouralt (Indre), France, depicts Adam and Eve on either side of a tree of knowledge that is unequivocally a branched *Amanita muscaria.* The warrior Viking *berserkers* were, according to one authority, under the influence of *Amanita,* as evidenced by their capacity for tremendous feats of strength and their own report of visions and high sexual activity. In western Siberia, among the Finno-Ugrian peoples and among the Lapps, hallucinogenic use of this fungus is widespread. Women of the Koryak tribe in Siberia prepare the mushrooms by chewing them (without swallowing) and then pass the softened pellets to the men, who are allowed to swallow them. The hallucinogenic alkaloids are excreted in urine, which the women are permitted to drink. Their urine, in turn, is taken by the children. Russians refer to these hallucinogenic fungi as *paganki* (little pagans) and believe that they have Powers of the Dark. There are scattered reports in the English literature of the late nineteenth and early twentieth centuries that hallucinogenic properties of *Amanita* were exploited by upper-class opium eaters, and one wonders whether the Rev. C. L. Dodgson of Christ Church College in Oxford was aware of this when, in *Alice's Adventures in Wonderland,* he wrote about our heroine nibbling at a mushroom to change her size (Figure 6.41).

SOMA

When the Aryan people entered India from the north about 3500 years ago, they brought with them a cult involving a plant called *soma.* The *Rig-Veda* (X, 119:7–9) glorified soma in the following couplets:

> Heaven above does not equal one half of me.
> Have I been drinking soma?
> In my glory I have passed beyond earth and sky.
> Have I been drinking soma?
> I will pick up the earth and put it here or there.
> Have I been drinking soma?

Figure 6.39 Guatemalan mushroom stone dated from 3000 B.P.

Botanists and anthropologists have suggested many plants as being soma, but none seem to meet all the criteria found in the ancient poems and legends. The possibility that soma is a drink made by steeping *Amanita muscaria* in milk or wine has been suggested by several people.

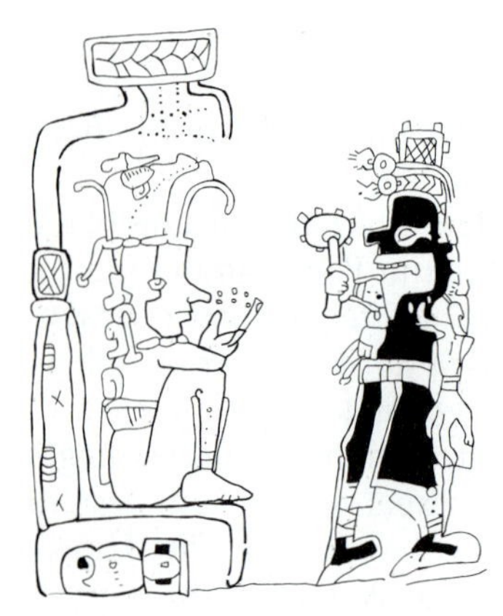

Figure 6.40 Messenger from the Gods offering the sacred *Amanita* to a ruler. Taken from a wall painting in Mexico and copied in the *Madrid Codex.*

They argue that in tribal societies a psychedelic experience is a journey into the realm of the Supreme, a transcendent opportunity to become part of the Supreme. Soma, according to the ancient writings, could send one's soul to the abode of Deity and, equally important, could allow the traveler to return. John Allegro extended this argument in *The Sacred Mushroom and the Cross* and concluded that *Amanita* was a focal point in the Christian tradition. The distinction between the mycophobe and the mycophile may be basically religious, based on acceptance of the divine mushroom concept. These speculations are not generally accepted by other scholars and are most assuredly not shared by organized Western religions.

In a book on hallucinogenic plants, *Flight from Reality,* the late Norman Taylor asked why human beings have searched so avidly for psychedelic plants. Since use of hallucinogens has great religious ritual significance, was religion itself a flight from reality? Admittedly, this is a cold, cruel world we live in, and it was much colder, harder, and crueler in the past. Today we are exhorted to face life, to stand up to reality, to become masters of our fates and captains of our own souls. In our Western European milieu, we condone the use of only one drug, alcohol, as our vehicle to flee reality and have interdicted virtually all others, some with many fewer physical and mental consequences. Did the Aztec noble, the Siberian peasant, and the Vedic priest know something that we don't know?

ADDITIONAL READINGS

Allegro, J. *The Sacred Mushroom and the Cross.* Garden City, N.Y.: Doubleday, 1970.

Brodie, H. J. *Fungi: Delight of Curiosity.* Toronto: University of Toronto Press, 1978.

deRios, M. D. *Hallucinogens: Cross-Cultural Perspectives.* Albuquerque: University of New Mexico Press, 1984.

Findlay, W. P. K. *Fungi in Folklore: Fiction and Fact.* Eureka, Calif.: Mad River Press, 1982.

Gray, W. D. *The Use of Fungi as Food and in Food Processing.* Parts 1 and 2. Boca Raton, Fla.: CRC Press, 1971.

Harner, M. J. *Hallucinogens and Shamanism.* New York: Oxford University Press, 1973.

Ott, J. *Teonanacatl: Hallucinogenic Mushrooms of North America.* Seattle: Madroña, 1978.

Puharich, A. *The Sacred Mushroom.* Garden City, N.Y.: Doubleday, 1970.

Rolfe, R. T., and R. W. Rolfe. *The Romance of the Fungus World.* London: Chapman and Hall, 1925.

Rumack, B. H., and E. Salzman (eds.). *Mushroom Poisoning: Diagnosis and Treatment.* Boca Raton, Fla.: CRC Press, 1978.

Savonius, M. *Mushrooms and Fungi.* London: Octopus Books, 1973.

Scheiffer, H. *Sacred Narcotic Plants.* New York: Macmillan, 1973.

Figure 6.41 Alice speaking to the green caterpillar, who is seated on a mushroom. Taken from the drawing by John Tenniel in the first edition of *Alice's Adventures in Wonderland.*

Singer, R. *Mushrooms and Truffles: Botany, Cultivation and Utilization.* New York: Wiley, 1961.

Smith, A. H., and N. S. Weber. *The Mushroom Hunter's Field Guide.* Ann Arbor: University of Michigan Press, 1980.

Wasson, R. G., G. Cowan, F. Cowan, and W. Rhodes. *Maria Sabina and Her Mazatec Mushroom Vedala.* New York: Harcourt Brace Jovanovich, 1975.

Wasson, R. G. *Divine Mushroom of Immortality.* New York: Harcourt Brace Jovanovich, 1972.

Pot—Grass of the Living Room

The only part of conduct of anyone, for which he is amenable to society is that which concerns others. In the part which merely concerns himself, his independence is of right, absolute. Over himself, over his own body and mind, the individual is sovereign.

JOHN STUART MILL
On Liberty

It has been suggested that hallucinogens permit people to enter the "real" world, closed off from childhood by the many layers of culture that surround us from birth. Hundreds of plants containing mind-altering drugs have been used at some time in human history, but aside from alcohol, none has had as much social impact in North America as the hemp plant, *Cannabis sativa.* Botanically, *Cannabis* is a genus of several species in the Moraceae, a family containing hops used in beer, the mulberry, figs, and the Osage orange. It is a tall annual weed that sometimes reaches four meters in height and can grow in most temperate and tropical climates and in virtually any soil. Cannabis is dioecious, with separate male and female plants that can be distinguished even when young. The male plant tends to be taller and less prone to branching; its leaf form is slightly different from that of the female plant (Figure 6.42). The genus is native to China, although it has been disseminated to all continents.

SOOTHER OF GRIEF

Although the uses of the plant as a fiber and food crop dominated its production in the Orient for a long time (Chapter 9), its medical uses were known and exploited. In a medical book ascribed to the legendary Emperor Shen-Nung in 4737 B.P. (but probably written in the Han Dynasty about 2100 B.P.), ground, dried leaves of cannabis, known as *ma-yo,* were recommended for malaria, beriberi, and constipation, as

Figure 6.42 Hemp plants *(Cannabis sativa)*. Female flowers produce hashish gum or resin (left). Male plants do not form hashish.

an anaesthetic in surgery, and for that disease of old age—absent-mindedness. Shen-Nung noted that its primary medical value was in calming hysterical women. In India and throughout Southeast Asia, cannabis was also used medically, and its capacity to ease anxiety was noted in both Hindu and Buddhist writings, where it was referred to as the "soother of grief." It seems that cannabis entered India about 3500 B.P. from Assyria, since seeds have been found in archeological sites in Iran dated from before any records of Indian use.

It was in India that cannabis was first formally exploited as an hallucinogen, although the ancient Chinese knew that it could be "the delight giver." In India, cannabis had its greatest use among those religious groups whose norms include introspection, meditation, and bodily passivity. Indra, Lord of the World, was believed to drink it, and the plant thus became a gift of the Gods, imbued with mystical properties; those who used it were partaking of the sacred. This attribution of drugs to the Godhead is found in many societies. Certainly, anything that could deaden pain and "remove the noise of the world" had to be holy. To the Hindu peasant, it was "the joy giver, the sky flyer, the heavenly guide, the poor man's heaven," although to the Brahman it could only be used religiously, and even its fabrication into rope (Chapter 9) was forbidden as a sacrilege. Nevertheless, as with other mind-altering drugs, it could not be restricted to magic-religious functions, and it became a product of everyday use.

The refinements accompanying the use of cannabis and much of our basic nomenclature are derived from India. There are three basic grades. The poorest of these is *mj*, consisting of leaves and twigs of both male and female plants. Also called *kif* or *dogga,* it is used in India by the poorest of the poor—it is also the most common grade in North America. *Bhang* is a better grade; its name is derived from a supplication to Siva, the Hindu God who taught that the word should be prayerfully chanted over and over during the period of cultivation to ensure a good harvest. Bhang consists of the upper leaves and cut tops of female plants harvested before the flowers open. The plant material is sun-dried, coarsely ground, and taken either as a decoction with milk or smoked. Better grades in North America are basically bhang, although we have substituted the Portuguese word *marijuana (marihuana),* which came into North America via Mexico in the late nineteenth century. The third grade is *ganja,* composed entirely of the upper leaves of carefully cultivated female plants. In India it is smoked, mixed with wine or milk, or incorporated into candy. The Indian government considers cannabis cultivation a licensed agricultural industry and, until recently, licensing was solely to regulate quality; it was not taxed, for, it was asked, "Who can dare tax a necessity of life?"

A special grade is called *charas,* consisting solely of the unadulterated resin of cultivated female plants at the flowering stage. This resin, which we call *hashish,* is synthesized when the female flowers open. It was collected in many ways; at one time naked men ran through the open fields and the resin stuck to their bodies. Since the odor of sweat offended the nostrils of the elite, the men were later clothed in leather vests and pants. In Persia, female plants were cut down and beaten onto carpets, the rugs were washed, and the dissolved resins concentrated.

The active hallucinogen is delta^1-tetrahydrocannabinol (THC) (Figure 6.43), but several other closely related compounds are also present in resin, and these may interact with THC in the body. THC was first synthesized in 1965 and is of chemical interest because no other hallucinogen has a chemical structure even remotely resembling it. THC is synthesized in special gland cells (Figure 6.44) found on the leaves and in the female flower parts; male plants have much lower concentrations than do female plants. The THC content of stable genetic strains of cannabis, grown under good conditions, will vary from season to season. In spite of subjective judgments, chemical analyses indicate that THC levels are greatest in plants grown under high temperatures and with a long growing season, conditions not usually obtained in dormitory closets or in the colder parts of the United States and Canada. The dramatic, well-publicized burning of a field of cannabis in northern areas is basically an exercise in publicity. It is also fairly well established that the form in which the material is taken is a factor in the response,

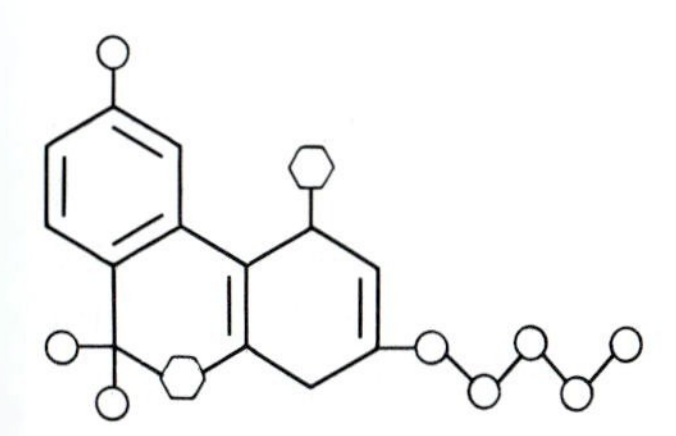

Figure 6.43 Molecular structure of Δ^1trans-tetrahydrocannabinol (THC), the major hallucinogen in *Cannabis*. Hydrogen atoms and carbon atoms in the rings are not shown. Carbon atoms are represented as circles, oxygen atoms as hexagons.

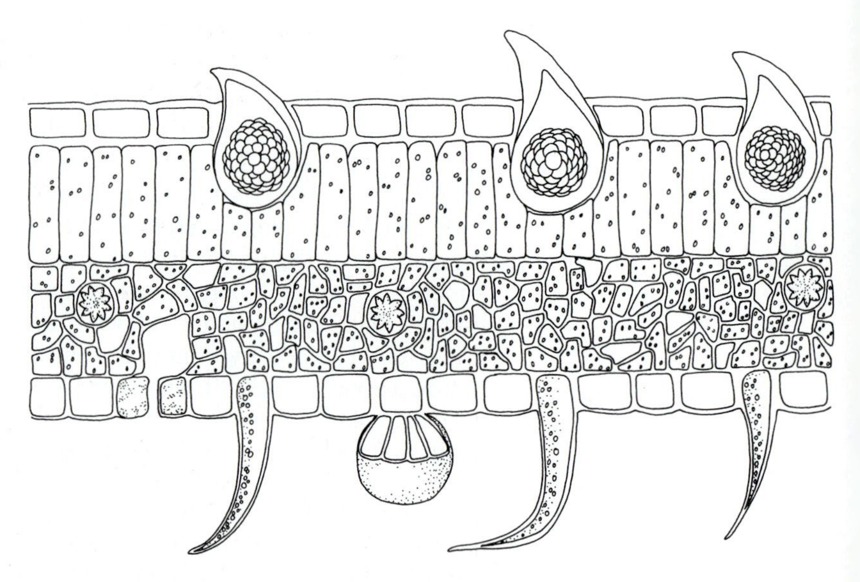

Figure 6.44 Diagram of a *Cannabis* leaf showing structure of the glands that synthesize THC.

even with synthetic THC. Smoking provides the quickest and most long-lasting euphoria, but subjective attitudes, peer-group pressures, and other factors are clearly involved. Clinical, psychological, and pharmacological effects can be found in references cited herein and in many other reputable reports. It is, however, worth noting that use of cannabis does not lead to sexual orgies as detailed in many bad films, to murder, or to other lurid activities lovingly detailed by the communication media.

Examination of cannabis is an excellent way of looking at the flow of a plant through various cultures. Knowledge of its hallucinogenic properties spread from India throughout Asia and Asia Minor about 2000–3000 B.P. Herodotus described its use by the Scythians in 2500 B.P.; they made a woolen tent, burned hemp in a dish, and stood in a circle with their heads under the tent. The ancient Thebans mixed hemp with sweetened wine, and Galen reported that the Greeks baked it into little cakes. Under the influence of Mohammedanism, whose faith proscribed alcohol, cannabis spread throughout the Arab world. The word *hashish* is derived from the name of a Persian prince, Hasan-i-Sabbah, whose private army, the hashishins, were paid off partly with resin; our word *assassins* is derived from these mercenaries whose avowed purpose was to kill all enemies of the faith. The sect was not completely suppressed in Syria and Persia until the thirteenth century. At least part of the legends about the dangers of cannabis relate to this sect, and the story (probably false) that the hashishins committed their ritual murders while under its influence has been cited as reason enough to ban it. It then entered the lands of Africa below the deserts. In the Zambesi valley, piles of fresh leaves are simply thrown on a fire and the celebrants inhale the smoke. When a Bantu tribe, the Baloubas, conquered much of what is now the Congo River Basin, their chief,

Kalamba-Moukenja, wished to unite the various groups. To do so, he developed a cannabis ceremony now common throughout central Africa. Hottentots, Bushmen, and Kaffirs have used cannabis for at least 300 years. In the Middle East and in Africa a variety of reed tubes, pipes, and the famous waterpipe were developed. Preparation of cannabis have been mixed with tobacco, with other hallucinogens, or with alcohol in order to increase physiological response.

HASHISH CANDY

Although it is likely that the hallucinogenic properties were known in Europe during the Middle Ages, we have no good evidence that it was commonly used. The introduction of cannabis to "polite" society came when Napoleon's army brought hashish back from conquered Egypt. In 1844 a private club, the Club de Hachichins, opened in Paris, dispensing hashish in candy or mixed with wine. Its founders included Gautier, Baudelaire, Dumas, and others of the socially elite, literary set. Other clubs sprang up and introduced smoking. At the same time, French medical authorities began recommending it as a calming agent for the hysterical, justifying the recommendation of Emperor Shen-Nung.

Marijuana entered the New World via Mexico when that nation's French rulers introduced its cultivation for hallucinogenic purposes in the nearly ideal hot, dry lands around Mexico City. Its use spread very slowly to the native peoples because the peon preferred native hallucinogenic plants that were sanctified by use dating back to Aztec times—marijuana was good for a mild smoke after a long hard day in the fields or mines. The product entered the United States through the Southwest and by 1860 was introduced to the eastern seaboard via immigrants from the Caribbean Islands. In New York, it was used by the black poor and by the socially chic. *Harper's New Monthly Magazine* reported in 1883 that "there is a large community of hashish smokers in this city . . . and I can take you to a house uptown where hemp is used in every conceivable form and where the lights, sounds, odors and surroundings are all arranged to intensify and enhance the effects." Not unexpectedly, lurid journals and newspapers, like the *Police Gazette,* had a field day (Figure 6.45). In the late nineteenth century, Tilden's Extract of Cannabis was medically prescribed for nervousness and even for lockjaw, and marijuana cigarettes, made by Grimault & Sons, New York, were a prescription item for easing asthma. Marijuana entered the American heartland via the Mississippi River, along with New Orleans jazz, and was popular among the black poor who were then moving to Chicago, Kansas City, and St. Louis. From all these centers, it moved into the major Canadian cities.

Immediately after World War I, as crime and violence seemed to take a quantum leap upward, political and police authorities were quick to blame marijuana rather than to evaluate shifts in population and other

Figure 6.45 "A New York Hasheesh Hell" wherein innocent young women of the upper classes were lured into a life of drugs and shame . . . and heaven alone knows what else. Taken from a cartoon published in the *Police Gazette* of the 1890s.

social changes—it was a convenient scapegoat. In 1937, prodded by the same temperance groups who had outlawed alcohol through the Volstead Act, Congress passed the Federal Marijuana Tax Act, ostensibly to limit and control distribution, but which actually made cultivation or possession a felony punishable by a $2000 fine and up to five years in federal prison. Most states passed their own marijuana acts with penalties about as severe as for narcotics. A Canadian commission, on the other hand, stated, "If we confuse the issue of whether an individual should use a substance with that of whether society should forbid his doing so, we may be laying the foundation for tragic errors in public policy."

Over the past ten years, the use of marijuana has increased at a rate close to 5 percent a year. Much of the dried leaf and most of the hashish are imported, usually from Mexico and the Caribbean, with Jamaica being a prominent producer (Table 6.2). Although about 90 percent of the North American supply is still imported, the years from 1979 to the present have seen a significant increase in the domestic production. The requirement for hot, sunny weather for the synthesis of the active hallucinogens can be met in many areas of the United States, including the southern states, California, and Hawaii. In descending order of production, Oregon, Kentucky, Missouri, Oklahoma, and Washington produce significant amounts of marijuana. Many of the illicit fields are started with a strain called *sinsemilla,* a hybrid type that contains up to ten times as much THC as the types grown in the tropics and in Colombia, long a major source of dried plants. Sinsemilla cannabis will produce highly potent leaves even under poor growing conditions, such as are found in upstate New York, New Hampshire, and Vermont. Garages, attics, and basements are being used, with powerful lights such

Table 6.2 SOURCES OF MARIJUANA

Country	Production	
	Estimated metric tons	Percent of U.S. supply
Colombia	8500	59
Jamaica	2300	13
Mexico	5500	11
United States	1600	9
Other	2000	8

Source: U.S. State Department estimates for 1984.

as are common in baseball fields. It now seems to be almost impossible to interdict the huge volume of both domestic and imported marijuana.

There has been considerable agitation to modify or repeal certain aspects of the laws relating to possession and personal use. Penalties for possession of relatively small amounts, together with evidence that the suspect was not a dealer, have resulted in less severe fines or jail sentences than were imposed a few years ago. Nevertheless, there is a concerted federal effort to punish dealers and to destroy plants as they are found. To a large extent, the North American public is ambivalent about marijuana, partly because its use is so pervasive and partly because it has become so common that presumed bad effects are questioned. No consensus has been reached about the contraindications for use. It is not pharmacologically addictive, although it may be psychologically habit-forming. Certainly there are effects on perception, on reaction time, and on some aspects of behavior. This is seen in the increasing incidence of automobile accidents in which drivers show impaired capacity as a result of marijuana. Obvious reductions in attention span, cognitive ability, and ambition are sufficiently serious to cause concern, if not to justify continued legislation.

The medical evidence linking moderate to heavy use of marijuana with levels of the male sex hormone, testosterone, with damage to sperm production, or with alterations in female reproductive function is scientifically inadequate, the studies are poorly controlled, and statistical confirmation is unsatisfactory. This does not mean that the suggestions of physical damage are incorrect, but they do appear to be premature in the light of available evidence. There is, moreover, an increasing body of information indicating that there are certain medical conditions for which marijuana may be helpful. It appears to be a drug of choice in the treatment of glaucoma, a condition in which the internal eye pressure is elevated. Studies also reveal that THC is effective in combatting the nausea and vomiting produced by chemotherapy used for cancer treatments. Attention is also being given to the antianxiety effects of marijuana, which seem to be due to another compound, cannabadiol, rather than to THC. In any event, there is a growing realization that, because personal use has potential physical and psychological dangers, the public requires legal protection against the known dangers of driving while under the influence of the substance. Like alcohol and driving, pot and driving don't mix.

ADDITIONAL READING

Abel, E. L. *Marihuana: The First Twelve Thousand Years.* New York: Plenum Press, 1980.

Abel, E. L. *Marihuana, Tobacco, Alcohol and Reproduction.* Boca Raton, Fla.: CRC Press, 1983.

Andrews, G., and S. Vinkenog. *The Book of Grass.* New York: Grove Press, 1967.

Mendelson, J. H., A. M. Rossi, and R. E. Meyer. *The Use of Marijuana: A Psychological and Physiological Inquiry.* New York: Plenum Press, 1975.

Nahas, G. *Marijuana—Deceptive Weed.* New York: Raven Press, 1973.

Chapter 7

Medicinal Plants

Herbs, Herbals, and Herbalists

Here begynneth a newe mater, the whiche sheweth and treateth of ye vertues and proprytes of herbs, the whiche is called an herball.

RYCHARD BANCKES
Title page of Banckes's Herball, 1525

Medical archeology has shown that ills to which flesh is heir have been with us since we have made the mistake of climbing down from the trees and ruining our backs by standing upright. Bacterial disease, injuries, and organic ailments have been diagnosed from bones, mummies, and cave paintings. Some of the earliest written records have been of epidemics. Early tribal societies began to develop systems of medicine based on causation by supernatural forces and on treatment by a combination of religion and herbal medications. As settled communities developed, priests of the Great Mother received knowledge from Her of plants which could ease human misery. As written records supplemented and then replaced oral traditions, the healing arts developed.

For us in the Western world, medicine can be dated from ancient Greece. Homer sang of healing roots. The father of medicine, Hippocrates of Cos, organized a college of medicine in 2490 B.P., the Temple of Aesculapius. Hippocrates taught medicine derived in part from the Code of Hammurabi of Babylon (4100 B.P.), the Elbers papyrus of Egypt (2500 B.P.), and records of herbs used by peoples of the Near East. The Temple of Aesculapius continued after the death of the founder, with its philosophy modified by the intellectual ferments provided by Socrates (2470–2399 B.P.), his pupil Plato (2427–2347 B.P.), and Plato's pupil Aristotle (2384–2322 P.B.). Aristotle was first and foremost a naturalist. As tutor to Alexander the Great, Aristotle traveled with military expeditions, gathering herbs and learning of their properties from captured physicians. His star pupil, Theophrastus (2372–2287 B.P.), expanded the master's work in the first book devoted to useful plants. The *Enquiry into Plants* described the plants of the Mediterranean Basin, classified them by use—as herbs, trees, and shrubs (Chapter 1)—and discussed the medicinal values of these plants.

As leadership passed from the Greeks to the Romans, medicine moved west. Superb engineers, the Romans drained swamps, built a water supply system that brought 300,000 gallons of pure water each day to the city, and instituted medical services for all citizens. The Valetudinaria followed the model of the Temple of Aesculapius, and physicians expanded their herbal healing arts to include plants of North Africa, Gaul, and Britain. In 77, Pliny published the first parts of his massive *Natural History,* which eventually included 37 volumes, several of which were on plants and their uses. Pedanius Dioscorides (50–?) was a widely traveled army surgeon whose book, *De Materia Medica,* discussed over 600 medicinal herbs, their preparation, and their use. This book formed, as we shall see, almost the sole basis for clinical medicine for 1500 years, supplemented only by the works of Claudius Galenius (129–199). Galen, a Greek, served as court physician to Marcus Aurelius. He asserted that "disease is contrary to nature and can be overcome by that which is contrary to disease." His several volumes included botanical materia medica, pharmacy, and therapeutics.

As the Roman Empire fragmented and the Dark Ages descended on Europe, observations of nature were replaced by the concept of authority. Dioscorides and Galen, although heathen, were accepted as authorities by the Roman Catholic Church. Thus their writings formed a canon and a dogma, and no person who wished to remain a communicant could do other than follow the directions in these books. From about 200 to 1100, no books on herbal medicine, or on any science, were written in the Western world. Scribes copied Galen and Dioscorides, taking care to avoid introducing any new information.

MEDIEVAL MEDICINE

Figure 7.1 Hound's-tongue *(Cynoglossum officinale).* The rough leaves bear some resemblance to the rough tongue of a dog. Witches' brews containing "tongue of dog" and "adder's tongue" actually contained plant leaves.

In searching for the basis of the disappearance of natural philosophy, historians have suggested several factors. A series of wars followed the dissolution of the Roman Empire. The attendant failure of agriculture due to a scarcity of people to till the land led to recurrent famines and general squalor, and fostered the spread of epidemics that wiped out any residual intellectual values. Infant mortality exceeded 60 percent, and the average life span was less than 30 years. England experienced the Black Death in 664, 672, 679, 683, and 728, while equally frequent and equally deadly pandemics occurred on the continent. In today's terms, many people of Europe were insane—melancholia, depression, alcoholism, and suicide were the norm, and critical evaluation of natural phenomena was impossible. Europeans spent their short lives in abject fear and terror with demons, witches, and evil ones as their constant companions (Figure 7.1).

Disease was a manifestation of evil or of God's displeasure and could be cured by prayer, exorcism, visits to shrines, flagellation, and recourse to curse lifters, who used herbs with virtues known to them through oral records or through the writings of Galen and Dioscorides. Intercession to heaven was formally ascribed to specific saints: leg diseases to St. John, cancer to St. Giles, toothaches to St. Appolonia, and so on, and their assigned plants formed the medical treatments. The hierarchy of the Roman Catholic Church, composed of men subjected to the same stresses as were the rest of the population, echoed the same superstitions. Knowledge was useful insofar as it exemplified doctrine and dogma. Medicine, too, had to serve only God, and His priests assumed the duties of attending the sick. Herbs for healing were cultivated in monastery gardens, and hospitals were adjuncts of nunneries. Innovation or even observation of the natural world had no place in dogma, and, besides, death was a release from misery and offered the possibility of salvation and eternal bliss.

As conditions worsened in the eighth and ninth centuries, people of Europe developed a folk medicine that was a complex of garbled Dioscorides, ill-translated Galen, ancient superstition, and religious fervor stirred together and imbibed with a prayer. Since few could read or write and those who could wisely refrained from challenging authority, herbal remedies were passed orally from generation to generation, becoming altered in the process.

DOMINANCE OF THE ARABS

The responsibility of preserving intellectual thought passed to the Saracens. Medical practice was the province of Arabian physicians such as Rhazas (919—965), Avicenna (981—1037), Avenzoar (1113—1162), and their Jewish colleagues like Moses Maimonides (1135—1204). Greek and Roman medical tomes were translated into Aramaic, Syriac, or Persian, which permitted the introduction of some Indian herbal lore to enter the West. Commentaries on Galen and Dioscorides written by later physicians included descriptions of drugs of North Africa, Byzantium, and other lands conquered by the followers of Mohammed. A medical college was established in Alexandria in the early eighth century, and by the tenth century, medicine was centered in Salerno, a former Greek and then Roman health resort. Alexandria and Salerno attracted heretic Nestorian Christians and dispersed Jewish physicians, who then traveled throughout Europe serving the medical needs of the noble courts. These physicians were known throughout Europe; Chaucer noted the debt owed them in the "Prologue" to the *Canterbury Tales.* Among the first herbals to appear was the hand-written *Paradise* [Garden] *of Wisdom* in 850 and the *Knowledge of Drugs* in 1025. An important work was the *Minhaj al-dukkan* (Handbook of Drugs), published by an Arabian Jew, Kòhên al-attar (Cohen the druggist), in 1259. Precise directions on collection and storage of drug plants, on methods for distilling, compounding and testing, on weights and measures, and on the values of each compound were included. Arabian medicine was deliberately going beyond the mere parroting of the wisdom of Dioscorides.

Figure 7.2 "Nymfea," the lotus *(Nymphaea).* Taken from a ninth-century manuscript copy of *Herbarium Apuleii Platonici.*

During this period, few European herbals appeared, most of them copies of Galen and Dioscorides, although some did "comment" by adding local lore. The *Leechbook of Bald* was an early Anglo-Saxon manuscript. *De Morbis Acertis* by Caelius Aurelianus and the *Herbarium Apuleii Platonici* were of immense value and utility (Figure 7.2).

Albertus Magnus of Cologne (1193–1280) published *De Virtutibus Herbarium* (Healing Values of Herbs), and Hildegard von Bingen, a Benedictine abbess, wrote *Liber Simplicus Medicina,* one of the first books to discuss the role of the nursing sister. As priests and nuns became less worldly, medicine passed to uneducated laypersons and to the barbers who became surgeons. Gerald of Cremona (1114–1187) wrote *Canon Medicinae,* which translated into Latin the Saracen medical writings captured when Toledo was freed from Moorish domination. In 1317 at Oxford, John of Gaddesdon published *Rosa Medicinae* which incorporated the Arabic information of Gerald into Dioscorides's treatments plus, for possibly the first time in Europe, his own clinical experiences: "For nothing is set down here but what has been proved by personal experience either of myself or of others and I, John of Gaddesdon, have compiled the whole in the seventh year of my lecture." The impact of this book on the medical profession was immediate; members noted that John was not subjected to the inquisition for his writings and concluded that they, too, could speak more freely—albeit circumspectly.

BOTANICAL RENAISSANCE

The invention of the printing press in Germany in 1448 served both God and Man. The Gutenberg Bible permitted reading of the Scriptures by people other than the priests, and the printing of John of Gaddesdon's book ushered in the era of the herbal. Father Barthalomaeus Anglicus published *Liber de Proprietibus Regum* in 1470, and Cunrat published *Das Püch der Natur* in 1475, both volumes containing more clinical herbal medicine than did Galen. The *Herbarius zu Teutsch* of 1485 was a mixed Germanic-Latin copy of the *Latin Herbal of 1484* in which Galen and Dioscorides were accorded second place to the lore of contemporary physicians (Figure 7.3). The Latin *Ortus* [Hortus] *Sanitatis* of 1491 went even further than its predecessors in reducing the importance of the ancients. Within a few years, *Le Grand Herbier* was published in France, demonstrating that the vernacular—the language of the people—was replacing the Latin that the Church and its universities had insisted on for centuries.

Figure 7.3 "Nenufar," the water lily *(Nuphar)*. Taken from a revision of the *Latin Herbal,* published in 1499.

The early sixteenth century saw the publication of two English books that specifically used the name *herbal* in their titles as an indication that the medicines were almost solely derived from plants. Rychard Banckes, apparently a printer of London and not a medical specialist, published *Banckes's Herball* in 1525. The book proved so popular that successive editions and translations appeared all over Europe. Its provenance is unknown. *The Grete Herball* was published the following year by Peter Treveris (Figure 7.4). Treveris prepared an excellent translation of *Le Grand Herbier* of France that was a version of the *Herbarius zu Teutsch,* but Treveris included a number of British

Figure 7.4 "Nenufar," the water lily *(Nuphar)*. Taken from *The Grete Herball* by Peter Treveris, published in London in 1526.

remedies which are still in use: horehound for coughs, henbane as a narcotic, and rosewater and glycerin for soothing the skin. The numerous remedies for melancholy give us some idea of the unhappy state of mind of many people in the early sixteenth century. The book is very medieval; remarkable powers were ascribed to magic stones, to the unicorn horn, to "powdre of perles," and to spells and incantations, but great weight was also placed on the efficacy of "wedys of ye feldye." Treveris, deeply concerned that the common people were at the mercy of doctors and apothecaries and the ignorance of "ye olde women," stated: "Wherefore brotherly love compelleth me to wryte thrugh gyftes of ye Holy Ghost shewynge and enformynge how man may be holpen with grene herbes."

Succeeding years saw publication of herbals in many languages. In order to transmit information about useful plants, the herbals began to include plants of food or of horticultural value. Otto Brunfels published *Herbarium Vivae Eicones* (Kinds of Living Plants) in 1536 with masterly woodcuts by an artist who worked from living material and did not conventionalize or stylize the drawings. With this as a model, Hieronymus Bock published the *Kreütterbock* in vernacular, even ribald German and included plants used by the people of the Palantine. Leonard Fuchs (for which the genus *Fuchsia* is named) was a doctor of arts and professor of medicine at Tübingen. His *Historia Stirpium* of 1542 was in Latin, but he arranged the plants alphabetically by common name and discussed over 400 German and 100 foreign plants including, for the first time, New World plants like Indian corn and "the Grete pompion." Pierandrea Mattioli published *Commentarii in Sex Libros Pedacii Dioscorides* (A Six Volume Commentary on the Works of Dioscorides) with magnificant illustrations. The publication went through several editions, selling 30,000 copies at ten ducats each, due in large part to the fact that the commentary was the major part of the book and Dioscorides's writings were subtly downgraded. William Turner's *A New Herball* came out in 1551 with plant names in "Greke, Latin, Englishe, Duche and French wyth ye cummune names that herbaries and apotecaries use." Since England was almost completely removed from the authority of the Roman Catholic Church, Turner could afford deliberately to debunk Dioscorides and could also object to the superstitious witchcraft that had infiltrated clericalism.

The latter part of the sixteenth century was made botanically notable by the publication of several herbals that tremendously influenced books published in the seventeenth century. Mattias de l'Obel (for whom the genus *Lobelia* was named) received his medical education at Montpellier, but settled in England, where he was a go-between for the now-heretic English and the Catholic botanists of Europe. His *Stirpium Adversaria Nova* of 1570 is noted for excellent woodcuts. De l'Obel modeled his study on the 1560 edition of Konrad Gesner's *Historia Planatarum,* an encyclopedia of all known plants, much less medically oriented than any previous herbal. The Portuguese and Spanish herbals of Garcia de Orta, who wrote about the plants of India, and of Nicolas Monardes, whose 1569 book was translated into English as *Joyfull Newes Out of the Newe Founde World,* reflected the success of Latin countries in exploring in both east and west. Sunflowers, tobacco, chili peppers, and other plants were described and illustrated, and the European scientific community avidly bought and studied these books. One other sixteenth century herbal is a landmark in botany. Charles d'Ecluse (Clusius) was a Montepellier-educated physician who served as court physician at Vienna and as professor of medicine at Leyden. He obtained New World plants from Sir Francis Drake and, among other books, published *Rariorum Plantarum Historia* in 1576, when he was 76 years old (Figure 7.5). Not only were many American plants illustrated, but Clusius also included the first comprehensive overview of the fungi.

During this period, herbals were the only permanent record that physicians and natural historians had of plants. Animals could be kept in royal zoological parks and their skins and skeletons stored for examination. Medicinal plants were grown in "physick gardens" (gardens of medicinal plants), but they seemed impossible to preserve in any condition so that the dried specimen could be identified or, even more important, could be used to determine whether another dried plant was the same. Herbaria, collections of dried plant specimens, did not exist until an Italian physician and professor of botanical medicine at Bologna, Luca Ghini, conceived the idea about 1550 that plants could be dried between blotters under mild pressure so that their form was retained. By the end of the sixteenth century, herbaria were developing in most of the universities of Europe, usually in the medical colleges, where they were used to compare plants received as possible drug sources with plants known to contain effective medicines. As botanical science developed independently of medicine, herbaria were located in botanical gardens and in departments of natural science.

These early herbaria are exceedingly valuable, for they contain specimens which were used to name a plant for the first time—these are the "type specimens." World-famous herbaria, containing literally millions of herbarium sheets, are found at the Royal Botanical Garden at Kew in Surrey, the Gray Herbarium at Harvard University, the New York Botanical Garden, and in several other centers throughout the world. Each specimen is labeled with the name of the collector, the date collected, where collected, the taxonomic diagnosis (family, genus, species), and information on the ecology of the location, thus providing a wealth of information for all fields of botany.

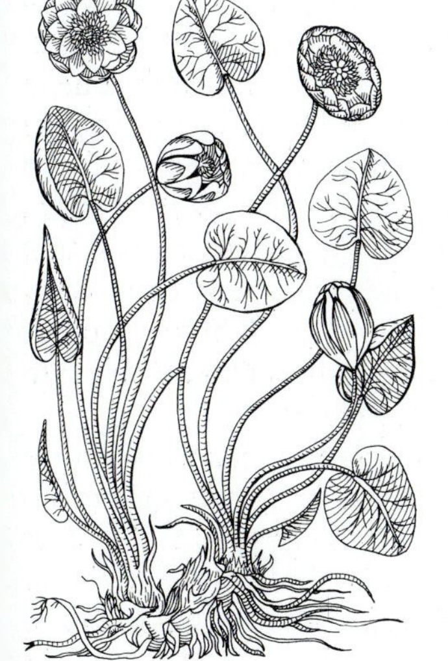

Figure 7.5 Egyptian lotus *(Nyphaea alba).* Taken from *Rariorum Plantarum Historia* by Carolus Clusius, published in Antwerp in 1576.

THE AGE OF HERBALS

The early seventeenth century has been called the age of herbals because of the numbers of important publications and the impact of these books on both scientific thought and medicine. Certainly among the more valuable was *The Herball or Generall Historie of Plantes,* published in 1596 by John Gerard. Gerard (1545-1612) was superintendent of the estate gardens of Lord Burghley, Master in the Company of Barber Surgeons, and an apothecary with his own physick garden. The text and many of the illustrations were taken from 1554 *Crüyderboeck* of Rembert Dodoens, but Gerard so amended and commented on Dodoens that his herbal was essentially a new book.

Gerard was a man of his times. He believed in trees that bore geese and others that had barnacles as fruit. Nevertheless, he took a firm stand against witchcraft and magic in herbal medicine, intelligently discussed plants from the Americas, and attempted to standarize the prescriptions then in use. To a great extent, Gerard's fame—and he was very famous—rested not on *his* herbal but on the revised editions by Thomas Johnson published in 1633, 21 years after Gerard's death (Figure 7.6). Johnson was an herbalist, a Free Brother of the Worshipful Society of Apothecaries, and a doctor of physick. The second edition had over 2500 woodblock pictures made in France, and these illustrations, the accuracy of the plant descriptions, and a clearly written text caused the book to be a financial success.

Two other herbalists are worthy of mention. John Parkinson wrote two books, the first *Paradisi in Sole: Paradisus Terrestris* appeared in 1629. It was a strange book for the time. Not only was at least half of it devoted to gardening, but the herbal portion contained a religious approach to even the names of plants. Parkinson assumed that Adam had received from God knowledge "of all naturall things" and that all

Figure 7.6 Title page of *The Herball or Generall Historie of Plantes.* This is the revision by Thomas Johnson of John Gerard's herbal. Courtesy Library, New York Botanical Garden.

plant names "descended to Noah afterwards." Parkinson's second book, *The Theatre of Plants or An Herball of Large Extent,* published in 1640, described almost 40,000 plants. This was probably the first book that could properly be called a *flora*—a systematic description of plants. Only in the nineteenth century did floras become, out of necessity, restricted to descriptions of the plants of a circumscribed region—for example, flora of Saskatchewan or plants of Thailand.

Nicholas Culpepper's *The English Physician* of 1652 is notable for its brusque, dogmatic, and combative tone, the author's pen dripped in vitriol when discussing the medical profession and anyone else with whom he disagreed. Culpepper was interested in bringing medical treatments out of the hands of doctors and directly into the control of the poor. He believed, not incorrectly, that the physicians of the day were more interested in financial reward and social prestige than in healing. Unfortunately, Culpepper's education in medicine consisted of having been apprenticed to a London apothecary, and the book is a mishmash of herbal medicine, a reversion to the almost discredited Doctrine of Signatures (Chapter 5), plus the addition of a large amount of astrological medical superstition derived apparently from Paracelcus. An example of Culpepper's medicine can be seen in his discussion of the burdock *(Arctium lappa):*

> Venus challengeth this herbe for her own, and by its leaf or seed you may draw the womb which way you please, either upwards by applying it to the crown of the head, in case it falls out; or downward in case of fits of the mother by applying it to the soles of the feet; or if you would stay it in its place, apply it to the navel; and that it is a good way to stay the child in it.

Independently of trying to figure out the various antecedents of "it," it seems that this plant, applied externally, can guard against miscarriage, can induce abortion, and can encourage conception.

BOTANICAL ART

For herbals to be of practical use, illustrations of the plants discussed were necessary. Consider the description of the Venus's-hair fern *(Adiantum capillus-veneris)* given in *Banckes's Herball* of 1525:

> This herbe is called maydenheere or waterworte. This herbe hath leves lyke ferns, but ye leves be smaller, and it groweth on walles and stones and in ye mudde.

Contrast this with the written description of the same plant in the eighth edition of *Gray's Manual,* published in 1950:

> Fronds 1–5 cm long with a continuous main rachis, often pendulous from a slender horizontal chaffy rhizome, the polished dark stipes 0.3–3.0 cm long; the blade ovate-lanceolate, 2–3 pinnate at base.

Clearly, once a vocabulary of gross plant structure had been established, there was no difficulty in writing a precise identification of the plant. But botany in the sixteenth century, and for more than a century thereafter, was not sufficiently well advanced to provide accurate descriptions and the writers of herbals turned to illustrations. In the early herbals, illustrations were so crude that identification was almost impossible. One problem was the lack of an artistic or esthetic standard by which to compare the drawing. In addition, techniques necessary to make good drawings suitable for incorporation into books were still being developed. Early illustrations were made as crude scratchings on soft copper plates. These were little better than the verbal descriptions that accompanied the drawings (see Figure 7.4). The concept of naturalism in botanical art that developed in the sixteenth century was foreshadowed by the work of the Flemish artists, including Hans Memling, the brothers van Eyck, and especially Hugo van der Goes. Van der Goes's Portinari Alterpiece of about 1450 provided a wonderful model for delineating plants that were easily identified. Dürer's *Das Grosse Rasenstück* (A Piece of Turf) of 1503 provided a ground-level view of grasses and flowers that were drawn with accuracy, naturalism, and sensitivity. Jan Bruegel (1568–1625) expanded on the concept that accurate botanical detail could go hand in hand with aesthetic considerations. Somewhat earlier, writers of herbals, like Leonhard Fuchs, began to approach these ideals, but he and his illustrators were frustrated by the available materials on which to prepare drawings for publication.

The solution was found in the woodblock. Invented in China, it had been used extensively in both China and Japan for a thousand years before it was reinvented and introduced into Europe. The carved woodblock allowed the flexibility and the precision of line that the copper plate did not. The woodblock was sufficiently rugged so that it was not destroyed by the printing presses, and the blocks could be stored and reused when the books were reprinted or when the illustration was wanted for another book. The audience for such books included not only the botanist and the medical practitioner. The volumes were avidly collected by the wealthy, who already had been educated by painters. The books were also purchased by the growing numbers of universities whose libraries were supported by nobility and by those with both wealth and artistic appreciation.

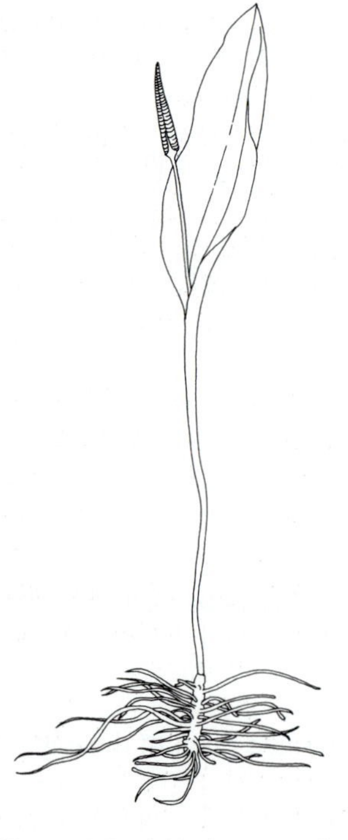

Figure 7.7 Adder's-tongue fern *(Ophioglossum vulgatum)*. Taken from *Historia Stirpium* by Leonhard Fuchs, published in Basel in 1542.

Woodblocks for herbals were made by people who were themselves artists. Hans Weiditz illustrated Brunfels's *Herbarium Vivae Eicones.* The team of Albrecht Meyer (artist), Heinrich Füllmaurer (transcriber of the drawing to a woodblock), and Viet Speckle (block cutter) was responsible for the illustrations in Leonhard Fuchs' *Historia Stirpium* of 1542 (Figure 7.7).

By the end of the sixteenth century, the woodcut was supplanted by line engravings and etchings, techniques previously developed by Dürer and Rembrandt. *Hortus Floridus,* published in 1617 by Crispaen van den Passe, shows the degree to which accuracy and beauty could be combined (Figure 7.8). Work such as this established a tradition and a standard of excellence equaled, but not surpassed, by water colorists who dominated botanical art in the eighteenth and nineteenth centuries. Even color photography has rarely come up to these standards and, in most instances, is not as botanically useful as are well-executed drawings.

Figure 7.8 Star-of-Bethlehem *(Ornithogalum umbellatum)*. Taken from *Hortus Floridus* by Crispaen van de Passe, published in Arnheim in 1617.

BEGINNINGS OF MODERN MEDICINE

The European herbals were almost passé by the middle of the seventeenth century. One of note was William Coles's *The Art of Simpling,* published in 1657. This book has as its subtitle *An Introduction to the Knowledge and Gathering of Plants* and was the forerunner of the modern pharmacopoeia. In clear, unambiguous language, Coles told his readers which herbs were medicinal and how they were to be gathered, preserved, compounded, and administered. There was emphasis on doctrine-of-signature plants, but Coles firmly rejected Culpepper's astrological medicine. Johnson's revision of Gerard, Culpepper, and Coles formed the basis for medical practice throughout Europe, and copies were taken by colonists to North America and Australia.

Although some of Europe's physicians were formally educated at schools of medicine, many obtained what little training they had through the apprentice system, and many more simply set themselves up as doctors. Henry VII signed a statue dealing with this matter that stated: "The moste persones of the said crafte of surgeons have small cunnynge," and the personal physician of Elizabeth I noted that "medicine was in the hands of tinkers, horse-gelders, rogues, rat-catchers, idiots, bawds, witches, sow-gelders and proctors of spittle houses." Britain organized the Royal College of Surgeons to supervise examination and licensing of physicians. Women were forbidden to become members because "they partly use sorcery and witchcraft." The hospitals, originally operated by religious orders, were secularized.

Although Theophrastus had a shop in which he sold dried herbs—apparently the first apothecary or drugstore—subsequent herbalists and physicians either gathered their own plants or grew them in private physick gardens. In the late Middle Ages and even through the eighteenth century, a brisk trade in home-remedy plants was maintained by hawkers of herbs who set up booths at markets and fairs, extolling the virtues of their "simples," dried animals, and magic stones to local people who could not afford the services of physicians. They would, for a small additional fee, give the purchaser the proper incantation to be used with the potion. James I, son of Mary Stuart of Scotland, ordered that the selling of herbs should be restricted to apothecaries because "grocers are merchants, but the apothecary's trade is a mystery [an art or craft]." The Worshipful Society of Apothecaries was formed in 1617 and, to protect its reputation and financial position, distinguished between herbalists, defined as gatherers of plants, and apothecaries, who compounded and sold herbal remedies. The society was faced with several serious problems. Medicinal plants were arriving in London from all over the world. Samuel Pepys could stroll to the docks and see sarsaparilla roots from Jamaica, sassafras leaves from the Massachusetts Bay Colony, senna and myrrh shipped home by the East India Company, and rhubarb from China. There was, however, no good way to insure the purity or the potency of the herbs. As alchemy evolved into physics and chemistry, the apothecaries began to develop crude methods of analysis.

Apothecaries also recognized that the drugs from their plants varied from year to year, and, in an effort to secure standardized material, they started to grow their own—a return to the physick garden concept developed in the Middle Ages, when medicinal plants were grown on hospital land. Italy took the lead; the first secular physick garden was in Pisa in 1543, with plants arranged according to their assumed medicinal properties. Padua's medical school followed suit in 1545 and had the first professorship of materia medica and director of the physick garden. Paris, Rome, Heidelberg, Leyden, and Montpellier had physick gardens in 1590. In England, Oxford apothecaries purchased land across High Street from Magdalen College to develop a garden, while the Worshipful Society of Apothecaries organized the famous Chelsea Garden along the Thames River in 1673. One of the first physick gardens in North America was that of Dr. John Bartram of Philadelphia, whose garden was planted in 1782. The most famous American garden was established in 1801 by Dr. David Hosack of the Columbia University College of Physicians and Surgeons in New York. By 1890, the garden was completely surrounded by the growing city, and Columbia University rented the land for the eventual construction of Rockefeller Center. The garden was moved up to the wilds of the Bronx, where it became the New York Botanical Garden.

Another problem recognized by seventeenth-century apothecaries was the great variation in the formulation of prescriptions. Each person learned a different recipe, and every home compounder used the one passed down from his or her ancestors. As early as 1542, the guild of barber surgeons in Nüremberg prepared a volume which it called a pharmacopoeia; it gave direction for mixing herbal medicines. Other such books appeared during the sixteenth century to supplement the herbals. It was at this time that the apothecaries reinstituted the practice of marking each prescription with the sign of Jupiter, ♃, to indicate that—by this ancient oath—they had neither adulterated the medicine nor included any poisons. This has come down to us as ℞. In 1618, England's apothecaries produced the London Pharmacopoeia, in which, in addition to herbal remedies, included medical preparations of cockscombs, woodlice, and the hair balls from goat stomachs. Pharmacopoeias of many countries have gone through numerous editions, adding new remedies, removing others which proved ineffectual, and modifying directions for gathering, storing, and compounding. Chairs of pharmacognosy, materia medica, and medicinal chemistry were set up at medical schools. Boards to examine and license apothecaries were all part of the medical scene by the late eighteenth century. The era of modern pharmacy had dawned.

CHINESE MEDICINE

The civilizations of China and India were flourishing when only modestly sophisticated cultures were developing in Europe. Not unexpectedly, writings on medicinal plants and the aesthetics of vegetation were numerous. The legendary emperor Shen Nung discussed medicinal herbs in his works—which were probably written 2500 years B.P. instead of their traditional date of 3500 B.P. By 150, the standard Chinese study of medicine, *The Essentials of the Golden Cabinet,* was available; it was followed in 290 by the *Record of the Investigation of Things. Remedies for Emergencies*" appeared in 340, and *Records of Famous Physicians* was published in 510. The *Chiu Huang Pen Tsau* (The Herbal for the Prevention of Famine) was written by Prince Chou Ting Wang in 1406, antedating by 70 years the *Book of Conrad of Mengenberg* on the same topic. Li Shih-chen published, in 1596, the *Pen-tsao Kang-mu,* an important record and herbal of medicinal plants, their storage, and their uses. This book had a profound effect on Oriental medicine, because it summarized previous works on the topic and provided notes on the effectiveness of herbs for specific diseases that were described in clinical detail understandable even today. By 500, woodblock illustrations were used. By 1200, the blocks were of a quality unmatched in the West for another 400 years.

Western medicine entered China with the Jesuits, but found little favor with the Chinese court for a number of years. The sixteenth-century medicine of Europe was no more effective than the mixture of herbal medicine, acupuncture, and other traditional techniques. Gradually, traditional medicine began to merge with Western practices. The two types are now used together in China; medical students receive instruction in both.

ADDITIONAL READINGS

Arber, A. *Herbals: Their Origin and Evolution.* New York: Macmillan (Hafner Press), 1970.

Duke, J. A. *Medicinal Plants of the Bible.* New York: Trado-Medic Books, 1983.

Duke, J. A. *Handbook of Medicinal Herbs.* Boca Raton, Fla.: CRC Press, 1985.

Gerard, J. *The Herball or General Historie of Plantes.* Edition of 1633, revised by Thomas Johnson. Reprinted Edition. New York: Dover, 1975.

Griggs, B. *Green Pharmacy: A History of Herbal Medicine.* New York: Viking, 1981.

Glimm-Lacy, J., and P. B. Kaufman. *Botany Illustrated.* New York: Van Nostrand Reinhold, 1984.

Lehner, E. *Folklore and Odysseys of Food and Medicinal Plants.* New York: Farrar, Straus & Giroux, 1973.

Talbot, C. H. *Medicine in Medieval England.* New York: Elsevier, 1969.

Trease, G. E. *Pharmacy in History.* London: Brilliere, Tindall & Cox, 1964.

U.S. Public Health Service. *Bibliography of the History of Medicine: A Continuing Series of Reports.* Washington: Superintendent of Documents, Government Printing Office.

Wheelwright, E. G. *Medicinal Plants and Their History.* New York: Dover, 1974.

Nostrums and Cures

Purge me with hyssop and I shall be clean.

Psalms 51:7

When the settlers of North America reached Jamestown, Plymouth, or French Canada, they brought with them the herbal remedies common to their home countries. Some were effective and others were ineffective—the nostrums of modern times. They also brought their valued copies of Gerard, Parkinson, Culpepper, and Coles, although they realized that plants listed in these books did not grow in the new country. Undaunted, they selected from the wilderness those plants which seemed to match the ones pictured in their herbals and chose others by the Doctrine of Signatures. They also learned about plants from the Indians. Although the plants were sometimes very different from those they knew, they took comfort from the fact that the concepts underlying Indian medicine were the same as in Europe.

AMERICAN INDIAN MEDICINE

Indian medicine was closely allied to religious belief, and the Doctrine of Signatures was as well developed in North America as it was in Europe, Asia, or Africa. Yellow plants were good for jaundice, red ones were used when blood was involved, and threadlike roots were indicated for intestinal worms. Belief that illness was caused by improper behavior, by failure to honor and placate a superior Being, or by inadequate attention to sacred objects or spirits led to psychosomatic guilt that could be cured by exorcism, by analysis of dreams, and by hypnotic dancing, drumming, and chanting. Tranquilizers, hallucinogenic substances, and foul-tasting or disgusting medicines to induce cleansing vomiting were used by most tribes, and white settlers were gratified to observe that Indians believed as firmly as they did in purges and enemas. Settlers remarked on how healthy the Indians were, not having the biological sophistication to realize that many tribes practiced euthanasia, and that it was a rare individual who could boast of having seen 700 moons. The full spectrum of disease and disability seen in any diverse population was found among the Indians, and lacking immunity to European communicable diseases, they were easy prey to smallpox, scarlet fever, tuberculosis, and venereal diseases introduced by the white man.

Nevertheless, Indians used many plants which were effective and which have been adopted into modern medical practice. Early Spanish explorers of the West Coast found Indians using bark of a small tree *(Rhamnus purshiana)* as a laxative. Rediscovered in 1895 by Lewis and Clark, *Cascara sagrada* is still in the U.S. Pharmacopoeia. Jalapa *(Ipomoea purga)* from Mexico, the common May apple *(Podophyllum peltatum)*, and senna *(Cassia spp.)* were effective cathartics. Salicylic acid was named for the willow *(Salix spp.)*, from which it was first isolated, and Indians chewed willow bark to relieve headaches and ease sore muscles; aspirin is a derivative of salicylic acid.

COLONIAL MEDICINE

There were few physicians in the colonies, and most of them were barber surgeons skilled in bloodletting, bone setting, and the use of the silver tonguescraper. Herbal medicine was in the hands of goodwives, and cures were in the hands of God. An exception was Dr. John Josselyn, who, armed with the 1633 edition of Gerard, dosed the Massachusetts colonists and the farmers of the outlying districts. In 1672, Josselyn wrote *New England's Rarities,* in which he compared English herbs with their North American counterparts. The Virginia and Charlestown colonies had herbalists who relied more on Culpepper than Gerard, Parkinson, or Coles. Lawrence Hammond, for example, recommended for the *mergrums* (migraine headache):

> Mugwort and Sage a handfull each, Calomel and Gentian a good quantity. Boyle it in Honey and apply it behind and on both sides of ye head very warm and in 3 or 4 times it will take it quite away.

In the late seventeenth century, these remedies were supplemented with *The New Jewell of Health* by George Baker, "wherein is contayned the moste excellent secretes of Physick and Philosophy."

In the forefront of medical botanical work was a group of men, all physicians, who used gathered herbs and who started botanical physick gardens of their own. The lieutenant governor of New York, Dr.

Calwalader Colden, introduced, through his daughter, the Linnean system into the colonies. Dr. John Clayton (for whom the spring beauty, *Claytonia caroliniana,* was named by his friend André Michaux), Dr. Alexander Garden (the genus *Gardenia),* and Dr. Michael Sarrazin *(Sarracenia purpurea*—the pitcher plant) were active physicians in the early eighteenth century and were equally active botanists who collected plants for European herbaria and physick gardens. Dr. Caspar Wistar (the genus *Wisteria)* was president of the American Philosophical Society and one of the founders of the medical college at Philadelphia. Dr. John Bartram of Philadelphia was appointed Plant Collector to His Majesty George III at £50 per annum to supply plants to the Royal Botanical Garden at Kew. Bartram botanized all along the Eastern Seaboard, from the dangerous Mohawk country around Lake George in New York down through swamps of the Carolinas, and exchanged pressed plants and fresh material with his friends and professional acquaintances in Europe. This tradition of the physician-botanist was maintained throughout the nineteenth century. John Torrey, whose flora of the eastern United States is classic, was professor of medical botany at the Columbia University College of Physicians and Surgeons. His star pupil and later collaborator was Asa Gray, who, after receiving a medical degree, was called to Harvard University as professor of botany and director of the Harvard herbaria and the botanical gardens.

As Americans moved west after the Civil War, settlers were very attracted to Indian medicines. Sweatings to rid the body of "poisons," purgings, and the use of strong-tasting mixtures appealed to the people. The stronger-smelling and the more vile-tasting the concoction, the better—some medical historians have called the 1860–1890 period the age of heroic cures. Castor oil was one favorite. So were sulfur and molasses tonics or mustard plasters capable of severely burning the skin. Many nostrums were guaranteed to cure after one spoonful—no normal person could stand any more. Children went to school wearing around their necks cloth bags containing powerful-smelling herbs which, their mothers insisted, would keep away "germs." Most frontier doctors received their medical degrees by correspondence from diploma mills and some, who could not scrape up the $200–$300 for a degree, called themselves naturopaths. Many credited their skills to secrets obtained while they were captives of the Apache or the Sioux. The settlers understood the medicine the Indians were using. The plants were familiar, and appeals to the wisdom of the "noble savage," the "unspoiled creatures of nature" (who had been pushed onto reservations and were then imbued with a romantic aura) struck responsive chords in the minds of people who had moved west to escape the educated authority of the East. Samuel Thompson, a New Hampshire farmer, promoted the idea that Indian healers discovered herbs and natural remedies far superior to the unknown materials included in prescriptions written in Latin. Quacks and self-appointed saviors almost took over the health care of rural North America.

MEDICAL HUCKSTERS

Although originally based on herbal medicine derived from Gerard et al., "Indian cures" became big business. Medicine shows, featuring "real" Apache, Sioux, or Blackfoot chiefs in full costume, were among the few diversions in small towns. Literally drumming up trade, the medicine-show "doctors" would come out and talk about health and then tell the audience that they had just the thing for what ailed them. Kickapoo Indian Sagwa Remedy, Ka-Ton-Ka, Dr. Morse's Indian Root Pills, and Dr. Lerox's Indian Worm Eradictor were sold to the accompaniment of drums, war whoops, and "How!"

Patent medicines were not the sole province of the traveling medicine shows; newspaper advertisements featured Ayer's Sarsaparilla for complaints of liver and kidney complete with case histories:

> J.W. of Lowell, Massachusetts was troubled with want of appetite, oppressive weakness and severe pains in the small of the back; all indications of serious derangement of the kidneys and liver. Ayer's Sarsaparilla made him a well man.

Ayer's contained extracts of yellow dock, sassafras, senna, pokeweed, wintergreen, cascara, quinine bark, prickly ash bark, glycerin, potassium iodide, and iron sulfate in alcohol. Lydia E. Pinkham's Vegetable Compound was the nation's favorite for "female weakness" as evidenced by "indisposition of the female reproductive apparatus amongst other female complaints." Containing a diuretic and 18 percent alcohol ("This is added solely as a solvent and preservative"), it was nipped and sipped by women all over North America. Paine's Celery Compound for nervous diseases had 21 percent alcohol and Hotsetter's Bitters was 41 percent alcohol—this is 82 proof, and was a blessing for the Americans of the Bible Belt who had taken the teetotaler's pledge.

> "Men!! Are You Slipping? Are you LESS ACTIVE than you used to be? Are you A GOOD HUSBAND? If you have doubts, don't delay!!!! Use . . ."

Not all patent medicines were innocuous. Some contained concentrations of senna, cascara, or jalap sufficient to cause diarrhea leading to dehydration, and others contained extracts of lobelia, which caused a "cleansing vomiting" dangerous enough to injure the patient. Tuberculosis was called catarrh and, said one ad, "Childs' Catarrh Specific will effectually and permanently cure any case of catarrh, no matter how desperate." Other cough and catarrh medicine contained enough morphine or even heroin to cause addiction. Cancer, heart disease, diabetes, ulcers . . . you name it and the friendly grocer had a bottle for it on the shelf. There were, in fact, medicines to cure diseases which never existed except in the fertile brain of the inventor (Figure 7.9). With pressure from the American Medical Association, the U.S. Public Health Service, and drug companies that prepared ethical drugs (sold

Figure 7.9 A somewhat scandalous advertisement for one of the many "cures" for tuberculosis. Courtesy New York Historical Society.

by prescription only), legislation in the form of the Pure Food and Drug Act of 1906 was passed to curb the more flagrant abuses, but loopholes in the laws still permit the sale of compounds of dubious value and preserve a multimillion dollar industry.

CAUSES AND CURES

If we compare the herbal remedies given in any of the large number of recent books on "using nature's ways" with those given in Gerard, Parkinson, or even Dioscorides, we are struck by similarities that border on plagiarism. Even the terminology is the same: *pectorals* relieve chest congestion, *carminatives* calm the stomach, *discutients* dissolve tumors, and *vulnaries* heal internal wounds. These terms date back beyond the tenth century, and some were used by Hippocrates. People who didn't feel "quite right" took tonics or, in increasing order, *aperients, laxatives or cathartics;* if draconian measures were necessary, they *purged* themselves. *Astringents* tightened the skin, *emolients* soothed it, *demulcents* smoothed it, and *maturating agents* cleaned its pores of boils and pimples.

Based on the Greek four elements, fire, air, water and earth, Hippocrates derived the four cardinal humors which made up the body:

blood, phlegm, white choler, and black choler. From these, Galen derived the four temperaments: sanguine, phlegmatic, choleric, and melancholic. Temperaments could be modified by being hot, cold, moist, or dry. Seizing on these dicta, for there was no other philosophy of human physiology, herbalists searched out plants to control the imbalances among humours that resulted in disease. *Alteratives* changed the character of the blood, *depuratives* purified the blood, and *rubifacients* increased the flow of blood. *Chalgues* increased the flow of bile, and *expectorants* aided in eliminating phlegm from the chest. Hot, cold, moist, or dry states could be modified by *diaphoretics, diuretics, emetics, febrifuges, refrigerants, sialogogues* (promoting salivation), *sudoriferics* (increasing perspiration), and *hepatics* (activating the liver). But from today's perspective, the entire medical system for which these herbal remedies was developed is difficult to rationalize with knowledge now available on the etiology of disease. A broken leg or a bleeding wound was the same problem to Dioscorides as to the medical faculty of the University of Chicago, but a *pectoral* will not cure tuberculosis. Essentially, herbalists of the fourteenth through eighteenth centuries—and those of today—were treating symptoms. Modern medicine is based on two fairly recent concepts, that of the controlled experiment developed in the eighteenth century and the discovery of bacteria in the nineteenth century.

DOES IT REALLY WORK?

Garlic, a most excellent and necessary food seasoning, has been recommended by one or another writer on plant medicinals for falling hair and baldness (usually genetic in males), sore throats (caused by bacteria), sprained ankles (sports injuries), peptic ulcers (from the stress of modern living?), the blahs, and over 20 other conditions, none of them showing any functional or structural relationship. A comparative examination of a dozen of the more popular modern herbals reveals that the same plant has been recommended for a wide range of normal and abnormal states of health. To limit their liability, writers or compilers of these books use the same weasel words that Gerard or Culpepper used in the seventeenth century: "This plant is reputed to . . ."; "it has been reported that a poltice of the leaves of this plant relieves . . ."; "teas made from the roots of this plant are said to . . . ." Of significance is the fact that, from the sixteenth century, when the foundation of modern science was established, until today, none of the herbalists have provided their readers or the scientific community with a shred of evidence supporting the nebulous claims that are made. While it is certainly true that some, perhaps a fair number, of the herbal remedies are made from plants which contain physiologically active and pharmaceutically effective chemicals that do work in specific maladies, the information supplied by the herbalist does not give much of a clue to the treatment, and most of the claims that the material works for the herbal practitioner can be taken with a very large grain of salt.

If I feel run down, think I have tired blood, take a herbal tonic and feel better, there is no reason to assume that I had tired blood (whatever that might be), that the botanical cocktail cured the tired blood, or that this particular herbal remedy would work for *your* tired blood. There was, you see, no experiment in the formal sense.

There has been a great deal of nonsense written about the concept of the controlled experiment. Certainly the philosophers of the Golden Age of Greece didn't think of it; they believed that knowledge could be obtained by pure reason. The scholastics of the Middle Ages were so involved with the authority of the ancients that they, too, rejected direct experimentation; Galileo's studies were proscribed. Only slowly did the idea develop that one could obtain direct information about how things worked only when the experiment had a built-in control, or check. If I think that photosynthesis involves the production of oxygen, it is not sufficient for me to place a plant in the light and determine whether oxygen is liberated. At the very least, it is also necessary for me to put a precisely equivalent plant under conditions where photosynthesis does not occur, but where all other physiological and biochemical processes do occur, and then see whether I can detect oxygen formation. As my experiments become more sophisticated, the controls will correspondingly become more rigorous. If I believe that a particular bacterium causes sore throats, I can prove this only by obtaining a pure culture of the suspected bacteria, infect *a large number* of test organisms with the bacteria, and have an equally large number of identical test organisms which have not been inoculated with this particular live bacterium. To complete the proof, I must determine that only the infected test organisms get sore throats, and then I must reisolate the bacterium from the diseased animals and repeat the experiment again with the isolated bacteria. The requirement for a large sample size is even more recent than the idea of a control. It is always possible that some of the inoculated organisms may be resistant to the bacteria and, were I to use only one inoculated mouse, it might be immune, and I would conclude that this bacterium doesn't cause sore throats. To make the study even more rigid, I should use inbred strains of mice to provide genetic uniformity, and all the mice—experimental and control—should have been raised under precisely the same conditions.

The application of the concept of a controlled experiment is nowhere seen as clearly as in medicine. It is impossible to obtain a large population of genetically identical individual humans, and it is also impossible to maintain even a few people under absolutely identical conditions for even a few days. If the treatment is believed to be valuable, the experimenter is faced with the moral dilemma of deciding who will be treated and who will be a control. This is dramatically illustrated in *Arrowsmith* by Sinclair Lewis. Fortunately, plant scientists avoid these problems.

MODERN PHARMACEUTICS

This polemic should not indicate that plants used in folk medicine cannot be effective. The pharmacists' shelves contain plant products derived from the herbal remedies of Europe and many other cultures. The Chinese invented vaccination 2000 years before Jenner learned about cowpox; they understood the circulation of the blood 1000 years before William Harvey; and they used advanced surgical procedures, including acupuncture, for centuries. Although the Chinese coated their herbal medicine with the same mystical-religious gloss as did the Europeans, rhubarb roots for laxatives, *Ephedra* as a source of ephedrine, and many other effective chemicals have been extracted or synthesized from herbs used in Chinese medicine.

So, too, have we benefited from the herbal medicine of another great Eastern culture, that of India. Precise reference in the Rig-Veda to a plant called *chandra* can be found. Chandra is a Sanscrit word for moon and an oblique indication that the plant was used in the moon's disease—lunacy. The plant, *Rauwolfia serpentina,* was known to Westerners as an effective reducer of fevers and a cure for dysentery and for many other apparently unrelated complaints. The common denominator in the uses was that the plant seemed to be an excellent sedative. This escaped the attention of Western physicians. Only in the 1940s was the fact that extracts of *Rauwolfia* were amazingly effective in reducing blood pressure brought to the attention of the West, although it was known by everybody in India. Pills made from the dried roots were equally effective in reducing the violence of the insane to a degree and with a safety unmatched by barbiturate narcotics. An alkaloid, reserpine, was isolated and characterized in 1952–1953, allowing standardized dosages to be administered. Schizophrenics, who make up over 50 percent of the patients in our mental hospitals, could be treated with reserpine and could resume virtually normal lives in society. The concept that mental and nervous disorders can be successfully treated and controlled with chemicals has developed to the point that many synthetic chemicals have been made and tranquilizers are a way of life.

The recent history of pharmaceutical botany lends support to the idea that plant-derived medicinals will assume even greater importance in the future than they have today. The list of plant-derived medicinals is continuing to enlarge. But in contrast to earlier centuries, modern usages are based on isolation of the active substance, determination of chemical structure, careful testing on laboratory and human subjects, and extensive clinical trials prior to general use. Products from some plants that have been a part of herbal medicine for centuries have passed these tests and are fully accepted. The true aloe, *Aloe vera,* a reputed remedy for insect bites, ringworm, boils, rashes, and other skin ailments, has been in use since the time of the Egyptians. It was noted in the Bible (John 19:30) that juices of its leaves were applied to the skin. The gel-like leaf sap is, in fact, an excellent first aid for simple burns, decreasing pain and promoting healing. It is under testing as a useful treatment for X-ray burns. An arrow poison of the South American Indians *(Chondrodendron tomentosum)* contains curare, a muscle relaxant used in surgery. The Penobscot Indians of Maine used the May apple *(Podophyllum peltatum)* to treat tumors; the active ingredients are now useful adjuncts in cancer therapy. So too are vinblastine and vincristine, isolated from the periwinkle *(Vinca)*. Nonprescription cough medicines still include balm of Gilead from *Populus x gileadensis,* extracts of cherry *(Prunus serotina),* and horehound *(Marrubium vulgare)*. Such listings can be extended. But because the necessary determinations of effectiveness are expensive and time-consuming, the introduction of a new medical treatment is a step taken with considerable caution.

HUMAN FERTILITY

Nowhere in pharmaceutical botany is more attention given than to the regulation of fertility in humans. Substances that can increase fertility or serve as antifertility agents will play a more and more important role in all of human society. Fertility control has economic, political, social, religious, and cultural significance; Rome, like many modern societies (and families), viewed sterility as a punishable offense. China, where sons were a religious necessity, mandated that a barren wife could not be permitted even to die in her own home. In modern South American countries, where *machismo* is a cultural factor, a man without children is "not quite a real man." In almost all societies, infertility is rarely considered the man's "fault." Herbs that could increase female fertility were prized and are described in medical writings throughout the world. Many of the recommended plants are examples of the doctrine-of-signatures concept. The Chinese recommended the peach fruit because of its resemblance to the female genitalia; Greeks had their women eat nuts and pomegranates; Romans included breast-shaped fruits in the diets of their wives. The Mohawks, Cherokee, and Plains Indians fed phallus-shaped roots to brides, Near Eastern civilizations favored the mandrake (Chapter 6), and the Japanese and Chinese chose the morphologically similar ginseng. At this time, no plant product has been demonstrated to be effective in enhancing fertility or curing sterility.

Repression of fertility has long been a matter of great interest. A search through the ethnobotanical literature reveals that many plants have been used as contraceptives. Viewed from the perspective of modern pharmacology, most of these are nostrums, depending upon supernatural powers for effectiveness. Several do, however, show some promise. The Cherokee used the spotted cowbane *(Cicuta maculata),* whose tuberlike roots contain coniine and other alkaloids similar in structure and action to the ones in the hemlock *(Conium spp.)* that Socrates drank. Indian women of the western and southwestern plains drank an infusion of the roots of stoneseed *(Lithospermum ruderale)*, which contains chemicals that abolish the normal menstrual cycle, decrease the size of ovaries, thymus, and pituitary glands, and interrupt the secretion of other endocrines. Similar chemicals are found in one of the sweet peas and *Lycopus*, the bugleweeds of the mint family. Estrogen-like hormones are also found in a clover *(Trifolium subterraneum)*, in stems of soybean plants, and in tulip bulbs, all of which have been used for contraception by various cultures (see section on cotton in Chapter 9).

The trend in modern medicine is either to extract the effective medicinal from the plants in which they are synthesized or to find out what the active compound is and then make it in the laboratory. Not only does the latter method eliminate the possibility that other constituents of the plant can cause undesirable side effects, but it permits precise standardization of dosages. In many instances, it is still cheaper to have the plants gathered in the wild or grow them on herb farms, then isolate the effective compound, than it is to attempt synthesis of the molecule.

HUMAN POISONING

As part of the trial and error process by which food and medicinal plants were discovered and eventually incorporated into nutrition and medicine, it is very likely that many of the experimenters died, in some instances rather horribly. The fact that a given plant had a harmful effect—whether leading to minor discomfort or to death—became clear enough. But it remained for chemists and pharmacologists of the present century to work out the chemical nature of the poison contained in a plant or plant part and to develop methods of counteracting the toxin. These tasks are not completed; as new species are brought into cultivation and escape from gardens into which they have been introduced (that is, they grow wild), the list of potentially harmful plants is lengthened. Adults are, of course, at risk as well as children, but the majority of reported poisoning cases are among children. Not only do they tend to put just about anything into their mouths, but their body weight is low relative to the amount of toxin necessary to cause overt symptoms. Each year, there are close to 15,000 fatal or near-fatal plant poisonings of children under 5 years of age in the United States. The majority of such incidents are caused by plants that are commonly found in homes, yards, and public parks.

Many classes of chemical toxins are present in such common plants. Some are alkaloids that affect the central nervous system; others interfere with heart action; and others, usually classified as causes of dermatitis, include irritants, allergens, and toxic oils (such as those of poison ivy). Pollen and plant dusts are usually the causes of allergies such as hay fever. The best control is avoidance, but since the range of potential poisonous plants is so broad, education and control of access is

Table 7.1 COMMON POISONOUS PLANTS

Plant	Plant part	Symptoms
	Skin irritants	
Poison ivy, oak, and sumac	All parts	Redness of skin, blistering, itching
Nettles	Leaves	Itching and burning of skin
Trumpet creeper	Leaves	Reddening of skin
Poinsetta	Milky sap	Itching
	Internal poisons	
Castor bean*	Seeds	Nausea, diarrhea, vomiting
Cherry*	Seeds, leaves	Convulsions, difficult breathing, coma
Dieffenbachia	All parts	Swollen tongue, loss of speaking ability
Foxglove*	All parts	Cardiac abnormalities
Jimson weed*	All parts	Hallucinations, convulsions
Lantana*	Leaves, fruit	Muscular failure
Larkspur*	All parts	Difficult breathing
Mistletoe*	Fruit	Weak pulse, coma
Monkshood*	All parts	Nausea, irregular heartbeat, coma
Nightshades*	All parts	Hallucinations, vomiting, coma
Oleander*	Primarily leaf	Respiratory disturbances
Philodendrons	Leaves	Burning mouth, swollen throat
Poison hemlock*	All parts	Nausea, vomiting, weak pulse, respiratory failure
Rhubarb*	Leaf blade	Convulsions, coma
Water hemlock*	All parts	Convulsions
Yew*	Seeds	Dilated pupils, coma

*Known to be fatal.

necessary. Again, this is most important for children, who must be carefully watched and rigorously instructed to keep all unknown plants out of their mouths, not to make tea party beverages from leaves and flowers, and to avoid eating wild fruits. Table 7.1 provides an abbreviated list of some of the more common poisonous plants.

ADDITIONAL READINGS

Bolyard, J. L. *Medicinal Plants and Home Remedies of Appalachia.* Springfield, Ill.: 1982.

Brophy, P. J. "The History of Contraception." *History of Medicine* 1, no. 4 (1969):2–4.

Carson, G. *One for a Man, Two for a Horse.* Garden City, N.Y.: Doubleday, 1961.

Clarkson, R. E. *The Golden Age of Herbs and Herbalists.* New York: Smith-Dover, 1975.

Hardin, J. W., and J. M. Arena. *Human Poisonings from Native and Cultivated Plants.* 2nd Ed. Durham, N.C.: Duke University Press, 1974.

Kelly, H. A. *Some American Medical Botanists.* Troy, N.Y.: Southworth, 1914.

Kinghorn, A. D. (ed.). *Toxic Plants.* New York: Columbia University Press, 1979.

Krochmal, A., and C. Krochmal. *A Guide to the Medicinal Plants of the United States.* New York: Quadrangle Books, 1973.

LeStrange, R. A. *A History of Herbal Plants.* New York: Arco, 1977.

Lewis, W. H., and M. P. F. Elvin-Lewis. *Medical Botany: Plants Affecting Man's Health.* New York: Wiley, 1982.

Mettler, W. *History of Medicine.* Toronto: Blakestorn, 1947.

Mitchell, J., and A. Rook. *Botanical Dermatology: Plants and Plant Products Injurious to the Skin.* Vancouver, B. C.: Greengrass Press, 1979.

Oyle, I. *The New American Medicine Show.* Castro Valley, Calif.: Unity Press, 1979.

Stephens, H. A. *Poisonous Plants of the Central United States.* Lawrence: Regent's Press of Kansas, 1980.

Tampion, J. *Dangerous Plants.* London: David and Charles, 1977.

Vogel, V. J. *American Indian Medicine.* Norman: University of Oklahoma Press, 1970.

Ergot and Ergotism

St. Anthony, pray for me.

MASTER OF MARY OF BURGUNDY
Book of Hours, 1480

The drug known as *ergot* is defined pharmacologically as the dried sclerotium of a fungus, *Claviceps purpurea,* developed on rye plants (Figure 7.10). This purplish-brown, somewhat hornshaped structure has over 150 names, including spurred rye, horned seed, and *mutterkorn.* In the United States, the inclusion of ergot in materia medica dates back to a scientific paper presented before the Massachusetts Medical Society in 1813 when Dr. Oliver Prescott suggested that ergot could ease difficult births. Actually, Dr. John Stearns of New York had demonstrated five years earlier that ground-up ergot could quicken childbirth. Although unknown to Prescott and Stearns, their reports only repeated observations made for centuries. In the *Kreuterbuch* of 1582, Adam Lonicer (for whom the genus *Lonicera*—honeysuckle—is named) noted that ergot was used by German midwives. The ancient Chinese probably knew of ergot's ability to cause uterine contractions, and Avicenna, a Moorish-Arabian physician of the tenth century, used the material for the same purposes.

Figure 7.10 Head of rye bearing sclerotia of the ergot fungus.

Search for the active principal in the fungus has occupied the attention of many bio-organic chemists. The active alkaloids constitute less than 0.1 percent of the dry weight of the sclerotia. The isolation of the first (of many) alkaloids from ergot was made by a French pharmacist, C. Tanret, in the nineteenth century, but it wasn't until 1918 that A. Stoll, a Swiss, obtained the first chemically pure and biologically active alkaloid—ergotamine.

Most ergot alkaloid is isolated from sclerotia formed in the grain heads of rye plants. In order to obtain a large-scale infection of rye plants, a variety of techniques were developed in the 1930–1940 period. When the grain is just beginning to flower, large machines mechanically shake the plants, and, at the same time, a spore spray is directed at the flowers. When sclerotia are mature, which occurs simultaneously with maturation of the rye, the fungus is picked by hand. Yields depend on all the imponderables that attend the maturation of any crop, and deviations from yet-unknown environmental optima result in variable concentrations of the active alkaloids. Understandably, the variation in potency is of considerable concern, and harvests are appropriately mixed to provide a product of uniform alkaloid content. Cultivation is further complicated by host-parasite interactions, the genetics of which are unknown; races of the fungus may develop from a single gene mutation at frequencies of about 1 per million. Since a field inoculation consists of trillions of spores, genetic uniformity is very unlikely. The ergot industry in the United States, Germany, Russia, and Spain produces huge quantities. In 1945 the United States imported 44 tons of ergot (each sclerotium weighing less than ¼ ounce) and produced 100 tons in Minnesota, Nebraska, the Dakotas, and Illinois.

MEDICAL USES

Methods of isolation and purification of ergotamine and other active alkaloids were worked out shortly after the end of World War II, primarily in the laboratory of Albert Hofmann in Basel. Prior to the use of sophisticated chemical techniques, crude biological tests for activity were used. For example, a sample was injected into a White Leghorn rooster, and the time required for the rooster's comb to turn blue was measured. Alternatively, frogs were injected and constrictions of blood vessels in the toe webs were observed.

The color change and vasoconstriction reactions supply clues for the medical use of ergot. Basically, it causes blood vessels to contract—thus reducing the danger of hemorrhaging during and immediately after birth. It also causes contraction of muscles, serving to assist in shoving the unwilling child into the world. Blood vessel constrictions and uterine contractions are accompanied by nerve blockage, alterations in motor fuctions, and some brain dysfunctions. When provided in appropriate concentrations, all of these ergotamine actions are useful in easing childbirth.

The knowledge derived from reactions to ergot alkaloids noted in childbirth was applied in the 1925–1930 period by internists treating migraine headaches. A small but significant proportion of the population of every country, migraine sufferers experience periodic, persistent, and unilateral headaches, frequently accompanied by nausea and mild hallucinations. There is vasodilation of blood vessels in the brain, and hormonal balances are sharply altered from the normal pattern. Ergot, causing physicochemical changes in humans which are almost diametrically opposite the migraine syndrome, is widely used to control, but not to cure, the agonies of migraine. Unfortunately, ergot therapy is effective in only about 40 percent of migraine victims.

Whether used in childbirth or for migraine, ergot therapy is not without its deleterious side effects, although fewer than might be expected of a drug with such potency. Obviously, there are wide variations in tolerance, and reports of toxicity can usually be traced to differences in patients rather than to variations in the composition of the drug. Especially sensitive newborn children whose mothers were treated during childbirth will occasionally show reddening of the skin or breathing difficulties, but these symptoms usually pass without permanent damage. This is fortunate, because there is no countertherapy for ergot poisoning.

ERGOTISM

For uncounted centuries, ergot has been one of the cursed scourges of mankind. It has plagued the body and mind ever since we began to use grasses for their edible seeds. In Europe it was called the holy fire (*ignis sacer*), St. Anthony's fire, the ignis beatae Virginis invisibilis, or the infernalis.

Three major cereal grains have been used to make bread. The common bread was an unleavened product made from barley, but only wheat and rye flour make a raised or leavened bread. It is likely that Thrace and Macedonia, but not Greece, grew some rye, but the plant wasn't introduced into much of Europe until the Christian era. France began to grow rye about 300, and Britain obtained her starting seed when the Teutons invaded the island. Rye was not grown as a major cereal grain in Europe and European Russia until about the fifth century, and attempts to pinpoint just when historical records of ergotism began is difficult. Thus, an epidemic suspiciously like ergotism broke out among the Spartans in 2430 B.P., and a plague of 857 in the Rhineland also matches the clinical symptoms. A disease "like fire" was reported in Paris in 943, from Aquitaine-Limousa in 994 with 4000 deaths, and from Rheims in 1041 with 2000 deaths. From that time on, instances of ergotism have been recorded in sufficient detail so that we can be sure of its cause.

Ergotism results from the ingestion of sclerotia of ergot ground up in rye flour. Two major types of ergotism are known, gangrenous and convulsive. In the former, severe constriction of the blood vessels results in swelling as blood accumulates in the hands or feet, with burning sensations alternating with intense cold. Numbness follows within a few days and this, in turn, is followed by blackening of the limb, horrible odors, and eventually merciful, but unbearably painful, death.

Convulsive ergotism accurately describes the symptoms. Twitching of head, arms, and hands is followed by contractions of muscles thoughout the whole body. The afflicted typically roll themselves into a ball and then stretch themselves out at full length, the actions accompanied by terrible pains. Vomiting, deafness, blindness, and hallucinations usually follow. Feats of superhuman strength, and the conviction that flying is possible, have been noted. If the victim recovers, and about 30–40 percent do, hallucinations can continue aperiodically for up to a year. Domestic animals who eat ergot-contaminated grain or table scraps exhibit identical responses. It is said that dogs will tear bark from trees until their teeth fall out and that ducks will strut like roosters, attacking people and other animals. Depending upon the weather, the genetic constitution of the host and the fungus, the care taken to eliminate sclerotia before milling grain into flour, and the amount of bread eaten, devastating outbreaks of ergotism could occur. And they did occur in Europe on an average of once every five to ten years.

Innumerable people had ergotism before its etiology was recognized in 1673 by a Parisian lawyer-physician, Denis Dodart. Up to the middle of the eleventh century, over 20 massive epidemics were reported in France alone, and by the middle of the fourteenth century, over 50 epidemics had been reported from central Europe. Because they lacked any knowledge of its cause, it was reasonable for Christians to call upon the saints to intercede with heaven for succor. But which one? St. Anthony the Great was born in Egypt in the first century and established the idea of monastic life. Long the patron saint for erysipelas, a bacterial disease of the skin which causes swelling and burning, he could logically become the intercessor for this disease as well (Figure 7.11). In 1039, a French nobleman, Gaston de la Vollaire, built a hospital in the Rhone Valley, obtained relics of St. Anthony, and asked monks to serve in the hospital. These men formed the Order of St. Anthony and dedicated themselves to nursing the survivors of ergotism. In the *Book of Hours* by the master of Mary of Burgundy (1480), St. Anthony is asked for protection against the disease. The Holy Fire disease was soon called St. Anthony's fire.

Figure 7.11 St. Anthony, patron saint of ergotism, praying for a victim of the affliction. The gangrenous type of disease is seen. Taken from a woodcut in *Fendtbuch der Wundartzney* by Gersdorff, published in 1517.

Although Dodart's identification of the cause of ergotism was known among the few educated physicians of the seventeenth century, the direct connection between *mutterkorn* and St. Anthony's fire didn't become general knowledge until the eighteenth century. The dark, heavy, sour but very nourishing bread of central and eastern Europe contained so much ground-up weed seed that the dark sclerotia went unnoticed. Since bread was truly the staff of life, the persistence of the peasants in eating ergot-contaminated bread is not surprising. When a high probability of starvation had to be weighed against possible ergotism, people made the only logical choice.

Between 1580 and 1900 there were 65 major ergot epidemics, including 29 in Germany, 11 in Russia, 10 in Sweden, and 5 in the United States. The introduction of the potato to Europe undoubtedly reduced the death toll by substituting for bread. In 1722, Peter the Great mounted an invasion of Turkey to obtain for Russia the still-coveted ice-free port to the seas. His cavalry ate ergotized black bread and 20,000 men and horses were stricken; the invasion was called off. Between 1770 and 1780, epidemics raged through Germany and France with over 8000 documented deaths. In the winter of 1812–1813, Napoleon's troops and horses ate bread baked from rye commandeered from the Ukraine, and the resulting epidemic of ergotism contributed to his Russian defeat and turned the retreat from Moscow into a horror. In 1812, Austria passed a law stating that inadequately cleaned rye would be confiscated, and other European countries quickly passed similar legislation.

Although the gangrenous form of ergotism has almost disappeared—with no major outbreak since its last appearance in France in 1885—there is a growing suspicion that it is reappearing in central Africa from sclerotia developing on cereal grains other than rye. The convulsive form, however, has increased in severity and is responsible for outbreaks of ergotism in the twentieth century. There was a severe outbreak in the Soviet Union in 1926 and a smaller one in England in 1928–1929, when the Jewish community imported rye from central Europe. During the well-studied Soviet Union epidemic in 1926, flour containing 2 percent ergot was found to be enough to cause convulsive ergotism; some samples of rye flour contained up to 7 percent sclerotia. It has been claimed that convulsive and hallucinogenic ergotism struck a town in Provence in 1951, but the French government denied this, stating that there was an inadvertent contamination of the flour by an insecticide; this bureaucratic explanation does not conform to the symptoms noted.

As home milling decreased during the late nineteenth century, purity control became easier, and large sclerotia, which are bigger than the grain, could be removed by sieving. Removal of smaller sclerotia and fragments is facilitated by the observation that when rye is mixed with salted water, the sclerotia float and can be skimmed off. Nevertheless, rye grown for the organic foods market is not subject to federal control.

Although no records were kept prior to the nineteenth century on ergotism in animals, ergot alkaloids have long caused serious economic problems. Canada decreed in 1971 that no more than 0.1 percent by weight of ergot is permitted in feeds for poultry and other domestic animals, but even this level severely reduces the weight gain of cattle. Many other wild grasses are susceptible to infection by ergot, including those that form the food of migrating birds. The reports of abnormal behavior of wild plant-eating animals may be due to their feeding on ergotized grasses. The parallels between these behavioral modes, some aspects of the convulsive type of human ergotism, and the "spaced" appearance of highly sensitive women in labor who have been given ergotamine suggest that the ergot alkaloids may, in fact, be potent hallucinogens.

LSD

As part of a long-term chemical study of ergot alkaloids, two Swiss chemists, Stoll and Hofmann, prepared a large number of compounds

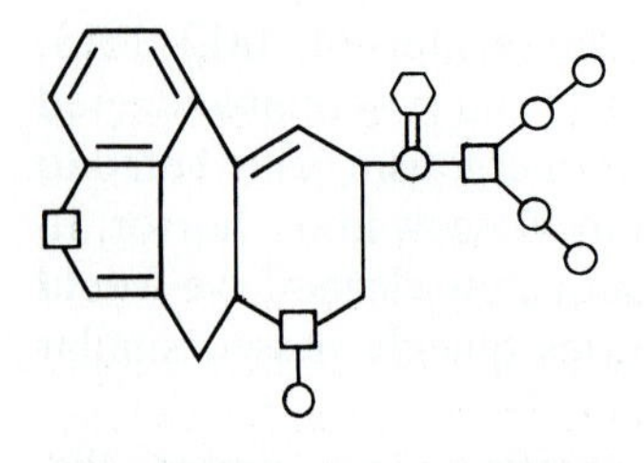

Figure 7.12 Molecular structure of lysergic acid diethylamide (LSD). Hydrogen atoms and carbon atoms in the rings are not shown. Carbon atoms are represented as circles, oxygen atoms as hexagons, and nitrogen atoms as squares.

from ergot. Among these were a number of lysergic acid derivatives, the most familiar of these being the diamine of lysergic acid—LSD (Figure 7.12). Hofmann apparently was the first to experience the effects of LSD when he inadvertently touched his contaminated fingers to his lips and then attempted to bicycle home. The trip took several hours, rather than the usual half-hour. Not only did one early experimenter experience initial hallucinogic effects of LSD, but he reported that for several months thereafter he would see blood coming from the faucet whenever he opened the tap. The subculture use of LSD and the somewhat hysterical reports of uses of LSD in psychiatry have had at least one useful spinoff. They have directed the attention of botanical and medical investigators to the effects of hallucinogenic materials among the Indians of South America.

In a compendium on the plants, animals, and rocks of New Spain, Francisco Hernandez reported in 1580 that the Aztecs used the seeds of the snake plant, *coaxihuitl,* in religious ceremonies. Infusions of the seed, called *ololiuqui,* deprived people of their senses; those who used it "believed in the owl, sucked human blood and could communicate directly with their gods." To make themselves fearless, Aztec warriors mixed ashes of poisonous insects, tobacco, and ololiuqui and rubbed the mixture on their bodies. Not unexpectedly, Jesuit upholders of the one true faith railed against this diabolical seed and drove the cult into hiding. The users hid their knowledge of the plant for close to 400 years. In 1897, Dr. Manuel Urbina concluded that ololiuqui was a morning glory. Most *norteamericano* botanists couldn't believe that the seeds of such a pretty plant could harbor hallucinogens, and the definitive authority of William Safford was invoked to discount both Urbina and an anthropologist, B. P. Reko, who in 1919 was so rash as to identify the plant as a morning glory, *Rivea corymbosa.* The adequacy of identification by Urbina was confirmed in 1939 by a botanist, R. E. Schultes of Harvard. At about the same time, seeds of another member of the morning glory family, *Ipomoea violacea* (a close relative of the true sweet potato), were found to be used by the Oaxacan Zapotec tribe for divinatory rites under the name *tlitlitzin* or *badoh negro*—the sacred black seeds. By 1955, it was clear that the hallucinogenic materials were alkaloids of the ergot type and by 1960 Albert Hofmann isolated several, including ergometrine, an alkaloid useful in childbirth. Other members of the same botanical family, the Convolvulaceae, also contain active substances.

The number of plants known or suspected to contain hallucinogenic compounds continues to increase, both as a result of careful anthropological studies of cultures throughout the world and as a consequence of careful chemical and pharmacological research. Although many hallucinogens are alkaloids, important ones are not. In addition to the marijuana hallucinogens, which are not alkaloids, psychoactive substances have been found in nutmeg oils such as safrole, myristicin, and elemicin.

Many plants have nitrogen-containing active principles. The hallucinogenic mushrooms, peyote cactus, and the solanaceous plants, in addition to opium and cocaine, are active solely or principally because of their alkaloids. Less well known in Western societies are the South American snuffs from leguminous plants such as *Piptadenia* and *Mimosa* and the various species of *Banistereopsis,* known collectively as *caapi* or *yaje,* whose alkaloids are used in the Rio Negro region of Brazil in male adolescent rituals. Tolerance limits and the nature of the physiological responses of these and many other hallucinogens have not been established, and most are capable of causing severe reactions including respiratory failure and death.

ADDITIONAL READINGS

Barger, G. *Ergot and Ergotism.* London: Gurney and Jackson, 1931.

Bove, F. J. *The Story of Ergot.* Basel, Switzerland: Karger Publications, 1970.

Fuller, J. C. *The Day of St. Anthony's Fire.* New York: Macmillan, 1968.

Hofmann, A. Ergot. In T. Swain (ed.). *Plants in the Development of Modern Medicine.* Cambridge, Mass.: Harvard University Press, 1972.

Schultes, R. E. "Hallucinogens of Plant Origin." *Science* 163 (1969):245–263.

Van Rensburg, S. J., and B. Altenkirk. "*Claviceps purpurea*—Ergotism." In I. F. H. Purchase (ed.) *Mycotoxins.* New York: Elsevier, 1974.

Youngken, H. W., Jr. "Ergot—A Blessing and a Scourge." *Economic Botany* 1 (1974): 372–390.

Jesuit Powder

That in marshes there are animals too small to be seen.
But which enter the mouth and nose and cause troublesome diseases.

VARID, *gentleman farmer of Rome, 2100* B.P.

Throughout Man's troubled history, few diseases have played so tragic a role as malaria. It has killed or incapacitated more people than all plagues, wars, and automobiles. More than 10 percent of the U.S. overseas armies in 1943 had malaria. In 1938–1939, there were over 300,000 reported cases of malaria worldwide, including 20,000 deaths in the Brazilian province of Rio Grande do Norte. The Spanish-American War of 1898 saw four times as many troops inactivated by malaria as by wounds. The Army of the Potomac under General George McClellan lost at Chickahominy because it didn't have enough healthy troops to oppose General Lee. The Pilgrims' decision to settle in New England was made, according to William Bradford, because, "hott countries are subject to greevous diseases . . . and would not so well agree with our

English Bodys." In 1596, the earl of Cumberland captured Spanish Puerto Rico but couldn't hold it because his forces were decimated by malaria. The city of Florence was depopulated 2200 years ago by malaria. Alexander the Great died of it in June 2323 B.P. Among the 12 labors of Hercules was the slaying of the nine-headed Hydra, the monster that brought misery and ruin to the people, and the elimination of the man-eating birds of the Stymphalian marshes—allusions to malarial infections. Untreated malaria may kill about 1 percent of those infected. The survivors, prone to relapse, may suffer from anemia, weakness, sexual impotence, chronic abortion, or secondary infections—all of which lower the value of the individual to self, family, and community.

ETIOLOGY OF MALARIA

The name *malaria* was coined in the seventeenth century by Dr. Francisco Torti by combining the Italian for *bad* and *air,* and it has been called the shakes, the fevers, the ague, and many other things, none affectionate. Hippocrates reported several clinical types of malaria. He believed that imbalances in the ratios of the four humors—blood, phlegm, black bile, and yellow bile—caused the disease. It was asserted for a thousand years that invisible worms were carried on the dank night air into the body and caused changes in the humoral ratio.

Yet it was also know that swamps and mosquitos were involved, for malaria was rarely found in dry and windy areas, and it disappeared during the winter. Empedocles of Acragas (Sicily) 2500 years ago had marshes drained to rid local towns of malaria. From India, still an endemic area, a lyric poem from the fifth-century *Susrata* said:

> The green and stagnant waters lick his feet
> And from their filmy, iridescent scums
> Clouds of mosquitos, gauzy in the heat
> Rise with his gifts: Death and Delirium.

Only during the middle of the eighteenth century was the relationship of mosquitos to the fevers accepted, and it was not until 1880 that a French physician, Charles L. A. Laveran, found microscopic parasites in red blood cells of human victims of malaria. For this discovery he received the Nobel Prize in 1907. In 1902, the British physician-bacteriologist Ronald Ross received the Nobel Prize for his discovery, some years before, of another stage in the life cycle of the parasite in the gut of the *Anopheles* mosquito. Giovanni Grassi proved that the *Anopheles* mosquito carried the parasites to the blood stream of humans. Thus by 1899, the complete life cycle of the parasite, called *Plasmodium,* was known (Figure 7.13). There are four major species of malarial parasite, each causing different clinical types of the disease.

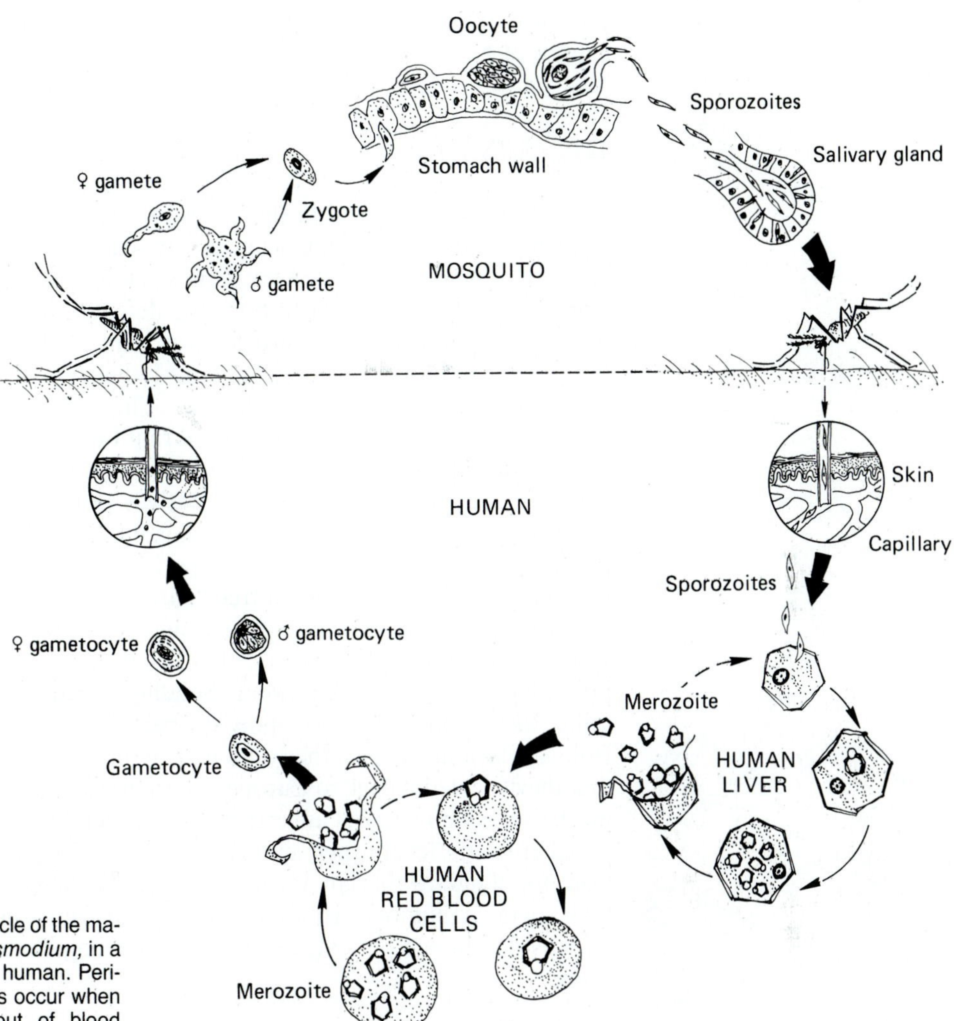

Figure 7.13 Life cycle of the malarial protozoan, *Plasmodium,* in a mosquito and in the human. Periodic chills and fevers occur when merozoites break out of blood cells.

The need for mosquito control became so obvious that marsh draining and other control measures were adopted throughout the world.

Since mosquito larvae develop in water and must come to the surface to breathe, draining was accompanied by pouring oil on water to alter surface tension and prevent the mosquito larvae from "holding" to the surface, but this procedure was ineffective. DDT came into general use as an insecticide in the 1940s, and in spite of almost criminal misuse, it saved millions of live. But it is impossible to kill all female mosquitos. Even a few people with malarial parasites in their blood can serve as a reservoir for the insect and can become the base for a whole new epidemic. As far back as the fifteenth century, physicians were dreaming of some medicine which could cure the disease, what we today would call a *chemotherapeutic agent* specific for the malarial parasite.

One was found in a plant, and its history starts in Lima, Peru, the capitol of New Spain.

THE QUINA TREE

After Lima was founded in the 1520s, it became a proud and wealthy city with almost feverish financial activity. Riches were quickly amassed; young Castilian gallants vied for permission to make an assured fortune from minerals so easily taken from the defeated Indians. The turnover of these hopeful merchant princes was high; malaria was endemic in Peru. Yet, as Church fathers noted, Indians in the bush were not excessively bothered by the disease. Through converts and slaves (usually one and the same) they heard that on the eastern slopes of the Andes Mountains was a tree whose bark, powdered and mixed with water, would cure the fevers. The natives called the tree *quina*—the fever bark tree (Figure 7.14). An Augustinian monk, Father Calancha, published a book in 1639 about his experiences in New Spain in which he noted that the bark of the fever tree "cures the fevers and tertians." The Augustinians were a contemplative and less worldly order than the Jesuits, and it was the Society of Jesus that recognized the political potential inherent in this powder. Sending expeditions into the mountains, they soon had a monopoly on bark from Peru, Colombia, and Bolivia. By some means, they obtained an official monopoly by 1650. At first they trickled it back to Europe and then systematically built up their markets through discreet advertising of the fact that they and they alone could cure the fevers then sweeping Europe. They had little competition, for standard therapy for malaria was still based on the Hippocratean dicta of humors. Treatment was a combination of bleeding the already debilitated patient, bed rest, and cooling cloths placed over the body.

Figure 7.14 The "tree of the Jesuit's powder." It is doubtful whether the illustrator ever saw the tree; the drawing was done from verbal descriptions. Taken from *Rariorum Plantarum Historia* by Carolus Clusius, published in Antwerp in the edition of 1601.

RELIGIOUS MEDICINE

The Society of Jesus formally decided that this wondrous, God-given plant should be used wisely—some would say jesuitically—for "the greater Glory of God and for good and useful Christians." Rome, especially the Vatican, had first call on it, for the need was especially great. Summer fevers were raging in Italy, and travel through the Campania to Rome was an invitation to death. The princes of the Church remembered a previous conclave which resulted in the death by malaria of eight cardinals and 30 attendants, and postponed their conclaves until late fall. Secular and religious bishops, and members of the nobility—then down through the hierarchy—were ordered in priority lists. The Chinese emperor, the great K'ang Hsi, had several bad malaria attacks in 1693, and Jesuits in attendance at his court introduced the bark and saved his life; K'ang Hsi was grateful but never became a convert.

It was obvious that regulation of the bark's use and study of the best method of treating malaria with it needed to be centralized. Control was vested in Juan de Lugo, S.J., native of Madrid, cardinal upon the insistence of Pope Urban VIII, eminent theologian, and sometime professor of philosophy at Gregorian University. De Lugo had a superb organizational mind, one capable of both administration and research; the Society could not have chosen a better man. He devoted over ten years of a busy administrative life to evaluating dosages, attempting to extract the active principle, and in 1651 oversaw the publication of *Schedula Romana,* a book in which instructions on the use of the powder were detailed. With the imprimature of the Society and the tacit approval of the pope, no Catholic physician would dare openly to obstruct the use of the powder, although many were, for a variety of reasons, opposed to it.

Meanwhile, Spanish galleons were lumbering back to Santiago and Cadiz with gold and silver bullion and with bales of bark bearing the mark of the Order. With malaria decimating French and English colonies on the eastern seaboard of North America, buccaneers who plundered the Spanish main were as welcome for their free-spending ways as for their loads of Jesuit bark. Some of the bales were transshipped in good English bottoms back to Britain, where the London ague was endemic. The bark, commonly called "Jesuit's powder," was available by 1658 at John Crook's Bookshop, but because of fears of a popish plot of curses placed on it by "those Jesuit Devils," Oliver Cromwell died of malaria the same year without being treated.

Cardinal de Lugo died in 1660. His death, and changing religious politics, pushed the powder further into disrepute even in Catholic lands. Distribution and control of the powder by the Jesuits was waning, several prominent persons died after taking the powder, an unidentified fever in Rome in 1655 was not controllable by the medicine, and the power of the Roman Church to demand obedience and conformity had decreased. Physicians had never liked the deft hand of the Jesuits slipping into their pockets, and with the Jansenite apostacy challenging the authority of the Church, attacks on Jesuit powder intensified. Popular prejudices against the Society were fanned by reformists and physicians, supplies were short because of piracy and slave revolts in Peru, and people didn't like swallowing a bitter draught without knowing more about what was in it.

THE FEVEROLOGIST

Rumors were, however, flying in England that a former apothecary's assistant in the county of Essex could cure the ague. To quell fears, he announced that he was not a physician, but a *pyretiato,* a feverologist, who by long and arduous study, "by observation and experimentation," had "a certain method for the cure of this unruly distemper." He further announced, "Beware of all palliative cures and especially that known by the name of Jesuit's powder, for I have seen most dangerous effects following the taking of that medicine." The pyretiato moved to London in 1668 and set up a very lucrative practice under the horrified noses of the Royal College of Physicians. When Charles II contracted malaria, he called for this interesting man, Robert Talbor. Because of suspicions that Talbor was really using the hated popish remedy on the monarch, loyal English subjects carried placards warning that the Jesuits were trying to poison the king. Happily Charles recovered and Talbor was knighted. Charles also forced the College of Physicians to make Sir Robert a full member and warned this outraged body that "you should not give him any molestation or disturbance in his practice."

Through Robert Talbor, Charles saw a way to improve the always touchy relationships between England and France. The dauphin, son of Louis XIV, suffered from fevers, and Talbor was sent to Paris as "royal envoy and physician to the king of England." The entire royal family was cured, and Louis, in a magnanimous gesture, sent Talbor to cure the queen of Spain. Although Spain's colonies were the source of the curative bark, there wasn't any in the mother country because the Jesuits had recently been expelled from Madrid and they had taken their powder with them. Upon his triumphal return to France, Talbor changed his name to Talbot—a distinguished French name—and became a celebrity in Paris. Madame de Sévigné referred to him as *un homme divin,* and his bedside manner was most favorably commented on by ladies of the court. As the Chevalier Talbot, and with Paris rife with malaria, he amassed a fortune. In 1680 he decided to return to England with his secret, but Louis paid Sir Robert 3000 gold crowns and the promise of a life pension for a sealed envelope containing the remedy formula, on the promise of the king that it would not be opened until after Talbot's death. Talbor, again English, returned home in glory, became a fellow of St. John's College, Cambridge, and composed his own epitaph before his death in 1681.

> The most Honorable Robert Talbor, Knight and singular physician. Unique in curing fevers of which he delivered Charles II of England, Louis XIV of France, the most Serene Dauphin, princes, many a duke and a large number of lesser personages.

In January 1682, Louis opened his expensive envelope and in a book entitled *The English Remedy or Talbot's Wonderful Secret for Curing Agues and Feavers* the secret was revealed. It was, of course, Jesuit powder mixed with wine to disguise the bitter taste. The discomfiture of the physicians who had denounced Talbor was great, but even knowing that the bark was effective didn't make the British medical profession happy using it. When Charles had another bout of malaria in

1682–1683, he literally had to beg his doctors for Talbor's remedy; they simply would not accept the fact that they had been outdone by a mere apothecary's apprentice.

PLANTATION CULTIVATION

By the end of the seventeenth century, quinine powder, no longer Jesuit powder, was the standard treatment for malaria. Spain still controlled trade through its exclusive mandates in Peru and Bolivia. As demand increased, it became obvious that there weren't enough trees available to assure supplies. Bark collectors had to go further and further into the mountainous bush to find the trees, getting lost and dying of dysentery or from the poisonous darts of the head-hunting Jívaro Indians. This danger had been envisioned by the Jesuits 100 years previously when they tried, unsuccessfully, to require that a tree be planted for every one cut down and stripped of its precious bark. In the middle of the eighteenth century, a group of French botanists studied this problem and concluded that there were at least four species of the tree. This was confirmed in Sweden when botanical specimens were sent to Uppsala for classification by Linnaeus. Linnaeus gave the genus name *Cinchona* to the trees (Figure 7.15) to honor the viceroy of Peru in 1628–1639, inadvertently dropping the "h" following the initial "C" in Count Chinchona's family name.

By the middle of the nineteenth century, the fundamental facts about the fever bark tree were well known. The trees are native to mountains of Peru, Bolivia, and Colombia, growing best in areas of abundant rainfall, at altitudes above 1000 meters and up to 8000 meters. The genus is in the Rubiaceae, the family which also contains the coffee tree. The compounds that destroy the malarial parasite are all closely related alkaloids which, when purified, form white, crystalline powders with very bitter taste. Quinine alkaloids are found in the bark and may represent over 7 percent of the weight of the bark. Two French chemists isolated the quinine alkaloids in the early nineteenth century. By 1870, quinines could be isolated and purified and chemical determination of active materials was a routine laboratory technique. With this information, standardization of treatment for malaria was accomplished by basing dosage on the alkaloid content rather than on the amount of bark mixed with water. There appeared to be no barrier to plantation cultivation and freedom from malaria seemed merely a matter of time.

Figure 7.15 Flowering branch of the quinine-bark tree *(Cinchona officinalis).*

The South American monopoly was broken in 1865. Charles Ledger, an English bark trader, established his business in Puno, Peru, across Lake Titicaca from Bolivia. Knowing that the best bark was found only at high altitudes, he secretely sent his servant-translator, Manual Incra Mamini, into the Bolivian Andes near the headwaters of the Rio Beni, where *Cinchona calisaya* trees with high-yielding bark grew. Mamini left Puno in 1861 and returned in 1865 mumbling about bad growing years and leading a shy young Indian woman with two small children. Ledger paid Mamini £150 for the seed he had obtained, and Mamini foolishly went home to Bolivia, where he was arrested and charged with the traitorous crime of smuggling cinchona seeds out of the country. In jail for less than a month, he was severely beaten and allowed to die. The precious seeds were, by this time, in London in the care of George Ledger, Charles's brother. The seeds were first offered to the British government, which refused them. George went to Amsterdam and managed to sell some to the Netherlands government for 100 gulden, with a promise of an additional payment if the seeds were viable.

The seed packet arrived in Java in December 1865, was opened, and smelled so bad that it was assumed that the seeds were rotted. Fortunately, they germinated well and the Ledger brothers received their promised 500 gulden. Over 10,000 trees were outplanted the following year, and by 1873, several plantations were established. The Dutch East India Company sent an alkaloid chemist to Java who methodically tested samples of the bark of each tree, marking those with quinine yields exceeding 10 percent. These trees were used as grafting scions for fast-growing root stocks. By 1874, the superintendent of the plantations reported that within five years there would be over 2,000,000 high-yielding trees . . . and there were. In 1881, South America had exported 9 million kilograms of bark, but by 1884 there were less than 2 million kilograms exported, and by 1890, the Dutch had a world monopoly on quinine. After some years of overproduction which led to reduced prices, controlled harvesting with ruthless stabilization of the world's supply was achieved by 1910, not to be altered until the Japanese overran the plantations in World War II.

With quinine under monopolistic control, attempts were made to grow high-yielding trees in other parts of the world, but none were economically viable operations. As World War II extended into the jungles of Asia, to the Pacific islands, and into the parts of North Africa and southern Italy where malaria was very prevalent, supplies of quinine became worth more than their weight in gold. During the 1930s, the Winthrop Chemical Company in the United States had developed an antimalarial called atabrine, used as a substitute for quinine. It was effective in wartime, but dosage was poorly understood, and it turned a patient's skin yellow. As the quinine shortage became a serious problem for warring armies, a crash program for the total synthesis of quinine started in the United States and England, and several effective antimalarial synthetics were developed by the mid-1940s, a bit late for the troops but still a most worthwhile effort. Quinine isolated from bark is still cheaper than synthesized quinine (Figure 7.16) and is still being used to cure one major type of malaria that is resistant to synthetics. A small but growing market for quinine has developed since the 1960s, when tonic water with quinine was found to mix nicely with gin. Natural quinine, synthetic antimalarials, and other mosquito-control measures continue to provide us with, if not complete freedom from malaria, at

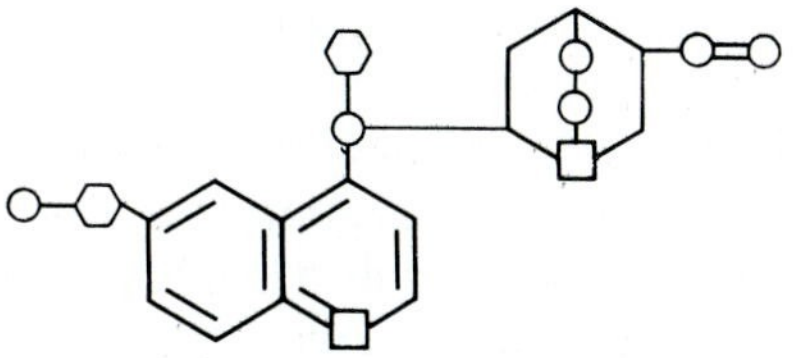

Figure 7.16 Molecular structure of quinine. Hydrogen atoms and carbon atoms in the rings are not shown. Carbon atoms are represented as circles, oxygen atoms as hexagons, and nitrogen atoms as squares.

least reasonable control. And, ironically, the people and their children whose lives have been saved are a major part of the world's population problem.

ADDITIONAL READINGS

Duran-Reynals, M. L. *The Fever Bark Tree: The Pageant of Quinine.* Garden City, N.Y.: Doubleday, 1946

Haggis, A. W. "Fundamental Errors in the Early History of Cinchona." *Bulletin of the History of Medicine* 10 (1941):417–459, 568–592.

Stanford, E. E. "The Story of Cinchona and the Mosquito," *Natural Magazine* 21 (1933):65–68.

Taylor, N. "Quinine to You." *Fortune Magazine* (February 1934).

Taylor, N. *Cinchona in Java: The Story of Quinine.* New York: Greenburg Press, 1945.

Urdang, G. "The Legend of Cinchona." *Scientific Monthly* 61 (1945):17–20.

The Purple Foxglove

The foxglove's leaves with caution giv'n
Another proof of favoring heav'n
The rapid pulse it can abate
And blest by Him whose will is fate.

WILLIAM WITHERING

Among the ills to which flesh is heir is cardiac insufficiency, a condition in which a weakened heart fails to pump enough blood through the body. Heartbeat is irregular and fluids collect in the arms, legs, and abdomen because the kidneys cannot perform their normal function. The swelling is known as *dropsy* or, more formally, as *edema.* This disease syndrome is not new. Ancient physicians knew of it, but because they lacked knowledge of the circulation of the blood—it was discovered by William Harvey in 1628—and information on the function of the kidneys, treatment was limited to usually unsuccessful attempts to reduce edema with medicines which increased urine production (diuretic agents). Today, millions of people pop a small pill that regulates and strengthens the heartbeat and allows the kidneys to expell excess fluid quickly; cardiac insufficiency kills few people since the discovery of digitalis.

BOTANY AND MYTHOLOGY

Digitalis is a mixture of several naturally occurring cardiac glycosides synthesized by *Digitalis purpurea* and related species in the figwort (Scrophulariaceae) family (Figure 7.17). Native to Europe, western Asia, and central Asia, it is grown all over the world. The plant may be an annual—flowering the first year after planting—or a biennial—flowering the second year. The leaves are large, up to 30 centimenters long, covered with fine hairs (pubescence), and borne on a condensed stem producing a rosette rarely taller than 10 centimeters. When the plant flowers, it sends up tall stalks on which develop a raceme of thimble-shaped, white to pink to red flowers. The floral lip is structurally adapted as a landing platform for its pollinator, the bee, and colored spots on the corolla "light" the landing strip and direct the bee to the nectar-secreting cells at its base. Because foxglove is cross-pollinated, there is great genetic variability in the 1 million seeds formed in the flowers of a single plant—a fact of some significance in evaluating the quality and quantity of cardiac-stimulating chemicals formed.

The Latin generic name, *Digitalis,* is something of a puzzle. It means "little finger," from the Latin *digitis,* and is directly derived from the German name for the plant, *fingerhut.* The Latin term was applied by Hieronymus Tragus in 1539 and was repeated in Leonhard Fuchs's *Historia Stirpium* in 1542. Several species are native to Europe, and all bear the common name foxglove or variants on the same theme: foxbell, fairy bell, fairy glove, fairy fox, fairy thimble, or its equivalent in French, German, and other languages. According to one legend current during the reign of England's Edward III, a band of bad fairies gave *Digitalis* flowers to a fox so that he could put them on his toes to muffle the sound of his footsteps when he raided hen houses. Foxbell and fairy

Figure 7.17 Flowering heads of foxglove *(Digitalis purpurea).* Courtesy W. Attlee Burpee Co.

bell are obviously derived from the shape of the flowers in their natural habitat of hollows and wooden glens favored by wee folk as well as by foxes. The common names Our Lady's glove or Our Lady's thimble are examples of the attempts by Church leaders to relate everyday sights to the holy family. Some have suggested that the common names are derived from old Middle English *Fuche's glew*—fox music—from a likeness of the flowers to a Saxon musical instrument that consists of a series of bells hung on an arched support.

Digitalis was a medicinal herb for centuries; Dioscorides praised it as a plant whose leaves, applied to the skin, could cure many diseases. Juice pressed from the leaves became an ingredient of salves applied to cuts, bruises, and the leg ulcers common in an era of inadequate diets and lack of soap. Rural people made hot water infusions of leaves and drank foxglove tea to experience an inexpensive but dangerous intoxication. One herbalist said that "it will scoure and clense the brest of the thicke toughness of grosse and slimie flegme and naughtie humours," and another reported that it would cure "pimply bodie, sore head, sore ears, heat of the maw and churnels"—whatever they were. John Gerard wrote that "foxgloves are bitter, hot and dry with a certain kind of clensing qualitie joined therewith; yet they are of no use, neither have they any place amongst medicines," but John Parkinson disagreed, believing that digitalis ointments would cleanse out sores and ulcers and could cure epilepsy. Nicholas Culpepper reported that an ointment containing the juice of foxglove leaves would cure the king's evil (tuberculosis of the lymph glands) and that "it is one of the best remedies for a scabby head that is." In the New London dispensatory of 1687, William Salmon wrote:

> Fox-glove expectorates thick phlegm, if drunk with thick mead [it] clears away obstructions of the liver and spleen, is an extraordinary good wound herb, prevalent against King's Evil and may be used instead of gentian. Two handfuls of the herb taken with polypody helps the epilepsy.

ENTER WILLIAM WITHERING

The modern history of digitalis in cardiac insufficiency is completely tied up with the life of a single person. William Withering was born in 1741, the son of a Wellington, England, apothecary. A bright, middle-class boy, he played the flute, bagpipes, and harpsichord, was a skilled archer, a good golfer, and participated in local dramatics. In 1766 at the age of 25 he received his medical degree from the University of Edinburgh, distinguishing himself in several courses, but, as he wrote his parents, despising the required botany course: "The offer of a gold medal by the professor would hardly have charm enough to banish the disagreeable ideas I have formed of the study of Botany." He established a general practice in the town of Stafford in Shropshire where one of his first patients was the gentle and delicate Helen Cookes. William courted her by collecting flowers for her to paint. Through the influence of a good woman, he became devoted to the "lovable science," married Helen Cookes in 1772, and, in 1776, published *A Botanical Arrangement of All the Vegetables Naturally Growing in Great Britain According to the System of the Celebrated Linnaeus, With an Introduction to the Study of Botany.* The book so interested the distinguished Erasmus Darwin that Withering was invited to share the lucrative practice of Charles's grandfather in Birmingham. Withering maintained a clinic in Stafford, making the 60-mile trip each week. On one trip, as he stopped to change horses, he was asked to visit an old woman suffering from the dropsy. Dr. Withering concluded that she could not survive and was astounded when, a few weeks later, he found his patient alive and apparently healthy. In his own words:

> In the year 1775, my opinion was asked concerning a family receipt for the cure of the dropsy. I was told that it had long been kept a secret by an old woman in Shropshire, who had sometimes made cures after the more regular practitioners had failed. I was informed also that the effects produced were violent vomiting and purging; for the diuretic effects seemed to have been overlooked. This medication was composed of twenty or more different herbs; but it was not very difficult for one conversant in these subjects to perceive that the active herb could be no other than Foxglove.

The real story is slightly different. Mrs. Hutton, locally known as a witch, grew herbs in her garden and also collected plants from which she brewed various concoctions and sold them to the local people as remedies. Villagers believed that she conspired with the Devil to make people sick so that she could cure them. She would wave her arms over the person's head, mumble incantations, stare intently at the person, and then run out into the garden, pull up plants, and brew tea which she sold to the now thoroughly frightened patient. The "family receipt," purchased by Dr. Withering for "several gold sovereigns," included the critical information that only the foxglove had any effect on dropsy and that the other plants made the patient throw up to prove how powerful the medicine was.

Still maintaining a successful general medical practice in Birmingham and his clinic in Stafford—plus the responsibilities of a growing family and the public service that is the lot of a prominent member of a respected profession—Withering spent close to ten years studying digitalis therapy in dropsy. Chemistry was not sufficiently advanced to permit the isolation of the active ingredients: biology in general and human physiology in particular were just in their infancy. Questions included the following: Which part of the plant was most active? (the leaves); could the leaves be dried? (they could); what was the best solvent for the active material? (alcohol was good, but the extract had undesirable side effects—cold water was best); should one pick leaves in early or late summer? (mature leaves from 2-year-old plants were the most active); and, most important, what was the optimum dose and how frequently should it be administered? In 1785, he published *An Account of the Foxglove and Some of Its Medicinal Uses,* a clinical study so detailed and so accurate that if one had no other information than is contained in the *Account,* one could use digitalis effectively and safely:

> Let the medicine therefore be given in the doses and at the intervals mentioned above; let it be continued until it either acts on the kidneys, the stomach, the pulse or the bowels; let it be stopped upon the first appearance of any of these effects, and I will maintain that the patient will not suffer from its exhibition, nor the practitioner disappointed in any reasonable expectation.

The report was something of a bombshell in England, where the standard treatment for dropsy was to puncture the waterlogged tissues with a scalpel (unsterilized) and stretch the patient over bedsprings to allow the fluid to drip into buckets. Withering was elected to the Linnaean Society and the Royal Society of London, received honorary degrees in Germany and France, and had a genus of tropical American plants, *Witheringia,* named in his honor. His medical practice grew apace, but he found time to botanize and even discovered a new crystalline form of barium carbonate, which was named *witherite.*

Nevertheless, foxglove treatment of dropsy was still viewed with suspicion since it had a history of association with old herb women, and many doctors, following the fifteenth-century herbalists, were using it for king's evil and epilepsy. If, they said, it could cure dropsy, it should certainly cure these other diseases, and when it couldn't, it was clearly useless. Furthermore, if a little was good, a lot is better, and when some patients died of overdose, the fault clearly lay with Withering. Yet in a time when concepts of witchcraft and of divine punishment in the form of disease were still extant, when the Doctrine of Signatures was still believed, when the relationship between cause and effect was unclear, and when the idea of a controlled experiment was scarcely understood, a paragraph in Withering's *Account* is worth reading:

> As the more obvious and sensible properties of plants such as colour and taste and smell have but little connexion with the diseases they are adopted to cure; so their peculiar qualities have no certain dependence upon external configuration. Their virtues must be learnt either from observing their effects upon insects and quadrupeds; from analogy deduced from the already known powers of some of their congeners, or from the empirical usages and experiences of the populace. This last lies within the reach of every one who is open to information, regardless of the source from whence it springs. It was a circumstance of this kind which first fixed my attention on the foxglove.

William Withering died of tuberculosis in 1799 and was buried in a vault on which a *Digitalis* plant was engraved.

Although he noted that digitalis "has a power over the motion of the heart, to a degree yet unobserved in any other medicine, and that this

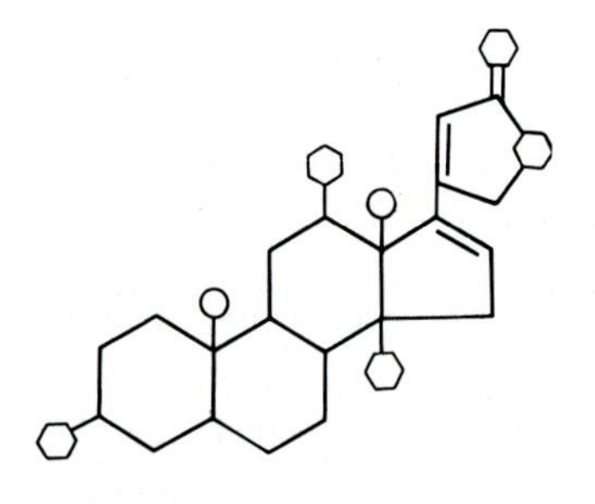

Figure 7.18 Molecular structure of digoxigenin, a cardiac stimulant from foxglove. Hydrogen atoms and carbon atoms in the rings are not shown. Oxygen atoms are represented as hexagons, carbon atoms as circles.

power may be converted to salutary ends," there is no evidence that he related dropsy to cardiac insufficiency. With the discovery of the stethoscope by René Laennec in 1820 and the development of medical pharmacology in the middle of the nineteenth century, the effects of digitalis on the heart became clearer, although much still remains to be studied. Indeed, we now know that digitalis is not a single compound but a mixture of four or five cardiac glycosides (Figure 7.18) with interacting effects on the heart. Recent clinical studies have shown its effectiveness in treating glaucoma (increased fluid in the eyeball), in neuralgia, asthma, and several other physiological diseases.

Digitalis purpurea arrived in North America after the Revolutionary War when Withering sent seeds to Hall Jackson of Portsmouth, New Hampshire, in 1787, to be cultivated in Jackson's medical garden. Escapes from this initial planting and subsequent introductions of ornamental plants have distributed the plant throughout New England, Oregon, Washington, Pennsylvania, and New York; a few hundred acres supply the medical needs of the world.

ADDITIONAL READINGS

Dowling, H. F. *Medicines for Man.* New York: Knopf, 1970.

Estes, J. W., and P. D. White, "William Withering and the Purple Foxglove." *Scientific American* 212, no. 6 (1965):110–119.

Fisch, C., and B. Surawicz. *Digitalis.* New York: Grune & Stratton, 1969.

Gordon, B. L. *The Romance of Medicine.* Philadelphia: Davis, 1944.

Kranz, J. C., Jr. *Historical Medical Classics Involving New Drugs.* Baltimore: Williams & Wilkins, 1974.

Marks, G., and W. K. Beatty. *The Medical Garden.* New York: Scribners, 1971.

Trease, G. *Pharmacy in History.* London: Balliere, Tindal & Cox, 1964.

New-Mown Hay

Doc, what's killing my cows?

EDWARD CARLSON OF WISCONSIN

One sure sign of spring is the snarl of the rotary lawnmower chewing through the sacred turf of suburbia. Another is the smell of the new-mown grass. This pleasant odor is produced by a quinonelike compound, coumarin (Figure 7.19). Coumarin was chemically known as early as 1820 and its structure worked out before 1900. Because of its agreeable odor, it was widely used as one of the ingredients in perfumes. A man's cologne called, for example, Field and Forest, or a woman's scent named Breath of Spring, certainly would contain coumarins. Bergamot oil, another coumarin, is a fixative for perfume. These substances were used to flavor chocolate or to give a "manly" scent to pipe tobacco, they are found in vanilla flavorings and in many other products, including Earl Grey tea.

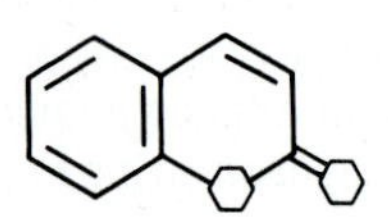

Figure 7.19 Molecular structure of coumarin. Hydrogen atoms and carbon atoms in the rings are not shown. Oxygen atoms are represented as hexagons.

Coumarin is only mildly toxic and the amounts necessary to cause more than a slight discomfort are very high. And yet there had been reports for at least a hundred years of cattle dying when fed on clover hay or silage containing clover. Livestock that eat clover hay can die of massive hemorrhages within a few days, with blood spouting from their mouths. This phenomenon was reported in detail in 1919 by Dr. F. W. Schofield, a noted Canadian veterinarian. Schofield wrote that in all cases, the hay was moldy. He precisely described the clinical symptoms and then ran a brilliantly simple experiment. Carefully, he separated stalks of clover hay into two piles, one which contained plants that were visually free of mold and a second consisting of moldy stalks. Three laboratory-raised rabbits were fed clean hay, and three received moldy hay. Those fed the infected clover died within a few days with typical clover-disease symptoms; those that ate the clean hay were normal.

Schofield then took the next experimental step. He isolated fungi from the moldy clover and infected fresh hay. These stalks also killed his test rabbits. Extracts of moldy hay, but not those from clean hay, were injected into rabbits; there was a dramatic drop in the ability of their blood to coagulate. Schofield correctly concluded that moldy hay contained something produced by the interaction of mold and plant that caused the bleeds. He and others tried, unsuccessfully, to find out what this substance was. It wasn't the coumarin in the hay, they were sure about that.

During the depths of the Depression, Edward Carlson, a Wisconsin dairy farmer, lost two heifers to the bleeds, and a milker died after receiving a scratch on a barbed wire fence. In early February 1933, Carlson found that his prize bull was bleeding from the nose. He loaded his truck with a dead cow, a hundred pounds of his hay, and several milk cans filled with the blood of a cow that had died the day before. In a midwest blizzard, he drove the 200-odd miles to Madison, hoping to find the veterinarian of the State Agricultural College. The vet was out of town and Carlson, by chance, went into the agricultural biochemistry building to find out where he could leave his gruesome load. He met Professor Karl Paul Link, who was familiar with the clover bleeds but had had no direct experience with it. Link urged the farmer to stop feeding the moldy hay, but he couldn't hold out much hope for the cattle that were already sick.

After Carlson left, Link opened one of the milk cans and saw that the blood was still liquid. Obviously, this was what the bleeds condition was all about; any simple injury could rupture a blood capillary and, with the failure of the blood to clot, internal bleeding would soon cause death. This is what happens to sufferers of hemophilia, including some of the male members of royal families in Europe.

Why Professor Link decided to get involved in this problem even he didn't know. He just decided. He also made another decision of great importance. If you are trying to find some specific chemical out of the thousands present in a plant, you must have some way of identifying it.

This development of a biochemical assay system can be very critical. Earlier workers on the bleeds disease used rabbits, injecting them with various chemicals and then waiting to see if typical symptoms developed. Link decided that since the clover-bleeds substance works directly on the blood, some biochemical assay could be developed that would measure the clotting time of blood, a much simpler and more direct approach than using whole rabbits.

The trouble was that the standard method of measuring blood clotting time was imprecise and somewhat unreliable. So Link's first job was to improve and standarize the blood clotting assay. This took over a year, but it was worth the time. The assay he developed not only worked well for domestic animals but supplanted the one used in hospitals to assess surgical risks. With this assay, it became possible to evaluate which of the hundreds of chemicals extractable from moldy clover might be the chemical agent of the bleeds. Tons of moldy hay were chopped up and extracted with water, alcohol, and other solvents; each extract was tested to see if it reduced the ability of blood to clot. By 1937–1938, he had a concentrated extract that had over 200 times the anticoagulating activity of hay.

These extracts were laboriously purified and the components tested. On June 28, 1939—a full five years after the project started—Dr. H. A. Campbell, a postdoctoral fellow in Link's laboratory, obtained a few tiny crystals on a microscope slide. Not only did the compound prevent blood clotting, but a highly diluted solution produced the typical bleeds symptoms in rabbits. By 1940, the research team had 1800 milligrams of the crystals and were able to determine its chemical structure. In 1971–1942 they synthesized it. The compound, named *dicoumarol* (Figure 7.20), was essentially two coumarin molecules linked together, the link being accomplished by the fungi that infected the hay.

Although this was a fine job of laboratory biochemistry, it didn't seem to be of much use to Ed Carlson or other dairy farmers. And so the Wisconsin biochemists started putting together the available information on dicoumarol with data in the medical literature on blood clotting. In essence, blood clots normally develop because fine, thread-like protein molecules, fibrin, form a mesh that hardens into a clot. This traps blood cells, and the wound is sealed off. For fibrin to form, many chemicals are needed, one of which is vitamin K (Figure 7.21). It had long been known that people with vitamin K deficiencies bleed easily. The Wisconsin scientists found that dicoumarol prevented vitamin K from acting in the formation of fibrin. The cure for the cattle bleeds was at hand; animals showing the first signs of the bleeds could be saved by a massive injection of vitamin K. Ed Carlson's trip had paid off.

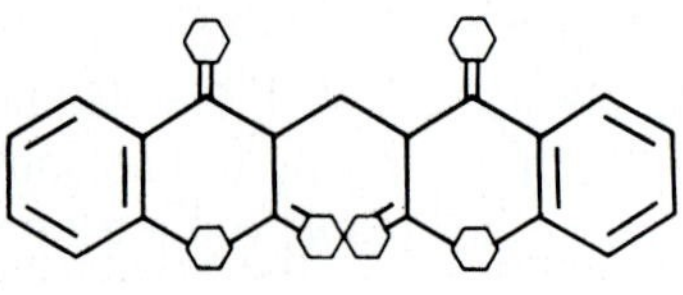

Figure 7.20 Molecular structure of dicoumarol. Hydrogen atoms and carbon atoms in the rings are not shown. Carbon atoms are represented as circles, oxygen atoms as hexagons.

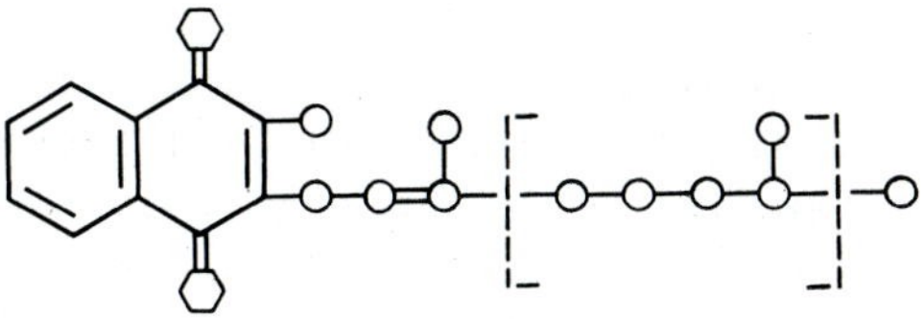

Figure 7.21 Molecular structure of vitamin K_1. Hydrogen atoms and carbon atoms in the rings are not shown. Carbon atoms are represented as circles, oxygen atoms as hexagons.

BLOOD CLOTS IN HUMANS

The publication of the Wisconsin studies caused medical scientists to perk up their ears. The fact that a pure compound was available to prevent blood clotting was of considerable medical interest. In many human diseases and injuries blood clots can form, move through the body to the heart or lungs, and cause a serious or even fatal blockage of blood flow. If an anticlotting agent were supplied, many deaths could be prevented. Heart specialists at the Cornell University Medical School in New York found that dicoumarol was just what they were looking for. Coronary thrombosis, caused by massive clots in heart blood vessels, could now be controlled. Indeed, President Eisenhower was one of thousands of people who spent the rest of their lives using small, controlled doses of dicoumarol to prevent clots. In phlebitis, as President Nixon found out, the danger of damaging clot formation in the legs is greatly reduced by dicoumarol. If, by miscalculation, too much dicoumarol is taken, its effects can be quickly determined by a blood clotting test and the patient can be removed from danger with an injection of vitamin K.

RAT POISON

End of the success story? Not quite.

In 1946, Link was hospitalized and, to keep his mind occupied, he began to read up on natural history and on the role of rodents in agriculture. Based on his own laboratory studies, he knew that rats and mice were fairly insensitive to dicoumarol because their blood contains considerable amounts of vitamin K. But suppose dicoumarol were chemically modified? Could it be a rat poison? Link's train of thought was set off by one of the many strange things about rats. Apparently packs of rodents choose some wise old rat as official taster. He will eat some new food and the rest of the pack won't touch this food for several days until they are sure that the taster has remained healthy. Since dicoumarol can take several days to have its anticoagulating effect, maybe the pack could be fooled into eating a slow-acting poison. Obviously coumarin itself would not work, but perhaps it might be possible to modify the dicoumarol molecule. This task took over a year to accomplish, but by 1947, "Compound 42" met all the requirements for a rodenticide. The research was supported by the Wisconsin Alumni Research Fund—an endowment built up from the royalties on patents at the University of Wisconsin for the irradiation of milk with ultraviolet light that increased its vitamin D content ("Vitamin D Added"). Link named Compound 42 "Warfarin," and it is now one of the standard rodenticides in the world.

LEUCODERMA

Actually, coumarins had been known medically for centuries. In India, where human skin pigmentation is heavy, individuals occasionally were found with patches of pure white skin somewhere on their bodies. This condition, called *leucoderma,* resulted in a piebald, unsightly appearance. The condition also existed in Egypt, as seen in tomb paintings. It is not restricted to dark-skinned people, however; about 1 percent of the North American white population has leucoderma, also called *vitiligo.* It is not dangerous to health or longevity, but is psychologically unpleasant.

In the *Atharva Veda* of 1600 B.P., Brahman physicians recommended that certain herbs be mashed into a paste and applied to the pigmentless regions of the skin. Chinese Buddhists of 1400 B.P. noted that the pastes were most effective when, within a few hours following application of the plants, the patient stood in the sun for an hour or so. The recommendations of the Egyptian physicians were exactly the same. In all cases, the treated areas would darken to virtually natural skin tone in a few days, and the treatment didn't have to be repeated for several months. In Western medicine, the first clue came in 1916, when Dr. E. Freund, a German dermatologist, found that some of his very light-skinned patients displayed pigmented areas on part of their bodies that had been exposed to both plants and sunlight—knees and elbows were the most common areas. He also found that these same people were frequently sensitive to eau de cologne, with darkening of the skin on the back of the neck and around the wrists likely to result. Once his findings were reported in the medical literature, other physicians observed such occurrences of photodermatoses in light-skinned patients. In all cases, the exposure to plants had to be followed by exposure to sunlight.

The connection between the medical observations and the ancient Indian treatment for leucoderma was made by a Indian dermatologist in Boston. By 1940, Dr. M. Pathak found that the active substance in the plants was a chemical related to coumarin; when activated by solar ultraviolet radiation, it caused changes in the pigment-producing layers of the skin. The Egyptian and Indian plants, identified with the help of ancient medical texts and the knowledge of plant taxonomists, were found to be primarily in the genus *Amni.* A North American representative is our bishop's-weed (*Amni majus*). By the early 1950s, a standard treatment for leucoderma was developed. Plants didn't have to be mashed up, instead, the pure chemical, psoralen (Figure 7.22), was painted on the skin or taken as a pill and the patient exposed to radiation from an ultraviolet lamp. This treatment is called PUVA

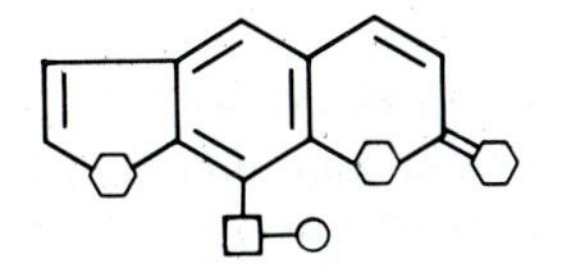

Figure 7.22 Molecular structure of 8-methoxypsoralen. Hydrogen atoms and carbon atoms in the rings are not shown. Carbon atoms are represented as circles, oxygen atoms as hexagons.

therapy (Psoralen UltraViolet A). Most of the available evidence indicates that leucoderma is a genetic abnormality; while the treatments will not change that, the patches of white skin will be gone. The treatment can be repeated as necessary, and the psychological lift experienced by the individual is gratifying.

Since we in the Western world equate a tanned skin with health (nothing could be further from the truth), cosmetic manufacturers soon exploited these medical findings by developing preparations for a quick tan. Such preparations are now banned because there are side effects that cannot be controlled with casual use.

DERMATITIS

It was first reported in 1897 that some vegetable growers were prone to blistering of their hands and arms. Those who grew parsnips seemed to be the most severely affected, but carrot growers were also at risk. Fig pickers in Italy developed eruptions, scaliness, and pigmentation of their hands; in 1926, celery workers in Belgium and citrus pickers in Morocco began the same symptoms. Botanical dermatologists started putting these reports together and, by 1930, it became obvious that the farm and produce workers were all handling plants that naturally contained coumarins. The condition was more severe when the vegetables and fruits were infected with fungus diseases. The worst cases were caused by members of the Umbelliferae, which includes parsnips, carrots, celery, and fennel, and by members of the Rutaceae, which includes the common citrus fruits.

Celery dermatitis was particularly virulent. In 1948, Dr. J. S. Wiswell examined workers in the Boston market where the 'Pascal' celery was shipped. The workers told Wiswell that they came down with the "trouble" only when they were handling crates with rotted plants, but, of course, no one listened to them. It was 1963 before someone found that a pink rot fungus (*Sclerotinia*) was changing the coumarin in celery into a fucocoumarin, 8-methoxypsoralen, precisely the same compound that is used to treat leucoderma. Obviously, careful washing of exposed skin before going out into the sun solved the problem. A number of the rashes that people, especially children, get in the summertime are the result of photodermatitis caused by accidental contact with a coumarin- or psoralan-containing plant. Cow parsnips (*Heracleum*), Queen Anne's lace (*Daucus*), the angelicas used in gin, oil of caraway, poison hemlock, and many other wild plants are sources of psoralens.

PSORIASIS

End of story? Well, almost.

There is a very annoying disease called *psoriasis,* in which red, scaly patches appear on the scalp, elbows, knees, and back. The name is derived from the Greek word for itching, and the condition can drive the sufferer to distraction. Almost 8 million North Americans have psoriasis. The usual treatments are not too effective and are potentially dangerous. Dr. Mahdukar Pathak and his colleague Dr. Thomas Fitzpatrick, both of whom had worked on the treatment for leucoderma, found that oral treatment with one of the coumarin-derived psoralens, combined with ultraviolet irradiation, is highly effective and with proper supervision is an excellent therapy for psoriasis. More recently, these same researchers have suggested that some persistent fungal diseases of human skin respond well to the same regimen.

One final note. If you do any Indian, Mexican, or Chinese cooking, prominent seasonings include cumin and coriander, which impart a coumarin flavor to the food. And the spicy taste of common parsley or the dillweed we sprinkle liberally on boiled potatoes is also a coumarin flavor.

ADDITIONAL READINGS

Kadis, S., A. Ciegler, and S. J. Ajl (eds.). *Microbial Toxins.* Vol. III. New York: Academic Press, 1972.

Lampe, K. F. *Plant Toxicity and Dermatitis.* Baltimore: Williams & Wilkins, 1968.

Link, K. P. The Harvey Lectures 1943–1944. 39:162.

Mitchell, J., and A. Rook. *Botanical Dermatology: Plants and Plant Products Injurious to the Skin.* Vancouver, B. C.: Greengrass Press, 1979.

Pathak, M. A. "Phytophotodermatitis." In M. A. Pathak et al. (eds.). *Sunlight and Man,* Chapter 32. Tokyo: University of Tokyo Press, 1974.

Chapter 8

Booze

Wine

Not only does one drink wine, but one inhales it, one looks at it, one tastes it, one swallows it . . . and one talks about it.

EDWARD VII OF ENGLAND, 1905

In a book on wine published in 1923, the author described wines he had drunk. He spoke of one as "a wine of senatorial dignity with the mysterious haunting appeal found only in very old wines that may be compared with the sound of the harmonic on the violin." And of another: "This wine breathed forth a perfume worthy of the Gods. It was compounded of a multitude of subtle fragrances, the freshness of sunripened grape, etherealized by the patient work of Nature into a quintessence of harmonic scents. The full organ swell of a triumphal march might express its appeal." These statements, however overwrought, do not tell one just what wine is. As a start, wine is the fermented juice of grapes. But wine is considerably more than alcoholized grape juice, and its preparation is worthy of study as part of the liberal education of the intellectual (Figure 8.1). If one were to state that the world's finest wines are French, the uproar from other countries would shake the walls of Jericho, but techniques developed in France over 2000 years of winemaking are the standards by which all wine is judged. To discuss the science of *oenology,* the French model will be used.

Figure 8.1 Punishment for drunkenness: the stocks. Taken from *Spiegel Menchlicher Behaltnis* by Berger, published in 1489 in Augberg.

LE VIGNE

The wine grape, *Vitis vinifera,* originated in the Caucasus Mountains at the southern end of the Caspian Sea as a self-fertile (monoecious) mutant from the normally two-sexed (dioecious) wild plant. It was domesticated before 7000 B.P., was grown in Greece by 5000 B.P., and "the blood of the grape" was known in Egypt by 4400 B.P. (Figure 8.2). Homer sang of wine in the *Iliad.* According to *The Bacchae* of Euripides, Dionysus gave wine to Man: "Filled with that good gift, suffering mankind forgot its grief." As libation and beverage, wine was known to the Israelites; there are over 160 references to the vine and its product in the Testaments. Rome, too, was well acquainted with wine. Pliny observed that it refreshed the stomach, sharpened the appetite, blunted care and sadness, and was conducive to slumber. Vines were planted in France about 2600 B.P., and Rome introduced wine grapes to the Rhine Valley in 2100 B.P.

The wine grape is one of 50 species of *Vitis* in subtropical and temperate zones around the world. It is a climbing vine, capable of attaching to upright supports by tendrils. Flowers are small and

Figure 8.2 Steps in winemaking. Taken from an Egyptian tomb painting.

inconspicuous. The fruit is botanically a berry, produced in clusters, and contains two or more hard, bitter seeds. Through selection of natural mutants (what horticulturists call *sports*) and by crossbreeding, there are now several thousand recognized cultivars or varieties, although fewer than a hundred provide the bulk of the world's fine wines.

In France, two dozen cultivars are used. In Burgundy, the pinot types and the gamay, brought to France as tangible results of the crusades, are the primary wine grapes. Within each of the cultivars there are subtypes, the French *cépages,* so that a black pinot grape of one grower is not exactly the same as the pinot noir of another vineyard. The tokay of middle Europe is a pinot, as are many others under different names. The white-skinned merlot and cabernet sauvignon grapes are used for most white wines in Bordeaux, including both dry wines and sweet sauternes, while the dark-skinned merlot and cabernet sauvignon grapes are used for rich, red clarets. The reisling and sylvaner grapes produce rhine and moselle wines, while two red-skinned grapes, the sangiovese and the malvasia, predominate in the chianti of Tuscany.

Of the six or eight North American species of *Vitis,* the most important is *V. labrusca,* the northern fox grape. It differs from *vinifera* in that the skin is easily separated from the pulp because of a mucilaginous layer just below the skin. Wines made from labrusca grapes have a characteristic taste due to the presence of methyl anthranilate. The fox grape so impressed Leif Ericsson in the tenth century that he called the shores of Newfoundland "Vinland." Labrusca was first bred by the Rev. W. W. Bostwick of upstate New York to yield niagara, and the famous concord was a mutant of niagara isolated in Massachusetts. The catawba cultivar was isolated from the wilds of North Carolina. Labrusca cultivars are disease resistant and cold-tolerant, but the grapes are low in sugar, and produce soft, fruity wines frequently fortified with sugar and additional alcohol; kosher wines are examples of this processing.

LE CLIMAT

Climate is almost as important as is the cultivar. Vinifera grapes can be divided into two major groups. Those that do well in the hotter, drier climates of southern Europe are called Mediterranean types and include grenache, syrah, muscat, and cabernet grapes. Temperate climate grapes include gamay, semillion, sauvignon, all pinots, the merlots of Bordeaux, and the white grapes used for rhine and moselle wines. The success of California's wine industry is due to the wide variations in climates found in the state. Dry, very hot Mediterranean climates are found in semidesert valleys with cooler, moister areas of mountain slopes planted to temperate zone cultivars.

Climatic preferences are, however, more subtle than seen on a temperature chart. Vines grown on north- or south-facing slopes of the same valley will produce grapes varying in sugar, acid, and pigmentation. The wine of Château Latour and that of Château Léonville–Las Cases are produced from the same grapes grown in Bordeaux vineyards separated by a narrow gully. Latour's slopes face south, receiving perhaps an hour more sunlight a day than do the vines on the north-facing Léonville–Las Cases, but this adds up to several hundred more sunlight hours for the photosynthesis of Latour's vines. Latour is a great wine, while Léonville–Las Cases is merely a very fine wine.

Variations in weather may strongly affect the character of wine. Low night temperatures favor the conversion of sugar into acids, resulting in a more long-lasting wine. Rain, or lack of rain, a too hot or too cold period of only a few days, hail, or a multitude of slight changes may so alter the grapes that the best skills of the vintner are required to utilize the fruit properly. Differences in weather are reflected in the vintage dates that appear on wine labels, and "perfect" years are few and far between, resulting in prices which make a wine lover weep.

LA TERRE

Soil composition and structure are important (Figure 8.3). Soils used for wine grapes are generally rapidly draining, poor in organic matter, and relatively poor in nutritive minerals. Even slight differences in soil can greatly alter the character of the wine. In the St. Emilion district, the highly rated wines of Château Cheval Blanc are from grapes grown on a stony soil overlying solid limestone. The neighboring Château La Gaffelières–Naudes has a slightly better soil with more humus, but its wines are not as highly rated. Fertilization of the vines is done cautiously. Addition of too much nitrogen fertilizer results in more growth of the stems and leaves, and the photosynthate that would otherwise go into the grape is used for vegetative growth.

Figure 8.3 Breaking the soil to plant grape vines. Taken from *Opus Ruralium Commodorum* by Piero Crescentio, published in 1493 in Venice.

LA CULTIVER

In general, five years are needed to bring a new planting into full production, while up to ten years may be required to determine whether a planting will have the potential to produce a superior wine. All grapes are propagated by cuttings or graftings and remain true to type. Vines are trained along wire trellises, with row distances calculated to allow maximum exposure to the sun. Only rarely are there more than 1500 plants per hectare. Pruning (Figure 8.4) is done after the harvest to control the number of flower-bearing branches laid down the previous summer and early fall. The grower must balance maximum grape production against grape quality.

Figure 8.4 Pruning vines. Taken from *Opus Ruralium Commodorum* by Piero Crescentio, published in 1493 in Venice.

Although it was their own fault, the French have never forgiven Americans for the introduction of *Phylloxera* into French vineyards. Europeans took grapevines from North American to France in 1855–1860 as breeding stock, unaware that the roots carried this tiny insect. Wingless females of *Phylloxera vastatrix* insert their mouth parts into roots, sucking up the sweet sap and causing disruption of root structure. Females produce more wingless females and also lay eggs which develop into winged females. In less than 15 years, phylloxera had almost destroyed the vines of Europe. Charles V. Riley, a U.S. entomologist, noted that the phylloxera then ravishing Europe's grape vines was relatively benign on American species of *Vitis* and suggested that the vinifera grapes be grafted to *V. labrusca* rootstocks. This saved the wine industry of Europe.

LA VENDANGE

There are usually 80–100 days between fertilization and fruit maturity. Sugars accumulate only slowly in young fruit, as we know from Aesop's fable of the fox and the young, sour grapes. In the few weeks before full maturity, sugars accumulate rapidly. When the grapes are ripe, 70 percent of the weight of the fruit will be the juice (cell sap) and between 20–24 percent will be the two hexose sugars, glucose and fructose. Organic acids such as malic (also found in apples) and citric (in oranges) plus tartaric acid, vitamins, proteins, and other cell constituents are present. Acid levels are important; if the acid constitutes less than 0.5 percent, the resulting wine will be insipid; levels above 0.7 percent will allow the production of a tart, long-lasting wine. Aroma and flavors are determined by many different compounds, including aldehydes, esters, and alcohols, and decisions on when to pick the grapes include evaluations of all these factors.

Figure 8.5 Harvesting the grapes. Taken from *Opus Ruralium Commodorum* by Piero Crescentio, published in 1493 in Venice.

Depending on the grape and the wine to be made, picking may be done all at one time or may be spread over several weeks (Figure 8.5). The sauvignon grapes used to make sauternes are picked at full maturity to yield a relatively dry, fruity wine. At Château d'Yquem, the grapes are allowed to become overmature to the point that they shrivel and begin to rot. The consequent reduction in water content and increase in sugar plus the flavors imparted by the rot fungus interact to produce a sweet dessert wine. In the Rhine wines, grape overmaturation determines the name of the wine. *Spätelese* wines are made from grapes picked late; *auslese* wines are made from individually selected bunches of overmature grapes. *Beerenauslese* wines are made from the juice from individually selected grapes from bunches showing both drying and some rotting, while the most expensive Rhine wines are the *trockenbeerenauslese,* made from individually selected, rotted grapes which have dried out almost to the raisin stage.

LE PRESSOIR

Following mechanical removal of stems, grapes are pressed to yield their juices. The process was traditionally done with human feet. The screw press of the fifteenth century and its modern descendants are now used (Figure 8.6). Pressing must be thorough enough to crush the grapes, but not so harsh as to break the seeds and release compounds that can ruin the wine. For white wines, the grape skins are removed immediately after pressing to avoid getting any skin pigments into the fermenting vats. Both white-skinned and red-skinned grapes can be used to make white wines. For rosé wine, the skins are allowed to remain with the juice during the initial stages of fermentation.

LA FERMENTATION

Grape juice, plus a pomace of pulp, seeds, skins, and some stems, is transferred to large fermentation vats traditionally made of oak (redwood in California), but now of concrete, ceramic, or stainless steel (Figure 8.7). Temperatures close to 30° C are best for the initial fermentation of red wines; 18° C is optimum for white wines. Yeasts are killed at temperatures above 38° C, and, since active fermentations can reach this point, cooling the vats is important. Although a few vineyards still utilize natural yeasts which form a whitish bloom on the surface of grapes, the bloom also contains bacteria, other yeasts, and other fungi that can spoil the fermentation. It is now standard practice to kill or inactivate all microorganisms in the juice (now called the *must*) in the vat and then to inoculate the must with a pure culture of the desired strain of *Saccharomyces ellipsoides* (Figure 8.8), the wine yeast. This concept of using a pure culture of yeast was initially developed by Louis Pasteur, a loyal wine drinker whose scientific life revolved around the wine industry. Pasteur also found that when fermentations went "bad,"

Figure 8.6 A windlass wine press. Taken from a sixteenth-century woodcut.

Figure 8.7 Primary fermentation of wine. Vintner checking temperature of red wine. Courtesy Wine Institute of America.

resulting in bitter, acidic, or evil-smelling wines, they could be restored by heating the very young wine briefly to kill all microorganisms and starting again with active "good" yeasts. This pasteurization process was subsequently adapted to kill pathogenic and spoilage bacteria in milk, an adaptation the French would not have anticipated.

Fermentation (see Chapter 1) of the must starts almost immediately and continues at a high rate for several days. The amount of alcohol formed depends on the strain of yeasts used, the temperature, the level of acid, and the initial concentration of sugars.

Within a few days, the rate of wine fermentation in the vats slows down and the pomace of pulp, skins, and seeds floats to the top. This thick layer reduces the penetration of oxygen into the young wine and lessens the danger of conversion of alcohol to vinegar by aerobic bacteria. Fermentation stops when the alcohol reaches 15–18 percent by volume, a concentration that inactivates yeasts. This level is usually reached in about two to three weeks. In the preparation of sweet wine, fermentation is stopped by adding sulfur dioxide before all the sugar is converted to alcohol, by pasteurizing the wine, or by adding alcohol. The wine is allowed to remain in the fermenting vats for several more weeks to allow solids to settle out. At this point, the young wine has a distinct "grapey" or fruity flavor and aroma.

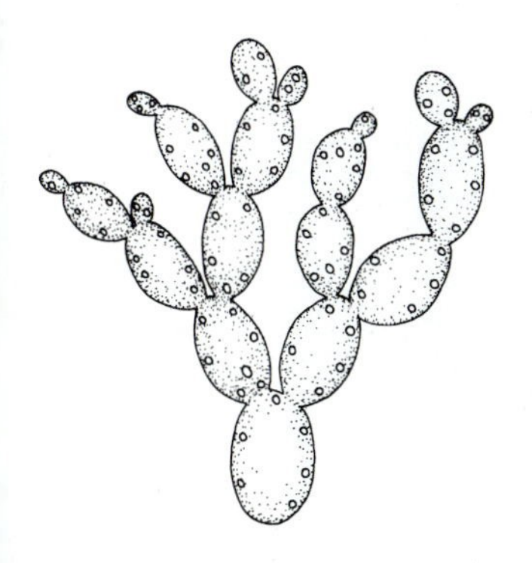

Figure 8.8 Yeast *(Saccharomyces ellipsoides)*. Asexual reproduction by budding with more than one daughter cell per mother cell.

RACKING

Racking consists of drawing the wine from the fermenting vat into clean barrels (Figure 8.9). These barrels are held in cool cellars (*les caves*) over the winter. Fermentation may continue at a very slow rate until the

Figure 8.9 Cooperage. Taken from *Libro della Agricultura*, published in 1511 in Venice.

last trace of sugars is converted into alcohol. Sugar-free wine is called "dry." By spring, the wine has a precipitate of dead yeast cells, proteins, and other components of the grape, and the wine is again racked into clean barrels. For red wines, racking is repeated the following fall to allow removal of precipitated tannins and tartaric acids (*lees*). The final racking may last one to three years, depending on the wine. This period, *la marié,* is the period of the marriage of the flavors and aroma in the wine (Figure 8.10). Many subtle changes occur, including complex chemical unions of acids and alcohols to form aromatic esters that

Figure 8.10 Testing racked wine. Courtesy Taylor Wine Co.

provide bouquet to the wine. The time of racking is critical: bottling too early will delay wine maturation, and bottling too late will result in a dull-tasting wine.

LE COLLAGE

Prior to bottling, wines are clarified (*fined*) to remove suspended solids. For premier wines, collage is done by adding egg whites, gelatin, or clay to racked casks. These additives slowly sink to the bottom of the cask, carrying impurities with them. Fining takes two or three months.

Clear wine is drawn from the top of the barrel into a cask, where it is chilled to precipitate tannins in solution. A final "polishing" or filtration is done to make the wine very clear. Inexpensive wines may be pasteurized or are sterilized by passage through bacterial filters, but these wines will never show any further development in the bottle.

MIS EN BOUTEILLES

Finally, the wine is bottled (Figure 8.11). Different types of wine are placed in bottles of traditional shapes, each major wine-making area having evolved its own bottle shape; the raffia-covered *fiaschi* for chianti and the tall slender bottles for Rhine wine are immediately recognizable. The bottles are capped with cork, the bark of the cork oak tree (Chapter 1). Corked bottles should be stored on their sides to prevent

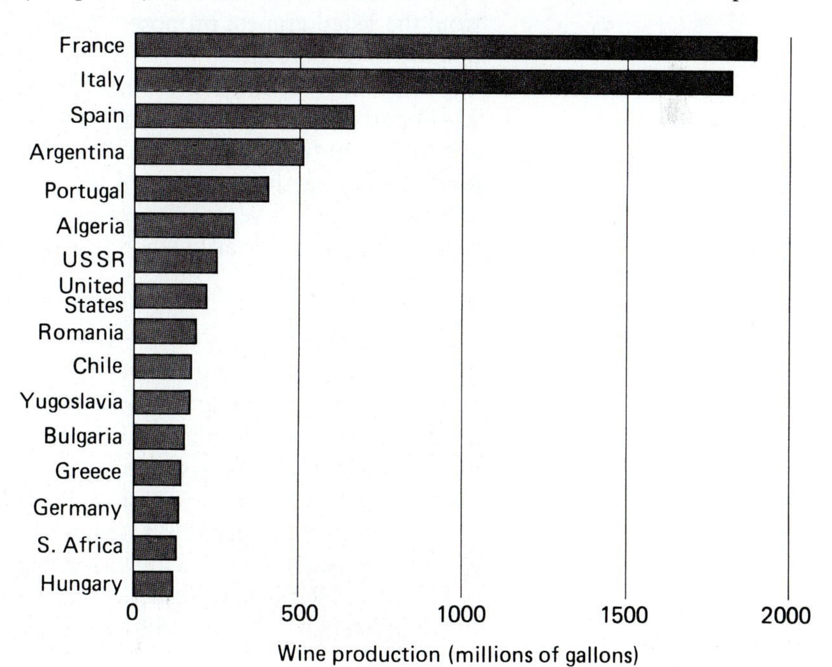

Figure 8.11 Major wine producing countries of the world.

the cork from drying out and allowing oxygen to enter the bottle. It is also a good idea to open a bottle of wine an hour or so before drinking to allow the bouquet to develop; part of the pleasure of wine is its aroma.

Where the label indicates *mis en bouteilles* followed by the name of a château, estate, or domaine (the German equivalent is *original abfüllung*), it means that the wine was bottled where it was made. In many instances, wines from several vineyards are blended by the vintner or by a shipping company, and the label can bear only the designation of the district from which the wine came. This is the *appelation controlée* in France, and wines simply called médoc or burgundy are usually blends. This is not necessarily bad. In years when none of the wines are outstanding, judicious blending can provide an excellent wine. In 1855, the French government classified all the vineyards in Bordeaux according to a complex rating scale that included assessment of climate, soil, cultivars, and so on. Although this classification, later extended to other districts in France, is debated endlessly, classifications are still generally accurate. Outstanding vineyards are called *grand cru* and, in descending order of excellence, wines from specific vineyards are called first, second, or third growths. Until recently, most California wines were not regulated, and the consumer was told only that the wine was a "burgundy" or a "rhine." Since California is not in French Burgundy nor in Germany's Rhine district, such names were misleading and, under international law, were illegal. Reputable vintners in California are now providing the name of the area plus the name of the grape cultivar used. Such "varietal" wines must contain 51 percent of the wine from the listed grapes; proposed federal legislation is designed to raise this to 75–85 percent.

Most red, rosé, and white wines are "still" or table wines and contain 12–14 percent alcohol by volume. If not sterilized, they will continue to age and mature in the bottle. Red wines, usually high in acids and tannins, may age for 30 or more years, but white wines rarely age well for more than 2–4 years. The vintage year refers to the year the wine was made, not necessarily when it was bottled.

LA BOISSON

Wine drinking is a highly culturated activity. In some countries, almost no wine is consumed, while in others, wines are a standard feature of meals as well as casual refreshment. The ritual use of wine in Judeo-Christianity is not paralleled in other major religions. Although wine drinking in Europe certainly antedates the Christian era, North America had little acquaintance with wine during the colonial period, primarily because of the lack of suitable climatic and soil conditions along the Atlantic seaboard. The wealthy plantation owners of the southern

United States imported wine for personal use, while people of the northern states and Canada were content with rum and whiskey, both of which could be made locally. Some wine was prepared from native grapes, but it did not compare in quality with the wines of Europe. By the late sixteenth century, California, still separate from the United States, was growing wine grapes from European grapes imported by the Catholic missions.

Only during this century, when transportation from the western part of the country has become reliable, and when there has been a sufficiently high level of affluence among the population, has wine drinking taken on economic and social significance. With an assured market, the number of California vineyards increased rapidly, especially after about 1950. The Finger Lakes region of New York State also developed extensive plantings, although the European cultivars frequently were killed by the severe climate. Hybrids between the European *Vitis vinifera* and native North American species, including *V. labrusca, V. ruprestris,* and *V. riparia,* were developed in France in the late nineteenth century. The grapes are highly disease-resistant and, most important, very cold-hardy, but the wines produced from these hybrids are considered to be inferior. Most of the vines in France have, by law, been replaced by European cultivars grafted to American rootstocks. These hybrids are extensively planted in eastern North America.

Until the 1970s most of the wine consumed in North America was red wine; neither white nor rosé wines had much of a market. This situation has changed drastically; white wines now far outsell red wines, while the rosé wines have maintained a fairly steady rate of use. White wines, which can withstand or may require chilling, have become appreciated as a summertime casual drink (Figure 8.12).

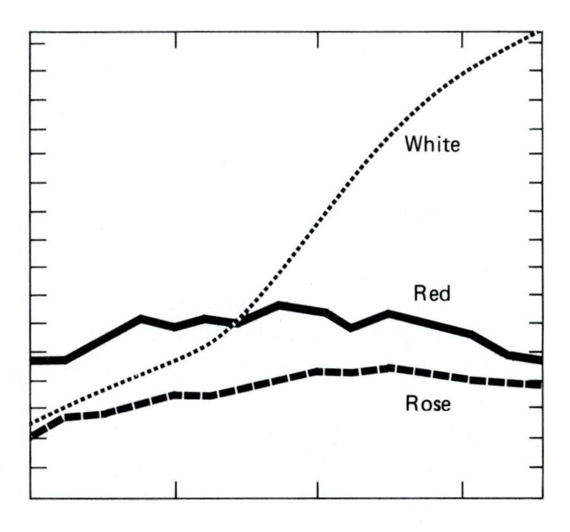

Figure 8.12 Consumption of table wines in the United States: gallons per adult per year.

VIN AROMATIQUE

Figure 8.13 Drinking champagne. Taken from *Opus Ruralium Commodorum* by Piero Crescentio, published in Venice in 1493.

Thus far, our discussion has centered on table wines. There are, however, many other grapevine products, and three categories are so common that brief discussion is warranted.

The manhattan or martini cocktail is an important part of the social life of North America. The manhattan consists of whiskey plus sweet Italian vermouth; the martini, of gin or vodka plus minute quantities of dry French vermouth. Both vermouths are made from white wines in which herbs and spices including coriander, wormwood (*Artemesia*), cinnamon, cloves, nutmeg, and many other flavorings are steeped. Sweet vermouth has sugar plus caramel coloring added. For both vermouths, the alcohol level is raised to 20 percent by volume. May wine—best served chilled with a garnish of fresh strawberries—is a Rhine wine flavored with woodruff (*Asperula odorata*).

VIN MOUSSEUX

Champagne (Figure 8.13) is a sparkling (carbonated) wine produced only in the Champagne district of France from both white and black pinot grapes, but other wines can be made by the same process (sparkling burgundy, crackling rosé, cold duck, and so on). Traditionally, the invention of champagne is ascribed to Dom Perignon, a Benedictine cellarmaster in the monastery of Hautvilliers. The vineyards surround the towns of Épernay, Ay, Gramant, and Avise. Growers sell their grapes to large companies that press, ferment, and bottle the white wine. Each of the two or three grape cultivars is pressed, skins immediately removed, and each is fermented separately. The young wines are racked over the winter and, the following spring, the cellarmaster blends the different wines to form a *cuvée,* the success of which depends on the cellarmaster's skill and vision. The blends are bottled and undergo a second fermentation over the hot summer. The wine rests for two or three years during which time each bottle is slowly turned and eventually upended to allow the sediments to collect in the neck of the bottle, the *remuage* process. The neck of each bottle is then frozen and the cork removed. The plug of sediment gushes out of the bottle, and the wine is ready to be made into champagne. A small amount of sugar solution is added to each bottle, it is corked, and the cork wired in place. Since yeasts are still present, a third fermentation occurs, with the carbon dioxide remaining in solution under pressure. The amount of sugar added determines the final dryness of the champagne. *Brut* champagne has less than 0.2 percent residual sugar; *extra dry* has less than 0.5 percent sugar; *sec* may have up to 2 percent sugar; and *doux* is a cloyingly sweet 3–4 percent. Vintage champagne is made from wine produced in a single year; *sans année* champagne may have wine from several years' pressings. A Frenchman, M. Charmat, who should have had more native pride, invented the "bulk champagne" process in which the third fermentation is done under pressure in large tanks, and the sparkling wine is bottled under pressure. The Russians, who enjoy a very sweet sparkling wine, invented a continuous process in which, from grape to bottle, the wine is made in only three weeks.

VIN CUIT

Wines to which alcohol is added are generally very long-lasting, do not develop in the bottle, and are called *aperitif* or *dessert wines.* The most famous of these are the sherries. The vineyards are dispersed around the city of Jerez de la Frontera near Cadiz, in southwestern Spain; the primary grape is the palomino. The first fermentation lasts the unusually long time of three months. During the latter part of this fermentation, a film of yeasts forms on the surface of the vats. This *flor,* or flower, of natural, wild yeasts confers a special flavor to the wine. For heavy, sweet *oloroso* sherries used as dessert wines, the fermentation is stopped by the addition of grape brandy when there is still considerable sugar left in the wine. For the drier, lighter *fino* sherries, fermentation is stopped later, and the final product is used as an aperitif. This fortification raises the alcohol concentration to 19–21 precent by volume. All sherry is aged by the solera process, in which the young wine is racked into the top row of stacks of barrels. As older, aged wine is removed from the bottom row of barrels, younger wine is racked down and fresh wine is placed in the top row. The hot weather of southern Spain causes some residual sugar to carmelize, providing the brown colors and the "nutty" flavors characteristic of sherry. Finos (amontillado and manzanilla sherries) are less caramelized than the oloroso and cream types. Our name *sherry* is an English corruption of Jerez, and when Falstaff, in *Henry IV,* spoke of "a good Sherris sack," he corrupted not only the name of the city but also the Spanish word *saco,* which means for export.

In the 1930s the agricultural experiment station in Geneva, New York, invented a quick method of obtaining a fortified, brown wine sold as sherry. Wine from labrusca grapes is fortified with sugar and alcohol and aged at temperatures of 40–60° C for several months to caramelize residual sugars. Madeira is a fortified wine aged at an elevated temperature for three to six months. Port, muscatel, tokay, and others are fortified wines whose fermentation was stopped before all the sugar was converted to alcohol.

RAISINS

Sun-dried grapes—raisins—have a history almost as long as does wine. Undoubtedly, dried wild grapes were utilized as storable food even before the vines were cultivated.Their use as food has been documented in dynastic Egypt and ancient Persia. The Old Testament reports that King David's subjects paid taxes in cheese and raisins. The Book of Samuel noted how raisins "returned the spirit." Armenia and Asia Minor were the centers of raisin trade by 1500 B.P. Homer sang of the raisins that sustained Odysseus. The Romans bought slaves with raisins; Nero included them in his Bacchanalian feasts. Pliny recommended them "against dysentery, quarten ague (malaria) and fits of coughing."

Because of the distance between sources and markets, raisins were a luxury item in Europe until the fourteenth century, when grape cultivars adapted for drying were grown in Italy and France. Introduction of raisin grapes into California, the primary producer in North America, occurred in the sixteenth century when Franciscans established missions with vineyards along the California coast. In 1851, the first vineyard devoted solely to raisin grapes was started near San Diego. The Egyptian muscat grape is still grown in this area. Raisin cultivation soon was extended to the Sacramento and San Joaquin valleys. The muscats were replaced by seedless white grapes developed by William Thompson; today 90 percent of the California crop is from the Thompson and cultivars derived from it.

Harvesting begins in late August and early September. The grapes are usually sun-dried for two to three weeks until moisture levels are reduced to 15 percent. The dried fruit is then placed in "sweat boxes" where mild heat allows an equalization of moisture content. After destemming, cleaning, and inspection, the raisins are packed. Four types of raisins are generally available from California: sultana, the Thompson seedless, golden seedless (Thompson grapes dried out of the sun), and muscats. The black zante currants are grapes; true currants (*Ribes spp.*) are in another plant family (Saxifragaceae), which also contains the gooseberry. A host for the fungus that causes white pine blister rust, *Ribes* are not grown extensively in North America.

ADDITIONAL READINGS

Amerine, M. A., and E. B. Roessler. *Wines: Their Sensory Evaluation.* San Francisco: Freeman, 1983.

Johnson, H. *The World Atlas of Wine.* 3rd Ed. New York: Simon & Schuster, 1985.

Schoonmaker, F. *Encyclopedia of Wine.* 6th Rev. Ed. New York: Hastings House, 1976.

Simon, A. L. *Wines of the World.* New York: McGraw-Hill, 1972.

Wagner, P. M. *Grapes into Wine: The Art of Winemaking in America.* New York: Knopf, 1976.

Weaver, R. J. *Grape Growing.* New York: Wiley, 1976.

Webb, A. D. "The Science of Making Wine." *American Scientist* 72, no. 4 (1984):360–367.

Winkler, A. J., J. A. Cook, W. M. Kliewer, and L. A. Lider. *General Viticulture.* 2nd Ed. Berkeley: University of California Press, 1974.

Whiskey

See what the boys in the backroom will have.

American Ballad

On May 4, 1964, cognizant of its awesome responsibilities in guiding the nation, the Congress of the United States of America passed Senate Concurrent Resolution 19, introduced by Senator Thruston Morton and Representative John C. Watts, honorable gentlemen of the great state of Kentucky:

> That it is the sense of Congress that the recognition of Bourbon whiskey as a distinctive product of the United States be brought to the attention of the appropriate agencies of the United States Government towards the end that such agencies will take appropriate action to prohibit the importation into the United States of whiskey designated as Bourbon.

It was said the Congress adjourned immediately after this exhausting act of statesmanship to demonstrate its wholehearted support for the product under consideration. This was far from the first act referring to distilled beverages. One of the earliest was a law enacted by the first general assembly of Virginia in 1619:

> Against drunkenness be it also decreed that if any private person be found culpable thereof, for the first time he is to be reprooved privately by ye Minister, for the second time publically, the third time to lye in boltes 12 howers in the house of ye Provost Marshall and to paye his fee.

These two quotations demonstrate the sinfulness and profitability of distilled spirits—twin themes that run throughout the entire liquor experience.

A BRIEF HISTORY

Although production and consumption of low-proof alcohol predate recorded history, Aristotle mentioned that one could obtain the spirit of wine. Greek and Egyptian alchemists in the second century obtained rectified liquids and apparently stumbled upon ardent spirits. Distilling entered the thinking of Europe only when crusaders learned the technique from North Africans. A treatise on distilling appeared in France in 1310 and, by 1500, there were books and practical manuals on preparation of brandy from wine, *aqua vitae* from beer, and the

Adapted from "A Nation of Moonshiners." *Natural History 85* (1976):23–31. Copyright American Museum of Natural History.

Figure 8.14 A brick still with a metal alembic. Taken from a woodcut of the sixteenth century.

fabrication of medicinal cordials from herbs infused in grain alcohol made from wheat, barley, and millet. Our English word *whiskey* is derived from the Gaelic *uisa beatha* or water of life (aqua vitae). Originally, it was medicine containing anise, cloves, nutmeg, ginger, caraway, licorice, sugar, and saffron—an all-purpose cure.

Preparation of alcohol is very simple. Basically, a carbohydrate source is fermented with yeast, and the 8–16 percent alcoholic solution is rectified by distilling the alcohol away from the water. Both in Europe and in its colonies, the medieval alembic (Figure 8.14), a closed vessel to which heat was applied, was filled with the ferment; the vapors, high in alcohol, were passed through a pipe in the top of the alembic to a twisted metal "worm" immersed in cold water to condense the vaporized alcohol. In addition to alcohol, distillates contain a wide variety of volatile compounds which confer specific odors and flavors to the brew. When grain is the carbohydrate substrate, fermentation is initiated either by adding sugar or supplying an enzyme (amylase) which converts starch to the sugar fermented by the yeast. Barley is an excellent source of amylase. The grain is wetted, allowed to sprout, dried with gentle heat, and the seedlings ground into malt. Scotch whiskey's distinctive taste is due to the fact that the malt is dried over peat fires. Grain is ground into meal, boiled into a gruel to solubilize its starch, malt added, inoculated with yeast, and allowed to ferment. Most yeasts stop working when the alcohol content reaches 15 percent, and the alcohol can be distilled away from the beer. The resulting mash can be refermented several times and spent mash used as animal food. It was a source of much innocent amusement among the colonists to see how drunk their pigs would get on used mash.

SOLACE OF THE AMERICAN REVOLUTION

For reasons lost in the fogs of history, the early settlers of North America failed to bring the alembic and worm with them and thus were restricted to beer, ale, and wine. Thomas Hariot of the Roanoke Colony reported that barley malt was unavailable and that he had some success with malt made from Indian maize. George Thorpe, a kind-hearted soul, gave the Indians back their corn in the form of strong beer and was killed by them for his trouble. Jamestown was making corn beer by 1608, derived from maize planted by Captain John Smith in 1607. With Yankee ingenuity, just about all the colonists' produce could be used. Pumpkins, potatoes, plums, Indian corn, carrots, and even turnips were made into beer. New Englanders used honey from wild bee trees and sap of the maple. Along with other necessities, seedling fruit trees were carried from Europe and orchards were planted as soon as the land was cleared. "Hard syder," about 4–5 percent alcohol, was made from "crabb trees." The alcohol content of these ciders was increased to about 25 percent by putting the jugs out in the winter cold and carefully removing the ice. In the Virginia colonies, peaches were favored because they fermented into a somewhat stronger product with a delightful odor and taste. Elderberries, currants, and wild *V. labrusca* grapes were used for wine. Ardent spirits were not used routinely. Brandy was standard in medicinal ship's stores, and rum made from molasses on the sugar plantations of the West Indies had to be transshipped to the colonies via England.

Of course, distilling apparatus was soon in use. At first, a blanket was put over a kettle and the condensed vapors simply wrung out into a pail. Each village had a mechanically inclined inhabitant who constructed an alembic and worm from bits and scraps of wood and metal. Applejack and grape, peach, and pear brandies were in use by 1630. The fruit brandy of New Jersey was considered the most potent, producing an "apple palsy" after just two drinks. Wilhelm Keift, director general of the Dutch colony on Staten Island, was distilling grain in 1640 with imported barley malt mixed with imported rye and native maize. Brandy was made from imported grapes by 1650 and aged for two to four years in barrels made from native white oak instead of the limoge or gascon oaks of France. Gin, too, was made in America by 1660, but never attained the popularity it has in England. A mixture of gin and applejack—called "strip and go naked," because this behavior was observed after several mugs—was used by the poor in northern cities. It was tamed for children and women by adding beer and blackstrap molasses.

RYE

Wheat and barley grew poorly in the colonies; wheat was needed for bread and barley was needed for making beer malt. Rye, not a native plant, formed the base for the first whiskeys of the colonies. It flourished in western Maryland and in eastern Pennsylvania, land settled by Scots who had previously lived in Ulster and by Germans from Moravia. These settlers had the skills and the equipment to make whiskey, and they set to it with a will and a fervor true to their heritage. Rye whiskey was originally called tiger spot, but soon became known as either Maryland or Monongahela. There were good economic reasons for the industry of the distillers. A pack horse could carry no more than four bushels of grain, but the same horse could carry two kegs of whiskey representing 24 bushels of rye. Spoilage was minimal unless the horse or driver stumbled, and the two kegs brought the cash equivalent of 40 bushels of grain. The fermentation and distillation of rye and barley malt produced a whiskey with a heavy, intense flavor, in great contrast to the blended "rye" whiskey that the modern unenlightened mixes with ginger ale. We have been sold a bill of goods; such "rye" is little more than a small amount of whiskey mixed with pure alcohol, water, and caramel coloring.

BOURBON

In the South and on the western frontier, rye was not a successful crop. The weather was too warm with too much rain in midsummer. The settlers turned to a cereal grain that did grow well, Indian maize. As Virginians moved west, they established villages that were sufficiently remote to require their own legal structures. By 1770, Harrodsburg, Harwood's Landing, and Boosfort were settled agricultural communities and were growing corn for food, fodder, and fermentation. In 1780, Fayette, Jefferson, and Lincoln counties were established in the western Virginia lands, a judicial district of Kentucky was formed in 1783, and Bourbon County was cut out from Fayette County in 1786. The population of Kentucky at that time was over "thirty thousand souls," of which a goodly proportion were growing corn. Just who first started making "corn likker" in Kentucky is unknown, although there are several claimants for this exalted title. Marsham Brashears purchased land in 1782 for 165 gallons of whiskey, Bartlett Searcy willed his 96 gallon still to his son John in 1784, and the Kentucky District Court had nine cases of illegal whiskey retailing in 1783, including one presentment to the grand jury against a woman. For convenience, the Rev. Elijah Craig is credited with making the first bourbon whiskey. Elijah's brother Lewis, also a Baptist minister, was a whiskey dealer who supplied the cargo for the flatboats which followed the rivers down to New Orleans. There was serious debate on the propriety of ministers engaging in the liquor trade, but since most of the congregation was similarly employed, no pot could call any kettle black. The Baptist debate raged until about the time of the Civil War, when a firm antiliquor stand was taken. Drunkenness was, however, another matter and occasionally led to outright expulsion from the congregation; a man and his wife were supposed to be able to handle their liquor.

As Maryland and Pennsylvania rye and the corn whiskey from Bourbon County became widely distributed, liquor's role in social, business, and political affairs became all-pervasive. Payment for debts, mortgages, and purchases of necessities was often in liquor. Jugs of whiskey and distilling apparatus were formally parts of estates; births, weddings, and funerals were paid for and celebrated with whiskey; and no business arrangement could be considered consummated without pledges sloshed in a cup or glass. A sovereign remedy for summer complaints of children and adults was an infusion of rhubarb, caraway seed, and orange peel in whiskey. Copperhead or rattler bites were treated both internally and externally with bourbon.

Most distillers were the farmers themselves, taking advantage of the fine, pure limestone water and the abundant corn. None of the products were trademarked; they were all bourbon or rye. A James Beam and a J. W. Dant had moved to Kentucky from North Carolina by 1760 and a John Ritchie and a Jacob Beam were licensed distillers by 1780, but it wasn't until after the Revolutionary War that specific names were attached to the product of a distillery or a district. Prior to and even after the revolution, whiskey was collected at Louisville or Cincinnati, then barged down the Ohio and Mississippi rivers to reach the Southwest and Southeast via New Orleans. John James Audubon made a stake for his then-ignored painting when as a merchant in Henderson, Kentucky, he barged a load of kegs to Missouri, where he sold the whiskey at two dollars a gallon.

The whiskey of the colonists was sold and consumed in its God-given, natural, water-clear state. Occasionally a small amount of caramel coloring was added to give the amber color associated with European brandy. The colonists knew that aging in charred oak barrels imparted a golden-brown color and smoothed out the product by the marriage of various organic compounds, but this took valuable time, and no one cared a whit what the color was so long as the drink was strong. The complex of harsh, skull-popping volatiles (fusel oils, aldehydes, and wood alcohol) could be absorbed by barrel aging or by filtering the raw whiskey through a layer of maple charcoal. Peach brandy was similarly treated with charred peach pits, which also contained a minute amount of prussic acid. One serious problem was the tendency for the alembic to become too hot and scorch the mash, and some genius figured out a way to introduce live steam into the pot still to volatilize the alcohol. Only much later did the continuous Coffey still replace the pot method.

By the beginning of the eighteenth century, copper stills were in general use, imported from either England or Holland. Most native stills were iron, which imparted an off-flavor and sometimes colored the water-clear liquid. Because of British restrictions on local manufacturing, rolled copper for alembic, worm, and boiler was impossible to obtain. Paul Revere, silversmith and sometimes horseman, was best known for being the only manufacturer of rolled copper in the colonies, a monopoly that he held until 1802.

Proof was determined by the gunpowder method. Equal volumes of gunpowder and whiskey were mixed and set afire. If it flashed up, the whiskey was too strong, and if it didn't burn, it was too weak. When it burned evenly, the whiskey was 100 percent perfect; it was "proved out" and the proof was 100—that is, 50 percent alcohol within about 2–3 percent.

Rye whiskey was made by inoculating the mash with fresh yeast, but bourbon utilized the sour-mash process. Here the mash was charged with a small amount of the liquid from the previous batch. This spent beer contained a small amount of lactic and other acids that promoted development of yeasts and repressed the growth of bacteria which might otherwise spoil the taste of the whiskey. Colonists were more particular about taste than spelling, and the *-y* or *-ey* ending was used interchangeably until about 1850. Frances Trollope used both spellings in her *Domestic Manners of America* of 1839. After the Civil War, the *-ey* ending was restricted to the rye, Canadian, and bourbon whiskey of the Western Hemisphere.

George Washington learned the value of whiskey very early. He stood for the Virginia House of Burgesses from Frederick County in 1785 and won with 307 votes . . . at a cost of £38, of which £34 was for liquor. At the time, rum was 16 shillings per gallon, rye whiskey was 8s per gallon, and corn liquor was 4s per gallon. It was money well spent. James Madison refused to supply refreshments for the voters and consistently lost elections. Washington's plantation was noted for its peach brandy; as his liquor became more famous, he branched out into making whiskey from rye and corn and imported a Scotsman to oversee the business. During the war, Washington insisted that the army be supplied a liquor ration: "In many instances such as when they are marching in hot and cold weather . . . it is essential that it [liquor] not be dispensed with." He recommended to the Continental Congress that public distilleries supply liquor to the troops. The young navy had its daily tot of rum following the British model. With many men away from their farms during the war, grain was in short supply. Washington noted with approval the restrictions imposed by the states on excessive use of grain for whiskey making and insisted that the liquor ration be of peach or apple brandy or of rum. This didn't sit too well with southern and western troops weaned on corn and rye whiskey, but their complaints were subdued by the high proof rum passed out by supply sergeants. War time profiteering was evident as the price of rum rose fourfold in a year.

THE WHISKEY REBELLION

All governments have taxed liquor, and all citizens have attempted to evade payment. The young republic, desperately searching for means to pay debts resulting from the war and to cover the growing obligations of statehood, considered a whiskey excise. Secretary of the Treasury Alexander Hamilton and his Federalist party rammed through the Excise Tax Act of 1791, empowering the federal government to impose duties on all spirits "distilled within the United States, from any article of the growth and produce of the United States, in any city, town or village." Progressive duties on each gallon according to its proof and a yearly tax on each still were mandated, together with the onerous responsibility for each producer to maintain complete records and to permit inspections of distilleries, warehouses, taverns, and even private homes. Indignation ran high. Kaintucks, Virginians, Tennesseers, and Pennsylvania "Dutch" rye distillers were hauled into court for back excises; Reverend Elijah Craig had a liability of $140 assessed against him and nearly went bankrupt, as did many others. The excise was, they all agreed, an invasion of rights won with their blood; they fought the British for freedom from excise, and they'd be damned if they were going to go through that again. Although the excise was softened in 1792 and 1794, distillers joined with shippers and retailers to oppose the tax, and meetings and resolutions gave way to direct opposition, including tarring and feathering the collectors. It was to be a test of the power of the federal government to tax the people directly without the interposition of the states, and Hamilton convinced Washington that the whiskey rebels of the Monongahela must be utterly and permanently crushed.

With a penchant for military overkill that seems to mark the United States defense establishment, a massive army was gathered. In the autumn of 1794, 13,000 troops with artillery, mortars, supplies (including whiskey), and other appurtenances of a punitive force had been mustered. General Henry (Light Horse Harry) Lee, then governor of Virginia, volunteered to lead troops through Cumberland Gap, General Howell and the New Jersey contingent moved through Carlisle, and Washington himself made an appearance on a white horse to show that the government really knew what it was doing. Happily, there was no battle; the rebels dispersed. It cost about $1.5 million, but excises were collected. Jefferson, who never supported any of Hamilton's ideas, repealed the tax when he assumed the presidency, and it wasn't imposed again until the Civil War. Since that time, whiskey has become a favorite target for state and federal legislatures.

BRANDY

Although the oversized glass is often considered an affectation comparable to a lorgnette or handlebar moustaches, there are good reasons for using one when drinking fine brandy; half the pleasure of brandy is in its aroma. True brandy is defined as a grape wine, distilled when young, and then aged in oak casks for 5 to 50 years. The wine is distilled to yield a clear, slightly aromatic spirit at 25 percent alcohol (about 50 proof), and then is distilled again to produce raw "heart of spirit" at 70 percent alcohol (140 proof). During the aging process, the alcohol content drops to 85 proof, esters and other flavorings develop, and the brandy becomes a rich, reddish brown. Any white wine can be used; German brandies use Rhine wine and *V. labrusca* cultivars are used in North America. By law, only brandies made in France's Cognac district can be called cognac. Different brandies are aged separately and are then blended to produce a uniform product. Brandies from the Champagne areas of Cognac produce the finest brandies, a source of some confusion since these areas are named for the local chalky soil and have nothing to do with sparkling wine. Names like "Napoleon" or a system of stars means very little except to advertisers. VO (very old) cognacs are usually less than 10 years in the cask, while VSOP (very superior old pale) brandies have been aged for more than 10 years.

France also produces other fine grape brandies, among the most notable being armagnac, produced in the area near Spain where Dumas placed the birthplace of d'Artagnan and where Louis XIV completed the cathedral of Auch. In most wine-growing districts of France, wine grapes are pressed a second time to yield juice called *marc*. Too strong and too full of tannins to produce a table wine, marc is fermented separately, distilled, aged, and sold as marc brandy. It varies in taste depending on the grape cultivar used and the subsequent handling.

Distilled spirits are made from many fermentable products in addition to grapes. These water-clear alcohols contain flavorings of the fruit used. Kirsch is derived from cherry pits, fraise from strawberries, and mirabelle from plums. Wood-aged brandies are made from many fruits: calvados is from Normandy apple cider; slivovitz from plums grown in eastern Europe. Peaches, apricots, pears, and other fruits are fermented into a beer, distilled, and then aged. In addition, many liqueurs include brandy as the flavoring and alcohol base: Chartreuse is a brandy with herbal essences, creme de menthe has mint flavorings, Cointreau, Curaçao, and Grand Marnier are flavored with oranges. Many liqueurs have sugar added for taste and to provide the "heavy" syrupy flavor.

RUM

Rum is the yeast fermentation product of the molasses remaining after the refining of sugar cane, which is then distilled to higher proof. Blackstrap molasses, a very dark brown syrup, is the most common starting material. Its modern history dates from the early seventeenth century, when sugar plantations were established in the Caribbean, particularly in the West Indies (Chapter 4). Not particularly popular in Europe, where brandy was preferred, it was used primarily as a medicinal. The Puritans of the Massachusetts Colony usually served rum as a flip. A mixture of rum, beer, and sugar was stirred with a logger—a heated iron poker—and after a half-dozen drinks, the guests were frequently at loggerheads. Rum was issued to British sailors in the 1650s because it was cheap and did not go bad on long voyages, as did beer. By 1725, Sir Edward Vernon, a British admiral, ordered ships' captains to mix the rum with sugar and lime juice as a *grog* to prevent scurvy (a vitamin C deficiency disease) that decimated sailors on long voyages. Soon a daily ration of grog became common in most naval nations. Daily rum rations to American sailors were abolished in 1862 by order of President Lincoln and, most reluctantly, by the British in 1970.

The rum first consumed in the American and Canadian colonies was made in the West Indies, but soon most of it was made abroad. Martinique and Guadeloupe shipped it to France, and the British Isles sent it to New England towns like Salem, Newport, and Medford. Benjamin Franklin, president of the American Philosophical Society, devised a walloping drink for membership dinners, consisting of high proof Medford or Demarara rum, loaf sugar, and orange juice. Britain, preferring to keep the lucrative rum trade away from the French, required that New England rum had to be shipped to England or, if kept in the colonies, was to be heavily taxed. The Molasses Act had two effects: it increased smuggling, and it was as distasteful to the colonists as the stamp, navigation, and tea taxes. The British, and later the Americans, organized rum triangles, in which rum was shipped to Africa to purchase slaves who were transported to the Caribbean where the ships loaded molasses for more rum production in New England. In the eighteenth century, there were several hundred rum distilleries in New England; Boston alone had over 60.

Rum lost favor during the nineteenth century and was replaced by whiskey. During Prohibition (1919–1932), its popularity increased because bootleggers were able to obtain adequate supplies from Cuba that could be smuggled into Florida. When Prohibition was repealed, rum consumption again fell. The current popularity of rum dates from the period when Caribbean cruises gained social status.

Rums can be divided into two main categories. The light rums range in color from clear to light tan and are distilled at high proof and then cut with water. They receive little ageing and mix well with fruit juices and ice for warm weather consumption. The rums of Cuba, Puerto Rico, and the Dominican Republic are primarily light rums. The heavy, dark-brown, and sometimes high proof rums of Barbados, Jamaica, Trinidad, and the French islands of Martinique and Guadeloupe are more comparable to those of the eighteenth century.

Analogous to rums are the fermented and distilled honey products. "Old Metheglin" was a dark-brown, sweet liquor containing about 60 percent alcohol (120 proof). In Vermont, it was asserted that one glass was enough to allow—even in the depths of winter—the buzzing of the bees to be heard.

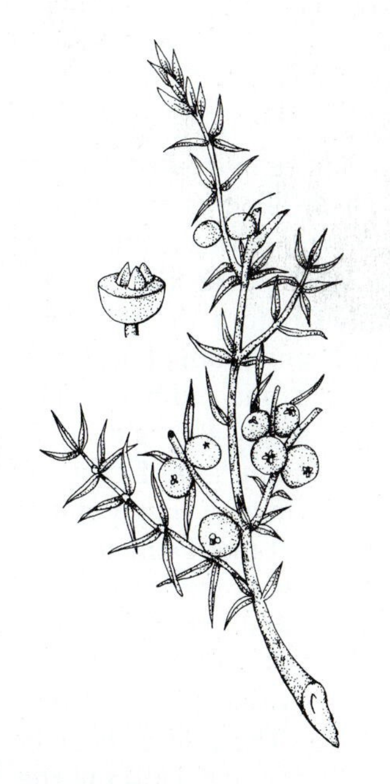

Figure 8.15 Juniper *(Juniperus spp.)*. The "berries" are a major flavoring in gin.

UNAGED SPIRITS

Of the unaged distilled spirits, the two most used are gin and vodka. For both, and for the Scandinavian akvavit, the base is usually grain, fermented into a beer, distilled several times, and purified of all other compounds by fractional distillation and filtration through charcoal. To produce gin, a variety of herbs are steeped in the 95 percent alcohol (grain neutral spirits) for several weeks; the procedure is followed by a series of distillations to yield a high proof alcohol containing essential oils and other flavorings. For gin, the major flavoring ingredient is the fruit of juniper (Figure 8.15). Other plant extractives include coriander seeds, angelica roots, and citrus peels. Akvavit uses caraway seeds, while some Slavic vodkas include buffalo grass, sorrel, or other herbs. Most should be taken neat and very cold; a bottle chilled in a block of ice is recommended.

Centuries ago, natives of Mexico discovered that if a hole just at the base of a flower bud of the century plant (*Agave spp.*) was gouged, golden sugary sap exuded. When allowed to ferment, *pulque*, a beer with 6–8 percent alcohol, was formed. When the Spanish introduced distillation apparatus to the town of Tequila, a new distilled product was formed and given the name of the town.

ADDITIONAL READINGS

Baldwin, L. D. *Whiskey Rebels.* Pittsburgh: University of Pittsburgh Press, 1939.

Carr, K. *The Second Oldest Profession: An Informal History of Moonshining in America.* Englewood Cliffs, N.J.: Prentice-Hall, 1972.

Crowget, H. G. *Kentucky Bourbon: The Early Years of Whiskeymaking.* Lexington: University Press of Kentucky, 1971.

Gray, J. H. *Booze.* New York: Macmillan, 1946.

Lender, M. E., and J. K. Martin. *Drinking in America.* New York: Free Press, 1982.

MacMarshall, W. (ed.). *Beliefs, Behavior, and Alcoholic Beverages.* Ann Arbor: University of Michigan Press, 1979.

Beer

Beer! Beer! Beer! cried the privates,
Merry, merry men are we.
There's none so fair as can
Compare to the fighting infantree.

Soldier's Song

From cans, bottles, and kegs, by the glass, stein, or pitcher, beer is sipped, drunk, chugged, and guzzled in quantities greater than that of any other alcoholic beverage. In contrast to distilled spirits, whose development can be dated from the Arabic invention of the still, beer has been a primary product of grain from the time people became farmers (Figure 8.16). Royal breweries existed before the Eleventh Dynasty in Egypt (4000 B.P.), and the Code of Hammurabi of 3750 B.P. regulated the quality and price of dark, light, pale, and red beers. The Chinese used both rice and millet beers, called *kui,* by 4000 B.P. In all cases, the people believed that beer was a heavenly gift. Egyptians credited it to Osiris, Babylonians to Siris, and the Chinese simply stated that it was a gift from heaven. Greek tradition has it that Demeter fled Mesopotamia in disgust because the people drank beer instead of wine and chose to settle in Greece, where the wine drinkers were more civilized. Rome dedicated beer to Ceres, Goddess of the Corn, and their name for beer, *cerevisia,* is the specific epithet for yeasts (*Saccharomyces cerevisiae*) used for both baking and brewing. The role of beer in ritual results from its attribution to a female deity. Women, closer to the Corn Goddess than men, were priestesses of these goddesses and the

Figure 8.16 Beer sippers. Taken from a Babylonian seal of 3900 B.P.

brewers. Rameses III modestly noted that he sacrificed 30,000 gallons of beer each year to Egyptian gods, and credited his country's prosperity to these libations. Norse Vikings asserted that beer was the drink of Valhalla.

A BRIEF HISTORY

From the beginning of the Christian era, brewing was a household art, with every girl instructed in baking and brewing, both processes requiring the same ingredients, both invested with the same mysteries, and both exalting woman as giver and sustainer of life. Beer was "liquid bread," and a meal could consist of bread, beer, and cheese—a combination not unfamiliar in North America today. Since water was suspect, beer was the thirst quencher, and people demanded high quality. Roman legions carried beer with them and mixed it with vinegar to avoid catching the diseases of the barbarians. When they conquered an area, they introduced Roman yeasts to ensure getting a decent drink. The *Senchus Mor,* a compilation of ancient Celtic or pre-Celtic law, contained regulations on beer quality. William of Orange invented British bureaucracy when, in 1067, he appointed an ale coiner to regulate quality and price. Official state breweries were established in Bohemia in 1256, in the town of Budweis, and by 1384, Pilsen's breweries were under the personal control of Charles IV. Thomas à Becket, then chancellor and later archbishop, brought casks of good English ale to France when he crossed the channel to conclude a peace treaty, and Elizabeth I sent courtiers to see if the local beer was drinkable before she visited towns. In Germany, if the local beer went bad, city leaders would import it from another town and sell it at cost in the basement of city hall—the *ratskeller.* Medieval marriages were toasted with brides-ale, now *bridal.* The name *ale* is derived from the medieval *hael,* meaning "good health."

Government regulated beer not only in the public weal, but also to ensure the collection of taxes. Since beer made at home was impossible to regulate, taxes were imposed on the ingredients, the containers, and on alehouses frequented by the public. Monasteries and nunneries were exempt on the grounds that they consumed what they brewed, but many bishops enriched the Church coffer by requiring that the community buy beer from the Church. The resentment to this practice in England was a factor in the overthrow of "popery."

BEERS OF NORTH AMERICA

The invasion of the New World by Europeans found beer already in situ. Columbus drank a corn beer on his fourth voyage in 1502. Beer was brewed from English barley in Jamestown. The Pilgrims carried kegs in the *Mayflower* and worried near the end of their voyage that "our victuals being much spent, especially our beere" whether they would survive the winter. The Dutch established the first commercial brewery in New Amsterdam in 1623. A Captain Sedgwick was licensed to brew by the Massachusetts Bay Colony in 1637. Roger Williams authorized a communal brewery for the Rhode Island Colony. William Penn brought Quakerism and a brewhouse to Pennsylvania, and every family maintained its own equipment for making "small beer," consumed immediately after fermentation. Heroes of the American Revolution were brewers and beer merchants: Samuel Adams managed the family brewery, General Israel Putnam was part owner of a brewery, and Thomas Jefferson imported a Bohemian brewmaster. Virginia planters all followed the advice of Dr. Benjamin Rush, who wrote of the "friendly influence upon life and health" of malted beverages.

Colonial tavernkeepers-brewers were respected; they had inherited the esteem of members of European brewers' guilds, to whom medieval people bowed. Usually retired military figures, they accumulated political and social power, and their taverns were the center of a town's activities. Because it was the town's gathering place, the British rightly viewed the tavern with suspicion; all too often seditious words were uttered. Patrick Henry frequented his father-in-law's tavern, and Captain Stephan Fay was the proprietor of the Catamount Tavern, where Ethan Allen and his Green Mountain boys met. When Great Britain required that Redcoats be quartered in private homes, indignation over this breach of privacy led to refusal of householders to supply beer to the hated "Lobsterbacks." Washington joined Patrick Henry and Richard Lee in the Declaration of the Association of Williamsburg, which stated "that we will not hereafter import . . . from Great Britain any one of the goods hereafter enumerated including . . . beer, ale or porter." Beer joined sugar and tea as symbols of the colonists' resolve to be free. Washington's farewell address to his officers in 1783 was, as would be expected, given in Fraunce's Tavern in New York's Battery.

THE BREWER'S ART

Basically, beer is merely fermented carbohydrate. Depending on temperature and other conditions, yeasts will produce carbon dioxide, ethyl alcohol, and their own energy supply from almost any carbohydrate. Wheat, rice, corn, pumpkins, beets, potatoes, and other starchy plant tissues can be converted into beer. But starting with the Mesopotamians, people noticed that the cereal grains produced better beer and also that barley was the best cereal. When barley grains were wetted, allowed to sprout, and then dried and ground to a powder, the resulting malt fermented more quickly than ungerminated grain. Barley malt seemed to cause the fermentation of other grain powders to occur more quickly. Only in the nineteenth century was it discovered that during germination of cereal grains, an enzyme is activated which breaks starch down into its constituent sugars—the primary substrate for yeast's fermentation processes. Not only did this enzyme (diastase or amylase) hydrolyze barley starch, but it could break down wheat, corn, potato, or any other starch, thus speeding up fermentation. Barley soon became the preferred malt because the resulting beer has a lighter, more pleasant taste. Today, especially bred strains of barley are steeped in water for several days, germinated on a temperature-controlled malting floor or in large drums for five or six days, dried, and ground into the light brown malt powder also used in malted milkshakes. Most modern beers mix the malt with "adjunct" starches from wheat, rice, or corn, but malt liquors use only barley malt as the substrate for the yeasts.

The malt and adjunct starch are mixed with water to make a mash. Pure water is essential; the best beers are from iron-free water, preferably from springs that flow through limestone. The mash is allowed to saccharify—that is, to undergo the enzymatic conversion of starch to sugar, resulting in the formation of the *wort,* an aqueous mixture of starch, sugar, and many soluble substances from the cells of the grains. The clear wort is transferred to a mash kettle where it is boiled with hops.

Figure 8.17 Hops *(Humulus lupulus),* showing the flowers used to flavor beer.

Were beer to contain only alcohol, wort, and the flavor of malt and grain, it would be insipid and slightly sweet, and would spoil quickly. By the eighth century, brewers in central Europe found that the addition of the unopened conelike inflorescence of the hop plant (*Humulus lupulus*) preserved the beer and provided a slightly bitter taste that increased the palatability of the product (Figure 8.17). Hops was not the only bitters used; *Tanacetum balsamita,* a chrysanthemum, is still used in England to make alecost; tannins from oak and ash trees were used in Scandinavia; cinnamon was used in southern Europe; and American colonists dosed their beer with sweet fennel (*Foeniculum vulgare*), licorice, or sassafras. Hops was first recorded in the *Kalevala,* the Finnish saga of 5000 B.P., but was unknown in the rest of Europe until the father of Charlemagne donated his hops garden to the Monastery of St. Denis. Addition of hops was accepted in central Europe, but since Henry VIII liked ale made without bitters, it was not until the end of the fifteenth century that Flemish brewers were brought to Kent to make hopped beer. Parliament denounced "this wicked weed which will spoil the taste of our drink and endanger the people." Bohemian hops from Czechoslovakia, Wisconsin, and Oregon account for the supply used in North America.

Boiling in the mash kettles transfers the bitter flavor of hops to the wort, sterilizes it, and causes the precipitation of proteins. Filtered again, cooled wort is placed in fermenting tanks where an appropriate strain of brewer's yeast is added to initiate alcoholic fermentation. Each brewmaster guards the yeast carefully; subtle differences in fermentation result in beers with different flavors. Beer differs from ale in several respects; the fermentation temperature is lower for beer (45° C) than for ale (60° C) and the yeasts used for beer tend to remain at the bottom of the tank instead of floating to the top as in ale brewing. The result is a higher alcohol content in ale, a tarter taste, and a paler color.

Young beer is separated from the yeasts and aged for several weeks before bottling. Some brewers add chips of beechwood to the aging tanks to serve as a nucleus for yeast growth. A second, slow fermentation occurs at low temperatures in which remaining sugar is fermented and carbon dioxide accumulates under pressure to yield a carbonated beer. This *lagering* process ("to store" in German) was discovered in the eighth century when it was noticed that the flavor of young beer was improved by aging it over winter in cold cellars in tightly stoppered wooden barrels. The second fermentation is sometimes eliminated and the beer carbonated with the carbon dioxide saved from the first fermentation. To increase shelf life, most beers are pasteurized following the recommendations of Louis Pasteur in his classic *Studies on Beer* in 1876. Draft beer is unpasteurized or sterilized by filtration to avoid the flavor changes that occur upon heating.

Since any of the steps in the brewing and aging process can be modified, there are many kinds of beer. Those most frequently seen in North America include the following:

> *Lager.* In this type of beer, strains of yeast that settle on the bottom of the vats are used. Lagers are aged in barrels or vats for up to six weeks. Almost all of the American beers are lagers.
>
> *Pilsener.* This is a general term for beers that are brewed in the style developed in the city of Pilsen, a Bohemian city which is now part of Czechoslovakia. The water used for Pilsener-style beers tends to be harder (higher calcium and magnesium content) than the water used for lagers. The color of pilsener beers is lighter than that of lagers. The so-called light beers, which contain less alcohol and hence fewer calories, are generally made in the Pilsener style.
>
> *Porter.* Porters are dark lager beers. The darker color and special flavor come from toasting the malt before brewing. Not only do porters have a stronger taste, but they usually contain more alcohol.

Bock. Traditionally, bock beers were made in Germany in the spring to launch the brewing season. They are dense, since they are made with roasted malts, and they have a very distinctive malty flavor and usually a high alcohol content.

Ale. Beers made with yeasts that float to the top of the brewing vats are called ales. They tend to be full-flavored and usually contain a somewhat higher alcohol content than lager beers. Ales are subjected to a different aging process from lagers.

Stout. Stouts are dark ales made with toasted malt and brewed for a longer period of time before aging.

Many cultures have developed beers. *Chicha* is a South American beer prepared from corn, manioc, bananas, or other starchy plant material. Since malt is not available, the enzymes needed to split or hydrolyze starch into sugar for yeast fermentation are supplied from human saliva (salivary diastase). Japanese *sake* and Chinese *samshu* are prepared from rice that has been inoculated with a culture of the fungus *Aspergillus oryzae*, which contains starch-splitting enzymes (amylases). Birch beer, ginger beer, a spruce beer from young needles and twigs of *Picea*, true root beers from sarsaparilla (*Smilax spp.*) and spikenard (*Aralia*) are still being made in various parts of the world.

Like other alcoholic drinks, beer has come in for its share of both taxation and vilification. Over $600 million are collected in the United States, not including local taxes, and the tax burden is proportional in other countries. Never popular, beer taxes have sometimes been collected at gunpoint and have occasioned revolts which have been put down violently as attacks on government itself. Public and private drunkenness have long been ascribed to beer drinking. Assuming an average alcohol content of 5 percent in beer, there are 50 milliliters of alcohol in a liter (roughly a quart). A 30 milliliter "shot" (roughly 1 ounce) of bonded bourbon contains 15 milliliters (half an ounce) of alcohol. It is possible to get very drunk on beer.

TEMPERANCE—THE NOBLE EXPERIMENT

People of Europe and North America have been modestly heavy drinkers for centuries, and the interaction between drinking and the moral standards of organized religions has been with us for as long. The phrase of poor Bill Sykes in *The Beggar's Opera* that "gin was mother's milk to me" was literally true in seventeenth-century London. The Holborn district boasted of 7000 homes in 1617, and 1300 of them sold gin. Gin mills advertised "drunk for a penny" and made good on their promise, even supplying straw piles on which to sleep it off. Between 1785 and 1787, the foreign exchange of the young United States was reduced $12 million for importation of heavy, dark rum with alcohol levels about 140 proof. In 1826, the American Society for the Promotion of Temperance was founded in Boston, the major port for rum importation. Proper Bostonians raised the rallying cry against "the demon rum" in spite of America's permanent love affair with whiskey and beer. The Society first approved of cider, beer, and wine, declaring that all they wanted to do was to "suppress the too-free use of ardent spirits." In view of the high level of American alcoholism, this was a social good. It was, in fact, 30 years before the Society decided that "the only logical basis for the temperance movement is entire abstinence from all intoxicants." The Society's leaders were religious people, mostly Baptists, Methodists, and Congregationalists. Catholics and Jews did not support the movement, but since members of these religious groups did not include political leaders, their voices were unheard in Congress.

Temperance was very popular in less urban parts of the United States in the mid-nineteenth century, when most of the population was rural, religious, and righteous. Three years after its founding, the Society claimed 100,000 members, and this number rose to over a million by 1833. The convention of the Society in 1836, when the doctrine of total abstinence was enunciated, also gave rise to the word *teetotaler,* because, as the story goes, one who fully abstained (even from beer) was marked down in the register of the Society with a T for total. If this story isn't true, it should be. Prior to the Civil War, the antis mounted a massive propaganda campaign. Edward Delevan, their spokesman, asserted that water used in brewing contained dead rats and also claimed that the delightful, nutty taste of Madeira wine was obtained by suspending a bag of cockroaches in the casks. Delevan also said that General U. S. Grant was a member of the Society, a most unlikely claim in view of Grant's ability to consume over a pint of bourbon a day and still fight a major war. In 1846, Maine passed a temperance law which, by 1852, became the model for laws in Rhode Island, Massachusetts, New Hampshire, and Vermont. Similar legislation was later introduced into the middle western states and the territories of Minnesota and Nebraska. Southern legislatures assiduously avoided even discussing such bills in committee. Canada, with its large French and Irish populations, also avoided the question in spite of active temperance groups in the Maritimes and Ontario.

The Civil War, subsequent expansion of settlements in the West, growing affluence in the eastern states, and grinding poverty in the South held the country's attention and prohibition laws were either repealed or ignored. Yet the postwar period of 1870–1890 was the golden age of American religiosity, when preachers on horseback moved their revival meetings and evangelical fervors from village to hamlet. "Sin! No Heaven for the tippler! What will happen to the drunkard's child?" Rum was, in literal terms, the drink of the demon, and Satan was a palpable reality. With tears streaming down their cheeks, men took "the pledge," and it frequently required several drinks to get over it. The country ignored its growing alcoholism rate, and the organization of the Anti-Saloon League in the 1890s went almost unrecorded. A small cloud in an otherwise clear sky, the League soon became a thunderhead. The antis were clever. They cut formal church ties, although their financial and moral support was still Baptist and Methodist, and, very wisely, they organized the women of America into a formidable coalition, the Women's Christian Temperance Union. With a flair for publicity which could be profitably emulated by modern television personalities, superstars like Carry Nation wielded their shiny hatchets.

Rather than a direct frontal assault on the last bastion of American malehood, the Anti-Saloon League and the Temperance Union decided to work quietly, to "nibble away at the Devil." Sunday blue laws were enacted in many states; even the ice cream sundae was banned. Candidates for office were sure to be asked about their "stand on morality," and when the antitemperance governor of Ohio was defeated in 1904 by an avowed prohibitionist—and this in spite of the fact that women couldn't vote—American politicians became very circumspect when it came to this burning issue. President Taft caught holy hell when he authorized the attendance of his Secretary of Agriculture at the International Brewers' Congress in Chicago in 1911. Taft was defeated in 1912 and, while the loss of the presidency was not the result of his stand on liquor, the Anti-Saloon League made political hay over its "victory." While a lame-duck president, Taft vetoed a bill prohibiting liquor shipments into or through the dry states of Kansas and Oklahoma, but Congress overrode the veto.

Woodrow Wilson, concerned about the possibility of war, came down hard on both sides of the issue, but American military intervention in 1917 provided the temperance groups with powerful arguments. With food in short supply, why use precious stocks of wheat, corn, and barley for liquor? If, as the President proclaimed, the role of the United States was that of a moral crusader, no moral issue was as immediate as temperance. After all, most brewers and distillers had German names, and it was not unlikely that they were poisoning the youth of America to prevent "our boys from defeating the kaiser." With society dislocated and preoccupied, with men in uniform effectively disenfranchised, and with weary and overworked legislatures wrestling with major economic and political decisions, the Eighteenth Amendment to the U.S. Constitution was passed by an act of Congress in December 1917, President Wilson's veto of the act was overridden, and the amendment passed by the states. The Volstead Act of October 28, 1919, defined an intoxicating beverage as one containing more than 0.5 percent alcohol by volume and authorized the formation of appropriate sections of the Department of Justice to enforce the law. Alcohol produced for medicinal or industrial use was *denatured,* a curious word which means that it was made unfit for drinking by the addition of poisonous compounds.

In spite of dire predictions that the agricultural economy of the United States would collapse, the drop in demand for wheat, corn, and barley for beer and whiskey was more than matched by the desperate needs of people in Europe. Farm income rose as barley fields became wheat fields and corn was fed to hogs and cattle. Brewers, distillers, and their employers were badly affected, and many survived economically by entering other fields or brewing "nearbeer" and industrial alcohol, which was in great demand. The effect on the mores of the people was, however, less happy. Bootlegging—the smuggling of liquor from Canada, Scotland, the West Indies, and South America—was soon dominated by criminal leaders like Al Capone of Cicero, Illinois. Disregard for government and law resulted in the wild twenties and thirties, when speakeasies, smuggling, home brewing, and public drunkenness were fairly common behavioral modes.

Europe's economy recovered in the 1920s with corresponding decreases in farm incomes in the United States. The American people became heartily sick of weak beer, expensive imported wines, adulterated spirits, a quantum jump in violence, and the massive, calculated federal disregard for civil liberties. President Hoover appointed a commission in 1929 to study the impact of Prohibition. The commission reported that the Eighteenth Amendment should be retained and Hoover agreed. With the stock market crash in late 1929, economic depression, the subsequent election of Franklin Roosevelt in 1932, and the feeling that the times had changed, Congress passed the Twenty-first Amendment to the Constitution, thereby repealing the Eighteenth Amendment. The states quickly confirmed the amendment. Liquor could again be sold, as long as it was in bars or taverns—the Anti-Saloon League was able to prevent the reintroduction of the hated name, *saloon.* No battle over a plant product, not even that raging about marijuana, has so polarized the people of any country in several thousands of years. The sanctimonious zeal of organized religion and its political power within the United States was broken by the repeal, leaving a legacy of distrust and suspicion toward government that is still in evidence. Prohibition was not a "noble experiment"; it was an unmitigated disaster.

ADDITIONAL READINGS

Astbury, H. *The Great Illusion.* Garden City, N.Y.: Doubleday, 1950.

Baron, S. B. *Brewed in America: A History of Beer and Ale in the U.S.* Boston: Little, Brown, 1962.

Braidwood, W., et al. "Did Man Once Live by Beer Alone?" *American Anthropologist* 55 (1953):515–526.

Cook, A. H. *Barley and Malt.* New York: Academic Press, 1962.

Jackson, M. *The World Guide to Beer.* Philadelphia: Running Press, 1984.

Sinclair, A. *Prohibition: The Era of Excess.* Boston: Little, Brown, 1962.

Chapter 9

Plants That Changed History

Paper—The Memory of Man

Papyrus sheets preserve the thoughts and deeds of man.

LEONARDO DA VINCI

Each of us acquires the morality and culture of our group through example, verbal signals, exhortation, and from the written record of the ideas and knowledge of the ages. The written word is paramount in education and is no less vital in other activities in which we participate. Writing is information storage, and reading is information retrieval. To meet the need for a cheap, permanent, storage-retrieval system, humans developed writing—a method of recording speech and thought—and, later, invented paper.

EARLY WRITING SYSTEMS

People were recording long before paper was invented (Figure 9.1). Cuneiform clay tablets of ancient Babylon, incised sheets of metal, and virtually any other surface from cave walls to public monuments graced with graffiti attest to the insatiable drive that humans possess to preserve their thoughts for a posterity that may not want them. But such surfaces are expensive and not easily transportable. Two of the forerunners of paper are noteworthy. Parchment was produced by rubbing skins of sheep and goats with lime instead of tanning them into leather. Today,

Figure 9.1 An Egyptian God of Life writing a person's fate on a leaf held by a servant. Taken from an Egyptian tomb painting.

this expensive surface is used for prestige diplomas—the sheepskin—but in the time of ancient Persia, it was the preferred material for edicts and religious tomes. The best parchment came from Pergamum, a Greek kingdom on the coast of Asia Minor. On the assumption that the method of producing parchment originated there, we can understand how the term *parchment* was derived from *Pergamum.*

The other writing surface which preceded paper was *papyrus* (Figure 9.2), a Latin word derived from an Egyptian word from which the English word *paper* comes. As early as the Third Dynasty (about 4500 B.P.), writing sheets were made by cutting the pith of an Egyptian reed (*Cyperus papyrus*) into long, thin strips, pressing them flat, and gluing them together into large sheets. As long as the length of papyrus could be handled on rollers, there was no limit to its size. The Great Harris papyrus, made during the reign of Rameses III (3198–3167 B.P.) is 400 meters long and is in better condition than last month's newspaper. Athens, and later Rome, used papyrus for consular and imperial edicts such as the law scrolls exhibited in the Agora and the Amphitheatre. It was used in Palestine, as evidenced by the Dead Sea Scrolls. Its surface was smooth enough and had enough body to serve as the vehicle for paintings. Rolls of papyrus, mostly of Egyptian manufacture, served the Western world as its primary writing surface until the eighth century, supplemented by parchment and, for lesser writings, by wooden tablets.

INVENTION OF PAPER

Paper, as we know it, consists of sheets of matted or felted plant fibers. Its inventor was the Chinese scholar Ts'ai Lun, who in 105 pounded the inner bark of the *Ku* tree (the paper mulberry, *Broussonetia papyrifera*) into short fibers, treated this with alkali (lye) from wood ashes to dissolve other material, mixed this slurry with water, and sieved off the water through silk. This left a sheet of paper which could be pressed flat and dried in the sun (Figure 9.3). Contemporary records proclaim, "Under the reign of Emperor Hi-Ti, Ts'ai Lun of Lei-yang conceived the idea of making paper from the bark of trees. . . . The paper was then used throughout the entire universe." For the Chinese of the Han Dynasty, China was "the entire universe." They guarded their secret process for over 400 years.

Figure 9.2 Papyrus *(Cyperus papyrus).* Plants grow up to three meters tall.

During the Han and the subsequent Wei (220–246) and Chin (265–520) dynasties and into the troubled Six Dynasties period (520–618), strips of bone, tablets of wood, and split stalks of bamboo were the usual writing surfaces. These elongated surfaces lent themselves nicely to the unique Chinese style of vertical writing, a style attributed to Ts'ang Chieh, who had invented calligraphy in 4700 B.P. Postal service was a government responsibility, and "letters" of wooden tablets in a clay envelope stamped with official seals have been found thousands of miles from the cities where they were written. And they

Figure 9.3 Papermaking in China. Bark of the paper mulberry *(Broussonetia papyrifera)* is pounded in a tree-trunk mortar and the pulp sieved from the water on a silk screen. Individual screens are dried in the sun and sheets of paper peeled off.

were probably delivered on time. Woven silk sized with rice starch or gelatin was another medium, but because of cost, silk scrolls were used only for important edicts and religious paintings. Silk and paper had surfaces that permitted calligraphy with camel's hair brushes and carbon-black inks made from animal bones.

The period of the T'ang Dynasty was marked by the development of foreign contacts and trade with the "barbarians," all people who were not Chinese. Trade routes were established to the west over the mountains and deserts to modern Afghanistan, into modern Iraq and Iran, and possibly into Greece and Rome. Although wooden and bamboo tablets of military records and business transactions have been found along these routes, no paper has been found; it was either not used for such mundane correspondence or was too fragile to survive a thousand years. Attracted by the flowering of Buddhism in China and the sophisticated and effective methods of governance, finance, and the fine arts, Japan sent many delegations to the T'ang court to learn all they could and to bring this new knowledge back to the Asuka and Heian emperors at Kyoto. The Silla rulers of Korea sent scholar-supplicants to China for the same purposes. Among the arts which went to Korea and Japan was that of making paper.

In China, Korea, and Japan, the art and technology of papermaking developed over centuries with other plant materials used to modify the character of the paper and the addition of coatings to improve its ability to hold various inks and paints. Fibers of precious metals were incorporated to yield papers of outstanding visual beauty, and the parallel development of calligraphy as an art form was exploited by scholars, poets, and even emperors. In Asia, these arts were not to be taken lightly; paper and writing were used to transcribe edicts and sacred thoughts for the benefit of posterity. Fibers of paper mulberry bark, bamboo, hemp (*Cannabis*), and rice straw were combined to prepare special sheets upon which were imprinted woodblock cuts of household dieties (Figure 9.4). Such printed pictures have the same religious significance as do the ikons of the Orthodox Russian and Greek churches. The "joss paper" used in funeral services was made from these ceremonial papers, the word *joss* being pidgin from the Portuguese *deus,* or God. Sheets of joss printed as "money" were burned so that the spirits of one's ancestors would have funds to buy their way through the beyond.

Figure 9.4 Household icon of the Chinese gods of life, health, and wealth. These were printed with the red ink of good luck on ceremonial paper.

MOSLEM PAPERMAKERS

Although the Chinese maintained trade routes to the west as early as the latter part of the Ch'u Dynasty (about 2300 B.P.), it was not until the T'ang dynasty that the major routes were extensively used. Caravans of camels and horses followed several routes laid out by Chinese engineers and policed by troops who built walls, forts, and watchtowers for thousands of miles. One route crossed the Gobi Desert, the desert of Takla Makan, and the formidable Tarin Valley to Samarkand, a city now in the Uzbek Soviet Socialist Republic just north of Afghanistan. The Chinese established a papermaking factory in Samarkand to utilize the fibers of flax (*Linum*) and hemp that grew in this area. The paper was shipped back to China under heavy guard. In 751 an army of the prophet Mohammed clashed with guard troops of the T'ang emperor Hsüan-tsung on the banks of the Tharaz River, a few miles east of Samarkand. The Moslems captured the city and with it the paper mill and its skilled artisans. Immediately recognizing that this product would provide a vehicle for spreading the faith, the conquerors continued paper production. Arabian artisans were trained by the Chinese, and new factories were established in Baghdad and Damascus by the beginning of the ninth century under the patronage of Harun-al-Rashid, he of *The Thousand and One Nights*.

The techniques used by the Moslem papermakers differed little from those developed by Ts'ai Lun 700 years previously. Today's processes are also basically the same, although many technical innovations have been developed in the past 1200 years. The inner bark (botanically called the phloem; Greek for bark—Chapter 1) of plants consists of several different kinds of cells. Some of these, the sieve tubes and companion cells, are structurally and functionally specialized to carry sugars made in the leaves down the plant to the roots, where they are stored as starch. Another cell type is composed of long, thin cells with thickened walls which serve to strengthen the phloem. These are called bast fibers, and in some plants the cells are long enough to be twisted into thread. Linen is a thread made from the very long and very tough bast fibers of *Linum*. Since these cells are tough, they are not destroyed by pounding or alkali treatments which can dissolve away the other phloem cell types. When freed of the other cell types and mixed with water, bast fibers tangle up and interlock as a felt. If this suspension is allowed to settle out on a fine-meshed screen, water is sieved out, and the fibers form thin sheets. Drying and pressing will complete the process. Ts'ai Lun used the bast fibers of the paper mulberry, a happy choice since its fibers are tough and resistant to the harsh alkali treatments he used. As Oriental technical workers continued to experiment, they added fibers from bamboo and other plants to give papers the characteristics of toughness, smooth or rough surfaces, ability to take ink, ability to be coated with oil for umbrellas or for windows. Neither the Chinese nor the Japanese solved the problem of pulping the plants; hand pounding was slow and inefficient. Arabian artisans invented a stamping machine worked by foot power, allowing them quickly to obtain uniform, clean fibers. They also found that old linen cloth gave a superior paper, one very similar to the linen paper used today for thank-you notes.

The Mayan and Aztec cultures independently invented paper using fibers from the century plant (*Agave*), the phloem of species of fig (*Ficus*), or bast from wild mulberry (*Morus niger*). The bark was softened by beating, treated with wood-ash lye to dissolve other cell types, washed with water, and laid on boards to dry (Figure 9.5). These

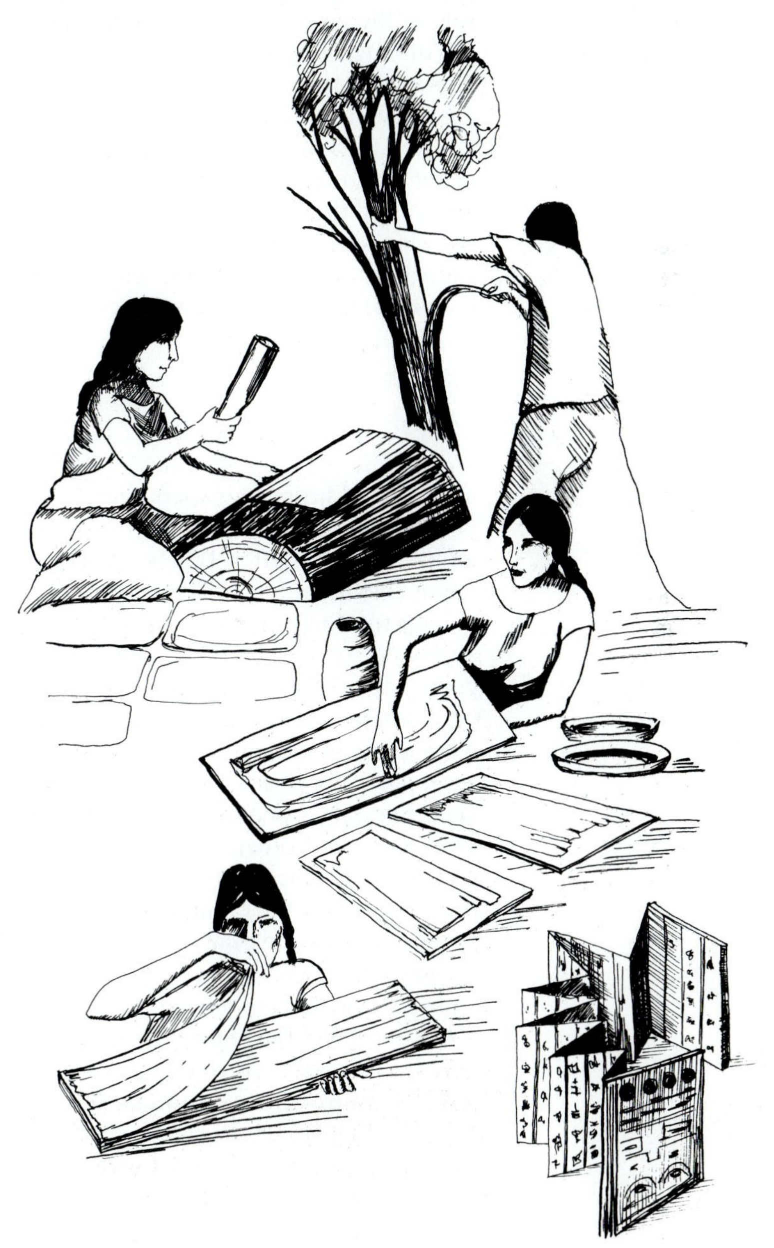

Figure 9.5 Papermaking in Central America. Tree bark is pulled off and pounded on a log into a thin sheet. Sheets are laid on boards to dry, peeled off, and pleated into an accordion-folded book.

papers, like those of the Chinese, were reserved for official documents and as currency. Paper was a precious commodity, and Montezuma II received sheets of papers as tribute. Apparently the Mayans were the first to make books by bending long sheets of paper into accordion folds and binding them between boards. Because paper can transmit information, it was involved in witchcraft, and Mexican peasants still cut out paper dolls of evil spirits, sprinkle the effigies with blood, and burn out the devils. The process and even knowledge of Mayan and Aztec papermaking was destroyed by the Spanish conquerors and played no role in the development of paper technology in the rest of the world.

Although Baghdad and Damascus attempted to maintain their monopoly on paper, the secret was bound to leak out. In the tenth century, a factory was built in Cairo, the Moors carried the technique throughout North Africa, and a major factory was built in Fez, Morocco. Moslem paper was made primarily from linen and cotton rags, including the wrappings of mummies when available. With the conquest of Spain by the Moors, the first paper factory in Europe was established in 1150 in Zativa in southern Spain. Papermaking had arrived in Europe a thousand years after its discovery in China.

EUROPEAN PAPERMAKERS

The early European papers were, however, not well received. The product was more expensive that the parchment used by monastery monks to hand-letter Bibles, it was much more fragile, and it was impossible to scrape away an error in transcription as one could do with parchment. Since paper was made by infidels, it was suspected of being a product of the Devil, a belief fostered by the parchment-makers guilds. Slowly, however, paper began to be accepted. Factories were built in Fabriano, Italy, in 1276, in Troyes, France, in 1348, and in Nuremberg, Germany, in 1390. The curse lifted suddenly, and paper's acceptance by the Church for religious writings made its general adoption only a matter of time. The development of movable type by Gutenberg and others and the printing of the Bible broke down any residual reluctance. Since the Latin word for bark is *liber,* books of paper were "on liber" and a collection of books formed a library. The pages of a book, bearing a clear resemblance to the leaves on a tree, are called *feuillet* in France, *blatt* in German, and *leaves* in English; equivalent words are used in most languages.

By the end of the fourteenth century, paper was in great demand, not only for imperial and religious use but also for the transmittance of business information and for a modest, but growing, secular literature. With demand so high, the papermakers in Ulman Strömer's mill in Nuremberg went out on what seems to be the world's first factory strike for higher wages in 1391. In general, the quality of European paper was lower than that made by the Moors, and there was a return to parchment during the fifteenth century. Indeed, it was not until the middle of the sixteenth century that better paper caused a return to its general employment, and it was only then that illustrations of the process began to appear. Under one of these woodblock prints is the following poem:

> Rags are brought to my mill
> There much water turns the wheel
> They are cut and torn and shredded
> To the pulp, water is added.
> Then the sheets 'twixt felts must lie
> While wringing them in my press
> Lastly, hang them up to dry
> Snow-white in glossy loveliness.

Ts'ai Lun's process had been supplemented with a wheel that replaced hand-pounding, the use of rags instead of raw bark, and drying felts to speed up the removal of water. Only this in 1500 years!

Two additional inventions in the sixteenth century accelerated the shift from parchment—now called *vellum*—to paper. Gelatin was added to the slurry of water and fibers before drying to prevent excess absorption of ink, and watermarking allowed identification of the producer. Watermarks were usually crosses, since most of the paper was used for the Bibles, but later the initials or names of the guilds or the factory were used—a practice that tended to raise standards and increase quality control.

Elizabeth I granted a royal patent to a John Spilman to build a paper factory at Dartford to supply the court and to promote the establishment of newspapers. As on the continent, the paper was made from linen and cotton rags, beaten to a pulp (a familiar expression whose derivation is now clear), filled with gelatin, and sized with glue to give a hard, opaque sheet, admirably suited for writing with quill pens and the thin inks then available. Printing presses were improved, and the increase in interest in politics led to the establishment of many newspapers. When William Rittenhouse built the first U.S. mill in Germantown, Pennsylvania, in 1690, rags were his sole source of fiber. His factory's output of two reams per day consumed such an enormous bulk of rags that he wasn't sure that he could find enough to remain in business. An issue of the Boston *New Letter* in 1769 announced that a "bellcart will go through Boston . . . to collect rags for the papermill at Milton . . . when all the people that will encourage the paper manufacture may dispose of them." New, more abundant sources of fiber were needed, and, although bast fibers made less satisfactory paper than rags, papermakers in many countries were experimenting with fibers from local plants. The French used the lime tree (our linden or basswood—a corruption of bastwood). In North America, corn husks, wheat straw, and several trees were tested. Bamboo, rice straw, sorghum stalks, fibers of the common stinging nettle (*Urtica dioica*), and other plants were tested as well, but most were not as good as rags and some were, given the technology of the period, almost useless. The Chinese developed a special paper from the pith or central part of the stem of the rice-paper plant (*Tetrapanax papyriferum*), which became the rice paper of commerce.

PAPER FROM WOOD

The modern paper industry, based on wood, originated with the acute and perceptive eye of a distinguished French scientist and inventor, René de Reaumer, who noticed that wasp's nests were paperlike and contained filaments of wood. In a report read to the French Academy of Sciences on November 15, 1719, de Reaumer suggested that wood might be a fine source of fibers, although he had not experimented with this possibility. Over 50 years later, Jacob Schaffer, a Lutheran minister, and Mathias Koop, a Dutch citizen living in England, finally were able to make paper from wood, straw, and rags. It was not until 1840 that Friedrich Keller of Germany made the first all-wood paper. Keller reasoned, quite accurately, that the fibers of wood, cell types botanists call xylem (a Greek word for wood), are sufficiently strong (Chapter 1) and long enough to form a felt which would mat well once the water had been removed. The problem was to find some way to separate the wood fibers. Keller patented a process in which blocks of wood were pressed against a slowly spinning, wet grindstone. The first factory using the ground-wood pulping method was built in Nova Scotia in 1841, followed by another in Maine in 1863. In 1870, the *New York Times* led the U.S. newspaper industry when it converted to newsprint made exclusively from wood.

Ground-wood paper pulp consists of whole or broken xylem vessels and fibers. At maturity, these are dead cells, their contents having been lost during maturation of the cell. The walls of xylem cells consist of a major component, cellulose, whose chemical composition is much like starch except for the arrangement of the chemical bonds. The cellulose molecule is long and threadlike, and a number of these molecules are twisted about each other to form a microfibril, and several layers of microfibrils form the xylem cell wall. Surrounding and coating the cellulose molecules is another compound, lignin, which serves to strengthen and waterproof the xylem cell wall, making these cells well adapted for their primary function of conducting water from the roots of the tree up to the leaves. Pure cellulose is white—cotton fibers are an example of almost pure cellulose—and the lignin is brownish in color. In paper, lignin is undesirable; in ground-wood pulp, strong bleaches are used to decolor the lignin, although this reduces the strength of the paper. The ground-wood process was a signal advance in papermaking, but it was time-consuming to grind the wood, and the resulting paper was suitable for cheap newsprint and little else.

Benjamin Tilghman, an American working in France, discovered that if wood chips are heated in the presence of sulfur compounds the lignin is dissolved away from the cellulose and the xylem fibers will separate easily from one another. In large-scale modern pulping mills, the bark of trees is removed by knocking logs together in a revolving drum. The peeled sticks are fed into a chipper, where whirling knives cut the wood into thin, short fragments. These are fed into a digester, where steam raises the temperature well above the boiling point of water. The quantity and composition of the chemicals added to macerate (separate) the fibers and remove the lignin will vary with the type of paper pulp wanted. In the soda process, lye (sodium hydroxide) is added; in the sulfite process, an acid solution of bisulfites and sulfurous acid is used under pressure. A third process uses a mixture of lye and sodium sulfide. The macerate or raw pulp, now almost pure cellulose cell walls, is washed with water to remove the chemicals, leaving a water slurry which can be further bleached or dyed. Water is removed and paper is formed by a modification of Ts'ai Lun's inventions. John Baskerville invented a wire screen in 1757 to replace the cloth screens, and in the same year Nicolas-Louis Robert invented a machine which put the wire screen on an endless belt so that, instead of making sheets of paper, continuous rolls would be formed. Because of the disruptions caused by the French Revolution, the continuous process was actually patented in England by Robert Gamble and Sealy Fourdrinier, and, by 1805, a papermaking mill was built in Hertfordshire which could turn out 100 times as much paper as did the hand process. When rolls of paper made from wood were available, in about 1835, the modern paper industry was ready, willing, and able to meet all possible demands for its product.

Demand increased at a tremendous rate. Kraft (German for strong) papers could be fabricated into bags to replace the burlap and cotton sacks in which grain and sugar were packed, cardboard replaced wooden baskets and crates, wallpaper decorated the homes of millions of families, and the availability of inexpensive paper allowed books, newspapers, and magazines to be within the financial reach of everyone.

North America is both the largest producer and largest consumer of paper. The United States consumes 16–19 million tons of newsprint paper, 13–15 million tons of pulp for other papers (bags, printing paper, tissues), and 3–5 million tons of container board per year with over 60 percent of the wood grown in Canada. Much of the rest of our paper is derived from trees grown in the southern states, both conifers like the southern yellow pine and hardwoods like the aspens (*Populus spp.*). Northern states and Canadian sources provide pulp from spruces, firs, and some pines. Western United States and western Canada utilize several species of fir, the Douglas fir, western hemlock, and ponderosa pine.

Because of increasing demand and rising costs, there is considerable interest in finding other sources of pulp and improving the use of tree processing. Breeding of species of poplar trees for southern pulp plantations has resulted in a tree that can be harvested in about half the time as for genetically diverse populations. There is more interest in other materials. Crushed stems of sugar cane (bagasse) remaining after the sugar is removed, bamboo and stems of the rapidly growing *kenaf* or Indian hemp (*Hibiscus cannabinus*) look promising. The recycling of used paper, particularly newsprint, now involves close to 25 percent of paper and paper products and the industry is capable of expanding to recycle double that amount. These alternatives are of increasing importance because forest lands are being lost to industry, roads, and housing at a rate that jeopardizes the future availability of wood for paper and for other purposes (see Chapter 1).

ECOLOGICAL CONSEQUENCES

As a consequence of the construction of huge, integrated papermills, severe ecological problems have emerged whose solutions are not yet in sight. The assault on one's nose as a mill is approached is well known to most people, but more serious is that some of the sulfur compounds are converted to sulfuric acid in the atmosphere, returning to the earth as acid rain. This alters the alkalinity-acid balance of soils with yet unknown, but potentially serious, effects on microflora in the soil and on plants growing in this soil. The efficient and time-saving method of clear-cutting large stands of forest trees for pulpwood and lumber has resulted in erosion of large areas, with consequent silting-in of streams having severe impacts on the insulted ecosystems (Figure 9.6). Just as severe are the problems of waste disposal in the mills. Until very recently, and in some mills today, mercury compounds have been used in the processing of the wood. These compounds were deliberately or inadvertently (mostly the former) dumped into streams and lakes along with sulfur compounds, bleaches, and dyes. They are all poisonous to the plants and animals in these waters. In Canada and in Japan, severe human mercury poisonings have been documented in which the mercury used in making pulp is concentrated in food chains, accumulating in the fish people eat. Finally, and not least among this catalogue of problems, is the lignin. Slowly degradable, with no known industrial use, it is routinely discharged into the water courses belonging to *all* of the people. It accumulates in large sludge beds that inexorably reduce the oxygen needed by aquatic animals and plants. The multinational paper companies have a great deal to learn about corporate social responsibility.

Figure 9.6 Erosion and silting of an Alaskan river after clear-cutting of the steep slopes of the drainage basin. Courtesy U.S. Department of Agriculture.

ADDITIONAL READINGS

Blum. A. *On the Origin of Paper.* New York: Bowker, 1934.

Butler, J. W. *The Story of Paper Making.* Chicago: Butler Company, 1902.

Clark, T. F. "Plant Fibers in the Paper Industry." *Economic Botany* 19 (1965):394–405.

Hunter, D. *Papermaking: The History and Technique of an Ancient Craft.* 2nd Ed. New York: Knopf, 1957.

Isenberg, I. H. "Papermaking Fibers." *Economic Botany* 10 (1956):176–193.

Perdue, R. E., Jr., and C. J. Kraebel. "The Rice-paper Plant—*Tetrapanax papyriferum* (Hook.) Koch. *Economic Botany* 15 (1961):165–179.

Smith, D. C. *History of Papermaking in the U.S. (1961–1969).* New York: Lockwood, 1970.

Sutermeister, E. *The Story of Papermaking.* Boston: S. D. Warren, 1954.

Weaver, A. *Paper, Wasps and Packages.* Chicago: Container Corporation of America, 1937.

Cotton—A Despotic King

Look away down yonder in the land of cotton . . .
To live and die in Dixie

D. D. Emmett

For a material to be used as fabric thread, it must meet several requirements. The fibers must be uniform in diameter, must have the ability to hold together when spun into a thread, must have high tensile strength, durability, and pliability, should individually be long enough to be twisted together, should be light in color, and should have the ability to take dyes. The material should also be relatively inexpensive and readily available. Wool meets most of these requirements; silk does too, but it is expensive. Thus it came to pass that men and women gratefully gave up wearing fig leaves or animal skins and started using specialized fibrous plant cells. Of these fibers, none has had the social, economic, or political impact of cotton.

THE COTTON PLANT

The genus *Gossypium* was established by Linnaeus within the mallow (Malvaceae) family, along with several garden plants (*Althaea, Hibiscus,* and the Rose of Sharon). The genus name is an ancient Greek word for cotton. There are over 30 species found native in Central and South America, Africa, Asia, and Australia. Four species are cultivated, and each has numerous subspecies, botanical varieties, and cultivars. *G. arboreum* and *G. herbaceum* form the taxa from which the cottons of India, China, and Africa have been derived. *G. barbadense* (sea-island, pima, and Egyptian cottons) and *G. hirsutum* (upland cotton) are both native to the New World and, because of the superiority of their fibers, are the major species planted throughout the world. The classification of cotton is complicated, not only because the plant has been selected and bred for superior fiber production but because the geographical distribution of the major species has peculiar patterns. All Old World cottons have clear affinities to one another, and all New World cottons also show affinities one to another. The difficulty is that there is no obvious geographical connection which would permit a link between the Old and New World species, but botanists are loath to believe that the two groups are not phylogenetically derived from some common ancestral type species.

COTTON PRODUCTION

In tropical climates, cotton can be a perennial, developing a thick, woody root and stem, and some wild species grow into small trees. The southern United States is too cool for overwintering, and the plants are grown as annuals. The flowers are large, bell-shaped, and look like those of hibiscus. Petal colors range from creamy white through spectacular reds and purples; cotton is a very fine annual garden plant. Cotton is both self-pollinated and bee-pollinated, and after fertilization the ovary develops into a large fruit, the boll, containing many seeds. The outer layer of the seed coat is covered with hairs (Figure 9.7). These seed hairs, the cotton fibers of commerce, are single cells that may grow to a length of 1 to 2 centimeters. During growth they elongate and their walls become thickened with additional layers of cellulose fibrils (Figure 9.8). At maturity, the cytoplasm is lost and the cell collapses and twists into a flattened filament, which is almost pure cellulose of considerable tensile strength and durability (Figure 9.9). At maturity, the boll splits open and the cotton fibers expand, covering the plant with white "snowballs."

Cotton is subject to a veritable jungle of insect pests, as well as bacterial, fungal, and viral diseases. Breeding for resistance to soil-born fungi and bacteria has been important in keeping production costs down, but insecticides and other pesticides are still used in massive control programs. The insect pests have developed increased resistance to pesticides, and there is a continued search for different and more

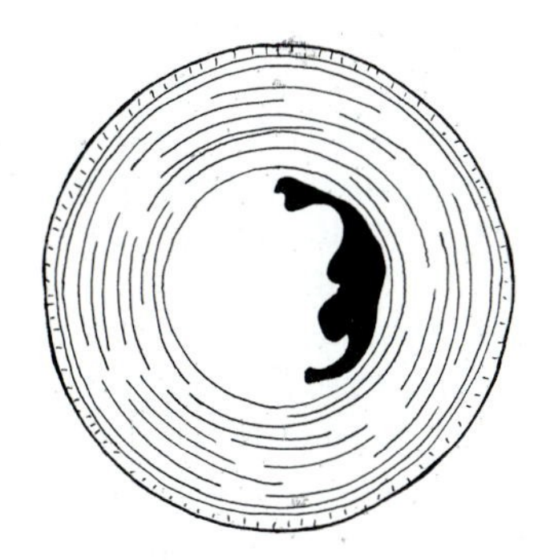

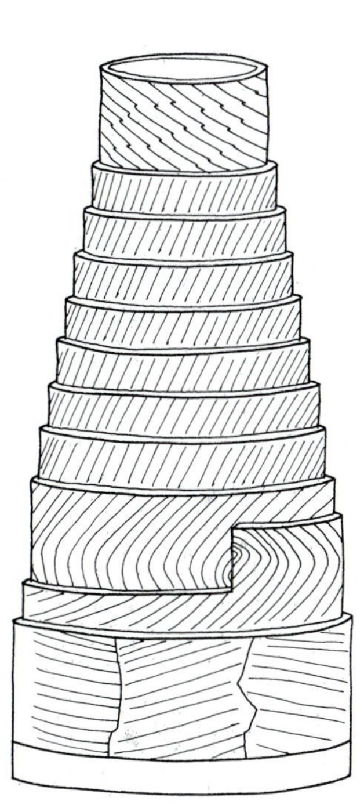

Figure 9.8 Diagrams showing the laminations of cellulose fibrils that comprise a cotton fiber. The dark mass in the center of the cross section is residual protoplasm.

Figure 9.7 Cotton seeds. Courtesy U.S. Department of Agriculture.

Figure 9.9 As cotton fibers mature, they collapse and twist.

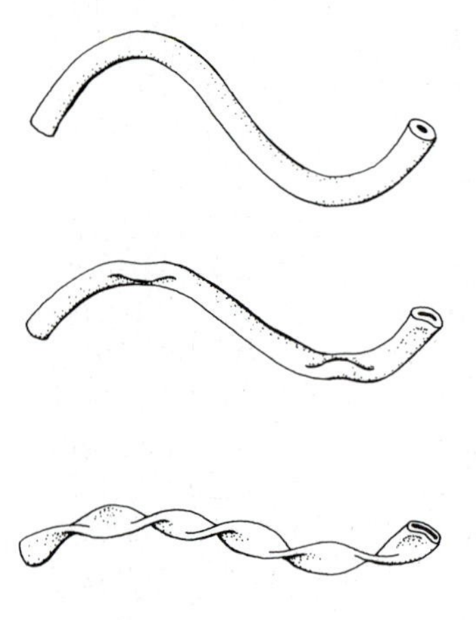

effective compounds. The boll weevil is still a fact of life in cotton fields, causing a 5 to 8 percent loss each year (Figures 9.10–9.11).

The growth rate of cotton is rapid, and it is very demanding of minerals and water. In order to supply these needs, cotton must be heavily fertilized and watered and kept free of weeds that compete with the plant. This has traditionally been done by hand-weeding (chopping cotton) but now is accomplished with mechanical cultivators and by application of selective herbicides. Picking cotton, too, was traditionally hand labor (Figure 9.12), since early picking machines damaged the cotton and indiscriminately collected bolls, twigs, leaves, immature bolls, and dirt which reduced the value of the crop. Efficient mechanical pickers were developed in the late 1940s when chemicals were found that could remove the leaves (defoliate the plants), leaving the bolls intact (Figure 9.13). Plant physiologists who studied the responses of plants to chemicals that later became herbicides and who investigated defoliation and leaf abscission as part of their research on plant development were responsible for this modern revolution in cotton cultivation. They were, inadvertently, responsible for the mass migration of the American black to a culturally inhospitable and economically unready North. Since field hands were no longer needed to chop or to hand-pick cotton, there was no longer a living to be made in cotton country, and black families moved North or into the cities of the South where industry was developing. In 1910, 89 percent of the black population was southern and rural; in 1960, three-quarters were urban. Poorly educated (a quarter functionally illiterate), without marketable skills, and subjected to climatic and cultural conditions unfamiliar to most of them, they were crowded into ghettos, paid minimal wages, and became the victims of business cycles that resulted in massive unemployment. King Cotton treated his subjects poorly.

Figure 9.10 Boll weevil on a cotton boll. Courtesy U.S. Department of Agriculture.

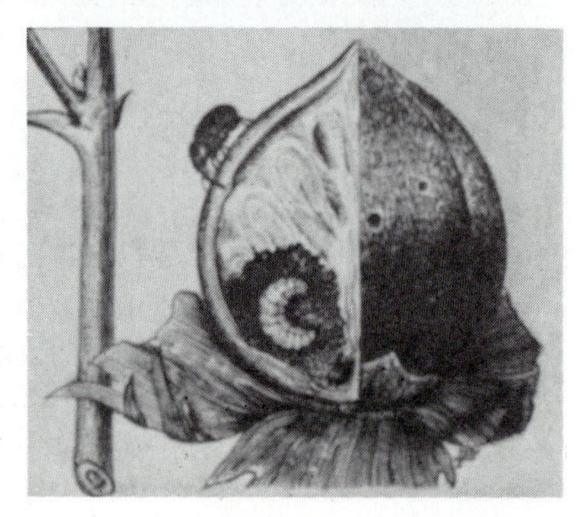

Figure 9.11 Infestation of cotton boll. Section of cotton boll with weevil puncturing boll and a larval stage forming its cell. Courtesy U.S. Department of Agriculture.

Figure 9.12 Hand-picking cotton in the 1930s. Courtesy U.S. Department of Agriculture.

COTTON HISTORY

The Western world's first knowledge of cotton was provided by the Greek historian Herodotus, who visited India in 2484 B.P. and reported that "there are trees on which fleece grows surpassing that of sheep and from which the natives made cloth." In 1350, Sir John Mandiville invented—out of whole cloth—the story of a tree that bore tiny lambs at the ends of its branches, the vegetable lamb (Figure 9.14), and the German name for cotton, *baumwolle* ("treewool"), reflects this legend. Alexander the Great introduced cotton to North Africa in 2330 B.P., and when the Moors conquered Spain in the ninth century, they brought cotton with them. A new type of cotton cloth, muslin (from *musselman*), was developed at about this time, and our word for cotton is from the Arabic word *kutun*. Spanish-made muslin was returned to North Africa and little of it was seen in the rest of the continent. Europe's interest in cotton cloth was aroused by reports of early travelers to China, Japan, and India who saw fabrics "so fine you can hardly feel it on your skin"; Indian poets sang of "webs of woven wind." Venetian merchants obtained a monopoly on Indian cottons and organized camel caravans to transport the cloth across the desert and mountains through Persia to Venice. Some of the cottons from Calcutta were so delicate that "when a man puts it on, his skin shall appear as plainly as if he were quite naked," but most of such cloth was reserved for the harems of the Middle East. These fabrics were a revelation to Europeans who wore heavy woolens or coarse-textured linens and could appreciate lighter

Figure 9.13 Mechanical cotton harvesting after chemical defoliation. Courtesy International Harvester Co.

fabrics for warm-weather use. Interest was so high that cotton became as valuable as the spices of the Orient, which also moved through Venice.

At the same time, explorers of the New World were astounded to find the natives of Cuba, Mexico, Peru, and Brazil clothed in beautifully woven cotton fabrics rivaling the best India had to offer. Columbus found the people of Hispaniola sleeping in *hamacas* loosely woven from strong cotton cordage. Montezuma wore a *tilmatl,* a girdle and cloak, of delicately dyed and embroidered cotton. Magellan found Brazilian mattresses stuffed with cotton batting, and Pizzaro reported that the peoples of Peru used spinning and looming machinery to weave differently colored cotton threads into complicated and elegant fabrics. Modern archeological research has shown that cotton cloth existed in Mexico and Peru by 5000 B.P. and was used in South Africa for at least 2500 years.

Cotton cloth was, however, a luxury fabric in Europe, more expensive than linen and wool, and it remained so until the beginning of the eighteenth century. In most of Europe, native wool was carded, spun into yarn, and woven into sturdy but rough cloth as a cottage industry, where parents and children worked in their own homes. Spinsters were legally unmarried women who were proficient in spinning woolen or linen yarns. Flemish protestors, feeling persecution by the Catholic majority in the fourteenth century, settled in Manchester, Lancashire, and introduced new techniques that helped England to achieve supremacy as the wool-cloth manufacturer for Europe. Small amounts of cotton were imported from Egypt and India to be used as bed linens and small clothes (underwear) and for candle wicks. As cotton cloth from India and Egypt became cheaper, Parliament passed a law in 1666 requiring

Figure 9.14 The vegetable lamb described and illustrated in *Voiage and Travayle of Sire John Maundiville, Knight,* published in London in 1568.

that every corpse had to be buried in a wool shroud on pain of a £5 fine levied, it is presumed, on surviving relatives. Later, the anticalico laws were amended to provide a fine of £200 on anyone using cotton "for domestic use, either as apparel or furniture." Yet, at this time, the British West India Company was instituting cotton cultivation in the Caribbean Islands and Indian seed was planted in Virginia. Between 1690 and 1750, the cloth manufacturing industry was transformed from a bucolic cottage system into the Industrial Revolution.

The Cotton Gin Many factors contributed to the change from a rural to a mechanized city economy in which people became factory workers instead of remaining in villages. One was the development of agricultural practices that allowed smaller numbers of people to supply the food for city dwellers and the exploitation of the potato as a cheap source of carbohydrates. But among the most important reasons were the series of inventions of cloth-manufacturing machinery which were too expensive for a single family and too productive to be managed by a few workers. Prior to the Industrial Revolution, raw fibers of wool, linen, and cotton were carded, spun into threads with the distaff or the spinning wheel, set onto hand looms, and woven with a single, hand-thrown shuttle. Between 1730 and 1760, British inventors devised carding, spinning, and winding machines with tremendous capacities. Arkwright's water-powered loom and the harnessing of James Watt's steam engine of 1769 to the power loom of Edmund Cartwright in 1785 completed the process of mechanization, allowing, indeed demanding, that factories be built to make inexpensive cotton cloth. The limiting factor in large-scale manufacture was the separation of the cotton fibers from the seed, a process that had to be done with a simple, inefficient churka roller gin (Figure 9.15) or by hand. This limitation was overcome in 1793 by Eli Whitney, a Yale-educated tutor to a wealthy Southern family (Figure 9.16).

It is impossible to overestimate the importance of Whitney's cotton gin. The export to England of cleaned raw cotton fibers from the United States increased 800 times between 1790 and 1800 and 2500 times by 1810 (Table 9.1). The parallel increase in the number of slaves is also noteworthy. Equally important were the direct social consequences on the people of England. Thomas Carlyle spoke of the owners of the cotton mills, the coal mines that fed the mills' steam engines, and the railroads and canals that transported the finished cloth as "captains of

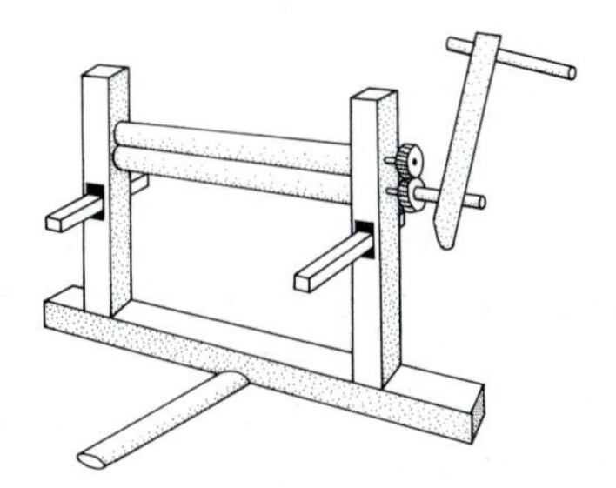

Figure 9.15 Indian churka or roller mills were suitable for separating small quantities of cotton fiber from the seeds.

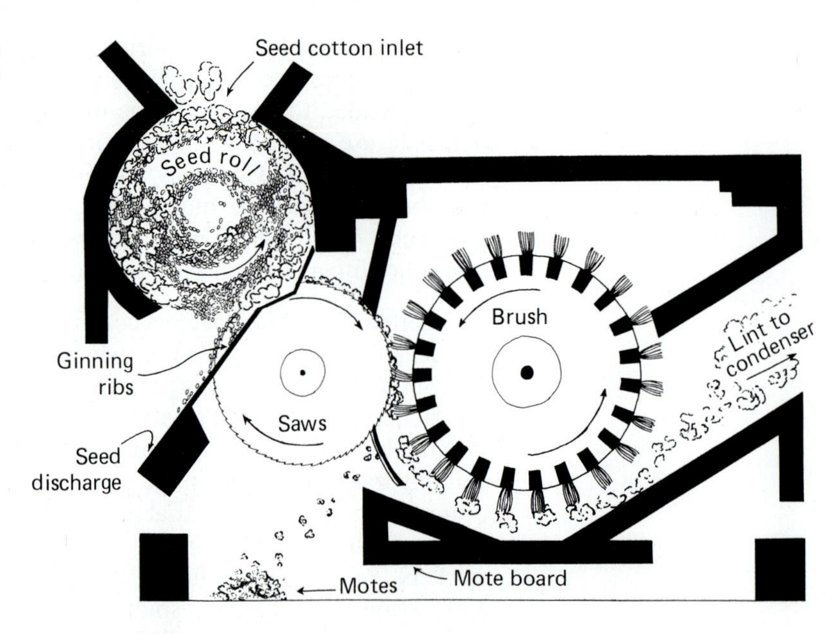

Figure 9.16 Diagram of a cotton gin. The condensed lint is the starting material for thread.

industry." These he defined as men whose only idea of hell was the idea of not making money. These men employed hands—not people—who were instruments for the production of capital and who were to be used as needed to make and remake profit. Factories and mills hired women and children who worked 16 hours a day. Most children entered the mills at the age of 7 and were frequently chained to the looms, for if they nodded and fell into the machinery, it meant a delay of up to a half hour to extricate their broken bodies.

Table 9.1 U.S. EXPORTS OF COTTON TO ENGLAND

Date	Pounds of cotton	Number of slaves in Mississippi and Louisiana
1785	5,000	8,000
1790	25,000	8,200
1800	20,000,000	10,000
1810	62,000,000	15,000
1820	125,000,000	32,000
1830	270,000,000	65,000
1840	500,000,000	193,000
1850	930,000,000	300,000
1858	1,100,000,000	450,000

Wages were pegged below cost of living so that the hands could never escape their just and legal debts. The injured and maimed were summarily dismissed. Rickets, due to lack of sunlight on growing bodies, was the least of a litany of diseases that included tuberculosis, lung cancer, and bladder or kidney infections. Mill hands married in their early teens and bore many children who could supplement the meager wages of the parents. Death at an early age was common and welcomed. "Charity," declared Lord Brougham, "is an interference with the healing process of nature which acts by increasing the rate of mortality." Adam Smith's *Wealth of Nations* and Karl Marx's *Capital* reached opposite conclusions on the morality of the Industrial Revolution. Thomas Malthus, observing the population explosion that accompanied the start of the Industrial Revolution, reached conclusions embodied in his *An Essay on the Principle of Population,* and this book strongly influenced the later writings of Charles Darwin. Elizabeth Gaskell's *Mary Barton* portrayed to a generally unsympathetic middle-class audience the distressed condition of the mill towns. Tories and Liberals fought each other in Parliament on the issues of poor relief and child labor. King Cotton treated his subjects poorly.

The U.S. Civil War The colonists of North America wore, as did their English and French cousins, clothing of linen and wool, originally European, but increasingly of local manufacture. By 1660, some cotton was grown in the Carolinas and by 1700 modest acreages were planted to cotton in Alabama and Mississippi. In 1732, seeds of *G. hirsutum* grown in the Apothecaries' Garden in Chelsea (Chapter 7) were sent to Georgia. This upland cotton formed the base for cotton cultivation in the United States. The fibers, although relatively long and smooth, are difficult to separate from the seeds, the Indian churka roller gin was ineffective, and the fibers had to be plucked by hand by slaves, for whom the cotton cloth was intended. Martha Washington wrote that she supervised the weaving of 300 yards of cotton cloth for the slaves of Mount Vernon, and Thomas Jefferson, the Madisons, and others followed the same practices. There were a few water-powered mills in Massachusetts by 1790, but cotton was a minor economic factor in the young United States. The major Southern crop, tobacco, had become a glut on the market. Slaveholders were being eaten out of house and home by idled field workers, and they were searching for some way to get rid of their slaves.

All this changed abruptly with the invention of the cotton gin. Land again became valuable, slave labor became "necessary," and cotton moved into all of the South and spilled over into the Southwest. As land was exhausted, planters simply moved to new land until the South was a sea of cotton. In spite of *Gone with the Wind,* there were relatively few large plantations with their white, pillared, antebellum mansions on hills overlooking fields where singing slaves picked cotton. Most planters had less than 100 acres, which they worked with the help of a few slaves. Nevertheless, the increased acreages devoted to cotton caused not only a resurgence in the value of slaves but a reactivation of the institution of slave-running. The profits of slave-trading were immense, and the laws against these practices—both in Europe and in the United States—were ignored or subverted. Table 9.1 shows the increase in the numbers of slaves, which paralled the increase in cotton sales.

The failure of the American South to industrialize was due partly to the huge profits that could be made in cotton. This was true in spite of ample supplies of coal and iron, abundant water for power and transportation, and the capital necessary to build factories. Slaves were, it was believed, "unfitted for any other work" and could not be trained to do other than lead a mule through a field. The northeastern states saw and grasped the opportunity to expand manufacturing, and the mill towns of New England date from about 1810. Arguments between an industrial North and agricultural South were fought in Congress, with the South favoring free trade and the North demanding a protectionist economic foreign policy. John Calhoun and Daniel Webster spent much of their legislative lives opposing each others views and helping to polarize the country. As their tobacco-planting forebears had been, Southern farmers soon were in thrall to cotton factors and to the bankers who loaned money for seed, slaves, and supplies in exchange for the right to market the crops. Much of the money was from the North, and Southerners resented this, believing that they were being cheated—and undoubtedly they were. Economic tensions continued to build as England and New England competed for each year's harvest and the North got richer and more powerful at the expense of the South. Opposition to slavery as a degradation of humanity was fanned by books like Harriet Beecher Stowe's *Uncle Tom's Cabin,* in which Eliza crosses the ice with bloodhounds baying at her little heels. Militant abolitionists like John Brown and the establishment of the underground railway for escaped slaves further worsened relationships, and the slave-or-free issue in Kansas stirred all segments of the country. By the time of the Lincoln–Douglas debates, there seemed little doubt that resolution of the differences was close to impossible and that the South would attempt to secede from the Union.

As Britain continued development of the cotton cloth industry, she became absolutely dependent upon American cotton. Sixty percent of all British exports in 1850 were cotton cloth, and close to 5 million people were directly or indirectly dependent upon the Lancashire mills; a quarter of the cotton cloth in the world was of British manufacture. It was clear by 1855 that trouble was brewing in the United States, and the British Parliament was warning of the "danger of our continued reliance upon the United States for so large a proportion of our cotton." In 1855, David Christy of Ohio published *Cotton is King or Slavery in the Light of Political Economy,* in which he proposed that the North accede to all Southern demands or allow secession since, he asserted, the entire economy of the United States was inextricably bound to cotton exports. When Fort Sumter was fired upon and war began in earnest, Jefferson Davis, President of the Confederate States of America, accepted the king cotton idea, believing that the North would soon collapse economically and that Britain would intervene on the side of the South; "Cotton is the King who can shake the jewels in the crown of Queen Victoria." Judah P. Benjamin, Davis's Secretary of State, led the move to prevent the export of cotton, a decision fully accepted by Southern cotton growers from the large plantation owner down to the farmer with 40 acres and a mule.

The South had, however, made a most serious blunder. Even as war clouds gathered in 1858–1859, millions of bales were shipped to the Lancashire mills, providing a reserve which would last for close to two years and allowing the mill owners time to obtain cotton from India, China, Brazil, and Egypt. The British and French upper classes and the mercantile princes favored the Confederacy, and most newspapers were overtly advocating intervention, but the working classes, familiar with exploitation, were so opposed to slavery that they vehemently favored the Union. Yet, as war continued in 1862 and 1863, raw cotton reserves dwindled, and many of the mills went on reduced schedules or shut down entirely. Linen and wool took up some of the slack, as did Indian and Egyptian cotton, but the quality of the Asian product was low and the cloth was expensive. Mill hands were, at best, working only part-time, and workers in subsidiary industries—machine makers, miners, stevedores—were furloughed or fired. Over half the population of Lancashire was destitute. Yet these people supported Lincoln's Emancipation Proclamation to the hilt. Even the captains of industry came around to a noninterventionist position as they found the value of their reserves of cotton cloth rising to unpredictable heights and their profits, undiluted by production and wage costs, continuing to increase. The British government, too, saw the wisdom of nonintervention as stores of American grain poured into a country experiencing a period of small harvests. Relief for the poor, late in coming, inadequate, and most grudgingly given, was finally instituted, but the profits reaped by the rich were bitterly noted by the unemployed. The gap between the ruling classes and the workers was widened and has still not been bridged. The present strength of the British Trades Union Council was at least partly based on king cotton.

As the horror of the Civil War (the War Between the States) rose to a crescendo with Sherman's march to the sea and Grant's bulldogging along the line, the South found that the press of the well-equipped Union was more than could be opposed. The end of the war saw a devastated South, stripped of human and material resources and plagued with carpetbaggers who ferretted out every bale of cotton to be confiscated and shipped with large profit margins to the mills in New England or overseas. The South began to rebuild with the only crops it knew—tobacco and cotton—and did so under unbelievably difficult circumstances. Lacking any capital themselves and finding that a vindictive North would not finance industrial expansion, Southern states remained rural, poor, and segregated. A system of tenant-farming or sharecropping developed in which the owner of land would supply seed, tools, and credit to farmers tilling small acreages of cotton land with some agreed-upon distribution of profits. This did little to create an economically viable South. Until the early part of this century, cotton was grown on worn-out soil by worn-out, undernourished, and undereducated people—black and white. A one-crop economy led to economic stagnation, and the monoculture of millions of square miles devoted to a single plant fostered the rapid spread of plant diseases and insect pests. The boll weevil moved from Mexico, where it was a minor pest on native wild cotton, into Texas by 1900 and into the entire South by 1910, where it reduced yields by over 50 percent in some areas The economic resurgence of the South, its industrialization and crop diversification, did not occur until World War II.

USES OF COTTON

Most cotton grown today is derived from cultivars of *G. hirsutum,* the upland cotton. Close to 12 million metric tons of cotton fibers are produced annually (about 50 million bales of 900 pounds each), with the United States among the major producing countries. Over 24 million metric tons of cotton seed are also produced. Almost all this cotton is grown as an annual, planted from seed each spring, and harvested in the early autumn. Cotton, like tobacco, has high requirements for nitrogen and phosphorus, and the newer, higher-yielding hybrids are even more dependent upon adequate fertilization for their rapid growth. Modifications in the ginning, thread-spinning, and milling processes over the two centuries since machinery was introduced have resulted in a wide variety of cotton fabric types now available. Additional modifications in cotton cloth have been occasioned by the introduction of sea-island cotton (*G. barbadense*). The fibers of upland cotton rarely exceed a centimeter, while those of sea-island cotton may be double that length, permitting the spinning of exceptionally fine, strong threads. Known as Egyptian or Pima cotton (it was first grown in Pima, Arizona), *G. barbadense* is the preferred plant for luxury yarn.

John Mercer (1791–1866) patented the process of soaking cotton threads in an alkaline solution that swells individual fibers, making them smooth and lustrous; mercerized cotton has a silky look and feel and slides easily through cloth when used as sewing thread.

The first of a host of synthetic fibers was rayon, patented in 1855 by a Swiss chemist who dissolved cellulose from wood chips and forced the viscous solution through small holes into a hardening bath. The filaments are then twisted together in groups of 13 to 270 to form thread or yarn. Commercial production of rayon began in France in 1891 and in the United States in 1911. Viscose thread is made by a similar process. When these and the petrochemical-based plastics were developed, synthetic fibers cut deeply into the dominant share of cotton in the world's fiber markets, and cotton growers did little except wring their hands and have their representatives in Congress vote larger subsidies. By 1930, cotton had slipped from 90 percent of the fiber market to 85 percent. By 1940, when the first noncellulose, synthetic fibers appeared, cotton constituted less than 80 percent, dropping to 56 percent in 1970. The development of mixed polyester-cotton blends stopped cotton from slipping even further. In 1970, North American consumers were using less than 16 pounds of cotton per year; the cost of producing fine cotton was almost more than its value on the open market. In the past few years, however, the king has found it unnecessary to abdicate. Cost-accounting methods for determining acreage and marketing, some increased affluence in underdeveloped countries, a resurgence of interest in the pleasant feel and flexibility of cotton compared to synthetic garments, and steeply rising petroleum costs have aided cotton production.

COTTONSEED OIL

Even when the fibers have been removed, the seeds of cotton are valuable. Dehulled seed contains about 35 percent oil that can be expressed by pressure. The press cake, containing almost 10 percent protein, is an excellent food for livestock. Some oil is converted into soap, but most of it is used as a food lipid in salad oils or is hydrogenated to make margarines and solid shortenings.

There is a great deal of interest in the possibility that one of the substances in unprocessed or raw cottonseed oil may be a male contraceptive. In the early 1940s, a report from the city of Jangsu, China, noted with surprise that not a single child had been born there in almost ten years. Attempts to find out why were fruitless until someone discovered that in about 1930, the inhabitants switched from soybean oil for cooking to readily available and inexpensive unprocessed cottonseed oil produced in the region. On the basis of these reports and follow-up studies, Dr. B. S. Liu proposed in 1967 that crude cottonseed oil might become a useful male contraceptive, although the chemical nature of the active substance was not well worked out. By 1972, the potential contraceptive was found to be a mixture of yellowish compounds, including gossypol (Figure 9.17). Gossypol was long known to cause liver damage, and the use of cottonseed oil in food was negligible until methods of inactivating and removing the compound were developed. Its structure was quickly worked out and it can now be synthesized. It has been successfully tested in China with 8000 volunteers; many of these men fathered children after they stopped taking the substance. Apparently, it interferes with the ability of the cells that form sperm to obtain the energy needed via respiration (Chapter 1) to permit sperm cell development. Investigations are now underway throughout the world to evaluate gossypol, with due consideration to the known side

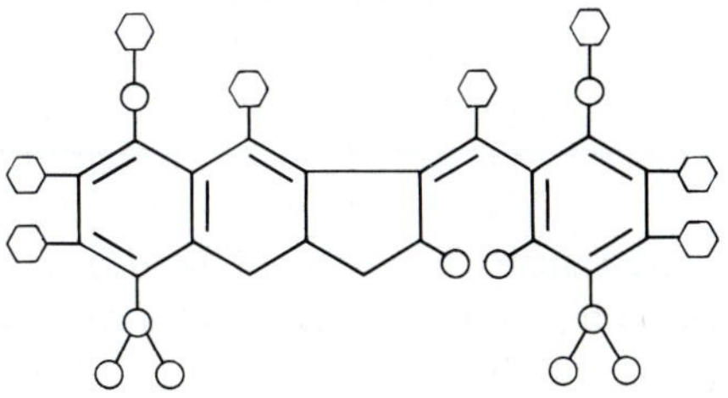

Figure 9.17 Molecular structure of one form of gossypol. Hydrogen atoms and carbon atoms in the rings are not shown. Oxygen atoms are represented as hexagons, carbon atoms as circles.

effects (stomach upsets, liver damage, and so on) and to the social and psychological responses of males whose cultures applaud procreation.

OTHER IMPORTANT FIBERS

Cotton and kapok (stuffing fibers derived from a tree, *Ceiba pentandra*) are known as surface fibers, borne superficially on seeds or other plant parts (Table 9.2). The other major fibers are internal and include the stem or bast fibers and those from leaves, the hard fibers. Bast fibers are obtained from a variety of dicotyledonous plants. They are long strands of several columns of cells cemented together by pectins and gums and are both strong and durable. In contrast to the hard leaf fibers, bast fibers can withstand bleaching and other harsh chemical treatments and can hold dyes well. Cloth made from the leading bast fibers can be pleasant against the skin, a characteristic they share with cotton. Some other bast fibers are less acceptable, although they have been used for apparel for centuries.

Table 9.2 MAJOR FIBER PLANTS OF THE WORLD

Fiber	Botanical name	Major producing countries ranked in order of production
	Seed fibers	
Cotton	*Gossypium hirsutum* *G. barbadense* *G. herbaceum*	Soviet Union, China, United States, India, Brazil, Pakistan, Turkey, Egypt, Sudan
Kapok	*Ceiba pentandra*	Indonesia, Thailand, India
Coir	*Cocos nucifera*	India, Indonesia, Sri Lanka
	Stem fibers	
Flax	*Linum usitatissimum*	Soviet Union, Poland, Czechoslovakia, Germany, Hungary, Italy, France
Hemp	*Cannabis sativus*	Soviet Union, Italy, Yugoslavia, Hungary, India, China
Ramie	*Boehmeria nivea*	China, Japan, Republic of China
Jute	*Corchorus spp.*	India, Bangladesh, China, Brazil
	Leaf fibers	
Manila hemp	*Musa textilis*	Philippine Islands
Sisal	*Agave sisalana*	African countries
Henequen sisal	*A. fourcroydes*	Mexico, Cuba
Bowstring hemp	*Sansevieria spp.*	Central African countries

Flax and Linen The social influences of flax probably antedate our use of cotton as a textile. The plant, *Linum usitatissimum* (Figure 9.18), is

Figure 9.18 Flax *(Linum usitatissimum).*

one of about 50 species in the genus found in both the Old and New worlds, although its exploitation occurred first in the Old World. Our early records of the use of flax to produce linen cloth come from ancient Egypt, where mummies were wrapped in linen cloth, and from the Swiss Lake Villages, where cloth and weaving equipment have been found. Some Egyptian linens were technologically very fine, with up to 500 threads per inch; a robe could be drawn through a finger ring. Linen was apparently common, with references in Genesis 41:42, in which "Pharoah took his signet ring from his finger and put it on Joseph's finger and arrayed him in fine linen," and in Exodus 26 and Matthew 28, among others. By biblical times, cotton was also available, but the superior qualities of linen and its fine texture made it a fabric for the wealthy and the priestly castes.

The Greeks and Romans were acquainted with linen, since it had been produced in Anatolia for centuries. The Romans introduced the plant and linen technology to France and Britain. Charlemagne in the eighth century established centers of linen weaving in Bruges, Courtrai, and other Flemish cities and, by the late Middle Ages, almost all of Europe's woven cloth was either wool or linen. Most European monasteries cultivated flax because linen was prescribed for articles of the Mass and most shrouds were of linen. The conquest of England by William of Orange in 1066 gave further impetus to the linen trade since Britain had the cool, moist climates ideal for flax cultivation. Some flax was grown in India, probably before cotton was cultivated, but linen never became an important fabric there.

By the sixteenth century, Henry VIII of England required "every person having in his occupation three score acres of land apt for tillage shall sow one rood [about a quarter of an acre] with flax or hemp seed." And no wonder. Britain's naval strength depended upon flaxen sails and hempen lines . . . and, of course, on iron men in wooden ships. In the early eighteenth century, the Flemish linen workers, Protestants in Roman Catholic Belgium and France, experienced sufficient religious persecution to force them to emigrate to Britain; they took their skills with them. Ireland became the major British center of flax and linen because the English actively discouraged the production of woolen cloth by the sheep-tending Irish, in order to preserve the mills and enrich the mill owners of the Midlands. While this stricture contributed to the poverty of the Irish, it stimulated the production of fine handmade laces, for which Ireland is still world-famous. Descendants of the Flemish immigrants and Irish linen weavers brought flaxseed to North America, where combined flax and wool fibers were blended into a sturdy cloth called linsey-woolsey. This was the major clothing material in the colonies. The Massachusetts Colony decreed that every homestead grow flax. In both North America and Europe, flax fibers were used as warp threads in hand-woven cotton until the nineteenth century because they were stronger than the cotton threads available.

The bast fibers of flax, like those of cotton, are almost pure cellulose and reach lengths of 2 inches. The small central holes in the cells account for the fiber's exceptional strength (Figure 9.19). Isolation of the fibers is time-consuming and has not been highly mechanized. Typically, the cut stalks are retted—that is, freed of the surrounding tissues by bacterial action. The stalks may be immersed in ponds or streams or simply left in the fields where they are wetted by dew. The retted stems are then broken and crushed (scutched) to remove the softened tissues surrounding the fibers, after which the fibers are combed out, a process called hackling. Our terms *flaxen-headed* and *tow-headed* refer to the yellow color of flax fibers before bleaching. Carding and spinning result in the formation of linen threads, which can then be bleached or dyed.

The flax used for fiber is a tall, almost unbranched plant, while the types used for seed production are shorter, with a large number of side shoots. Oil flax plants, whose seeds contain close to 35 percent oil, are used to produce the drying linseed oils. These are used in paints and soaps and for industrial purposes, although they have now almost been supplanted by synthetics and the industry is in decline. The Soviet

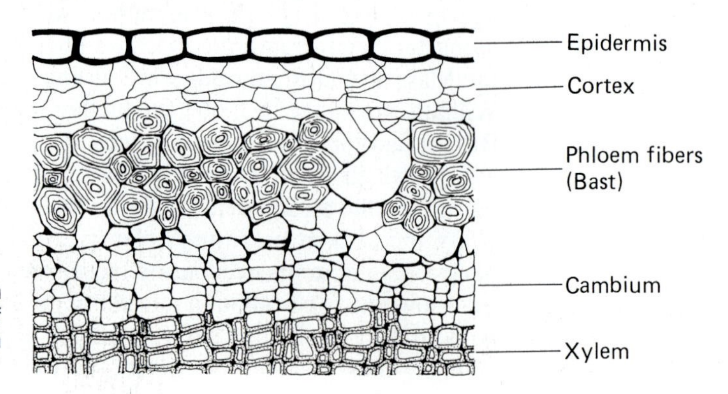

Figure 9.19 Cross section through flax showing position of the phloem (bast) fibers, from which linen thread is spun.

Union produces close to 50 percent of all flax. Like cottonseed oil, linseed oils contain a toxic compound which can be removed during refining so that the oil can be used for the production of margarine and other foods. The press cake remaining after oil extraction is used as feed for livestock.

Another major bast fiber is jute, derived from several species of *Corchorus* in the linden family (Tiliaceae). The fibers are of low cellulose content, are rough and weak, and are brittle because of the high lignin content. Nevertheless, production is the highest among the bast fiber plants because jute is used in making burlap and gunny sacks, furniture webbing, and carpet backings. Over 3 million tons are produced annually, much of it grown in Pakistan and Bangladesh, where it is one of the major cash export products.

Hemp The last of the major bast fibers is hemp (Figure 9.20), derived from the stem of *Cannabis sativa,* whose leaves and flowers are the source of marijuana. The Chinese traditionally ascribe the cultivation and weaving of hempen cloth (*ma*) to the legendary Emperor Shen Nung. Even making allowance for hyperbole, we know that hempen cloth and rope were in common use in the Orient at least 4000 years ago. The Greeks were weaving hemp fibers into cloth during the Golden Age of Socrates. It was employed as a fiber by most of the peoples of southern and central Europe as well as those of China, India, and Japan, where hempen cloaks are still traditional for mourners. Like the other bast fibers, it is isolated by a retting and scutching process and can be spun and woven to produce a durable and strong fabric, but the presence of lignin results in a textile that is less flexible than linen. Cultivation was mandated in England and in the American colonies for the production of cordage. Yankee shipbuilders also used it for caulking

Figure 9.20 Hemp *(Cannabis sativa)*, grown for fiber. Courtesy U.S. Department of Agriculture.

(oakum) hulls of clipper ships. It was grown extensively in Virginia and Kentucky; in 1762, Virginia gave cash grants to hemp planters and fined those who would not grow it. George Washington's plantations contained considerable acreages devoted to hemp; he exported most of his crop to England or had his slaves weave it into a burlap for baling tobacco leaves.

Hard Fibers The hard or leaf fibers are almost entirely derived from the leaves of monocot plants, where they form supportive and conductive strands of small, short cells arranged linearly. Botanically, they are fibrovascular bundles composed of both xylem and phloem or are collenchyma fibers. Although not used as a hard fiber, the prominent strands of celery stalks illustrate the collenchyma tissues that are used. The term *hard* refers to the fact that the cells are highly lignified—that is, their cellulosic cell walls are impregnated with the same compound that gives wood its strength. Isolation of hard fibers is difficult, time-consuming, and labor-intensive because they are freed from the rest of the leaf by scraping. The isolated fibers are strong but not particularly durable and are used primarily for rope and twine (cordage).

Manila hemp, also called *abaca,* is obtained from *Musa textilis*, a plant in the same genus as the banana. It is a tropical plant, grown where rainfall is heavy and temperatures are high. It is a Philippine native, and the islands are still the major source of the fibers. While some is used for textiles, most of it is spun into excellent cordage. The fibers are long, and because they are resistant to salt water, the rope made from them can still compete with synthetics for marine uses. Manila hemp is a fairly recent addition to cordage fibers, since it was not known to the West until the late seventeenth century and was not an article of commerce for almost a century thereafter.

The most important of the fibers of agave, *Agave sisalana*, is sisal, which constitutes half of the hard plant fibers grown throughout the world. Sisal agave was domesticated in central Africa and henequen sisal, *A. fourcroydes*, in Central America; there is little difference between the fibers. Both are known commercially as sisal, named for an old shipping port on the Yucatan Peninsula. Spanish explorers of Central America found henequen sisal in common use; over 90 percent of the sisal cordage is still produced in the hot, dry lands of Mexico.

SPINOFFS

The exploitation of all plants has involved a sequence of technological changes that permitted more efficient and more sophisticated use of each plant. People recognized the potential value of a wild plant, domesticated it, selected valuable strains, and invented the agricultural technology that increased yields. Over centuries, people developed the tools and the methods for the production, storage, distribution and use of the foods and materials that formed parts of human civilization. Nowhere is this better illustrated than in fibers.

It is beyond the scope of this book to discuss in any detail the development of the arts and sciences involved in converting raw plant fibers into the thread and textiles that are considered, along with food and shelter, one of the basic necessities of life. It is generally accepted that animal skins formed the first coverings. Pounded bark cloth, still used in Pacific islands, may have preceded cloth, but the use of woven plant and animal fibers must have developed quite early in human history. Baskets of flexible leaves and stems provided the idea of weaving, later to be extended to fibers. The first step had to be the development of spinning technology in which the matted fibers are straightened and twisted to form thread. This can be done by hand, but the manual procedure was supplanted by the distaff and spindle (Figure 9.21). which was apparently invented independently in many parts of the world and is still in use in tribal regions. The spinning wheel, possibly a Chinese invention, greatly increased efficiency and allowed a finer, more uniform thread to be spun (Figure 9.22). There was little improvement in the spinning wheel until the eighteenth century, when several inventions, noted above, revolutionized the textile industry and, not incidentally, also revolutionized our whole concept of production in

Figure 9.21 Spinning cotton thread. Taken from an Egyptian tomb painting.

Figure 9.22 An eighteenth-century spinning wheel. Redrawn from an old print.

Figure 9.23 Handloom weaving. Taken from a pre-Columbian Peruvian vase.

factories instead of in homes. The development of simple looms (Figure 9.23) and the final preparation of cloth involved many separate inventions in many places throughout the world. Other inventions—the cotton gin, the steam engine, and the electric motor—played seminal roles in textile manufacture.

So, too, was the nascent chemical industry employed in textile production. Natural wools are available in several colors, but the other major animal fiber, silk, comes in only one color. The plant fibers are white, off-white, or yellow-brown. Bleaching, first by simply putting the thread or fabric in the sun, and later by using lye derived from wood ashes, was supplanted by the broad array of chemical treatments available today. The human penchant for color and design was first expressed in woolen texiles, where different colors of yarn could be loomed together in an interesting display of patterns. Methods of dyeing thread and yarns that did not run when wet required that mordants be used. These chemical processes were developed by trial and error and finally by the direct application of chemical sciences that grew out of alchemy. Natural dyes, walnut shells, and so forth were replaced by aniline dyes, invented by William Henry Perkin in the middle of the nineteenth century. His discovery, in turn, gave rise to the giant chemical industries of the world. The chemical industry also developed the synthetics, starting from rayon and now including a bewildering variety of fabrics whose care is sometimes the despair of the wearer and the cleaner.

ADDITIONAL READINGS

Cohn, D. L. *The Life and Times of King Cotton.* New York: Oxford University Press, 1956.

Corbman, B. B. *Textiles: Fiber to Fabric.* 5th Ed. New York: McGraw-Hill, 1975.

Daniel, P. *Breaking the Land: The Transformation of Cotton, Tobacco, and Rice Cultures Since 1880.* Champaign: University of Illinois Press, 1985.

Dempsey, J. M. *Fiber Crops.* Gainesville: Florida University Press, 1975.

d'Harcourt, R. *Textiles of Ancient Peru.* Seattle: University of Washington Press, 1974.

Dodge, B. S. *Cotton: The Plant That Would Be King.* Austin: University of Texas Press, 1984.

Harper, J., and J. Harper. *The Early Development of the American Cotton Textile Industry.* New York: Harper & Row, 1969.

Hollen, N. R., J. Saddler, and A. C. Langford. *Textiles.* 5th Ed. New York: Macmillan, 1979.

Kirby, R. H. *Vegetable Fibres.* London: Leonard Hill, 1963.

Kohel, R. J., and C. F. Lewis (eds.). *Cotton.* Madison, Wis.: American Society of Agronomy, 1984.

Leggett, W. F. *The Story of Linen.* New York: Chemical Publishing Company, 1945.

Mann, J. A. *The Cotton Trade of Great Britain.* London: Frank Cass, 1968.

Qian, S.-Z., and Z.-G. Wang. "Gossypol: A Potential Antifertility Agent for Males." *Annual Review of Pharmacology and Toxicology* 24 (1984):329–360.

Scherer, J. A. *Cotton as a World Power.* New York: F. A. Stokes, 1916 (reprinted in 1969 by Negro Universities Press, New York).

Segal, S. J. (ed.). *Gossypol: A Potential Contraceptive for Men.* New York: Plenum Books, 1985.

Watson, A. *The Rise and Spread of Old World Cotton.* Toronto: Royal Ontario Museum Press, 1976.

Woodman, H. D. *King Cotton and His Retainers.* Lexington: University of Kentucky Press, 1968.

Rubber—Sulfur and a Hot Stove

I have seen a substance excellently adapted to the purpose of wiping from paper the marks of a black lead pencil.

DR. JOSEPH PRIESTLEY, 1770

In geology, a *watershed* is defined as a crest dividing two drainage basins. We also speak of historical and cultural watersheds, events that irrevocably alter the course of history. Certainly the development of the heliocentric concept of the then-known universe by Copernicus and Tycho Brache is a scientific watershed, and van Leeuwenhoek's peerings through a crude microscope represent another. The history and economic botany of rubber is so bound up with the invention of vulcanization in 1830 that one must discuss the topic in terms of pre- and post-Charles Goodyear, just as cotton can be evaluated in the light of the invention of the gin by Eli Whitney.

A BRIEF HISTORY

Like many other plant products, rubber was introduced to Europe by Columbus. On his return from his second voyage, he presented Queen Isabella with several small balls that "rebounded as if they were alive." Her Majesty wanted gold, silver, and spices, not children's toys, and was only minimally amused by the gift. A few years later, scribes of Cortez's army reported "there is a tree . . . much appreciated by the Indians. From this tree a kind of very white liquid flows. To obtain this, the trunk is struck with a hatchet and the liquid flows from these incisions." From *caoutchouc*—the weeping tree—latex was collected, smoked over a fire to coagulate it, and molded into balls. Mayans used these to play *tlachler,* a forerunner of basketball with stone rings as baskets. Since the ball weighed 14–15 kilograms and was moved about with head, legs, and shoulders as in soccer, players were selected for strength, not height. In 1735 Charles Marie de La Condamine, a botanist on a collecting expedition for the Paris Académie des Sciences, saw waterproof shoes in the Amazon valley. He brought some of this gum back to France, where, dissolved in turpentine, it was used to coat fabrics and as a rubber, or eraser, for pencil marks.

Nevertheless, before 1800 there was very little demand for rubber. Smoke-coagulated rubber could be dissolved in naphtha, and Charles Macintosh used this mixture to glue two layers of cloth together to make waterproof coats (hence the British *mackintosh* for a raincoat). Thomas Hancock, a London stagecoach builder, made rubberized cloaks for passengers and drivers and in 1819 opened an "elastic works" to make garters, tobacco pouches, and boots. He later expanded the factory to produce rubber bands, cushions, and even air mattresses. Hancock and Macintosh formed a company making hoses of rubber-covered fabric for breweries and attempted to make pulleys and belts for industry. The trouble was that smoked rubber was sticky when warm and brittle when cold; gum boots would literally dissolve in midsummer and crack in winter, and they smelled bad in both seasons.

CHARLES GOODYEAR

Recognizing the need somehow to stabilize rubber, many inventors turned their attention to the problem. Most tried mixing rubber with lime, magnesium sulfate (epsom salts), lead oxide (litharge), and elemental sulfur to dry out the stickiness. Charles Goodyear, born in New Haven, Connecticut, in 1800, was a rubber fanatic, convinced that stabilization would be a boon to humanity and the source of a personal fortune. He literally drove himself into physical and mental exhaustion, kept his family in abject poverty, and put himself into debtors' prison several times. In 1839, he discovered that coagulated latex, mixed with sulfur and heated, formed a stable product that possessed all the properties we associate with rubber. It was waterproof, elastic, resistant to abrasion, a nonconductor of electricity, would absorb vibration, and was relatively temperature-insensitive. Goodyear spent the rest of his life attempting to promote the use of rubber. He wore pants, coats, and hats of rubber; his calling cards were rubber; he had his portrait painted on rubber. He had the story of his life bound in a rubber book. Digging himself deeper into debt, he invented almost all the things for which rubber is now used. Rubber teething rings, bottle nipples, inflatable rubber boats and cushions, elastic shock cords, doormats, and other items were handmade. Goodyear tried to sell the processes to manufacturers and even peddled rubber goods from door to door. Although Goodyear patented rubber stabilization with sulfur, Thomas Hancock stole the idea and patented it under his name in England and in France. Hancock chose the name vulcanization (from the Roman God of Fire) and successfully beat off Goodyear's feeble lawsuits; Goodyear died a pauper.

Until close to 1850, rubber goods were limited to waterproofings, industrial pulleys, shock absorbers, drug sundries (rubber pants, hot water bottles), and overshoes. Thomas Hancock presented Queen Victoria with a set of solid rubber tires to replace the iron-banded wheels on the stage coach, and with the "patronage of her Majesty" he promoted this more comfortable wheel throughout Europe. In 1845, Robert Thomson coated cotton cloth with rubber and invented the pneumatic tire, later adapted by John Dunlop of Belfast for the bicycle. In 1898, tires for automobiles were made, and the subsequent history of rubber is the history of the automobile industry. The first gasoline-powered horseless carriage in the United States was handmade by the Duryea brothers in 1893, and in 1905 Ransom E. Olds started to produce modest numbers of one cylinder gas buggies. Henry Ford had an assembly line producing the Tin Lizzy by 1906. There were four automobiles registered in the United States in 1895, 8000 in 1900, 26,000 in 1902, and 2 million in 1908. America—and the rest of the world—was on rubber wheels. Akron, Ohio, a center for bicycle manufacture, became "Rubber City" in 1910 when Harvey Firestone invented the nonskid tire. In 1918, the United States had over 9 million cars and trucks on the roads (a Ford Model T cost $400) and was using 50 percent of the world's rubber production (Figure 9.24).

BRAZILIAN RUBBER

Virtually all of this rubber was collected in the jungles of South America and shipped to the United States through the port of Para (now Belém), Brazil. The center of the rubber-collecting area was the city of Manaos, 1000 miles from the coast. The rubber trees *(Hevea brasiliensis)* were wild, scattered over thousands of hectares of dense tropical rain forest land and were tapped by people who were essentially slaves, so deeply

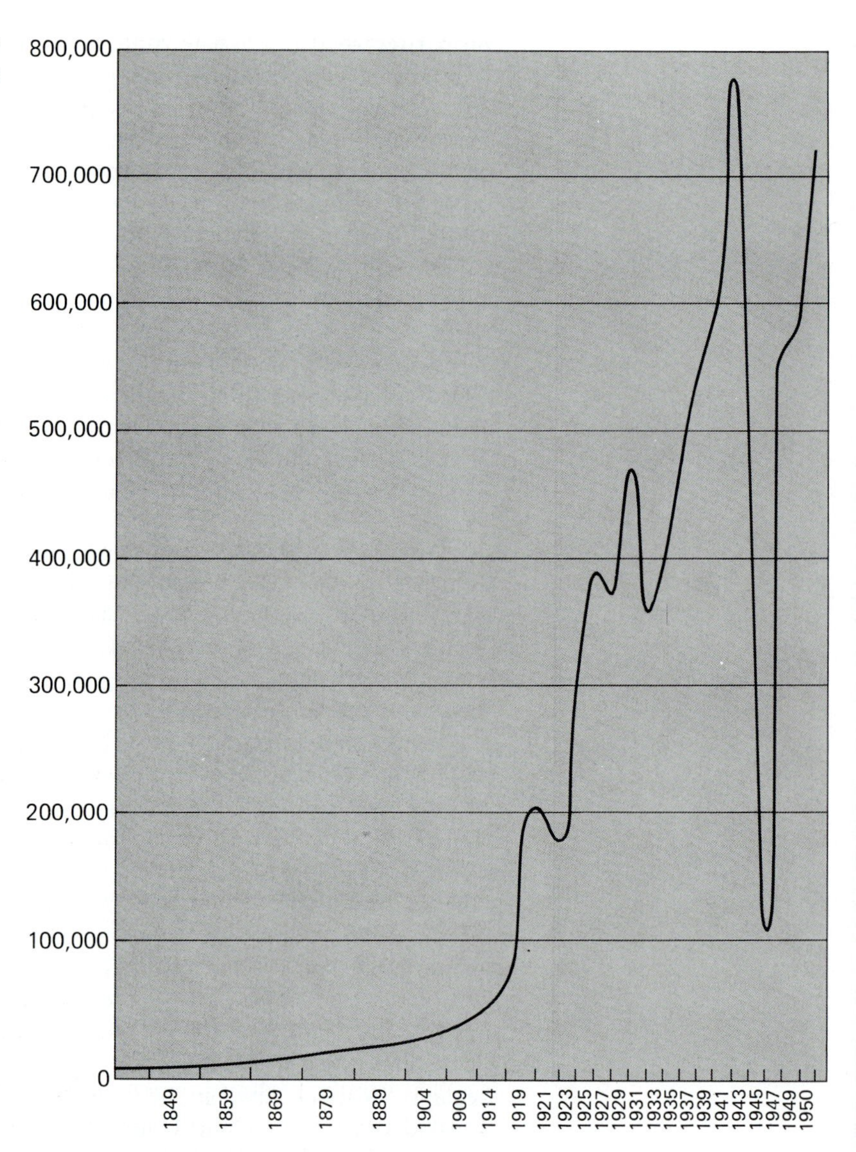

Figure 9.24 One hundred years' consumption of natural rubber in the United States.

in debt to the companies that they could never work themselves free. Malaria and yellow fever killed thousands, the Indians killed other thousands, and the companies also killed thousands who did not meet their daily quotas; it was said that each pound of rubber cost a pint of blood. As demand for rubber increased, unbelievable wealth accumulated in Para and Manaos. During the height of the rubber-tapping period of 1870 to 1890, Manaos built an opera house from imported Italian marble, and individuals built mansions filled with the luxury goods of Europe and America. Jewelry shops and stores filled with French perfume, champagne, and silk clothing sprung up and made fortunes for their proprietors. A few cassandras said that the trees were being overtapped and that prudent management was advisable, but rubber fever so gripped the exploiters of trees and of people that an end of the boom seemed impossible. Yet by 1900, Manaos was a city of decaying houses, an empty opera house, and smashed storefronts. The search for rubber moved elsewhere.

AFRICAN RUBBER

If the exploitation of the people of Brazil was horrible, the rubber experience in the Congo was infinitely worse. In 1850 the American missionary T. L. Wilson had pointed out that when *Landolphia* vines were cut down, a latex flowed from the segment. When rubber production began to fall in Brazil, and as automobile tire manufacture began to climb in 1900, interest in *Landolphia* rubber increased. Leopold II, king of the Belgians, read of the exploits and explorations of Henry M. Stanley (Dr. Livingston, I presume) and decided to "Christianize the Negro," spread trade, and exploit the treasures of west Africa. He formed the International Association of the Congo with the assistance of American missionaries, who saw souls to save, and the support of industrialists, who anticipated a tidy profit. Under the charter of the corporation, Leopold owned tropical west Africa, with one-fifth of all profits from ivory and rubber accruing to his personal account. Shares that sold initially for $100 were worth $2800 in 1900, and by 1903 dividends were several hundred dollars per year. These profits were obtained by outright enslavement of African natives, who were tortured or killed if they did not bring in enough rubber. By 1910, the Congo was depleted of rubber vines and of people. Supplies of rubber from other latex-synthesizing plants *(Funtumia elastica)* in Angola, Guinea, the Ivory Coast, and the Niger River basin were exhausted by 1914.

By 1918, the British realized that they controlled almost two-thirds of the world's rubber, with only trickles coming in from Brazil. There were plantations developed in the Dutch East Indies from seedlings given by the British to plantation owners in Sumatra, Java, and Malaya. Deciding to exploit this financial windfall, Winston Churchill, as Secretary of State for the colonies, suggested in 1922 that Parliament restrict rubber production to maintain prices; the United States alone was using 50 percent of the natural rubber, and the numbers of automobiles in North America had exceeded 12 million. The price of rubber rose to $1.25 a pound in 1925, and the American tire manufacturers decided that supplies under direct domestic control would defeat the British restrictive practices. Plantations were started in Liberia, a country founded as a home for freed American slaves, in Ghana, and in the Philippines. Henry Ford started a model city-plantation in Brazil which failed when the indigenous leaf blight fungi (*Dathidella ulei*) killed the trees. The African plantations were and still are successful; Liberia produced 50 million pounds of rubber in 1971. As the market for rubber increased,

Dutch planters intensified their efforts to develop superior plantings. Following their own example with quinine (Chapter 7), they grafted high-yielding stems to disease-resistant rootstock and increased production to the point that they were in direct competition with the British. The restrictive acts were repealed in 1927, but by that time the Dutch controlled almost two-thirds of the world's rubber, and when rubber became a glut on the market in the early 1950s, a British-Dutch cartel was formed to regulate rubber supplies.

RUBBER CHEMISTRY

The bulk of natural rubber is obtained from *Hevea brasiliensis,* a tall tropical forest tree native to the Amazon Basin. *Hevea* is a genus in the Euphorbiaceae, a family also containing the *Poinsettia,* the crown-of-thorns *(Euphorbia splendens),* and the tapioca plant *(Manihot utilissima).* Several members of the mulberry (Moraceae) family produce latex, including the ornamental rubber tree *(Ficus elastica)* which graces bank lobbies and barber shop windows and a vine *(Castilla elastica)* that sparked a short-lived rubber boom in Panama and Mexico. The mangabeira *(Hancornia speciosa),* a milkweed *(Asclepias),* and the Russian dandelion *(Taraxacum kok-saghyz)* have been used commercially; all told, over 160 species of plants product latex. As removed from the plant, latex is over 50 percent water, 10 percent resins, proteins, and sugars, and 30–40 percent rubber. The "very white liquid" reported by Spanish explorers is an emulsion, as is homogenized milk or mayonnaise. Rubber itself is a polymer of a 5-carbon compound called *isoprene:*

$$CH_2 = CH - \overset{\overset{\textstyle CH_3}{|}}{C} = CH_2$$

The polymer may be linear:

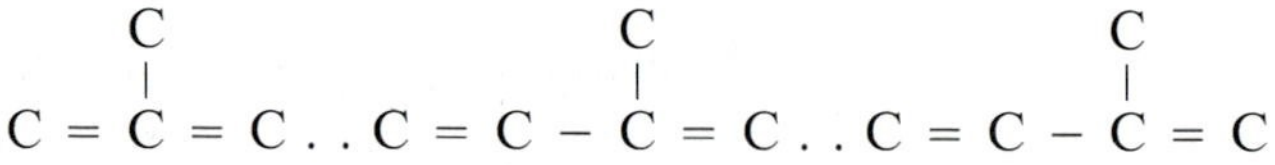

$$C = \overset{\overset{\textstyle C}{|}}{C} = C\,.\,.\,C = C - \overset{\overset{\textstyle C}{|}}{C} = C\,.\,.\,C = C - \overset{\overset{\textstyle C}{|}}{C} = C$$

or cross-linked:

$$\begin{array}{ccccccc} & & & C & & & \\ & & & | & & & \\ C = C & - & & C & = C & & \\ & & & | & & & \\ & & C = & C & - & C & = C \\ & & & & & | & \end{array}$$

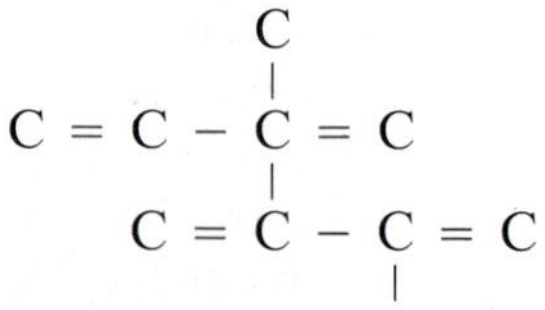

or it can exist in more complex configuration. Polymer chains may exceed several thousand monomers. Although it has been suggested that latex serves the plant as a sealant of natural wounds, its exact role has never been defined.

In *Hevea,* latex is produced in a series of latex vessels that anastomose through the phloem (Figure 9.25). Trees can be tapped

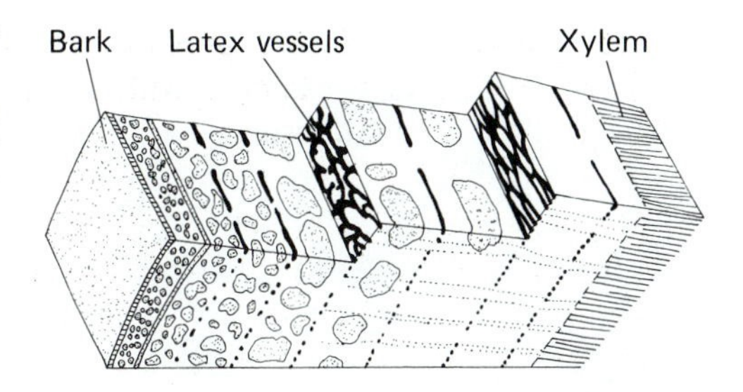

Figure 9.25 Diagram of part of the stem of the rubber tree *(Hevea brasiliensis)*, showing the location of articulated laticifers in the phloem.

several times a week during the growing season, each set of cuts made successively higher on the trunk (Figure 9.26). Each tree can produce 3 to 4 kilograms of rubber per year. Regeneration of phloem is rapid, and if trees are tapped on alternate years, a single tree can produce rubber for more than 30 years. Advances in technology have more than doubled yields with no apparent damage to the trees. A plantation of 1000 trees can provide close to 7000 kilograms per year. The latex drips into a cup and is brought to collecting stations, where the liquid is mixed with weak acetic acid to coagulate the rubber. Water and contaminants are removed by squeezing the rubbery sheets between steel rollers, and the crepe rubber is dried for shipment. At factories, dried rubber is thoroughly mixed, carbon black or other coloring matter added, and then it is vulcanized by the addition of sulfur, litharge (lead oxide), and heat. Used and reclaimed natural rubber and synthetic rubbers are added with additional mixing (Figure 9.27) and heating to produce rubbers for the many different products we use.

The enzymatic biosynthesis of latex illustrates the way that organisms expand and diversify biosynthetic pathways that follow the same sequence of chemical steps and require the same starting and intermediary materials. Once a biochemical pathway is established, organisms add on

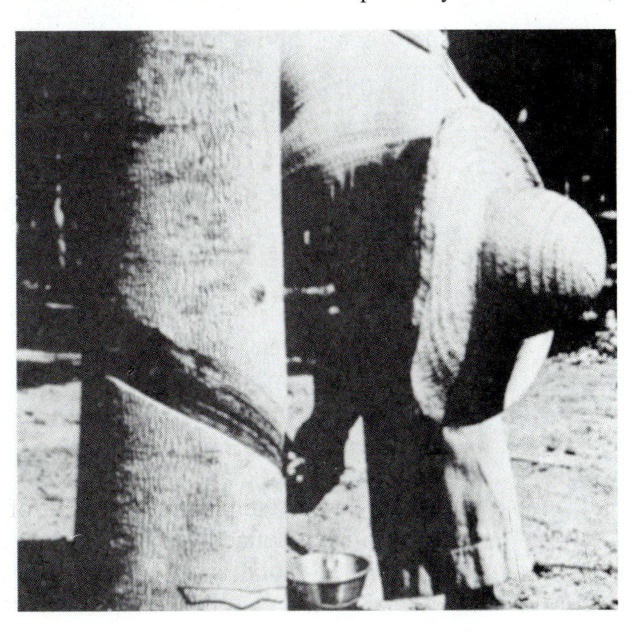

Figure 9.26 Tapping a rubber tree in Brazil.

Figure 9.27 Kneading rubber to make it pliable. Courtesy Firestone Tire Co.

to it by alterations in the sequences of genes to permit the formation of new end products. A 2-carbon intermediate (acetyl coenzyme A) in the respiratory pathway which links two portions of the energy-yielding respiratory pathway is shunted away from the respiratory route and is converted by several steps into a 5-carbon compound, mevalonic acid. Mevalonic acid is a branch or switch point from which a large number of important compounds are synthesized. One route leads to the formation of the carotene and xanthophylls, yellow to orange compounds that are part of the photosynthetic apparatus in the chloroplast and give carrots their orange color or Golden Delicious apples their yellow hues. A second route from mevalonic acid gives rise to the lycopenes, bright red pigments found in ripe tomato fruits. A third route leads to the synthesis of isoprenes, 5-carbon compounds that are themselves branch or switch compounds. From the isoprenes are derived the terpenes that give conifer needles their spicy odors. Isoprenes are also part of the complex of substances that we use as essential oils (see section on essential oils). One of the classes of plant growth regulating compounds, the GAs (gibberellic acids) are derived from isoprenes, as are sterols. Latex, too, is an isoprene-derived substance.

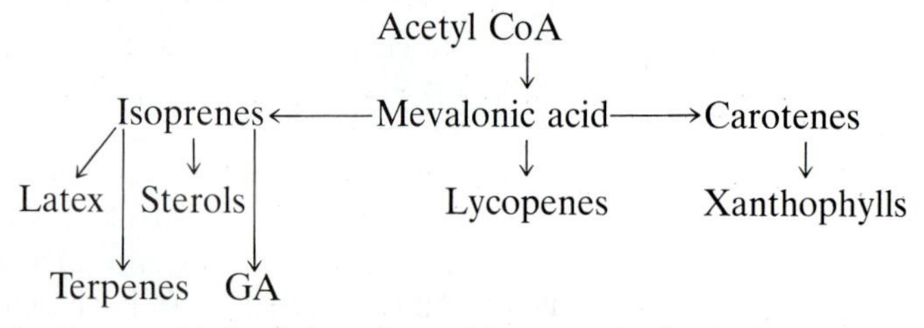

It is tempting to think of these branching metabolic chains as developing sequentially in evolutionary time with a respiratory pathway involving acetyl coenzyme A having evolved before the sequence of steps leading to mevalonic acid. However, plants do have alternate routes for the biosynthesis of various compounds; also, there is not enough data to justify such a linear concept of evolution. We are also unable to answer the question of how (much less why) plants in completely unrelated taxonomic groups will develop identical biosynthetic pathways.

SYNTHETIC RUBBER

In 1942, the Japanese invaded Southeast Asia and cut off over three-quarters of the West's supply of natural rubber. As early as 1939, perceptive governments and rubber corporations realized the dangers inherent in conflicts that would involve reduction of rubber supplies. Germany, the Soviet Union, Britain, and the United States began stepping up their efforts to produce synthetic rubbers. The first synthetic was based on a discovery of Julius A. Nieuwland at Notre Dame University. Nieuwland sold his patents to DuPont in 1925, and William H. Carothers developed DuPrene in 1931; DuPrene later became Neoprene. Isoprene, the monomer of natural rubber, is difficult to synthesize, and the attention of the German I. G. Farben group and of Standard Oil was directed toward rubber synthesis from polymers made from a related compound, butadiene ($CH_2 = CH - CH = CH_2$) easily obtained from coal tar, coal gas, or, as C. V. Lebedev of the Soviet Union found, from ethyl alcohol obtained from grain fermentation. The Russians called their product SKA and the Germans called theirs BUNA. Neither product was completely satisfactory; abrasion- and temperature-resistance needed for tires running at high speeds were low. In the early 1940s, nations found that addition of a plastic, styrene, stabilized synthetic rubber. The process of making synthetic rubber is fairly simple in outline.

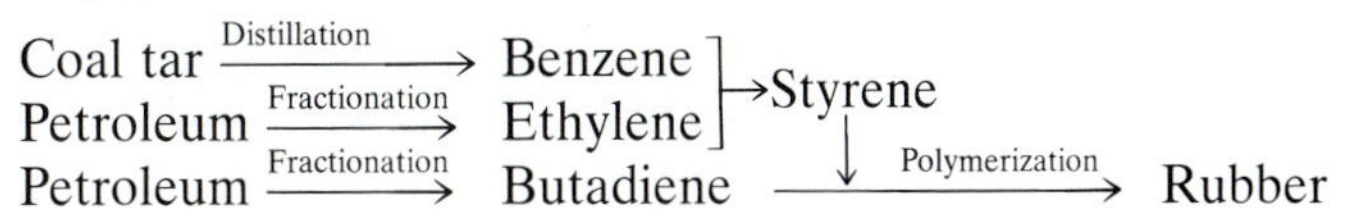

With the end of the war and reactivation of rubber plantations in Africa and Southeast Asia, natural rubber was again being collected. Most rubber products are mixtures of natural and synthetic materials.

As the cost of synthetic elastomers (elastic polymers) rises with the cost of petroleum, there has been renewed search for other sources of natural rubber. Attention has focused on guayule (why-oo-lay) a desert shrub *(Parthenium argentatum)* native to northern Mexico, Texas, and southern California. Guayule is a composite (Compositae) related to goldenrod. The latex is a true isoprene polymer. In contrast to *Hevea,* it is not found in vessels, but in the vacuoles of stem and root cells. Since 20 percent by weight of the plant is latex, its exploitation is not new; Aztecs used its rubber for balls and for chewing gum. In the early part of this century, processing plants were constructed in Mexico by an American consortium. Between 1900 and 1910, over 100 million pounds of guayule rubber were produced, close to half the rubber used in the United States. Since the plants were destroyed to obtain the latex, and little attention was paid to replacing the bushes, the supply was soon exhausted and the flood of cheap rubber from Asia made the operation superfluous. During World War II, the U.S. government spent $30 million to develop guayule rubber, and while 3 million pounds were produced in 1941–1943, the development of synthetic rubber made the plantations uneconomical. The federal government, in its wisdom, set fire to 50 square miles of guayule plantings. In 1977, interest was rekindled. A report by the U.S. National Academy of Sciences recommended that otherwise unproductive desert land should be planted with guayule to ensure a domestic supply of rubber and to develop an employment base in the American Southwest.

ADDITIONAL READINGS

Baum, V. *The Weeping Wood.* Garden City, N.Y.: Doubleday, 1943.

Collier, R. *The River That God Forgot: The Story of the Amazon Rubber Boom.* New York: Dutton, 1968.

National Research Council. *An Alternative Source of Natural Rubber.* Springfield, Va.: National Technical Information Service, 1977.

Polhamus, L. G. *Rubber, Botany, Production and Utilization.* London: Leonard Hill, 1962.

Wilson, C. M. *Trees and Test Tubes: The Story of Rubber.* New York: Holt, Rinehart and Winston, 1943.

Wolf, R., and R. Wolf. *Rubber: A Story of Glory and Greed.* New York: Covice, Priede Publications, 1936.

Tobacco—The Loathsome Weed

A woman is just a woman
But a good cigar is a smoke.

RUDYARD KIPLING

On October 13, 1492, one day after the *Santa Maria* made its historical landfall at San Salvador, the admiral noted in his journal:

> In the middle of the gulf . . . I found a man in a canoe carrying a little piece of bread . . . a gourd of water . . . and some dry leaves which must be a thing very much appreciated among them.

On November 2, Columbus landed at Cuba and sent his interpreter, Luis De Torres, and able seaman Rodrigo de Jerez with a letter and gifts to the "court of the Grand Kahn of Cathay" announcing the arrival of the fleet and expressing his interest in establishing commercial relations with China. The two men met many people on their fruitless journey "always with a firebrand in their hand . . . and a certain leaf after the fashion of a musket [tube]. They light them at one end and at the other they chew or such and take in with each breath that smoke which dulls their flesh and as it were intoxicates, so they say that they do not feel weariness" (Figure 9.28). Gifts of these cigars were given to the admiral, who almost casually added them to the growing pile of items to be returned to Spain. One story goes that Queen Isabella became the first Western woman to smoke a cigar. As other explorers invaded the Americas, a fuller picture of tobacco usage developed. Haitians called their rolls of leaves *tabacca.* Amerigo Vespucci visited Margarita Island in 1499 and saw natives with "a green herb which they chewed like cattle to such an extent that they could scarcely talk." Cortez saw Montezuma smoking a hollow reed filled with a mixture of liquid amber and tobacco; Jacques Cartier in 1534 described the ground tobacco of the Algonquins of the St. Lawrence Valley: "After taking some into our mouths, it seemed like pepper, it was so hot." Throughout the Americas nasal snuffers (Figure 9.29) for inserting powdered tobacco directly into the

Figure 9.28 The first illustration of tobacco. Taken from *Stirpium Adversaria Nova* by de l'Obel, published in 1570.

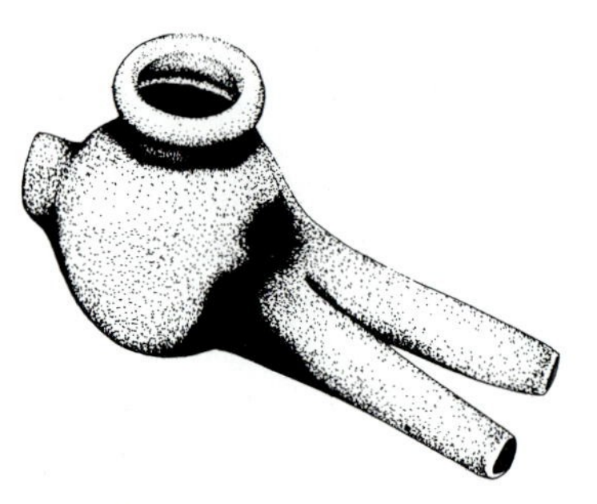

Figure 9.29 A clay nasal snuffer used in Mexico about 2100 B.P.

nostrils were in common use. Thus, all of the methods of using tobacco were well established long before the fifteenth century.

In use at least 1000 years before Columbus, tobacco was not only a medicine but served an important role in ritual. Aztec Gods smoked (Figure 9.30) to tell people of their decisions, and Mayan priests used smoke to intercede for their people (Figure 9.31). Tobacco was used by all North American Indians as money, and the leaf was so important that councils and diplomatic alliances were formalized by passing the calumet, the pipe of peace (Figure 9.32). The tomahawk pipe was a white man's invention, but it fit so well with established cultural patterns that the Indians immediately accepted it. Quids of folded leaves were held in the cheeks of men on hunting forays to dull hunger pangs and lessen the need for water. Women ate powdered tobacco to ease the pain of childbirth. When in the nineteenth century the settlers of the West spat out tobacco juice, the Indians were amazed—that was the best part!

IMPORTATION INTO EUROPE

Explorers noted that there were two different kinds of tobacco plant, a low-growing type bearing yellow-green flowers that was most common

Figure 9.30 Aztec Deity blowing tobacco smoke. Taken from a pre-Columbian stone carving.

Figure 9.31 Mayan priest blowing smoke as part of a ritual to bring rain. Taken from a bas-relief in Chiapas, Mexico.

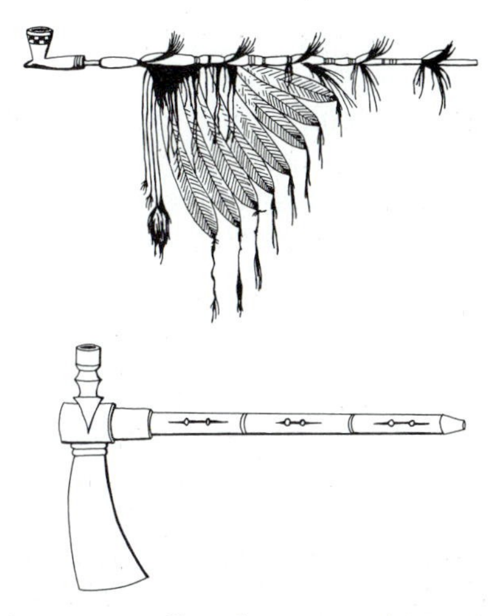

Figure 9.32 The calumet, or peace pipe, of the plains Indians and the tomahawk, or war pipe, of the prairie Indians.

in the temperate climatic zones of North America, and a taller, less coarse plant with white-to-pink flowers. The smaller plant, *Nicotiana rustica,* was decidedly less popular since its smoke was biting, harsh, and stupifying compared with the milder flavor of the taller species, *Nicotiana tabacum.* Columbus's cigars were probably *N. tabacum* (Figure 9.33), but the tobacco smoked by swaggering Spanish sailors in the barrooms of Cadiz was probably *N. rustica.* By 1531, the Spanish imported *N. tabacum* seed from the mainland and started cultivating it in the West Indies, and by 1540 both species were being shipped to Spain as dried, pressed leaves. The initial popularity of tobacco was its supposed medicinal virtues. Spanish ships' crews and the English, French, and Dutch freebooters who preyed on the Spanish fleet insisted that it was a potent aphrodesiac; the women in New Spain ports, it was

Figure 9.33 Indian of Hispaniola smoking a cigar. Taken from a woodcut by Thevet, published in Paris in 1558.

asserted, were heavy smokers and wonderful lovers. Spanish and Portuguese explorers such as Damiao de Geos reported that "the holy herb, because of its powerful virtue in many ways, of which I have experienced principally in desperate cases, will cure ulcerated abscesses, fistulas, sores, inveterate polyps, and many other ailments." That tobacco "improveth the lungs, cleanses the body of excess rheum," cured headaches, was an "excellent defence against bad air, eased convulsions," antidoted snakebite, and was a "sovereign medicine for female internal disorders" was avowed with the authority that only ignorant medical practitioners seem able to muster. As exploration of the Americas continued in the latter half of the sixteenth century, illustrated reports of medical and hallucinogenic uses of tobacco began to appear in Europe. Shamans in both Americas used tobacco smoke (Figure 9.34) to reach hallucinogenic states wherein they could "see the evil" and could prescribe remedies which, in many instances, involved purification of the patient with massive inhalations of tobacco smoke (Figure 9.35).

Although attempts were made to have tobacco dispensed only by the medical profession, its "pleasant stupification" spread widely throughout Europe. Andre Thevet brought seeds of *N. tabacum* to France from Brazil in 1556. Jean Nicot, ambassador to Portugal in 1560, diplomatically presented *N. rustica*—dubbed by him the queen's herb—to Catherine de Medici (1518–1589) and was subsequently honored by Linnaeus when the genus name was applied. Sir John Hawkins, privateer, introduced tobacco to England, but a person with more advertising flair, Sir Walter Raleigh, popularized it in the court of Elizabeth I. Tobacco grew poorly in Europe, and most of the leaf

Figure 9.34 Narcotism and hallucinogenic states induced by smoking the high nicotine *Nicotiana rustica*. Taken from *La Historia del Mondo Nuovo* by Benzoni, published in 1565.

Figure 9.35 Indians treating the sick with tobacco fumes. Taken from *Brevis Narratio* by DeBry, published in Frankfurt in 1591.

available was *N. rustica,* so harsh that it could comfortably be smoked only in pipes. Gallants in England and on the continent invented the smoke ring, the tobacco belch, and other important social graces (Figure 9.36). Although tobacco sold for its weight in silver, its use spread so rapidly that authorities tried to get rid of it, since money was leaving the country and entering the Spanish treasury. King James I published a *Counterblaste to Tobacco* in 1604, which ended on a high note:

Figure 9.36 Tobacco drinker and his visions. Taken from *The Melancholy Cavalier* by Cleaveland, published in London in 1656.

> A custome lothsome to the Eye, hateful to the Nose, harmful to the Lungs, daungerous to the Braine, and in the black stynking thereof, neerest resembling the horrible Stigean smoke of the Pit which is Bottomless.

Unsuccessful with this approach, James decided to impose a duty and excise tax on tobacco. The Portuguese introduced tobacco into China with so much success that the Ch'ing emperor K'ang Hsi ordered decapitation of anyone selling it; Murad II of the Ottoman Empire regularly executed smokers, and the Romanoff tsar Michael ordered that snuff-sniffers should have their nostrils slit. Pope Urban VIII issued a papal bull mandating excommunication of priests who smoked when saying mass, but it was not until the accession of Pope Innocent X in 1645 that the same stricture was placed on communicants who regularly filled St. Peter's with clouds of tobacco smoke. All this was to no avail; people smoked, in the words of one early seventeenth-century writer, "as an excellent defence against bad air . . . it is good for to warm one being cold and will cool one being hot." The French dramatist Molière wrote that "he who lives without tobacco is not worthy of living." The hold of tobacco on the population of England was tightened when in 1665 the Great Plague decimated the city and "ye precious stynke" was believed to be the only thing that could ward off the disease.

COLONIAL TOBACCO GROWERS

Although the obligatory purchase of tobacco from Spain or Portugal kept the smoker poor, enriched royal treasuries, and offended the churches, it was obvious that England had to develop its own source of supply. John Rolfe, who later married Pocahontas, started a plantation of *N. rustica* in the Jamestown Colony in 1610, and in 1612 planted seeds of milder *N. tabacum* stolen from the Spanish West Indies. In 1613, leaves were shipped to England and had immediate success. When, in 1620, Elizabeth granted the Virginia Company a monopoly to export leaves to the mother country, virtually every piece of open land was planted to tobacco; even the shoulders of the roads were used. Thomas Dale, governor of the company, observed that cropland had been planted in tobacco and being concerned that the colonists might starve, ordered that for every acre of land in tobacco, there had to be two acres of Indian corn planted. His order was rarely obeyed since a man could earn £50 sterling per year with tobacco and less than £10 per year from corn. George Calvert, Lord Baltimore, obtained a royal patent to settle Maryland in 1632, and, over the protests of the Virginia Company, planted tobacco as the main crop. The Maryland settlers developed from *N. tabacum* the cultivar known as Burley, which has become a major type of cigarette and pipe tobacco.

By about 1675, Spanish control of the tobacco market was ended, and all of Europe turned to England for its fine tobacco leaf. In order to control prices, the colonists of Virginia and Maryland agreed to restrict production on the understanding that the British crown would reduce their tax burden, an action which the crown used to its own advantage. The British navigation acts provided that all tobacco had to be shipped to England in ships of the British merchant marine. Obviously, the colonists objected to this; they saw that tobacco profits would be reaped by British merchants. They demanded the right to trade directly with other countries, insisting that they had not been consulted about the acts. This was denied, and the growers began selling their tobacco to New England merchants, who dodged his majesty's customs officials and sold the tobacco directly to the Dutch out of the port of New Amsterdam (New York). Pointing to this flagrant smuggling as a justification for direct trade, planters of Virginia demanded a new assembly for the colony, one that would stand up for their rights against the collectors of revenue and one that would work toward free trade instead of toadying to the king's governors. Shouting "God dame ye collectors!" and "No duty to ye king!" people rebelled, and it took over a year to put down the disturbance. Some of the demands were met, the rebels went unpunished, and the news that the Lion's tail could be tweaked with impunity spread throughout the colonies, to be remembered a hundred years later when the first Continental Congress met to demand free trade and eventually freedom.

TOBACCO AND THE AMERICAN SOUTH

Tobacco shaped the social, economic, and political life of the central Atlantic colonies from Maryland south through the Carolinas. As land was cleared and slaves became the labor force, tobacco was grown on huge stretches of land. The plantation concept was born, later to be extended to that other staple crop of the American South, cotton. The South was rural, life was gentle, somewhat monotonous, and flowed with the seasons. A small group of landed gentry dominated the area with a very few middle class merchants and a stratum of white people whose jobs consisted of overseeing black labor. Life on tobacco plantations was, if you were on top, very pleasant with overseers to order and slaves to work. The gentry spent much of their time hunting, gaming, dancing, and being active in politics. Because the navigation acts proscribed trade with other than Britain, virtually everything used in the Southern colonies was imported. Nails, glass, table service, books, brandy and fine wines, and clothing all came from the proceeds of tobacco sales. Tobacco could buy anything; 120 pounds of leaf bought passage of a marriageable woman from England. The local clergy prayed fervently for a good harvest, for hogsheads of tobacco rolled to the parish church in payment for their service. The sons of the planters were educated in England on tobacco money. William Byrd of Westover inherited 43,000 acres and by 1730 had 179,000 acres. William Fitzhugh of Bedford had an income of £6000 from 54,000 acres and 1000 slaves. Robert Carter had 300,000 acres and George Washington's slaves tilled 45,000 acres. Since months or even years flowed by between the time the tobacco was harvested and the accounts of the British and Scottish tobacco merchants were received, planters had no qualms about pledging their crops four, five, or more years in advance and having their bills paid by the merchants. Plantation life was on the installment plan, and as years went by, the planters fell further and further into debt.

Yet, in spite of, or perhaps because of, the easy life, the gentry took their political and public responsibilities very seriously. They served as governors, judges, leaders in the established Church, formed the local militia, and constituted the legislature. Their political and legal sophistication was high, and, as evidenced by the roster of United States Presidents from Washington to the Harrisons, they dominated the political affairs of the colonies and, eventually, of the United States. The Southern gentry were lords of all they surveyed, and they chaffed under the restrictions imposed by a distant Parliament which controlled taxes, duties, tariffs, and enforced the navigation acts. As debts mounted, they recognized that tobacco merchants offered them lower prices than they could obtain on the continent and overcharged them for the Georgian goods they wanted. They objected to being dunned for money by "mere tradesmen."

Yet the feeling that they were, in fact, aristocrats limited their ability to make common cause with people like Benjamin Franklin, a printer, or John Adams, a businessman, or Alexander Hamilton, a West Indian, a bastard, and a banker. Nevertheless, the desire for free trade was one common bond, confidence that they could manage their own affairs was a second, and a philosophical concept of dignity and freedom (if one were white) was a third. In 1777, the rebel Virginia House of Burgesses passed a law sequestering all British tobacco debts. Jefferson, the Lee family, and the others deposited in the State Loan Bank the sum of £275,000 in worthless paper money to cover these debts; they were gambling that the debts would be uncallable if the revolution succeeded. When, in 1783, the Treaty of Paris provided that the tobacco debts were still outstanding, many planters simply refused to pay them; George Mason rhetorically asked why the Virginians had been fighting if not to get out of the hands of bloodsucking tobacco merchants. The situation was so bad that the Congress finally authorized the use of federal funds to cover the debts. Thomas Jefferson negotiated privately and after insisting that tobacco burned by British troops on his Elk Island plantation should be applied to his credit, had a final bill of $10,000, about which he said,"Justice is my object, as decided by reason and not by authority or compulsion." This huge debt, his deep involvement in the affairs of the young country to the exclusion of his personal business, and, it must be noted, his tastes for fine wines and other amenities placed him in a desperate financial situation. When the British burned Washington in 1814 and the modest Library of Congress was destroyed, Jefferson offered Congress his personal library of 6500 volumes as the nucleus for a new library. The $23,000 received allowed him to pay his creditors, including the despised tobacconists.

THE BOTANY OF TOBACCO

Botanically, the genus *Nicotiana* is a complex of 65 species, most of them native to the New World. Based on its physiology, morphology, and chromosome complement, *N. tabacum* is likely to be a hybrid of several wild species; there is good reason to believe that *N. tabacum* is a cultivated species, not found in the wild. Although usually grown as an annual, it is capable of perennial development in tropical climates. For a general description, we can do little better than to present the one given by Dr. Nicholas Monardes of Seville in his *Joyfull Newes Out of the New-Founde Worlde,* which was "englished" in London in 1596:

> This hearbe hath the stalke greate, bearded [hairy] and slimie, the leafe large and long, bearded and slimie. The plant is very ful of leaves and groweth in height foure to five foot. In hot countryes, it is nyne or tenne monethes in the yeare laden . . . with leaves, flours and Coddes [fruit], ful of rype graynes. It sproutes foorth neere the roote muche . . . notwithstanding the graine is the least seede in the worlde, and the roote be like small threeds. The leafe of this hearbe being dried in the shadow, and handed up in the house . . . and bee burned, taking the smoke thereof at your mouth through a fonnel or cane. . . .

When Linnaeus established the genus to memoralize Jean Nicot, he included only *N. rustica* and *N. tabacum,* the only ones known in the eighteenth century. Since then, wild species have been found in Australia, and others have been collected in both North and South America. Many of these have been used as a source of genetic characters for cultivars of *N. tabacum.*

N. tabacum planting material is all derived from seed discovered by the Spanish in 1530 and transplanted to Haiti and Cuba. The bulk of North American tobacco includes two categories, the Bright tobaccos and the Burley types. Most Bright tobacco is grown in the Piedmont and coastal plain areas of the south Atlantic states characterized by sandy, well-drained soils with low organic contents, acid reactions, and very low reserves of inorganic nutrients. Burley tobaccos thrive on the richer, silt-loam soils of Kentucky and Tennessee and in the warmer areas of Ontario. Bright and Burley form the bulk of cigarette and pipe tobaccos. Cigar tobaccos are grown in Maryland, parts of Virginia, and in the Pioneer Valley of Connecticut, where the plants are grown under cloth tents which, by reducing solar intensity and increasing humidity, allow the plant to develop large, thin leaves (Figure 9.37). Lovers of fine cigars insist that the best cigars are made from tobacco grown in Cuba under conditions not reproducible on the mainland. Specialized tobacco types, such as the almost black perique cigar tobacco grown only in St. James Parish in Louisiana or the Latakia of Turkey, are blended with Burley or Bright.

Cultivation practices vary with plant type, soil, and weather conditions plus the use to which the leaves will be put. High levels of nitrogen fertilizers foster development of large, succulent leaves, while leaves for cigarette manufacturers are smaller and a bit woodier, controlled by restricting the nitrogen supply at predetermined stages in the growth period. Tobacco is, in the words of the trade, "a heavy feeder" requiring extensive additions of fertilizers. In the cigar-leaf areas of Connecticut, close to 2 tons of fertilizers are supplied per acre per year.

TOBACCO TECHNOLOGY

Tobacco cultivation is both labor- and capital-intensive. Plants are started from tiny seeds in seed beds and are transplanted to the field by hand. Because the plant is subject to many viral, bacterial, fungal, and insect diseases, a vigilant spray and dust program is necessary, involving millions of pounds of pesticides. During the growing season, the plants are topped to remove the terminal bud and apical meristem, to prevent flowering and fruit production, which tend to drain nitrogen and phosphorus from the leaves. Once topped, however, buds in the leaf-stem axil are released from dormancy since the growth-regulating chemicals formed in the apex are no longer translocated down the plant to repress bud growth. The resulting axillary shoots, called suckers, must be removed so that the plants do not become bushy and so that the

Figure 9.37 Tobacco growing under shade tents in Connecticut. Courtesy Connecticut Agricultural Experiment Station.

leaves on the single stalk will grow to the desired size (Figure 9.38). Plants are harvested either by picking individual leaves or, more usually, by cutting down the entire plant. Leaves are collected and either threaded on stringers or baled for curing.

Curing of tobacco involves a linked series of chemical events. The water content is reduced from 80 percent to 20 percent. At the same time, a complex series of biochemical changes occurs (called aerobic fermentations) wherein the leaf color changes from its bright green to various shades of yellow-brown as chlorophyll disappears and yellow leaf pigments become visible. Starch is converted to sugar, some leaf proteins break down to their constituent amino acids, and the alkaloids are modified. Originally, tobacco curing was accomplished by air-drying (Figure 9.39) or by smoking the leaves over a wood fire, and, for some specialized tobaccos, this is still done. Most tobacco is now cured by flue-curing. In modern flue-processing, leaves are heated to 70° C for a few days, resulting in a bright yellow leaf. Further aging may require several months. Leaves are classified by color, by projected use, by the method of curing and aging, and by the cultivar type and where it was grown. In cigarette manufacture, leaves are stripped to remove veins and petioles, shredded, blended with other tobaccos, flavored with sugars, menthol, moisturizers (humectants), and wrapped in special cigarette papers to which filters may be added. Although less popular than previously, plug (chewing) tobacco is cut and then blended with honey, licorice, and other flavorings before being compressed. Cigar making, originally a highly skilled hand operation, is now usually done by machine. In Europe, small cigars are baked; these are favored as an

Figure 9.38 Hand removal of sprouts (suckers) from tobacco. Courtesy U.S. Department of Agriculture.

Figure 9.39 Air-drying tobacco. Courtesy U.S. Department of Agriculture.

after-dinner smoke by women—a solace and pleasure still viewed with surprise in North America.

Although modern cultivars, curing practices, and methods of manufacture have been combined to give a cigarette with a relatively low nicotine content compared with the 35-40 milligrams of alkaloid per cigarette of the 1920–1950 period, there is still enough nicotine to cause physiological addiction in addition to habituation. Nicotine is synthesized in the roots and translocated up to the leaves, where it accumulates, reaching its highest concentration when the leaf is mature (Figure 9.40). Genetic manipulation has resulted in clones of tobacco which are free of nicotine, and these plants show no metabolic or reproductive disturbances; it is not known what, if any, role is played by nicotine in the economy of the plant. When nicotine is combined with sulfuric acid, the resulting nicotine sulfate is a potent insecticide. During the burning of tobacco, nicotine is volatilized and may constitute 2–3 percent of the smoke. It is readily absorbed by mucous membranes of the nose, mouth, and lungs, passing directly into the blood stream; unborn human babies can become addicted. Physiologically, nicotine alkaloids act directly on nerve cells, stimulating them at low concentration and paralyzing them at higher concentrations, interfering with the transmission of nerve signals. Whether this dual action can account for the hallucinogenic effects of tobacco is likely, but unproven. And, as with other addictive alkaloids, the pharmacological, biochemical, and psychological basis for habituation and addiction is unknown. The health hazards of tobacco

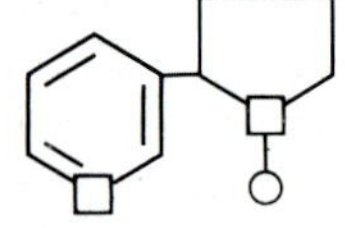

Figure 9.40 Molecular structure of nicotine, an alkaloid in tobacco. Hydrogen atoms and carbon atoms in the rings are not shown. Carbon atoms are represented as circles, nitrogen atoms as squares, and oxygen atoms as hexagons.

include not only cardiovascular alterations but emphysema and cancers of the lung, lips, and mouth caused by ill-characterized carcinogens, the "tars" formed during the combustion of tobacco. Since cigarette smokers do more inhaling of smoke, their disease rate is significantly higher than that of pipe and cigar smokers.

Since the end of World War I, the cigarette has dominated the tobacco market. The modern cigarette was invented in Spain, where cigar clippings were stuffed into preformed paper tubes and sold very cheaply. These "little cigars" became popular during the Crimean War of 1845, when soldiers learned to wrap Turkish tobacco in nitrate-impregnated cannon fuse paper which burned very evenly. The J. B. Duke Company of Durham, North Carolina, started producing machine-made cigarettes in 1883, but plug tobacco (peaking at three pounds per capita per year in 1890) and the masculine stogy (peaking at 86 cigars per capita per year in 1907) dominated the market for many years. Victorian gentlemen gathered after dinner for cigars and brandy, and their ladies were left to compare complaints about nursemaids. A farm woman could smoke a pipe, a European woman could smoke a thin cigar, but it was generally agreed that a cigarette-smoking woman was obviously of loose morals—for example, Bizet's Carmen or Puccini's Mimi in *La Bohème.* After World War I, chawin' tobacco fell into disfavor, and the brass spitoon became a collector's item. With the impetus of the most expensive (and successful) advertising campaign ever mounted, both men and women were urged to use the cigarette as an entry to wealth, health, and sexual magnetism, and cigar-chomping villians flickered on the silver screen. Seven-eighths of all tobacco consumed in North America today is in the form of cigarettes, the bulk being filter-tipped, which reduces the amount of tobacco per unit and increases industry's profits; filters are cheaper than tobacco.

Since tobacco has a high value per pound of leaf, many growers are small landholders with less than 10 hectares (25 acres). Because, in many cases, they use marginal farmlands unsuitable for growing any other cash crop, their livelihoods are absolutely dependent on a continuing market. All in all, 250,000 hectares are planted each year by over 60,000 farmers in the United States alone. North Carolina and Kentucky account for 30,000 tobacco-farming families and half of the U.S. production. Since the lands are generally infertile and have been drained of nutrients by almost 200 years of tobacco cropping, extensive fertilization and labor-intensive cultivation are economically justified by the relatively high dollar value of the crop. If the land were converted to raise food, dollar productivity would be reduced below that capable of supporting these farmers. Because of these economic and social considerations and the powerful financial and political pressures exerted through the manufacturing sector, producers of pesticides and fertilizers and other subsidiary industries, tobacco remains on the list of "basic crops" of the United States Department of Agriculture and considerable research effort of the USDA is devoted to tobacco. The U.S.

Commodity Credit Corporation finances price supports for wheat, peanuts, rice, soybeans, corn . . . and tobacco.

ADDITIONAL READINGS

Balfout, D. J. K. *Nicotine and the Smoking Habit.* New York: Pergamon Press, 1984.

Brooks, J. F. *The Mighty Leaf.* Boston: Little, Brown, 1952.

Corti, E. *A History of Smoking.* New York: Harcourt Brace Jovanovich, 1932.

Goodspeed, T. H. *The Genus* Nicotiana. Waltham, Mass.: Chronica Botanica, 1954.

Heimann, R. K. *Tobacco and Americas.* New York: McGraw-Hill, 1960.

Hewitt, E. J., and T. A. Smith. *Plant Mineral Nutrition.* New York; Wiley, 1975.

Levi-Strauss, C. *Introduction to a Science of Mythology.* Vol. 2. *From Honey to Ashes.* New York: Harper & Row, 1973.

Robert, J. C. *The Story of Tobacco in America.* New York: Knopf, 1960.

Tso, T. C. *Physiology and Biochemistry of Tobacco Plants.* Stroudsburg, Pa.: Dowden, Hutchinson & Ross, 1972.

Wagner, S. *Cigarette Country: Tobacco in American History and Politics.* New York: Praeger, 1971.

Essential Oils—Smells

OPIUM: Pour celles qui s'addonent à Yves Saint Laurent.
(Opium: For those who are addicted to Yves Saint Laurent.)

From an advertisement for Yves Saint Laurent perfume

You do not need to be reminded that many plants have odors. Some, such as the smell of skunk cabbage in spring, are almost offensive, much like rotting meat. These and other even stranger smells are insect attractants and are part of the floral display involved in pollination. The delightful odors of an old-fashioned rose, or lilacs still with the dew on them, or of gardenias at nightfall are also insect attractants. But there are other familiar plant odors which are not involved in pollination. The brisk smell of a clove or of cinnamon sticks, the musky odor of thyme or sage, the sharp pungency of pine or spruce needles, or the wake-up smell of perking coffee are as much a part of our concept of plants as the tastes we associate with a particular plant food. These odors are highly charcteristic; you can easily tell the difference among lemons, grapefruit, or oranges simply by their smell. If you have visited a citrus orchard, you can identify the species solely by the smell of the foliage, which is different from that of the fruit or the flowers.

Botanical chemists have studied the nature of odoriferous compounds for centuries. They have lumped the odor-producing chemicals with a broad range of other compounds to form a category called secondary plant products—secondary because they do not seem to be required for the life of the plant and also because they are usually produced in small amounts. The category is clearly arbitrary, since it includes alkaloids and some pigments (Chapter 5) in addition to the heterogeneous group of compounds that we can smell. Most of the odoriferous plant substances are placed in a subcategory that has been named the *essential oils*. This is not a particularly good name, because they are not oils like corn or soybean oil, but the historical roots of the name go back to the Middle Ages and it is too late to change it. Essential oils are defined as complex carbon-containing (organic) molecules that are not soluble in water but are soluble in alcohol, ether, and chloroform. These solvents can also dissolve oils and fats. The essential oils are volatile—capable of evaporating into the air at room temperature—hence the use of the word *essence*.

Within the past 50 years, many of the essential oils have been chemically characterized. It was found that they do not constitute a uniform group of chemicals but are very dissimilar. Some, like the *naval stores* obtained from pine trees, including turpentine, spruce gum, and resins, are called terpenes. The ratios of different terpenes produced by conifers are so characteristic that it is possible to distinguish one tree species from another simply by their odors. The smell of wet sheep's wool is the result of terpenes dissolved in the natural lanolin of the wool, and the smell of freshly cut liver is caused by a squalene, another terpene. Natural rubber is a polymerized terpene, as are the orange carotenoids and yellow xanthophyll pigments in leaves, flowers, and fruits. There are close to 20 different chemical groupings of essential oils; the odor of even a single flower is a blend of several to many oils.

A BRIEF HISTORY OF THE USE OF ESSENTIAL OILS

History notes the use of essential oils as far back as written records go. When the wood of pine and other conifers was burned, terpenes were released into the dwellings, providing a pleasant smell; pine cones and green branches were added to enhance these odors. Cave altars in the Indus Valley show ceiling stains that indicate that woody incense was burned to the Gods. Ritual incense burners have been excavated from Nineveh, Minoa, and Assyria. Baal Hammon (Lord of the Perfume Altar), a Phoenician deity, was placated by fragrant smoke; the Israelites used incense to veil the presence of the God of the Old Testament from human view. Even earlier, shavings of aromatic wood and gum resins together with cloves, cinnamon, and tars were included by the Egyptians in the wrappings of the dead to assist in their preservation. Among the cultures of the Middle East, two incenses were of particular importance. A resinous exudate from the olibanum tree in the genus *Boswellia* is biblically known as frankincense, and a similar exudate from bdellium shrubs in the genus *Commiphora* is recorded in Genesis 2:12 and Numbers 11:7 as myrrh. The spreading of smoke and the odor of the incense spiraling upward toward the heavens was, and still is, a form of prayer, and when Christianity developed, burning incense became part of ritual.

Both these aromatic trees are native to south Arabia and around the Red Sea. Prior to the first century, an extensive trade developed. Because the incense was an item of ritual, it was not heavily taxed, but the rarity of the plants and the small amounts that could be gathered in a year meant that it was expensive, more expensive than the precious pepper obtained from the islands of the Pacific. The gift of the Magi was, indeed, a kingly present. Beads, also called tears, of frankincense and myrrh traveled overland and by sea. Trading centers for these substances developed along the Incense Road in south Arabia. These same routes and towns became more important as commerce expanded to include materials from the Orient and from India. Some of the centers became deeply involved in the spice and silk trade.

Although the gum resins continued to be important religious items, other substances were also used for their odors. The solubility of essential oils in olive oil was known to the Egyptians, who prepared pomades, and to the Israelites, who incorporated fragrant spices, aromatic woods, and flowers into the holy oil used to anoint priests. Moses prescribed cinnamon, myrrh, and calamis (the sweet flag, *Acoris*) as ingredients in holy oils. The Old Testament notes the use of the white flowers of henna in oil as an unguent, a product derived from the *myron* of Egypt. The Greeks used oils containing flower petals to keep athletes' skin supple; Circe, the temptress, enslaved Odysseus by rubbing fragrant oils on her body. Essential oils held considerable medicinal interest; Hippocrates wrote about their use to cure skin diseases and to assist in the healing of burns and wounds.

In spite of the emphasis in Rome on manliness and the simple life of the soldier, Romans, like the Greeks they imitated in so many ways, found fragrant oils much to their liking. Edicts from military governors prohibiting their sale or use were ignored and warnings that Roman men were becoming effeminate were laughed at. The manufacture of perfumed products was centered at Capua, a city north of Naples with access to conquered Egypt and the Middle East. The essential oils of frankincense and myrrh, benzoin and labdanum (from species of rock rose—*Cistus*), were used as incense in temples and the homes of the wealthy. Our word *perfume* is derived from the Latin *per fumum*—through smoke.

The collapse of the Roman Empire effectively ended the civilized pleasures of scent in Europe, although interest and trade continued in North Africa, primarily in the cosmopolitan city of Alexandria, where

Arabian alchemists and physicians flourished during the Dark Ages. When Alexandria was conquered by Moslems in 641, the invaders found that alchemists had invented the alembic, a crude still. Although beverage alcohol was proscribed by the Koran, Moslem alchemists found that the distinctive odors of different flowers could be dissolved in alcohol, and our modern perfumes were invented. The most desired and hence most costly of these pure essential oils was the extract or *attar* of roses, reputed to have been isolated by Avicenna, an Arabian physician (980–1037). Attar of roses is obtained from a Mideastern species, *Rosa damascena,* and still costs over $1500 per pound. Bulgaria's "Valley of the Roses" near Sophia is the modern source of this important product. The attars or essences of many flowers soon followed. Jasmine *(Jasminum grandiflorum),* patchouli *(Pogostemon cablin),* violet *(V. odorata),* and many others were available by the sixteenth century.

Because pure essential oils or simple alcoholic solutions of these oils tend to turn rancid after a short time, a search was made for some substance that would fix or preserve the desired odor. The first of these fixing agents was musk, a secretion of a gland near the penis of an Asian deer. Natural musk had, for centuries, been associated with male sexuality. Indians and Chinese believed that it could attract either sex (human and suprahuman) and was so powerful that smelling too much could cause death. When added to the alcoholic attars, the odors were long-lasting, both in storage and on the body, with, of course, the singular odor and reputed virtues of the musk adding to the allure of the perfume. These essences were, however, toxic to some people; even in the thirteenth century it was recognized that some people would break out in a rash when musk was used. It was then believed that such individuals were not sexually desirable. We now know that some ingredients in musk are altered by sunlight or fluorescent lights into a skin irritant. Ambergris, a second fixing agent, was discovered in the eighteenth century. It is a waxy material secreted by sperm whales. While these are still the best agents, synthetic fixing agents have been developed in this century.

Alcoholic extracts of essential oils from flowers and other plant products, stabilized by musk, made perfumes in the modern sense available to a reawakening Europe. Eleanor of Aquitaine (1122–1204), queen (consecutively) to Louis IV of France and Henry II of England, had 2500 pounds of violets crushed to make a pound of violet attar, an extravagance of some scandal even then. The French guild of perfumers was established in 1190 by edict of Phillip Augustus, Eleanor's son. Because of the mild climate and the development of Venice as a port for trade with North Africa and the Middle East, Italy became the center for perfume manufacture in the early Renaissance. France supplanted Italy when, in the sixteenth century, Catherine de' Medici (1518–1589) married Henri II of France and included an alchemist and a perfumer in her entourage. Catherine found that the village of Grasse in southern France had the climate and soil needed to grow the flowers that produced large amounts of essential oils. The region also made wine that could be distilled into the alcohol that was used in making perfumes. Grasse is still a major source of natural essential oils, although Yugoslavia and California are also major producers.

While the price of attars was sufficiently high to restrict their use to the nobility, less affluent citizens employed native flowers soaked in oil, prepared sachets of dried flower petals, or made pomanders by sticking cloves into Seville oranges, which they wore around their necks. The extravagant use of scents caused laws to be passed to restrict their use. It was ordained that any woman who lured a man into matrimony by the use of odors and "other false objects" could be charged with procuring and the marriage annulled.

As the royal families of Europe became wealthier and more decadent, perfume use and the search for new scents reached a climax at the court of Versailles. A lady-in-waiting and her mistress (and master, too) would spend the equivalent of several years' wages of a working family for a single dab of a new perfume. Following the French Revolution, the remaining monarchies wisely decided to reduce their level of conspicuous consumption. Nevertheless, the tempestuous and passionate Marie-Louise, duchess of Parma, was so taken with the odor of violets that for intimate occasions she is reported to have had her bedchamber strewn with the flowers at any time of the year. When the petals were crushed, they suffused the air with attar of violet. As she aged, her favorite violet scent was formulated into a bath soap.

Once isolated, essential oils could be incorporated into many products. Soaps, after-shave lotions, colognes, bath waters, and other products were sold in Europe and the North American colonies. In the embryonic United States, Dr. William Hunter, the physician who introduced the study of anatomy to North America, opened a shop in 1752 in Newport, Rhode Island, to import and to formulate toilet articles. Originally called the "Sign of the Golden Mortar," the shop continued as the Caswell-Massey Company of New York. Early customers included George Washington, for whom Number 6 cologne was made.

ISOLATION OF ESSENTIAL OILS

Once people recognized that certain plants contained interesting smells, they set out to devise ways to obtain the essential oils in pure form (Figure 9.41). Some of the oils are easy to isolate. If you take an orange and peel it, your hands and the air around you will soon smell of the essential oils of orange peel. Oil glands in the peel are broken and release the odoriferous materials. Citrus peel can be ground up, subjected to pressure, and the oils collected. This technique is called *expression.* Citrus oils, used in many household products, are a byproduct of the orange juice industry.

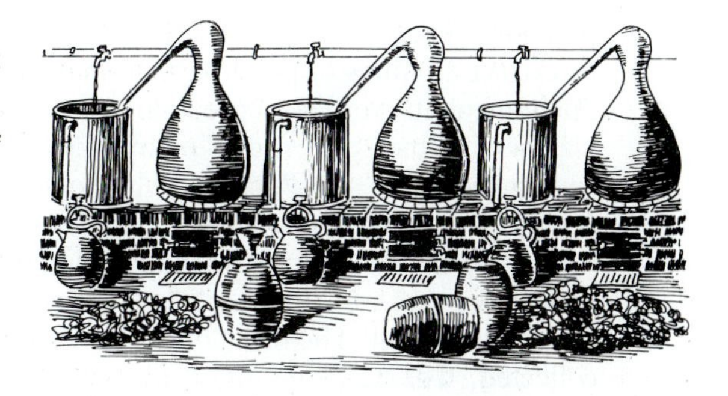

Figure 9.41 Capturing the essential oils of flowers by different techniques. From top to bottom: distillation, enfleurage, and maceration. Redrawn from *The Art of Perfumery* by C. H. Piesse, 1891.

A second procedure is *solvent extraction.* It was probably discovered when people started making wine and noted that the smell of the grape suffused the wine. Efficient solvent extraction of essential oils was accomplished in the late Dark Ages when Arabian alchemists developed the simple alembic still that is used to distill wine and collect the concentrated alcohol. Today, many fat-solvent chemicals are used, depending on the stability of the essential oil in different solvents. The plant material is usually finely ground and steeped in the solvent at a low temperature; the solvent is then evaporated. Although not used extensively today, the solubility of essential oils in true oils like olive or coconut allowed people of ancient civilizations to obtain unguents and pomades. These first appeared in India, where vegetable oils were used to extract mixtures of plant parts to make products with blended odors. Such unguents were very precious in Europe, having traveled overland via the silk and spice routes or by ships that hugged the coasts of the Indian Ocean to the Persian Gulf. Solvent extraction is still used for some essential oils because they are destroyed by heat.

The third method of extracting essential oils is called *distillation.* At first, flower petals or other plant parts were mixed with water and brought to a boil. The oils would float on top of the water and could be collected. The terpenes of turpentine were collected in this manner during colonial times. After simple boilers were developed, live steam under low pressure was substituted for the boiling water. The steam carried the oil into a collecting vessel, where it could be separated from the water. Steam distillation is used for some essential oils, but delicate compounds are altered chemically. A recent variation of the technique is called fractional distillation. Essential oils volatilize at different temperatures, and if a mixture of oils are subjected to successively higher temperatures, the pure compounds can be separated. The mint oils, used in a wide variety of items, are obtained by fractional distillation of the crude extract, as are the oils of many spices (clove, cinnamon, vanilla, and rosemary).

Cold fat extraction, called *enfleurage* by the French, was known to the Egyptians and Chinese at least four centuries ago. It had been known that essential oils will dissolve in true fats and oils. It was later discovered that if the volatile plant material was simply put in a tightly closed box with a fat, the essential oils moved from the plant into the still air of the box and were captured by the fat. Initially, sheep fat (lanolin) was used, but this has been replaced by a mixture of tallow and lard. Such unguents are rarely used directly. The essential oils can be dissolved out of the fat with alcohol or other solvents to obtain the pure essences of the plant. Essences are almost never single compounds, but contain all of the essential oils and some impurities which can then be separated by fractional distillation. The purified fat can be reused several times; eventually it is made into soap which retains the odors.

Even more recent is the synthesis of essential oils in the chemical laboratory. Synthetic methyl salicylate is chemically indistinguishable from true oil of wintergreen, and the smell of new-mown hay (a woodsy fragrance) is synthesized from coumarin (see Chapter 7). A vanilla scent, a frequent minor component in perfumes, is obtained from wood pulp.

THE SCENT INDUSTRY

The scent industry is large and profitable; essential oils are used in medicine, in household items, in foods, and in toiletries. Men's toiletries, an important business in the sixteenth to eighteenth centuries, dwindled to almost nothing in the nineteenth and early twentieth centuries. A small amount of bay rum (an alcoholic extract of leaves of *Pimenta racemosa*) as an after-shave astringent and some fairly smelly hair pomades were essentially the entire industry. After 1946, however, the industry was revitalized with an extensive advertising campaign. The industry now grosses over a billion dollars a year.

The women's scent industry is even larger; a yearly gross of $3 billion is a conservative estimate. Most of the name brands of scent are manufactured by conglomerates that find them exceedingly profitable. Borden Milk owns part of Jean Patou, Charles Pfizer Chemicals owns Coty, Norton Simon bought Halston, and the Squibb pharmaceutical firm runs Yves St. Laurent. Marketing has become so expensive that the introduction of one of the better men's products costs several million dollars and the best estimate of the start-up cost of a women's perfume is almost double that amount. The development of the perfume, the design of the container and the package, and the advertising are included in the initial costs. What's inside a bottle of cologne or perfume represents less than 10 percent of the retail price of the item. In an average year, 20–25 new perfumes are introduced with emphasis on more expensive formulations. Some of the more popular perfumes have, however, been around for a long time; "Chanel #5" came out in 1924, and "Arpège" was introduced in 1927.

Development of a new scent is a long and expensive process. Someone has to decide the overall character of the scent—sexy, sophisticated, liberated man or woman, day or evenings, or whatever. Since there are literally thousands of essential oils to choose from, many essential oils have to be mixed in different proportions. These are individually evaluated and the scaling up from laboratory to production is considered. The work is extremely difficult; the few people who have the expertise are called "noses," and their ability, a blend of heredity and training, is very well compensated; one manufacturer complained that his "nose" earned more than he did. Jean Louis Siezac, the nose for Yves St. Laurent, spent three years developing "Opium."

Many essential oils found in popular perfumes have been in use by different cultures for centuries. In addition to oils from flowers, other plant products and a few animal products are components of perfumes and colognes. Chypre is isolated from a lichen *(Evernia prunasti)* that grows on the trunks of oak and fruit trees. Originally from the island of Cyprus (hence its French name), it was discovered by the Crusaders by the twelfth century and brought to Europe, where it was used as a sachet. It now comes primarily from Yugoslavia. As a component of "Mitsouko" (Guerlain) and "Miss Dior" (Dior), it imparts a distinctive earthy scent. Oils of nutmeg, walnut, citrus, and other plants are included in most popular blends together with the floral oils. The fixing agents include musk, ambergris, and a yellowish, unctuous substance resembling musk that is extracted from a pouch in the genital region of the African civiet cat. The better perfumes contain 25 percent oils, with the fixatives and alcohol making up the rest.

There are several useful hints in purchasing a scent.

1. Perfume, cologne, and toilet waters are basically the same if they bear the same name, but the concentration of essential oils and fixatives decreases in that order. After-shave lotion is even more diluted. Buy the most concentrated form; it's cheaper.
2. Even when fixatives are included, light and oxygen from the air can destroy the oils. It is generally poor economy to purchase a large bottle unless you desperately need to impress someone. If you do get the giant economy, super-suburban size, transfer the scent to smaller bottles filled to the top and stopper the bottles tightly. To avoid photo-oxidation of essential oils and fixatives, don't keep the bottle on the dressing table, but store it in a cool, dark place.
3. It takes some time before the heat of your body and your particular blend of natural skin oils react with the essential oils to produce the aura that you want to project. At the same time, your own nose becomes fatigued so that it can no longer detect the entire fragrance. Thus, use good perfume or cologne sparingly and put it on at least an hour before you hope that it will have the desired effect on whatever companion you wish to enthrall.

Along with liquor and high-fashion clothing, the advertising of scents is among the most sophisticated in the world and uses some of the best photography around today. After all, millions of dollars are involved. As someone with an interest in science, you might want to see if you can classify (the science of taxonomy) advertisements for scents into categories. Included might be the "Liberated Woman (Man)" category, the "Gentle Romance" theme, "Sophistication," the "New Yorker Type" and various subcategories of sex in a sequence of increasing passion.

ADDITIONAL READING

Abercrombie, T. J. "Arabia's Frankincense Trail." *National Geographic* 168, no. 4 (1985):475–513.

Bailes, E. G. *An Album of Fragrance.* Richmond, Me.: Cardamon Press, 1984.

Bell, E. A., and B. V. Charlwood (eds.). *Secondary Plant Products.* New York: Springer-Verlag, 1979.

Cooley, A. *The Toilet and Cosmetic Arts of Ancient and Modern Times.* New York: Burt Franklin, 1970.

Groom, N. *Frankincense and Myrrh: A Study of the Arabian Incense Trade.* London: Longman, 1981.

Kaufman, W. I. *Perfume.* New York: Dutton, 1970.

Morris, E. T. *Fragrance: The Story of Perfume from Cleopatra to Chanel.* New York: Scribners, 1984.

Schery, R. W. *Plants for Man.* 2nd Ed. Englewood Cliffs, N.J.: Prentice-Hall, 1972.

Theimer, E. T. (ed.). *Fragrance Chemistry: The Science of the Sense of Smell.* New York: Academic Press, 1983.

Chapter 10

Plants in the Environment

Floods

For behold, I will bring a flood of waters upon the earth
Genesis 16:17

Among the disasters that have accompanied the human species through our history are devastating floods. The washing away of homes, arable land, and other human enterprises is far exceeded by the loss of life and the total, sometimes irrevocable, dislocations that are among the consequences of floods. The survivors, in addition to having to try to rebuild their lives, are at risk of disease and starvation. Some floods are beyond the ability of people to prevent or even to ameliorate—tidal waves, the flooding that accompanies volcanoes, or those that result from massive rains (Figure 10.1). Nevertheless, many floods need not have occurred, or the damage could have been much less severe than it was. It is this type of flood that we will examine. The floods in Florence, Italy, were the direct result of the failure of the people to use common sense in environmental management; they involved the abdication of even minimal responsibility by the civil authorities. To this extent, the Florence floods are a case study.

The announcement of the flood in Florence on November 4, 1966, aroused the immediate reaction of horror felt by most people when they hear of a natural catastrophe. Fragmentary stories on the wire services were followed by detailed descriptions of specific paintings damaged or destroyed, books soaked with water and oil, and accounts of death, hardship, and personal heroism. The word pictures of the odors of sewage, rotting animal and human flesh, molds, mud, and oil were graphic. In the succeeding days, as is usual in our overcommunicated world, the stories receded into the back pages of the newspapers and fresh stories of other tragedies took over the world's attention.

Yet there remained a nagging feeling that this flood should never have happened at the level of intensity and with such severe loss to the treasures of Renaissance Italy—indeed of Western civilization. As we read more about the history of the Florentine region, it becomes clear that the 1966 flood is an almost classic example of how to ensure that a flood will occur.

GEOLOGY

The area known as Tuscany is a well-marked and fairly compact geologic unit which rises gradually from the western coast of Italy to the heights of the Apennine Mountains. It is a harsh land of hills and valleys, of winter rains and summer droughts. The valley of the Arno lies on the northern margin of the African trade winds and on the southern margin

Figure 10.1 The aftermath of a flood in Bangladesh in June 1976. Over 2.8 million people were affected, crops were destroyed, livestock were lost, and soil nutrients were washed away. Photograph by T. Page, courtesy of the Food and Agricultural Organization, the United Nations.

of the prevailing westerlies that bring rain from the Atlantic Ocean. In winter, the warm air above the sea forms a low-pressure cap that draws winds over the water. As these warm, moisture-laden winds flow up the hills to the Casentino mountain range, they cool, and rain may fall furiously for days in succession. Of the 25 to 40 inches (63–120 centimeters) of rainfall that Tuscany receives annually, over 80 percent falls between October and January—much of this from late October through the middle of December. In the summer, there may be no rain for several months, because the hot, rapidly moving winds from Algeria do not pick up enough moisture from the sea.

Today, the hills surrounding Florence can scarcely support olives and grapes. There are occasional patches of herbs, including lavender, myrtle, rosemary, and thyme—aromatic, romantic, lovely to look at in flower, and completely useless as ground cover. The soil is leached-out clay with pockets of stone and sand. The sun beats down on impoverished and badly worn soil, changing it into a cementlike hardpan. In spite of, or perhaps because of, its aridity, the land above the city of Florence is very beautiful. From the surrounding hills, one can take in the peaks of the Casentino mountain range, bared and gleaming in the summer sun. First in abrupt descents and then more gently, the highlands fall away to the foothills. Sere, brown, and rocky, the hillsides catch the early morning or late afternoon sun, leaving the hollows in deep shadow.

The present city of Florence stands on land that has been a part of the civilized world for more than 3000 years. Excavations have unearthed artifacts dating back to the Bronze Age, and there is some evidence that Neolithic cultures flourished there even earlier. And well they might have, for the Tuscan valleys and hills were fertile and wooded, the climate was generally mild, with just enough seasonal variation to stimulate, but not debilitate, the people, and there was reasonable protection from marauders from the north. The major river, the Arno, was a freshwater stream, abounding with fish, flowing throughout the year, and was broad enough to permit travel, and hence trade, with the coastal peoples in the delta that is now Pisa.

ETRUSCA

The earliest complex civilization about which we have information is that of the Etruscans. They settled in what is now Tuscany about 2600–2800 B.P., coming from northwest Asia Minor via the Caucasus. In addition to being farming peoples, they were experienced merchants, and much of our knowledge of them is derived from their commercial activities. The Etruscans found, settled, and exploited a land of natural beauty. The Arno and its tributaries were clean, clear, fishable, and navigable. The hills were completely forested. In the valleys were found the Aleppo pine and two evergreen oaks, the Holm oak and the Valonia oak, forming open forests, with understories of small trees and shrubs. The midmountains contained excellent stands of beech, Spanish chestnut, and deciduous oaks, and the upper mountains were clothed with black pine and several species of fir. There was an abundance of animal life, for figures on Etruscan pottery depict deer and bear hunts. Theophrastus mentioned the beech groves of Tuscany as being of such excellent quality that the wood was bought from the Etruscans for ships' keels. Even by 2300 B.P. the forests were so dense that Roman legions hesitated to invade. Plant cover was still notable several hundred years later when Strabo reported that the central Apennines and the hills of Etruria provided wood for the construction of Rome. The Sila hill range yielded "fine trees and pitch" for shipbuilding activities centered near Pisa at the mouth of the Arno.

Etruscans were a vigorous people who used their land for many activities. They planted olive trees and grapevines on hillsides and farmed valleys for wheat and barley. They mined some copper, zinc, and tin to make bronze for tools and weapons, and they smelted mercury and some iron. Sheep were grazed on the lower hills to supply white, black, and brown wools, much prized for their softness, fineness, and colors. Thus, some deforestation and grazing injured the land, but the populations were too small to damage severely the fragile hills. By the end of the Etruscan hegemony, the Romans were lavish in their praise of this still-wooded, fertile, and well-watered land.

THE FIRST FLOODS

The first adequate records of floods date from the twelfth century, actually 1117, when the Ponte Vecchio, then the only bridge over the Arno, was swept away. Records show clearly that the Tuscan hills were denuded, the Arno was a seasonal stream, and the land was plagued by drought. If we date the demise of Roman rule at about 300, and we know that the land was ruined by 1117, we are left with approximately 800 years to account for in terms of physical geography. What happened to the forests, the clear streams, the good soil, and the fine climate?

The main city of the Etruscans was not modern Florence, but the now-small hill town of Fiesole. Present-day Florence was a suburb, about 15 miles downstream on the Arno. Florence itself was situated where the Arno issued from the mountains at about 180 feet above sea level, a few miles upstream from its present location. The upriver location was determined by a simple fact—the site of the present city was a marsh. Indeed, when winter snows melted in February, the river was not fordable. Livy noted that in the late winter of 2217 B.P., Hannibal and his armies floundered for four days in the muck of the valley floor. Yet the river bottomland was very fertile, producing an excellent crop of hard wheat, some millet, and good barley. Snowmelt served to supply nutrients, and the water moved out to sea by March, leaving time to plant. There was a second reason for the lowly political position of Florence, one related to the defense of the area. The Etruscan military wanted its citadel on the heights, not on the nearly indefensible valley floor.

Fiesole, as the political and defense center, gathered the greater share of the population, but it soon became apparent that this small, hilly area could not provide an adequate living for the populace. The hills about Fiesole had to be intensively cultivated, and the only practical way to do this was by terracing. The technique was certainly not invented by the Etruscans; the Greeks and other peoples of Asia Minor had been cutting horizontal ridges into their hills for eons. In the Tuscan context, terracing was necessary more as a means of fully utilizing all land within the protected reach of the fort than as a conservation measure (although it certainly served well in this respect). Cereal crops were not planted on the terraces but were grown on the Arno valley floor, and the crop was stored in the granaries of Fiesole. Vines were planted on the terraces, particularly on the drier, stony, south-facing slopes. Initially, the trees were native species, but these were replaced by the olive prior to the Roman invasions. Thus, even before the rule of Rome, some of the plant cover on the hills had been removed. Since the Romans left no records of flooding, we can assume that the major portion of the Tuscan hills was still covered with plants that sopped up the water and prevented the destructive run-off that resulted in floods.

ROMAN RULE

Tuscany, being on the route from the lands of the Franks and the Goths to that of the Romans, served as a staging ground for many armies. The mountains were no longer an effective barrier from the north as disciplined armies replaced wandering bands of military plunderers. Records are hazy and contradictory, but it seems that the Romans defeated the Etruscans several times between 2300 and 2200 B.P. and that by about 2000 or 1500 B.P. Roman rule was firmly established. The Roman road system was extended through Tuscany, and during its construction the engineers tried to stay in the piedmont except where there was a necessary valley river crossing. The Via Julia Augusta crossed the Arno at the site of the present Ponte Vecchio to connect with the Via Cassia going north. Even then, the Roman legions knew it as a bad crossing, because it was muddy and tended to flood in the spring.

Under the protection of the legions and the Roman governor Sulla, the population of Fiesole began to move down the Arno, to where life was easier. When Sulla was defeated by invading Franks, Florence was leveled and burned, and again the people retreated to the hills. By 2059 B.P., Florence, under the auspices of Julius Caesar, was built at its present location and given the name Julia Augusta Florentina, the latter word apparently being a good omen rather than having any botanical allusions. The city was a typical walled Roman city-fort, with temples, baths, aqueducts, and an efficient sewage-disposal system. There was a central marketplace, now the Piazza Vittorio Emanuele. Apparently there was some attempt to drain the swampy area then outside the city walls. Although flooding in the winter did not assume sufficient importance to record, the Romans were well aware that snowmelt was a danger, for they erected levees on the banks of the now-contained, but still free-flowing, Arno.

The ancients knew the value of woodland as flood protection. Plato deplored the reckless cutting of Greek forests and pointed out the cause-and-effect relationship between forest destruction and rapid run-off of rain from bared hills (Figure 10.2). Pliny quoted Plato when he admonished the Romans to maintain and to replenish their forests, and while we have no direct evidence, it seems likely that some form of forest conservation was practiced in the Tuscan hills.

MALARIA

With the population increasing, terracing also increased, but this stopped abruptly. Malaria, which had been introduced into Sicily from North Africa 24 centuries before moved onto the continent during the Second Punic War. By the beginning of the Christian era, it reached Tuscany, where it became endemic in the lower valleys and in the city. Within less than a century, Florence and the hill towns were abandoned. Terraced farms, no longer cultivated, were subjected to erosion that washed away precious soil. It is easy to visualize the stripping of the soil from the hills, the mud slides, and the consequent silting of the tributary rivers and, eventually, of the Arno itself. Although no records exist, we

Figure 10.2 Clear-cut area in California. Courtesy U.S. Department of Agriculture.

are fairly safe in saying that minor flooding occurred in the Arno basin from about the middle of the first century.

During the subsequent 300 years, the town of Florence continued to decline, local trade decreased, and this once relatively important Roman city became little more than a village. As the Roman Empire began to fall apart during the third and fourth centuries, tilling declined and plowed land and terraced hillsides were subjected to direct erosion. The upper hills, however, began to restore themselves, and flooding probably decreased.

The fortunes of the city took a turn for the better by the end of the fourth century as malaria began to disappear. The Lombards made Florence into a military center, and when the Franks replaced the Lombards in 774, the entire region assumed an increased role as a trade center. By then, Fiesole had declined to the level of a town, a condition that has never been reversed.

As populations grew, there was a greater demand for wood to be used for construction and for fuel. The hills were rapidly stripped, so that by the end of the seventh century, wood had become so scarce and expensive that stone became the primary construction material. From the perspective of the art historian, this had some importance. The artisans of the time were both good craftsmen and good architects. These early stone buildings served as models for the Renaissance masterbuilders, who made Florence a place of enduring beauty and, incidentally, laid out a city especially suited for flooding. As the city began to assume its present shape, virtual bathtubs were created there—piazzas and squares with a single inlet street and one or a few outlets. Water rushing into one of these squares could not easily pour out the other side, and the whole area filled up. Most of the listed high-water marks of the 1966 and earlier floods occurred in these piazzas.

THE WOOL TRADE

Although the Etruscans and the Romans engaged in the wool trade, and Florence had been a weaving center for many centuries, increased use of wool cloth by peoples less exalted than the aristocracy resulted in a great increase in woolen mill activity. As woodlands were decimated, the cleared land that resulted became pasture for goats and sheep. Both these animals nibble forage to the root level, effectively killing individual plants. Overgrazing became so severe that, by the eighth century, Florentine mills could no longer depend on local wool but were forced to import it from England and Spain. Parenthetically, it is likely that the denuded Spanish hills are partially the result of overgrazing by sheep to supply wool to Florence.

The effect of land clearing on the climate of Florence is not easy to document. Modern ecological and ecophysical studies show that forested areas serve as natural air conditioners. Heat is collected by trees during the day and released slowly at night. Modified by the cooling power of evaporation, heat never builds up as it does when captured by stone and paving; a walk from a city street into a park at dusk illustrates this dramatically. The presence of nothing but baked clay on the hills above Florence served to perpetuate its summer heat and drought. Forests served, as Etruscans and Romans knew, to ameliorate the heat of the summer. Not incidentally, the absence of the forests with their water-holding capacity resulted in loss of scarce summer moisture, and rivers and brooks ceased flowing in the summer.

By the beginning of the ninth century, Florence had about 30,000 people and the old walls had been breached and built farther out several times. The region had many different rulers, none particularly firm or effective. Weak rulers gradually let power slip from their hands, and authority, power, and ownership of the land became divided among Roman Catholic bishops and a few wealthy landowners. These interlocked concentrations of land ownership led steadily toward the development of a feudal system in which large areas were tilled by serfs for the production of a few economically important crops. Goats and sheep continued to graze, and meadowlands were cut from the forests farther up the hills. With the removal of the sparse grass, the summer droughts baked and caked the clay soil to the consistency of rock, and natural seeding in it became virtually impossible. Bound to squeeze the last florin of profit from the country, the Church and the landowners made no attempt to replenish their land. Tuscany, a land noted for good water, forests, and fine agricultural bottomland, inexorably declined into the arid, bare, and impoverished country that we see today. The greed of the Church, the avarice of the aristocracy, and the ignorance of the people have given us this legacy.

FLOOD RECORDS

The remainder of this sad tale can be told rather quickly. The first complete record of a major flood was provided by Giovanni Villani for the flood of November 1333. There was 4 feet of water, the city walls collapsed (never to be rebuilt), and three of the then four bridges over the Arno were destroyed; the Ponte Vecchio stood firm. Quoting Villani:

> Wherefore everyone was filled with great fear and all the church bells throughout the City were rung continuously as an invocation to heaven that the water rise no farther. And in the houses, they beat the kettles and brass basins raising loud cries to God of "misericordia, misericordia," the while those in peril fled from roof to roof and house to house on improvised bridges. And so great was the human din and tumult that it almost drowned out the crash of the thunder.

Not unexpectedly, the flooding was worse in the low-lying districts nearest the Arno. Three hundred people died in the hard-hit districts of San Piero Scheraggio and Porta San Piero. On the Via de Neri the flood crest was 13 feet 10 inches, only 14 inches lower than during the 1966 flood. Villani carefully listed the losses of grain and casks of wine, and in quite modern fashion he noted the cost of bridge repair, as well as municipal outlays for cleaning the streets and for minor compensations to citizens.

Villani noted the extraordinary rainfall over the Arno watershed, but then wrote: "On May 15 there was an eclipse of the moon in the sign of Taurus . . . and then at the beginning of July there followed a conjunction of Saturn with Mars at the end of the sign of the Virgin." The Church had the last word, stating that "by means of the laws of nature, God pronounced judgment on us for our outrageous sins." There was slight consolation from the King of Naples, who told the Florentines that "whom God loveth, He chastiseth."

A tiny note of biological intelligence was, however, struck by Gianbattista Vico del Cilento, who recommended a government sponsored reforestation plan. Two hundred years later—the plan never having gotten off the ground—Guistino Fortunato dismissed Vico's plan, stating that trees were ugly and useless.

We need not detail the economic and biotic history of Florence beyond 1333. In spite of accurate information on causes of floods and adequate technical information on flood control, nothing was done. Vico was probably not the first, and he certainly was not the last, to chastise the Florentines for their laxity. Giovanni Batista Adriana wrote that the severe flood of 1547 badly disturbed the people, who demanded that something be done. They were told that God wanted, with this act, to signify that worse calamities would occur, so they decided to wait. In 1545, Michelangelo cautioned a nephew about buying a house in the Santa Croce district, "every year there the cellars flood." His warning was apt, for the flood of 1547 was followed by another in mid-September of 1557, when the entire Santa Croce district was under several feet of water. Lorenzo de' Medici warned the population about the Arno: "Its arrogant anger spits and beats away at the weak banks." But in spite of his authority to order new construction, he did nothing, and there is no evidence that the people heeded the warning.

Leonardo da Vinci reported that the streams of Tuscany were muddier when they passed through populated districts and suggested that the mud was washed-out soil that clogged the streams and leached the soil of the hills. He conducted surveys and designed projects to develop water impoundments in the hills, to dredge the tributaries of the Arno and the Sieve, and to develop chambers under the city to hold excess floodwaters. These, too, were ignored.

The Tuscan flood situation had only one useful effect. In 1861, President Lincoln appointed the Vermonter George Perkins Marsh as American minister to the kingdom of Italy. Marsh's travels in Tuscany and his reports on the relatively minor floods of 1861 and 1863 undoubtedly contributed to his thinking in the preparation of *Man and Nature or Physical Geography as Modified by Human Action,* published in 1864. As a result of interest (indeed the controversy) created by this book, Marsh was asked to help compile the irrigation laws of Italy, laws never adequately enforced, which could have alleviated the floods that have occurred during the past 100 years. Marsh's book so stimulated Gifford Pinchot that he persuaded President Theodore Roosevelt to set in motion the legislation leading to the conservation policies of the United States.

PRESENT PROSPECTS

For what it is worth, the statistical picture is gloomy. Since 1333, there had been a moderate flood in Florence every 24 years, a major flood every 26 years, and a massive flood every 100 years. Of the massive floods, those of November 1333, November 1884, and November 1966 have been the worst. Since 1500 there have been more than 50 floods qualifying as moderate and 54 qualifying as major.

In 1797, a French engineer, Fabre, announced that alpine torrents which flooded the lowlands were a direct result of deforestation of the heights. One hundred years later, in 1890, the French government undertook the job of reforestation in the Hautes-Alpes. Engineer Surell reported that as reforestation continued, floods decreased in number and severity. This report was widely distributed in Europe, was acted upon in Germany and Switzerland, but was ignored in Italy.

Apparently no one in Florence could believe that water would cover their city in the twentieth century. But they, themselves, place marble markers on their homes and churches "commemorating" previous floods. They, themselves, proudly indicate the high-water marks of floods dating back 500 years. They, themselves, speak of their dry, bare hills which "shimmer in the summer sun," and they cannot fail to see that their "ribbon of shining silver," their Arno, is a muddy, silted-in, malodorous stream that almost stops flowing during the summer.

The *New York Times* of January 2, 1969, reported that during the week preceding Christmas of 1968—two years after the most damaging flood in Florentine history—the Arno crested toward the flood mark. "Only after the river began to subside at 9:00 P.M. did the crowds move homeward." The bed of the Arno is still filled with the debris of the 1966 flood and has never been dredged. It is estimated that the silt from 1966 alone is 2 meters thick.

There are two dams on the Arno, the Levane dam at Livorno, 35 miles upstream from the city, and the Penna dam, yet higher into the hills. These are power dams and do not provide for flood control. No money has been (or apparently ever was) appropriated for reforestation as recommended in 1334 by Vico, for the construction of water impoundments and for dredging the Arno as recommended by Leonardo, or for caring for the land as recommended by Plato.

It is beyond the scope of this book to discuss the storage of irreplaceable books and art treasures in the basements of Santa Croce, noted by Michelangelo as the district where the cellars flood every year. It is beyond the scope of this book to discuss the reasons for the lack of coordination between the two dam sites. It is beyond the scope of this book to discuss the lack of retaining walls bordering the Arno; even Lorenzo did nothing about them. As George Santayana warned in his *Reason in Common Sense,* "when experience is not retained . . . infancy is perpetual. Those who cannot remember the past are condemned to repeat it."

Few of the massive floods in North America are natural disasters in the sense that little or nothing could have been done to prevent them or reduce their severity. It has generally been assumed that heavy rains leading to swollen rivers and the flooding of low-lying areas are acts of God, but this is not usually true. The capacity of soils to absorb and hold water is greater than is generally believed. Where rivers are deliberately channeled for navigation, where the straight canals are periodically dredged and their banks shored up with levees, the unimpeded force of rushing water can undermine banks, levees, and retaining walls. The water, having no other route, will inundate the adjacent flood plain. Industries, municipalities, and summer-home owners who build on flood plains are asking for trouble—and usually get it sooner or later. Where rivers that drain highlands are dammed, the potential for backup of water is very high, and any failure in the dam structure or of the people who operate the dams is an invitation to disaster. Few engineering projects are "fail-safe." The record of dam failure is available and shows that failures have not been reduced sufficiently to view the situation with assurance (Figure 10.3).

Exploitation of mountainous terrain is one of the more important factors leading to floods, as seen in the Florence scenario. Most mountain soils are thin, shallow to bedrock, and their plant cover is very fragile. Clearcutting such forested lands can almost be expected to result in floods, but even selective cutting significantly reduces the ability of the land to hold water. Recreational trails, logging roads, and construction all play roles in altering the extent and pattern of runoff. These alterations can, under appropriate conditions, result in flooding. And the appropriate conditions are, too often, present.

ADDITIONAL READINGS

Dasmann, R. F. *Environmental Conservation.* New York: Wiley, 1959.

Leopold, A. *A Sand County Almanac.* London: Oxford University Press, 1949.

Owen, O. S. *Natural Resource Conservation: An Ecological Approach.* New York: Macmillan, 1971.

Skinner, B. J. (ed.). *Use and Misuse of the Earth's Surface.* Los Altos, Calif.: William Kaufman, 1981.

Wang, J. Y., W. B. Hemmer, and F. B. Tuttle, Jr. *Exploring Man's Environment.* Palo Alto, Calif.: Field Educational Publication, 1973.

Figure 10.3 Flood in Napa County, California. Courtesy U.S. Department of Agriculture.

Dust

At 10 A.M. the dust in the air was so dense
that objects could not be distinguished 100 yards off.
No one who could possibly remain indoors was on the
street.

REPORT OF U.S. WEATHER BUREAU,
Dodge City, Kansas, April 8, 1890

It was no accident that great civilizations of the ancient world developed in areas where the soil was renewed by flooding of rivers whose precious mud, high in organic matter, annually inundated the land. Egypt's dynasties were dependent upon the Nile, India's on the Indus, Mesopotamia's on the Tigris and Euphrates, and China's on the floodplains of the Yangtse and Yellow rivers. The Tigris and Euphrates valleys and the valley of the Indus have long since become arid, and Egypt utilizes a good deal of the electric power from its ill-conceived Aswan Dam to make fertilizers which the untrammeled river provided free for thousands of years. The great rivers of China are still serving people's needs. As the world's populations began to increase, less fertile and less desirable territory was farmed, including dry uplands and the semiarid and arid lands that constitute more than half of the world's arable area. It is these lands, called marginal by agricultural economists, which must be considered, for it is just these lands upon which the bulk of the people of the world must depend for food.

ARID LANDS

Among our most fragile lands are those which meteorologists classify as semiarid or arid. Here, rainfall is the limiting factor in plant development. For our purposes, we can call any area receiving less than 25 centimeters of precipitation per year an arid zone, and any area receiving less than 50 centimeters per year as semiarid; lands receiving less than 20 centimeters per year are deserts. Arid and semiarid land usually show extreme fluctuations in rainfall patterns, and the 25 or 50 centimeters limits are based on very long-term data. Rainfall is rarely "normal." The statistical average is based on a minimum of 30 years, but individual yearly totals, monthly averages, or daily precipitation recordings may differ markedly from this average, normal, or expected figure. Even using a 30-year average can be dangerous because of variations among successive periods. The midwesterner speaks of "spells" of rainy or dry weather, knowing that short-term changes in expected weather patterns are matters of chance and hence not predictable with any accuracy. Rainfall does, however, tend to come and go in cycles, perhaps with cyclic patterns of 11 or 22 years corresponding to sunspot activity, but deviations from such cyclic distribution patterns are frequently seen.

Semiarid and arid lands are further characterized by the plants which grow on them. With few exceptions, grasses dominate the landscape; early explorers of North America's grasslands spoke with awe of "seas of grass" with waves produced by the winds which swept over the land. Grasses are dominant because they possess characteristics which permit them to survive where other plants cannot. Most arid land grasses are perennial, with new leaves formed each year from shoot buds on the rootstock. Reproduction is mostly vegetative with extension of rootstocks from a center, producing mounds or bunches which may extend for several meters. Individual plant mounds may become confluent, with kilometers covered by the same species. The extensive root systems spread laterally as well as vertically, binding the soil into a tough sod. Mats of dead leaves and stems insulate the soil from heat, prevent wind and water erosion, reduce evaporation of moisture from the soil surface, and trap snow. Trees in grasslands are restricted to watercourses; most tree species cannot survive with the limited rainfall, and various sod-forming grass species tend to cover the ground so completely that tree seeds cannot get established. In contrast to many broad-leaved plants, grass leaves have the ability to curl, with the stomata-bearing lower epidermis on the inside to reduce water loss by transpiration. When undisturbed, grasslands can support huge herds of grazing animals—bison and prong-horn antelopes in North America and many species of antelope in Africa and Asia—which are preyed upon by carnivores like the wolf and the big cats. Mice, moles, birds, and their predators form interlocking food chains, all dependent upon the grass.

GRASSLANDS

Undisturbed grassland soils are usually very fertile, dark brown to almost black, and topsoil depths may exceed 2 meters. All soils are mixtures of mineral particles derived from rocks plus living and nonliving organic material, the humus, which provides the texture and structure of soil, its water-holding capacity, its biochemical and biophysical properties, and its ability to support the growth of plants. Without humus, plants will not develop properly unless carefully and individually tended. Over eons of time, as grass roots and leaves decayed or were digested by soil organisms, the life-sustaining humus developed. It provides suitable habitat for the animals whose tunnels and runs aerate the soils, with water-holding capacities which allow a buildup of a ground water supply which sustains the grass roots during periods of low rainfall.

Based on vegetation types, grasslands may be divided into *prairie communities,* characterized by tall-grass species, and *short-grass plains,* or *steppes.* Various species of grass are adapted to the climatic differences between Alberta and Texas, both of which have prairie and steppe communities. The tall-grass prairies, with 50 centimeters of rain per year, are dominated by the bluestems *(Andropogon spp.)* and Indian grasses *(Sorghastrum spp.)*, with grama *(Boutiloua spp.)* and buffalo grass *(Buchloe spp.)* found on the steppes where rainfall averages less than 25 centimeters per year. The best wheat lands—those of Kansas, Nebraska, the eastern Dakotas, and the prairie Provinces of Saskatchewan and Manitoba—were tall-grass prairies.

All semiarid and arid lands are drought-prone because of the deviations from normal rainfall patterns. Aridity carries the connotation of a permanent condition of low average rainfall, but drought is a temporary condition which occurs periodically in climatic zones where precipitation is usually adequate for the dominant plant cover. Drought is a relative term, depending upon life styles and land-use practices of the inhabitants. Conditions which a vegetable grower might consider drought would be fine growing weather for a wheat farmer. The sheep herder would be happy with rainfalls which would not permit wheat to develop. Prairie and steppe grasses are drought-enduring plants, selected over eons of time for their capacity to survive during those periods when precipitation is less than average. Leaves will wither and die, but root systems—protected from dessication by the leaves, tapping ground water supplies, and utilizing the moisture held by the spongy humus—can survive for several years, putting forth new leaves when rain again falls.

Survival and regrowth of plants is abruptly changed when the natural plant cover is replaced by crops or is destroyed by overgrazing of domestic animals. Semiarid lands, because of their fine soils, can be excellent agricultural lands, yielding reasonably well when the human populations dependent upon them are small, and when only small areas of grasslands are put into production. But when people possessing high levels of technology intrude upon virgin plant communities, they possess the tools to impose their wills upon the land. Land, in the modern view, is a commodity to be exploited, to be mined, and the natural plant cover must be completely removed to make way for crop plants. Opened to the onslaught of the wind, thunderstorms, and hot sun, humus levels begin to decrease, earthworms no longer open water channels, and the soil deteriorates. In Paul Sears's words, "Deserts are on the march."

THE AMERICAN DUST BOWL

From the Mississippi to the Rockies, the Great Plains flow in undulating waves north, from Texas to the tundra of Canada. The climate is continental, with hot summers and very cold winters. The long-term average rainfall is less than 50 centimeters per year, and much of the area receives less than 30 centimeters per year—in a good year.

In 1821, Stephen F. Austin led a group of immigrants from U.S.

territory into the Mexican province of Texas. First occupying the wooded river valleys of east Texas, settlers slowly moved into that portion of west Texas which was included in the Great American Desert. In spite of danger from the Comanche and from Mexicans, who realized that their country was being illegally taken over, settlers found that the semiarid grasslands supported the growth of long horn cattle. These were fattened on the grass and driven to Abilene, Dodge City, and other railheads to the eastern markets. The Indian Territory to the north contained a few whites until the end of the Civil War, but since the upper plains states had already become settled, covetous eyes were cast on these potentially valuable lands. Although the Territory was semiarid at best, it was covered with short grasses and the soil was fertile.

In 1889, the Oklahoma Territory was formally opened to settlement. In March and early April, thousands of people with horses, carriages, and wagons lined up on its boundaries. At noon on April 22, cannons were fired and the settlers rushed into the area to stake out homesteads. Those who jumped the gun were called "sooners," and Oklahoma, the "Sooner State," is proud of its law-breaking founders. Within a month the population increased from fewer than 5000 people—mostly despised Indians—to over 60,000 farmers. The sod was broken by John Deere's plow (see Chapter 3) and a wheat crop planted. Yields of the 1889 crop compared favorably with those of Kansas, but as a sign of what was to come, the rains of 1890 failed to materialize and the crop was almost a complete failure. Rainfall was about 75 percent of average, and, although the plants started out well, a hot summer and dry winds from New Mexico reduced the growth and seed production of the plants.

DUST STORMS

In spite of an occasional poor harvest, wheat farmers in Oklahoma prospered. The fertility of the soil and the water stored in underground aquifers and in the humus produced bumper crops. These were sold in 1914–1918 to Europeans so busily engaged in killing each other that they couldn't plant their own crops. Wheat more than doubled in price, and the areas cultivated increased rapidly. In the 1920s there were a few more bad years, each a bit more severe than the one before. The land was "dry-farmed," the farmers disk-harrowing a dust mulch on their fields to prevent soil-water evaporation and there was still some humus left. Much of the land was either tenant-farmed or farmed on lease, and neither the owners nor the farmers gave any thought to the possibility that fertility would run out, that ground water supplies would be used up, or that the humus would disappear. Wheat was still in short supply, and the government was encouraging cultivation of all available land. Loans for machinery and land purchase could easily be financed by banks whose officers were themselves speculating in land and whose financial stability was unregulated by the states or the federal government. American ebullience, boosterism, and an unquenchable optimism glossed over the few bad years; more sod was removed and everyone was mining the soil for all it was worth.

The inevitable could not have occurred at a worse time. Europe's agricultural recovery was complete, and American wheat prices were falling. The stock market crash of 1929 and the following depression dried up the money supply that was keeping the expansionist economy buoyant. The bank reorganization ordered by the New Deal administration of Franklin D. Roosevelt uncovered and made public an inadequate debt financing structure. Those banks that had not already succumbed to the panic withdrawal of funds began to call in their loans from farmers who could not pay them. The crushing blow was drought. From 1930 to 1937, the area that was to be known as the dust bowl received less than 65 percent of the expected, "normal" rainfall. This would have been enough for survival of tough, resilient prairie and steppe grasses, but it was not enough to allow wheat to grow. The prairie grasses were gone, and so was the humus, the ground water, and the soil fertility. As the hot, dry, bare soil radiated heat from cloudless skies, winds arose from the west and southwest and swept unhindered across the land. The first dust storm was reported on November 12–13, 1933, and was followed by so many others that complete records have not been kept (Figure 10.4). Custodians wiped Oklahoma topsoil from the desks of senators in the Capitol building, and homemakers in Boston and Atlanta complained about the dirt on freshly laundered sheets hanging on their clotheslines. At its peak, the dust storms affected 50 million acres centered in Oklahoma and the Texas panhandle, but

Figure 10.4 A dust storm in Texas in the 1930s. Courtesy U.S. Department of Agriculture.

drought and dust extended beyond the Canadian border. Dust filtered into feed stacks, it covered pastures, and filled in waterholes. Livestock died of starvation, suffocation, and thirst (Figure 10.5). Static electricity rendered automobile ignition systems inoperable, and countless stalled cars and trucks were abandoned. Trains were canceled because the engineers couldn't see the rails, and, if they did move, they spun their wheels on rails made slick and greasy by the crushed bodies of millions upon millions of grasshoppers who filled the air, eating anything that was still green. Driving sand scoured the paint from houses and sandblasted windowpanes. Jackrabbits were blinded by driving particles of sand, and children choked to death. When light rains did fall, they came down as mud showers.

Wheat yields averaged 2.5 bushels per acre, and, because of the depression, sold for 8 cents per bushel, less than the cost of the seed. Bank loans could not be extended and many farms were sold to pay the interest on these loans. Bread lines in the cities were paralleled by the long lines of cars and trucks carrying people away from their land. The "Okies" left for the golden dream of California, and Route 66 was the highway to a state where, it was thought, farming could again be a way of life. One jalopy had a sign reading:

1930—Frozen Out
1931—Dried Out
1932—Hailed Out
1933—Grasshoppered Out
1934—Dried Out
1935—Rusted Out

Figure 10.5 Aftermath of a 1938 dust storm. Courtesy U.S. Department of Agriculture.

1936—Blown Out

1937—Moving Out

California, beset by reductions in available irrigation water, increased salt accumulation in its irrigated fields, and its own depression, was in no mood to welcome the tattered immigrants to towns with 30 percent of their work force on relief (Figure 10.6). The immigrants were jailed for vagrancy, harassed, shot at, denied entrance to the state, and otherwise set upon, cheated, and hated as potential rivals for jobs. The record of the Okie migration is nowhere as movingly put as in John Steinbeck's *The Grapes of Wrath.* Today's migrant farm workers, still inadequately protected by minimum wage laws and lacking adequate housing and education, are the children and grandchildren of the Okies.

And the land? In 1935, the Department of Agriculture mobilized soil scientists, agronomists, and botanists into the Soil Conservation Service to restore the ravaged countryside. Irrigation and water-conservation practices were introduced (Figure 10.7), new methods of tilling the soil were developed, shelterbelts of trees were planted to break the force of the winds, and arid soil was covered with a mantle of prairie and steppe grasses seeded from trucks and planes. New crops were introduced to bind the soil, legumes were recommended to replace lost nitrogen and

Figure 10.6 An Oklahoma migrant family in a camp in San Jose, California, in the late 1930s. Photo from Resettlement Administration, U.S. Department of Agriculture (The Granger Collection).

Figure 10.7 Contour cultivation to reduce erosion. Courtesy U.S. Department of Agriculture.

humus, and silted-in ponds and streams were cleaned out. The more marginal lands were purchased by state and federal authorities and declared to be immune to further agricultural exploitation. The prosperity of the war years of 1941–1946 reestablished credit for the purchase of machinery and allowed those farmers who had not left the land to again become self-supporting.

The one thing which could not be done was to ensure more adequate and more dependable rainfall patterns. The former dust bowl is still, geologically, geographically, and biologically, semiarid to arid land, subject to the vagaries of weather and climate. A reduction of 25 percent in average precipitation in any given year has a reasonably high statistical probability, and, although much less predictable, there is absolute assurance that an extended period of drought can be expected. In 1954, after two years with subnormal rain, dust storms occurred in the former dust bowl and in western Kansas and eastern Colorado. Some dust blowing occurred in Nebraska and the Dakotas. Floods in Bangladesh, droughts in India, the Sahel, and other parts of North Africa, crop failures due to disease in Central America, frost in the Soviet Union, and excessively hot weather in China caused a rise in the value and price of wheat to the point that the restored grasslands of Oklahoma, Texas, and parts of New Mexico are again being plowed to plant wheat. And who is to say nay? In the United States, Canada, and other countries, land use is determined by the owner of the land. He or she is not its steward, but its exploiter. Government cannot tell the individual what use of the land is best for the future of that land. And everyone knows that there will again be years when hot winds will suck the moisture from the soils, when the rains will not fall, and when the topsoil, deprived of its grass, will blow and blow and blow.

IRRIGATION

Few areas in the world are blessed with adequate amounts of water. Even in the tropics, where rainfall can exceed 200 inches (500 centimeters) a year, there are dry seasons when rainfall is scanty and undependable. In order to bring land into intensive cultivation or to maintain lands already in cultivation, it is frequently necessary to provide extra water; this is the concept of irrigation. Irrigation is an essential part of agriculture; almost half of the world's plant production depends upon it. In 1984, 65,000 farms in 17 western states had 1.5 million acres under irrigation. Worldwide, 15 percent of cropland is irrigated.

While modern irrigation practices have benefited from advanced engineering technology, irrigation has been used for growing crops almost since the dawn of plant cultivation. The Nile Valley in Egypt and the valley at the conjunction of the Tigris and Euphrates rivers were irrigated. Productive lands of China have been irrigated since the Shang Dynasty. Mesopotamia used small-scale irrigation to grow its wheat and barley for at least 7000 years. Excavations in the lands of the Mayans demonstrated that these peoples had developed an elaborate system of canals, dikes, and bypasses that permitted controlled introduction of water into their maize, bean, and squash fields. These ancient irrigation systems were basically alike. They were furrow systems in which river water was diverted into a series of shallow channels or furrows running between rows of crop plants. The flooding of the rice paddies in China, India, and Southeast Asia, which represented a significant advance over simple furrow irrigation, has been practiced for over 6000 years. For both furrow and flooding irrigation systems, a major limitation was the necessity of raising water above the level of the river. The most primitive of methods, a person with a bucket, is still practiced today in some areas, particularly where plantings are small and irrigation is needed only occasionally. Many home gardeners use this method.

But as agricultural lands became larger, as more arid lands were put into cultivation, and as climate changed as a result of both natural and human alterations of the landscape, more sophisticated methods of raising water were developed in various parts of the world. Treadmills geared to paddle wheels or buckets were operated by foot power or with the assistance of domestic animals (Figure 10.8). Windmill systems were more efficient and required less labor. These were independently developed in the Orient and in the Middle East by 2000 B.P., but the furrow or flooding methods of delivery of water to the plants remained unchanged. Agricultural technology did, however, become modified when it was recognized that excess standing water injured the roots of many crop plants and reduced their yields. First with hand tools and later with plows, beds in which crops were planted were raised above the irrigation channels so that the standing water level was below the rooting zone. Raised bed cultivation in channel-irrigated fields is still practiced for many crops.

Raising water for large agricultural fields was made less labor-

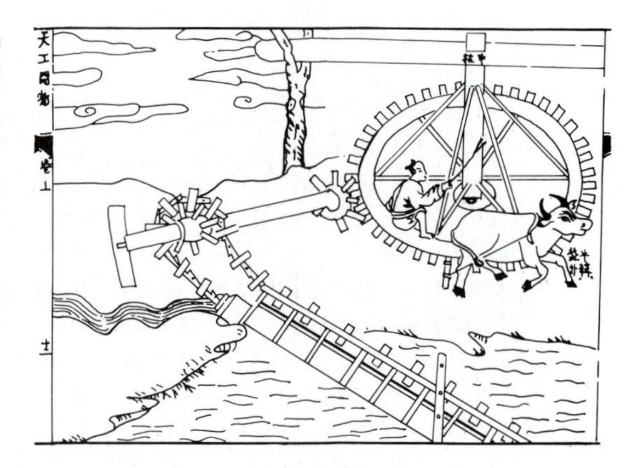

Figure 10.8 Raising irrigation water with an ox. Redrawn from the *Thien King Khai We,* an agricultural book published in China in 1637.

intensive with powered pumping systems which developed in the early nineteenth century, but the flood or furrow delivery systems were still used. Realization that much of the water in furrows simply soaked into the ground and was lost resulted in the development of flow systems lined with water-impervious layers, first clay and later cement and plastics.

Furrow irrigation is, however, still notably wasteful of water. Of the water supplied, less than half is taken up by plants, the rest being lost by evaporation and by soaking into the soil. To reduce these losses, other irrigation systems were developed. Most prominent of these are the overhead spray systems, essentially identical to a person sprinkling a lawn. The boom and nozzle rigs can be very large. In one system, horizontal booms on wheels are slowly moved across a field, watering a width of several hundred feet (Figure 10.9). Another system, developed in 1949 and used extensively on the dry lands of Nebraska, is the center pivot rig, a stationary boom that slowly revolves to spray-irrigate a circle

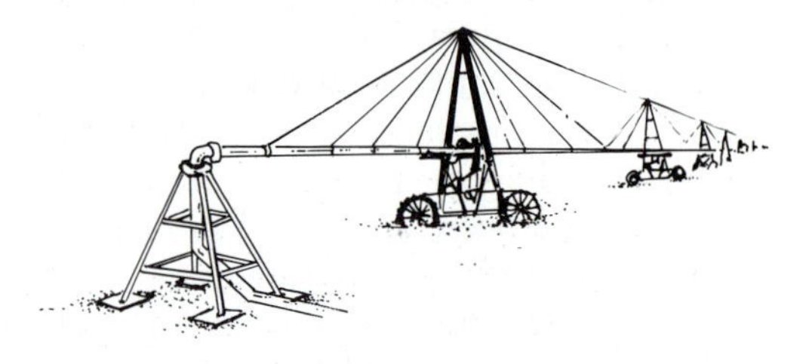

Figure 10.9 Rigs for large-scale irrigation. Top: center pivot system; bottom: side-wheel rolling rig.

of up to 160 acres (Figure 10.9). Water comes from deep wells that tap underground aquifers, particularly the Ogallala Aquifer, which extends down the Great Plains from Canada to the Gulf of Mexico. While these irrigation systems are expensive to purchase and to maintain, the costs are far outweighed by increased production. Corn yields have been increased threefold, wheat has almost doubled, and so has alfalfa for fodder.

The most advanced and water-conserving technology was developed in the early 1960s in Israel, a notably arid land. Initially, it was a modification of the furrow system in which perforated pipes replaced furrows. This reduced significantly the amount of water released at any one time and allowed a more or less continuous addition of water to the soil. Today, water is moved through plastic tubing that has a series of nozzles placed next to each plant to be irrigated. It was found that somewhat brackish or saline water could be used and that fertilizers and pesticides could be incorporated into the irrigation water. Drip irrigation reduces evaporation to a minimum, allows water to be supplied precisely when and where it is needed, and is highly conservative of water supplies. Impractical for field crops, drip irrigation is used primarily for high value horticultural crops and for fruit and nut trees. In field trials, yields of tomatoes or of melons increased by over 50 percent with only half the amount of water.

Of the water on earth, a surprisingly small percentage is available for land plants and animals. Most of the world's water is in oceans, which are too salty, and in permanent ice fields. Water vapor in the air is generally unavailable. Only 3 percent of the total water on earth occurs as fresh water. This water is cycled in rivers and ground water and returns as rain. Demand for fresh water is growing rapidly as direct human use increases proportionally to population expansion, as industrial use rises more than does population, and as the need for water for food production also increases. Since available fresh water is not uniformly distributed, decisions on who will get what share, how much, and when continue to occupy the serious attention of nations, regions, and individuals.

Droughts and acute shortages of water have been of deep concern for centuries. The removal of the forests from mountains of the Middle East dramatically altered the course of agriculture and even of civilizations; the valley of the Tigris and Euphrates is now a forbidding, arid land supporting fewer people than in biblical times. The movement of people to the American Sunbelt of Florida west to southern California has placed demands on water supplies that cannot be met locally and require the transmission of water for long distances and at very high environmental and economic costs. There has been greatly increased removal of water from underground aquifers. The water table in Texas, Arizona, and southern California has dropped by at least 100 feet (35 meters), and increases in pivot irrigation on the Great Plains have required that wells be deepened at frequent intervals. A center pivot can pump 1000 gallons of water per minute. The level of the Ogallala Aquifer, once thought to be inexhaustible, has dropped 60 feet (20 meters) in 30 years. The water is being pumped out at a rate 10 times greater than it can be replenished. It is now half depleted. Since recharge of ground water aquifers is measured in tens or hundreds of years, the drawdown of water is a critical matter that is not being considered in determining the cost/benefit ratios of large-scale irrigation. Overdrafts occur in many parts of the United States (Figure 10.10).

Another source of irrigation water is dammed rivers. The damming of major rivers throughout the world to supply human, agricultural, and industrial needs for water and for electric power has greatly altered climates and the landscape and has had considerable impact on the regions affected. Reservoirs in the United States can store 200 billion gallons of water, but this is less than the amount of water used in a day throughout the country. California takes water from dam sites and rivers in other states that believe that they have riparian rights that must not be lost. Decisions of water allocation in the western United States usually end up in the Supreme Court. Plans to dam rivers on the northern Great Plains affect the relationships between the United States and Canada.

Some note must be made here about water pollution. Human and agricultural water supplies are the recipients of many different pollutants capable of limiting the use of that water. Agricultural water, which constitutes over 80 percent of total water use in the United States, may be pure at its source, but is contaminated by many chemicals. Irrigation water is not used just once but is moved sequentially from one field to another and from one area to another and may be passed from agriculture to industry, to drinking water supplies and back to agriculture before the residual water enters the sea. Each user may pollute the water with wastes that can endanger the next user in the chain.

The pollutants of agricultural activities are numerous and include

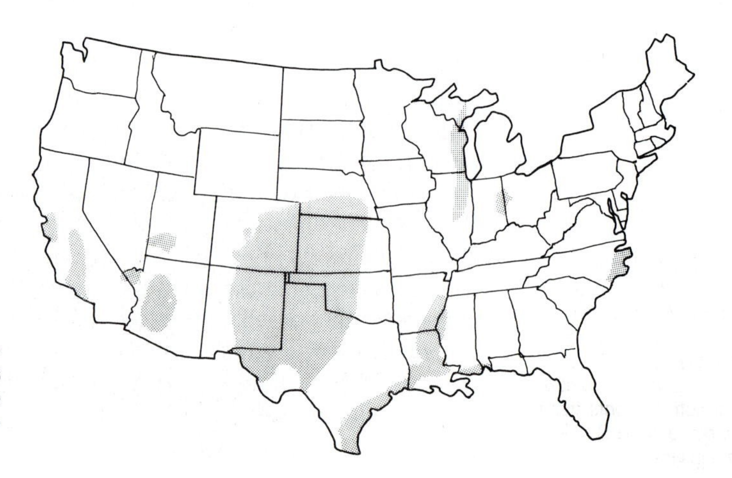

Figure 10.10 Areas of the United States where ground water is being used faster than it can be replaced. Courtesy U.S. Soil Conservation Service.

pesticides, excess fertilizers (especially nitrates and phosphates), and other agricultural chemicals. Crystal-clear water soon becomes a murky, smelly liquid as it picks up fine soil particles, organic matter, and other materials. It must be purified before it can be used industrially or as drinking water (Figure 10.11).

From an agricultural point of view, however, the most serious pollutants of irrigation water are various inorganic salts, including table

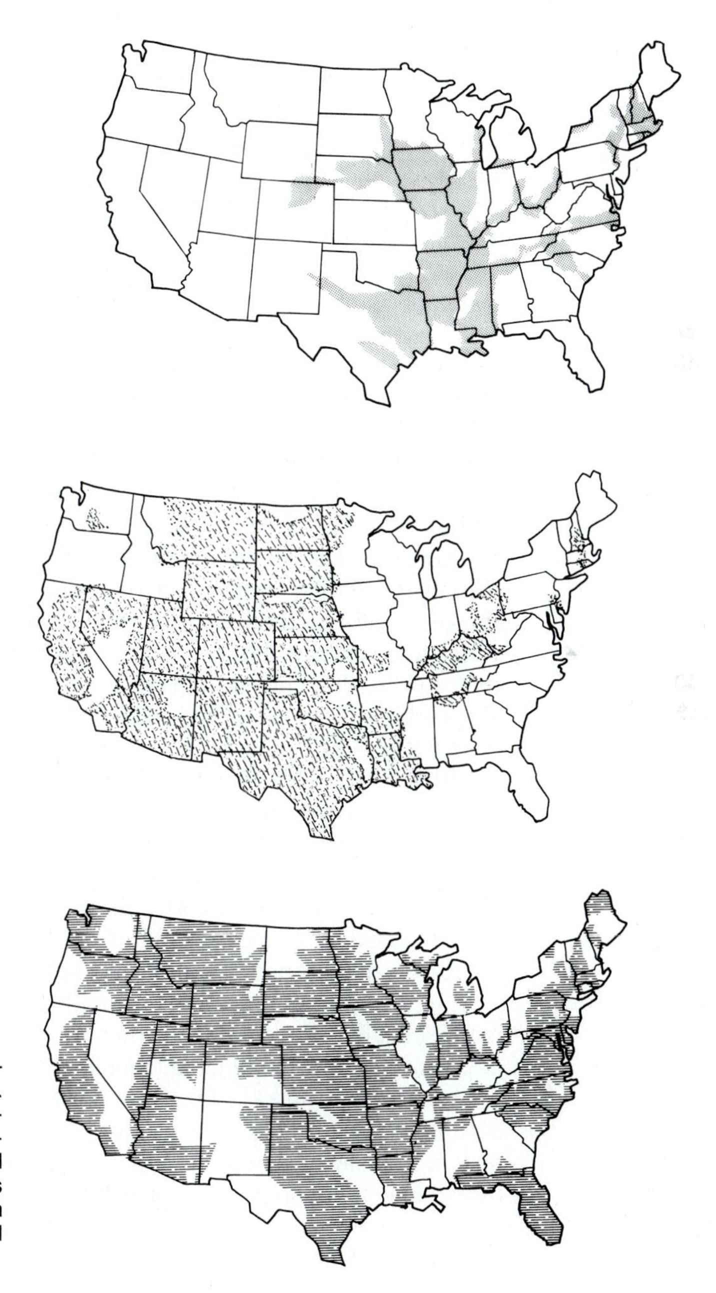

Figure 10.11 River basins receiving large amounts of pollution. Top to bottom: river basins polluted by pesticides; river basins polluted with dissolved solids (soil erosion and sewage); river basins polluted with runoff nutrients from agriculture. Courtesy U.S. Soil Conservation Service.

salt—sodium chloride. As evaporation from surfaces occurs, any salts initially in the water are concentrated. Additional salts are introduced as the water dissolves salts from the soil and as fertilizer salts accumulate. Serious reductions in yields are being experienced in soils that have become saline and alkaline from such water and from the capillary rise of salts to the surface of the soil from the subsoil. It is likely that the downfall of some of the ancient civilizations in the Middle East was due to salinization of their agricultural lands. Before the Aswan High Dam was built, accumulated salts in the soils of the Nile Valley were leached by the annual flooding of the Nile River, but the present sequential use of water through hundreds of miles of agricultural lands has resulted in salinization of the irrigation water and hence of the land itself. The goal of reclaiming large areas of desert land in Egypt with water from the Aswan Dam could not be realized because the irrigation water was too salty.

Salinization can occur in a relatively few years. Allocation of the water of the Colorado River dates from 1922, when irrigation and drinking water in the American West first became a matter of concern. A Colorado River Compact was worked out to divide the flow of the Colorado River equitably between the river's upper basin area of parts of Wyoming, western Colorado, eastern Utah, and western New Mexico and the lower basin area of eastern Nevada, Arizona, and southern California. The compact was based on expected water flow and envisioned a series of impoundment dams, many of which have been constructed. The states involved each thought that they were being shortchanged, and conflicts on allocation are still being fought through the courts. Since the Colorado is an international river that provides water for the vegetable-growing lands of the Mexicali Valley, Mexico, too, was finally included in the compact. The water rights of the Navajo and Hopi Indians were not considered until fairly recently, but they need water both for agriculture and for power plants being constructed on tribal lands. This dispute, too, is still in the courts.

As the Colorado River was divided and subdivided and its water used many times, the salt loads and toxic substance levels rose enormously. By 1961 it was found that users of the lower basin had water whose salt concentrations were well above those that are known to repress plant growth. Mexico, at the tail end of the flow, found that production of vegetables had dropped by almost 50 percent. After threatening to take its case to the International Court of Justice, Mexico was promised an increased allocation and construction of desalinization plants which have not yet been completed. Similar situations also occur with other rivers and the land served by these rivers. Ironically, the irrigation that permitted the transformation of the Central Valley of California into one of the most productive agricultural areas of the world is now threatening to render the land worthless. Production has dropped yearly since about 1970. Cotton production in the American Southwest, where the valuable long-staple cotton is grown, is also down, as is production of alfalfa and other forage crops. In spite of massive research efforts, land desalinization has not been successful and breeding studies have, to date, resulted in few salt-resistant cultivars.

Nevertheless, the need for irrigation water continues and water must be moved even greater distances. The Chinese plan a 700-mile (almost 1000-kilometer) diversion of the Yangtse (Chang Jiang) River to provide water for the north China semiarid plains in the hope of making China self-sufficient in grain. The Soviet Union has proposed to move water through canals from the rivers of western Siberia to Kazakhstan in central Asia, a distance of 1400 miles (2000 kilometers). The water would have to be raised over 500 feet (175 meters) along the way. Such water is desperately needed to make arid lands productive and to wash out the salt accumulated by centuries of irrigation. Long-distance transport of water is commonplace in the United States as well, and the end is not in sight.

DESERTIFICATION

Lands that are minimally productive or nonproductive are usually called deserts. While we think of hot, dry regions, the cold areas of Greenland and Antarctica are also deserts. The hot desert and semidesert areas need not be characterized by sand dunes against which camels are silhouetted; stony deserts and clay deserts also exist. About 20 percent of the earth's land surface is classified as desert (Figure 10.12). Temperature ranges in hot deserts are extreme, from 50°F at night to 120° during the day (10–50°C), with soil surfaces reaching 140°F (60°C). Water is the major limiting factor for plant growth. Desert areas usually receive much less than 5 inches (12 centimeters) of rain per year, but this average doesn't indicate the fact that there can be a series of years with virtually no rainfall. High rates of water evaporation further restrict the availability of water. The plants in such areas, through natural selection, are well adapted to the wide temperature swings and the low and irregular availability of water. Their life cycles are short, with bursts of growth and reproduction when water becomes available. They have structural and functional mechanisms to conserve water. The flora is generally simple, with few species and a low plant density per unit land area. Animals of desert areas also are adapted to the prevailing conditions and are in balance with the climate and the available plant food and shelter.

Natural deserts and semideserts are perilously balanced ecosystems whose biotic components may be periodically decimated by recurring droughts, fires, or plagues of insects, but they have the capacity to recover when the dangers are gone.

We are, however, now seeing extensive anthropogenic (human-caused) desertification: the conversion of semidesert lands into deserts incapable of supporting the plants, animals, and people who inhabit the

Figure 10.12 Areas of the world showing slight to severe desertification.

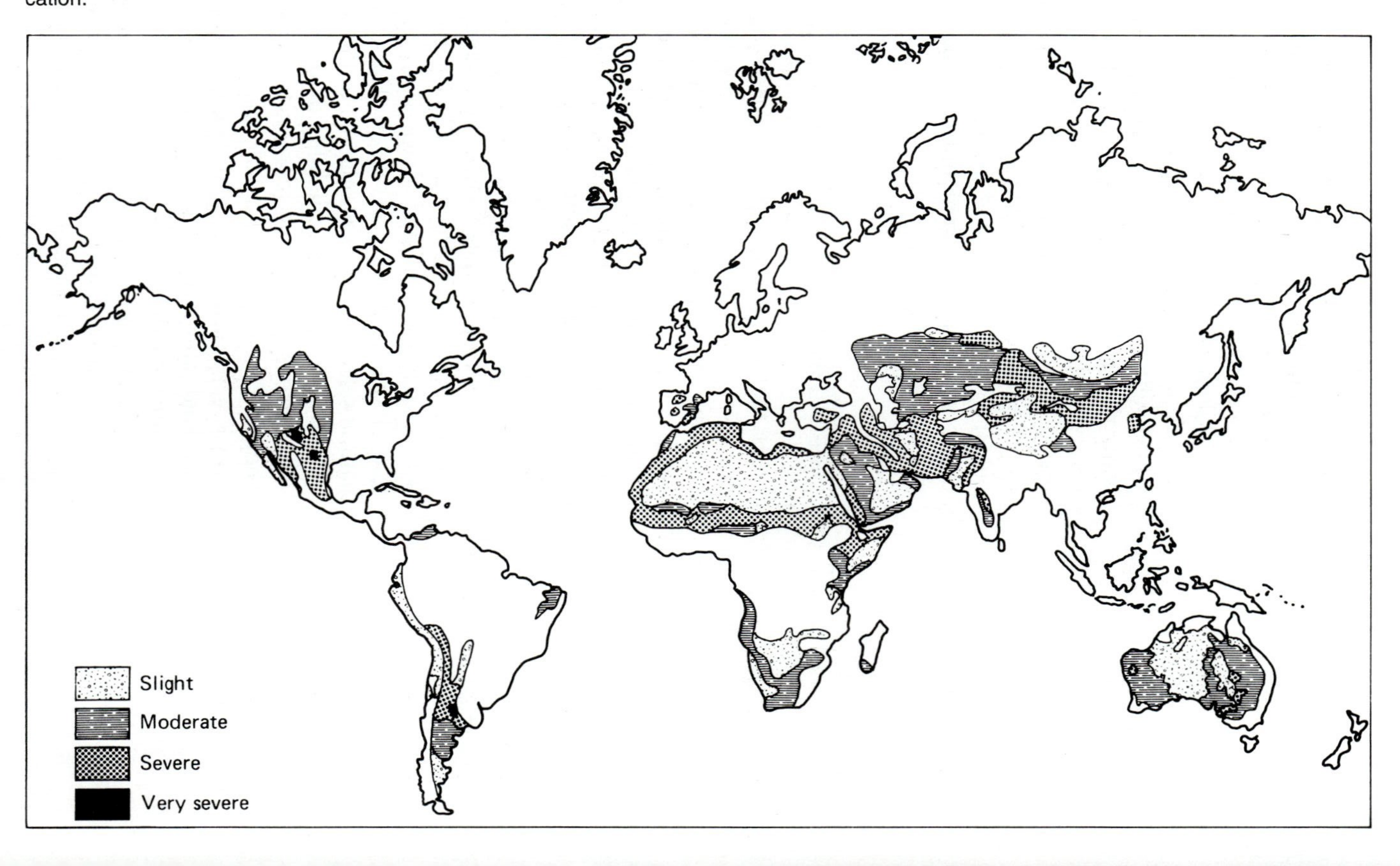

region. Actually desertification is not a new phenomenon. Historically, among the more fertile and productive lands of some ancient civilizations were the river bottoms surrounded by deserts. As human utilization and attendant insults were received by the land, the extent of desertification increased. The loss of the basin of the Tigris and Euphrates rivers of the Middle East involved the salinization of land by overirrigation and the destruction of their natural, adapted plant cover by timbering and overgrazing. Egyptians of 4600 B.P. began the decimation of Phoenician forests by harvesting the timber for ship and house construction and for the resins needed to preserve their mummies. Today, only a few sacred groves of the magnificent Cedars of Lebanon remain, totally inadequate to prevent desertification of the surrounding lands. Plato bemoaned the fact that Attica had become "a mere relict of the original country. All of the rich, soft soil has moulted away, leaving a country of skin and bones." Plato knew only his small area of the Mediterranean, but similar situations existed in Mesopotamia and in China.

Desertification has tremendously accelerated in the twentieth century as virgin lands are no longer available to serve increasing human populations and more intensive efforts must be made to wrest a living from the available, tired land. Our first modern experience with desertification was in the dust bowl of the 1930s in the American Southwest. Our second was the 1967 to 1973 drought in the African Sahal, where the resulting famines in Mauritania, Senegal, Mali, Upper Volta, Niger, and Chad brought the phenomenon forcibly to the world's attention. A third, less publicized desertification is underway in the Thar area of India.

It is convenient to blame droughts for desertification, but this is not the case. Admittedly, arid and semiarid lands exist independently of human activity, and people have no control over the vagaries of the weather. But where droughts occur frequently, desertification does not necessarily result. The destructive processes are halted and reversed when the drought breaks; the plants native to such regions recover. But when the degradation is accelerated by human activity before and during droughts, recovery becomes almost impossible, and the deserts become permanent. The consequences in human terms are devastating. The inhabitants, unable to sustain their agricultural base, either leave or face starvation. In recent times, both of these options have been realized.

The linked sequence of cause and effect that results in anthropogenic desertification is fairly clear. In the arid lands where rainfall is scant and irregular, loss of plant cover by overgrazing or through conventional agricultural practices eliminates the adapted, native species and exposes the soil to the direct impact of wind, rain, and runoff. Grazing animals eat the perennial, adapted grasses which are replaced, especially in drought periods, by less palatable and less well adapted species. The bare soil surface heats up, accelerating the degradation of soil organic matter (humus), a process that decreases water-holding capacity. Hot, dry soils are poor substrates for plants; the dead plants increase the area of exposed soil. Because any rain received runs off instead of percolating into the ground, gully erosion further depletes the soil of its organic matter. Wind erosion leads to dust storms, sweeping away any remaining topsoil. Eventually, in many cases within a few years, the soil becomes incapable of supporting plants or people. Where the land had been irrigated, salinization adds to the hastening decay of the soil. Heavily grazed lands are compacted by the hooves of domestic animals, and natural plant cover cannot regenerate. We can see this even in productive areas where cattle regularly follow paths to and from their barns.

It is not just semiarid lands that can be converted to deserts. Many areas that were forested are as barren. Removal of trees for timber or for firewood, by forest fires or by pollution, has exposed the soil, with the initiation of the same sequence of drying out, erosion, and lack of plant cover. And where forests were on hills or mountains above agricultural plains, the bottomlands are desertified when the water held by the forests is no longer available.

That desertification can be reversed has been seen in the United States (a wealthy country) and in China (a poor country). The methods used to restore the dust bowl of the 1930s and to regain lands lost in the far northern part of China—Inner Mongolia—are directly comparable. The first task is to stabilize the land to stop the erosional loss of whatever topsoil is left. This is most efficiently done by establishing a plant cover over the bare soil and binding the soil with plant roots. In Oklahoma and Texas, dust bowl lands were seeded from the air with hardy, drought-resistant grasses and lines of trees were planted as shelterbelts to deflect the drying winds. China, too, uses these techniques. It is now common to drive down almost any road in China and see a shelterbelt of three or four rows of trees planted on both sides. Signs exhort the people: "It is your duty to protect these trees," and people can be seen watering and tending the young trees.

The second task is to begin the process of soil restoration. As leaves are shed from grasses and trees, they are degraded by soil microorganisms into humus, essentially an organic mulch, that has high water-holding capacity so that evaporation of any available water is reduced. This, in turn, facilitates better growth of the protective ground cover and of the trees whose growth augments humus formation and begins the cycle of restoration. Reversal of desertification is not, however, a quick-fix operation. Restoration of the dust bowl required 30 years, even with extensive, energy-intensive efforts led by the Soil Conservation Service of the Department of Agriculture. China, lacking the capital for massive efforts, may have to keep at it for a century. Probably the most critical feature of restoration, and politically and socially the most difficult task, is restricting the use of the land during its slow recovery. A sparse green cover on previously barren land becomes an almost irresistible temptation to reintroduce grazing animals or to plow up and plant crops. Premature reuse is particularly difficult to prevent when the need—as in the Sahal—is so overwhelming. To date, not even a minimally satisfactory solution to this problem is at hand.

ADDITIONAL READINGS

Cantor, L. M. *A World Geography of Irrigation.* London: Oliver and Boyd, 1967.

Carter, V. G., and T. Dale. *Topsoil and Civilization.* Rev. Ed. Norman: University of Oklahoma Press, 1974.

Coats, R. "The Colorado River: River of Controversy." *Environment* 26, no. 2 (1984):7–40.

Dregne, H. E. *Desertification of Arid Lands.* Chur, Switz.: Harwood Academic Publications, 1983.

Frederick, K. D., and J. C. Handson. *Water for Western Agriculture.* Washington, D.C.: Resources for the Future, 1982.

Fukuda, H. *Irrigation in the World.* Tokyo: University of Tokyo Press, 1976.

Gibbons, B. "Do We Treat Our Soil Like Dirt?" *National Geographic* 166, no. 3 (1984):351–399.

Grigg, D. B. *The Agricultural Systems of the World.* Cambridge, England: Cambridge University Press, 1974.

Hall, A. E. (ed.). *Agriculture in Semi-arid Environments.* New York: Springer-Verlag, 1979.

Howarth, W. "The Okies: Beyond the Dust Bowl." *National Geographic* 166 (1984):323–349.

Hyams, E. *Soil and Civilization.* Rev. Ed. New York: Harper & Row, 1976.

Low, A. M. *Dust Bowl Diary.* Lincoln: University of Nebraska Press, 1985.

Maass, A. *. . . And the Desert Shall Rejoice.* Cambridge, Mass.: M.I.T. Press, 1978.

Nir, D. *Man, a Geomorphic Agent.* Hingham, Mass.: Kluwer, 1983.

Sears, P. B. *Deserts on the March.* Rev. Ed. Norman: University of Oklahoma Press, 1974.

Steinbeck, J. *The Grapes of Wrath.* New York: Viking Press, 1939.

Steiner, F. R., and J. Theilacker (eds.). *Protecting Farmlands.* Westport, Conn.: Avi Books, 1984.

Weatherford, E. D. (ed.). *Water and Agriculture in the Western U.S.: Conservation, Reallocation and Markets.* Boulder, Colo.: Westview Press, 1982.

Worthington, E. B. *Arid Land Irrigation in Developing Countries.* Oxford, England: Oxford University Press, 1978.

Shade Trees

I made me gardens and orchards
And I planted the trees in them.

Ecclesiates 2:5

A set of before-and-after photographs of roads leading out of cities can be called ecological pornography. What used to be a tree-lined, shaded highway has often become a maze of transmission lines, interspersed with hamburger joints, motels, and gas stations. Even residential streets and college campuses have had trees removed because it became "necessary" to put up an apartment building or a tower to house administrative functionaries whose role in education is unknown to faculty or students.

THE JOY OF TREES

All cultures have had a feeling of awe and reverence for trees (Chapter 5), and while the awe remains, reverence has been replaced by aesthetic appreciation. Explorers and early settlers of North America were both amazed and annoyed by the dense forests along the Atlantic seaboard and set forth with a will to remove the trees and bring the land into agricultural production. Trees became commodities, sources of fuel and timber needed by the community, and they were again worshiped, this time as sources of wealth. Paul Bunyan is a bigger folk hero than Johnny Appleseed. By the middle of the nineteenth century, most seaboard forests were gone, and the axe and saw were chewing their way through Ontario and the American Middle West. After the American Civil War, lumber barons cleared large parts of New England, upper Michigan, Wisconsin, and adjacent Canada, leaving the land naked to wind and rain and subject to erosion and flooding. From being barriers in the road to agricultural and urban expansion, trees suddenly became precious reminders of a paradise lost. As cities expanded, demand of their inhabitants for a piece of nature led to expansion of public parks—thus repeating Europe's experience—and tree planting became a social good. Establishment of national parks, national and state forests, and wilderness areas and demands for a sense of social responsibility from the lumber companies has given some assurance that trees will remain part of the North American environment. Activist groups, such as the Audubon Society, Wilderness Society, Nature Conservancy, and the Sierra Club, provide a vigilant and effective voice in the give and take of economics and conservation. It is now easier to mobilize a community endangered by tree removal than it is to obtain consensus for sewage systems or even public education.

With the increased urbanization of the past quarter century, a new assault on trees has become apparent, and new political and social issues have arisen. Private profit from public resources has intensified, and tract developers and their allies, the multinational lumber corporations, possess great political and economic influence with regulatory agencies and government. In addition, we are now seeing the decline and death of trees on a scale that has raised the consciousness of urban and suburban populations to unprecedented heights. Local governments, perpetually short of money, are seeing funds desperately needed for social services diverted to removal of dead trees. Funds are rarely available for replacement, and a skinny sapling, whose total cost may exceed $100, is no substitute for a mature specimen.

In North America, relatively few of the hundreds of tree species have been used extensively as shade trees. Some conifers, particularly Douglas fir *(Pseudotsuga taxifolia),* true firs *(Abies spp.),* spruces *(Picea spp.),* and the smog-resistant *Ginkgo* have been used. Deciduous trees and broad-leaved evergreens are more common. Elm *(Ulmus americana),* maple *(Acer spp.),* ash *(Fraxinus spp.),* willow *(Salix spp.),* the London plane (a hybrid: *Platanus x P. acerifolia),* some magnolias, and palms are extensively grown. All of these trees are in trouble. Since it is impossible here to discuss the full range of problems, conceptual understanding of the shade tree problem can be obtained by the case-history method in which a few examples are given.

CHESTNUT BLIGHT

The American chestnut *(Castenea dentata)* once covered 9 million acres in North America, with dense stands found from northern New England and adjacent Canada down to the Gulf of Mexico and west to the Mississippi River. As late as 1910, chestnuts were growing on 40 percent of the forested land of Connecticut. Valued for its decay-resistant wood used for fence rails, railroad ties, utility poles, and house foundation timbers, one out of every ten hardwood trees that went through the sawmills was a chestnut. Its nuts were staple food for wild turkeys, grouse, and squirrels. Roasted, ground into flour, or fed to hogs, the nuts were important carbohydrate and protein sources on farms. When woodlots were cut out and roads widened, chestnut trees were left standing, to become beautiful and useful shade trees. A shipment of Japanese chestnut *(C. crenata)* trees was imported into New York City. Since the trees looked healthy, they were outplanted in the Bronx Zoological Park, but no one was aware that they contained the spores of a fungus, *Endothia parasitica,* against which the American species had no resistance. The disease was seen in 1904. In less than 25 years, most of the mature chestnuts east of the Mississippi were dead. The fungus, an ascomycete, is exquisitely adapted to its parasitic role. Asexual spores, shed as jellylike yellow masses in wet weather, are easily carried in rain and by insects or birds to other chestnuts. The sexual ascospores are shed in dry weather and, being very light, are carried by the wind. Spores can germinate in wounds or in cracks in the bark, and the fungal mycelial threads grow rapidly, killing phloem and cambium cells, until the twig or trunk is girdled and dies.

By 1930, the American chestnut had been written off; attempts to discover a cure had failed, and the blight had spread inexorably throughout North America. It was noticed, however, that the fungus did not kill the root collar or the roots, and trees began to *coppice,* to regenerate new sprouts. Hopes that these new trees, sometimes 10–15 meters tall and occasionally bearing nuts, were resistant were dashed when, as bark began to furrow, they became infected and the cycle was repeated. To "bring back the chestnut," a nationwide search for seed was started, and nurseries of seedlings were planted with the thought of finding individual trees resistant to the fungus. Since this is a very long-range project, final assessment of its potential has not been made, but chances for success are low. In 1956, chestnut seeds, irradiated in the cobalt-gamma ray reactor at Brookhaven National Laboratory, were planted on the theory that mutations induced by the ionizing radiation would confer blight resistance. Over 10,000 trees were planted in five states, and, in 1973, seeds from these trees were germinated to see if any resistant strains were produced. The chances for success are slight.

It has long been known that Oriental chestnut trees are blight-resistant. The Chinese chestnut *(C. mollissima)* and the Japanese species were introduced into North America as substitutes, but they are less vigorous, require considerable maintenance, are not as handsome, and their nuts are not as tasty as those from the American species. They cannot compete with native trees and most forest plantings have died out. Hybrids of native and Oriental trees have been made, but none to date are fine specimens. Since the fundamental difference between a resistant and a susceptible tree is unknown, genetic engineering cannot proceed on a rational basis. A fungicide, Lignosan, can protect healthy trees and arrest the spread of the fungus, but it is expensive and hence impractical for any but individual specimens.

In the 1930s, *Endothia* was introduced into Europe, and it seemed that this same unhappy story was going to be repeated. About 1950, it was noticed in Italy and France that diseased trees of the European chestnut *(C. sativa)* were beginning to recover. French plant pathologists isolated a strain of the fungus which differed from the expected strain. Cultures of this hypovirulent or H-strain were sent to the Connecticut Agricultural Experiment Station, center for research on chestnut blight, and researchers confirmed the European reports that inoculation of diseased trees with H-strain caused regression of canker development and recovery of the trees. A cross-country skier in Michigan also noticed recovery of coppice sprouts near Ann Arbor, and other H-strains were obtained from samples of bark and wood. Although the evidence is incomplete, it appears that H-strains carry a virus that can invade and destroy virulent strains of *Endothia,* whose toxins are responsible for the death of phloem and cambial cells. H-strains are less vigorous than virulent fungi, and their spores are disseminated less widely; in France, H-strains spread only 10 meters per year, a fraction of the dissemination of virulent spores. There is, however, now hope for the return of the American chestnut.

DUTCH ELM DISEASE

The elm (Ulmaceae) family is a relatively small taxon whose trees are among our most important shade plantings. In Europe and in North America, elms grace streets and parks. Ulmaceous species of *Zelkova* are planted in Crete, the Caucasus, and Japan, and hackberry *(Celtis spp.)* cultivars are used as shrubby decorative accents. There are several North American species of elm, but the American elm is the most planted shade tree.

America's love affair with the elm is long-standing. In 1646, citizens of Boston planted American elms in the Boston Common for "the relief of travelers." A century later, on August 14, 1765, the "Free Men and the Sons of Liberty" met under one of these trees to hang an effigy of Lord Butte, the British parliamentarian who conceived the hated Stamp Act. A rallying place for rebels, the Liberty Tree was cut down by British soldiers in 1775, a less than futile gesture since hundreds of elms, all over the colonies, were immediately designated as Liberty Trees. A major centennial project in the United States was the planting of millions of American elms along roads and in parks across the country.

The elm is not a commercially valuable tree. Its wood is occasionally used for furniture, but any owner of a wood-burning stove will tell you how difficult it is to cut and split. It is, however, unexcelled as a shade tree. Rows of elms on each side of a street can grow so that their umbrella-shaped crowns meet to form a cool, green arch, and their thick, gray trunks line miles of town and city roadways. Ohio, very proud of its elms, was in 1930 the first state to experience the Dutch elm disease. Leaves of 60-year-old trees began to wilt, turn yellow, then brown, and fall. The blight was usually first noticed in hot midsummer; it looked as if one side or one main branch of a tree had been scorched by fire. The disease is progressive, moving from twigs to larger branches until the entire crown is devoid of leaves, buds were shriveled, and the tree dies within a year or two (Figure 10.13). New York was next, with the city being hardest hit, and dead elms were reported from Ontario and Quebec down to Louisiana and as far west as Nebraska by 1935. In 1975, the disease reached California. The disease is caused by a fungus, *Ceratocystis ulmi,* which grows rapidly in the vessels (Figure 10.14). The fungal cells bud, multiplying much like yeast cells, quickly plug up the water-conducting xylem cells. Leaves, deprived of water and minerals, soon wilt and lose the ability to carry on photosynthesis. The death of

Figure 10.13 Removal of an American elm dying of Dutch elm disease. Courtesy U.S. Department of Agriculture.

twigs, branches, and the stem becomes only a matter of time. There is some evidence that *Ceratocystis* also produces a toxin that kills tree cells.

The spread of Dutch Elm disease from tree to tree is due, at least partly, to two insects. Both the European bark beetle *(Scolytus multistriatus)* and a native bark beetle *(Hylugopinus rufipes)* bore into cracks in elm bark where they tunnel out extensive galleries in which to lay their eggs. The eggs hatch into larvae which feed on young bark tissues, opening up additional galleries. Although these tunnels and galleries injure trees, most healthy elms can withstand the damage. When the beetles mature into flying insects, they emerge from the tree. If this tree was infected with *Ceratocystis* fungi, the adult insect would carry both spores and mycelium to another tree, where the fungus would be introduced when tunneling begins.

For many years it was believed that these insect vectors were solely responsible for transmission of Dutch Elm disease, but more recent studies have demonstrated that root systems of adjacent trees form natural grafts, allowing infection to move directly from tree to tree. Until this phenomenon was discovered, control of the disease consisted of attempts to kill bark beetles with insecticides and the removal of dead trees which served as sources of infective fungi. Although these methods were partly successful, the ease with which the beetles can carry the fungus and the development of resistance to insecticides like DDT and methoxychlor merely slowed down the spread of the disease. Nevertheless, the ravages of the disease are essentially a man-made

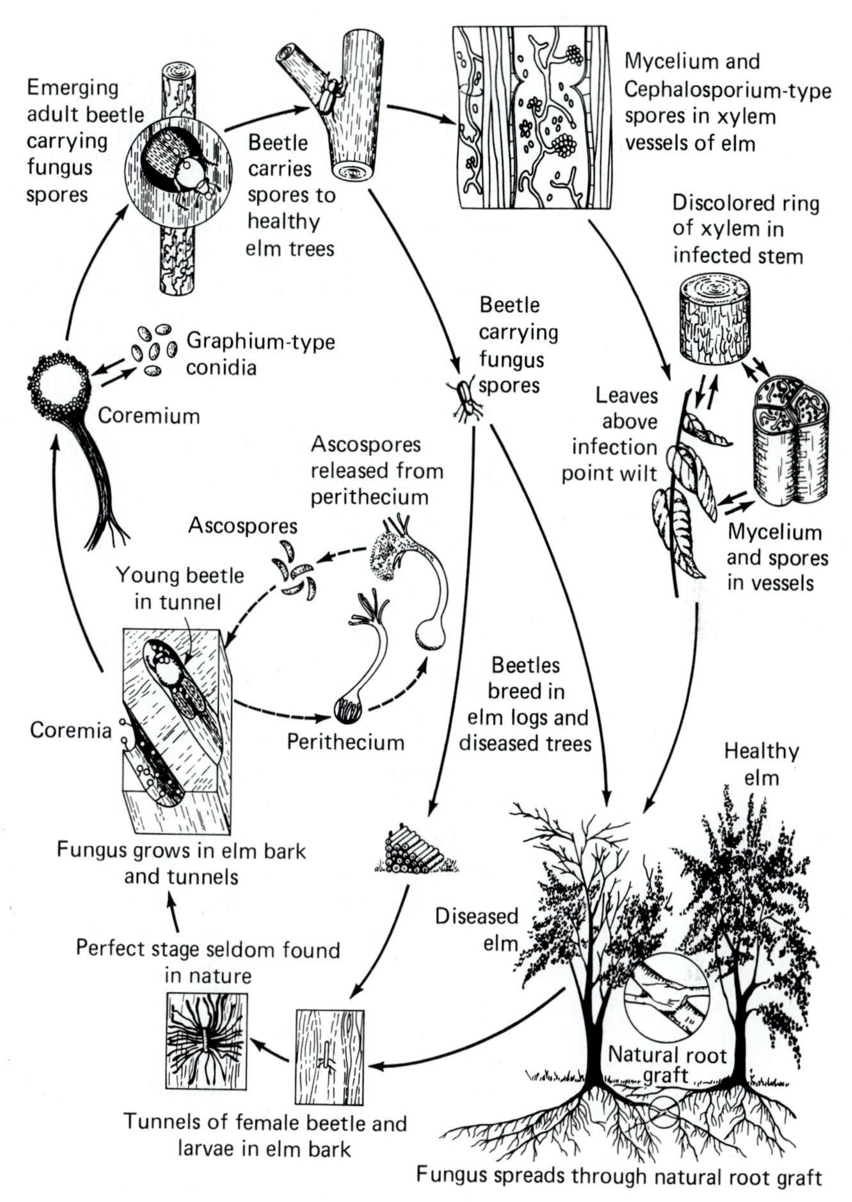

Figure 10.14 Disease cycle of Dutch elm disease caused by *Ceratocystis ulmi.* Redrawn from George N. Agrios, *Plant Pathology* (New York: Academic Press, 1969).

problem. By our extensive planting of elms, we developed a monoculture of *Ulmus americana,* in every way comparable to a field of wheat or corn, through which a disease or pest can sweep unhindered. A magnificent line of elms is a chain, with the weak link being one infected tree.

Demands for the elimination of Dutch Elm disease have not been met. To date, no resistant individuals have been found, and attempts to develop genetic resistance are not far advanced. The American elm has twice the chromosome number of the resistant European species and direct hybridization is impossible. It is possible to induce chromosome doubling in the European elm to make it genetically compatible with the American species. Once this is accomplished, another 15–20 years are needed before the tetraploid (double chromosome numbered) tree will mature and flower, and an additional 10–15 years are needed before the hybrids can be evaluated. A cross between an English species *(U. glabra)* and *U. carpinifolia* from France was made in Holland in 1963 and proved to be very resistant. Unfortunately, this tree rarely reaches 10 meters in height, too short to become a useful shade tree. Asiatic elms are resistant to the fungus, but are not particularly handsome trees and do not excite the imagination of urban planners and park departments.

Both species of bark beetle were found in North America in 1904, but the introduction of the fungus did not occur until 1933, probably on a shipment of logs sent from England. In 1935 the fungus was isolated and identified by Dutch plant pathologists, and Holland was inadvertently and incorrectly blamed for being the place of origin of the disease. There is little doubt that Dutch Elm disease was present for centuries in much of Europe, where it was a minor problem because European elms are relatively resistant. All this doesn't matter anymore. Quarantine restrictions have little meaning in an age when jet planes carry people and cargo around the world in a matter of hours. What does matter is whether the American elm, as a shade tree or even as a part of the forest environment, will be able to survive.

There are two current lines of attack on the problem. Instead of using insecticides which are less and less effective as beetles are selected for resistance and as evidence accumulates on the potential and manifest insults of chemicals on the environment, new methods of reducing the beetle population are being developed. Most hopeful is the use of *pheromones,* insect scents by which beetles can be lured to traps. It is also possible to sterilize male beetles of some species with radiation and release them to mate, unsuccessfully, with female beetles (the technique has not been tested with bark beetles). These potentially useful techniques cannot help infected trees or trees root-grafted to infected trees. Direct chemical assault on the fungus in situ was started in 1972 when the fungicide Benomyl (Lignosan) was developed for infected elms, and possibly as a prophylactic. Still experimental, Benomyl seems to be fairly effective, but it is expensive and its protective action is only moderately long-lasting. Its proponents correctly point out that use of Benomyl, prompt removal of dead trees, pruning, chemical control of insects, and other practices are "buying time"—time for genetic research to be completed, time to develop other treatments and preventatives, and time to permit the planting and establishment of other tree species to break the monocultural lockstep that resulted in the devastation of the elm.

MAPLE WILT

Native and imported maples are basic landscaping and shade trees. North American sugar maple *(Acer saccharum),* red maple *(A. rubrum),* and silver maple *(A. saccharinum),* the European or Norway maple *(A. platanoides),* the Japanese red maple *(A. palmatum),* and the Chinese Amur River maple *(A. ginnala)* are handsome shade trees whose spring and fall colors delight the eye. Box elder *(A. negundo)* is used as a foundation planting. Unfortunately, members of the genus are susceptible to a wide range of leaf, stem, and root diseases, attacks by insects, and abnormal reactions to environmental conditions. Leaf spottings and discolorations caused by various fungi *(Verticillium, Phyllosticta, Rhytisma)* or bacteria *(Pseudomonas aceris)* and insect depredations are common, but rarely fatal to the plant. Stem and branch diseases are more severe. Cankers (open, nonhealing wounds) progressively destroy the wood and weaken the trunk or may kill cambial or phloem cells to girdle and thus kill the tree. Sapsuckers chip holes in the trunks, the injured bark dies, and the holes serve as ports of entry for fungi, bacteria, and insects. Roots may rot following invasion of fungi, may swell and produce galls when attacked by nematodes, or have their meristems injured by insects. All in all, maples are delicate trees whose susceptibility is at variance with our image of a strong tree. Like the elm, the rapid dissemination of many diseases is due to the practice of planting these trees in rows.

Maples in suburban or even rural environments are not exempt from man-made predisposing factors. Economic considerations—the tourist industry, long-distance trucks which carry the goods needed for people, and inadequate public transportation—dictate that roads shall be ice-free. Salts, usually sodium chloride or calcium chloride, have been substituted for sand. Although most salt will leach down below the root-growing zone of shallow-rooted maples, the salts can cause some damage and will accumulate in ground water supplies and injure plants at some distance from the point of application. Altering the electrical charges on soil particles may change the character of soils. Automobile exhausts near heavily traveled highways cause leaf damage; sugar maples rarely reach their potential fall glory along such roads. In areas where winters are severe, one can see a zone of ice on roadside plantings from the water vapor in automobile exhausts.

MAPLE DECLINE

Construction work can cause severe damage to trees, and maples are among the most susceptible of our shade trees. Their shallow root systems spread out several meters from the trunk, usually beyond the circle of the outer twigs, gathering water and minerals for plant growth and development. Like all living cells, those of the roots metabolize, respiring to produce energy, synthesizing new cells, and growing in length. Oxygen requirements for roots are high, and a friable, well-structured soil is a necessity. Construction of a new road or modification of an existing road to carry more traffic usually results in direct injury to tree root systems, as well as indirect damage caused by compaction of the soil and alteration in the pattern of water percolation. Mature specimens along newly constructed roads frequently show very high mortality.

The seemingly inexorable march of "progress" can be measured by conversion of forests or meadowlands—usually of excellent quality—into housing developments. It is standard practice, because it is more "economical,"to chainsaw down all trees in the area, bulldoze the topsoil off for resale, and bring in heavy machinery that packs down the existing subsoil. Once the buildings are up, a 2–4 centimeter layer of topsoil is spread and planted with short-lived annual ryegrass *(Lolium spp.);* a few poor quality saplings are inserted into small holes. Lacking adequate mineral supplies, insufficiently supplied with water, and attempting to expand a root system into compacted subsoil, many saplings die as soon as the house is sold (never before), while others grow poorly and are susceptible to disease. Builders with some feeling for trees, whether aesthetic or economic, may leave mature specimens. Careless construction crews use the trees as guy standards or alkalize the soil with bits of building debris, resulting in decline and death. The more common cause of lowered vigor is the practice of back-filling or raising the ground level to promote drainage. As with compacting soil, this lowers the capacity of the soil to hold water and lengthens the path for oxygen penetration to the root zone.

Although first noted in the Northeast, maple decline has been reported in many states and provinces. The first symptoms include a premature yellowing or reddening of the leaves accompanied by browning of leaf margins. Leaves may fall in midsummer. The crown of the tree begins to show dieback of branch tips and fine twigs which continues toward the center of the crown and down toward the trunk. Bare branches with a tuft of leaves at the tip are seen the following spring. Growth is severely reduced, and, all too frequently, large portions of the tree deteriorate to the point that half or all the crown is dead. Attempts to isolate a causal organism have not failed; indeed, at least six pathogenic fungi have been found in dead or dying trees. Each can induce some of the observed symptoms. Maple decline is a *disease syndrome,* a complex of biotic and nonbiotic causes, each dependent on and related to the other. Salt injury, inadequate water, insect damage, compaction, back-filling, noxious gases, and other insults predispose the tree to reduced vigor which, in turn, facilitates the depredations of viruses, fungi, bacteria, insects, nematodes, and hence disease, decline, and eventual death. Obviously, there is no single or even multiple cure or control for maple decline. One cannot move a 20-meter maple back from a road or reroute traffic or individually water and fertilize a thousand maples along a 20-mile stretch of superhighway.

Diseases of shade tree maples are, like those of other plants, "caused" in most cases by definable, isolatable organisms. By suitable techniques, the virus, bacterium, or fungus can be obtained in pure culture, inoculated back into healthy specimens where it induces the typical disease symptoms, reisolated, and used again to infect the host. This lockstep of demonstrating cause and effect in disease was developed by the German bacteriologist Robert Koch in his studies on human disease. Koch's postulates provide excellent experimental evidence for the causal nature of disease. But cause and effect in the biological world is rarely as simple as the postulates would suggest. A sapsucker hole will not kill a sugar maple tree, but if the wound is infected with a wilt fungus, the resulting maple wilt may result in the tree's death. The sapsucker, trying to get a meal, is not the cause of the disease, but without its activity, the fungus would not have been able to invade the tree. The concept of predisposition is not a new one, but its relevance to disease of ornamental and urban trees is becoming more important. Maple decline illustrates this very well.

Consider the stress imposed on trees in an urban environment. In large cities, trees are planted in small holes chopped out of cement sidewalks to a depth determined by the enthusiasm of the employees of the street department. The hole is filled with poor soil; a 3-meter sapling planted in the hole is irrigated primarily with contaminated, alkaline water running off the pavement, and fertilized by dogs on leashes. Immediately adjacent to the tree is a busy city street where automobiles emit clouds of exhaust containing oxides of nitrogen and carbon monoxide. These block the respiratory process, and exhaust particulates clog the stomata and interfere with normal transpiration of water and uptake of carbon dioxide. The asphalt and the cement reirradiate heat in the summertime, raising ground temperatures to Death Valley levels, and hold cold in the winter, lowering temperatures to arctic permafrost values. When it rains, sulfur compounds emitted from factories and power-generating plants are converted to sulfuric acid which accumulates in the minimal amount of soil but cannot leach out because conduits, pipes, and foundations are just below ground level. The city street is no place for a plant. Photosynthesis, water relations, and respiration are far from normal; root growth is restricted; and bruises, cuts, and scrapes form openings through which disease-causing organisms can enter. A direct connection between low vigor and lowering of disease resistance is difficult to establish because the term *vigor* is difficult to define or measure quantitatively, but vigor, however defined, is certainly a component in predisposition to disease. And yet trees are necessary, indeed vital, for city living.

Maintaining shade and ornamental trees as part of our environment is, to date, an unsolved problem. The American Horticultural Society has noted that anthropogenic environmental stresses are increasing and pose threats to current and future installations of trees. Where a specific condition exists, it is possible to plant trees known to have the capacity to resist the condition. *Ginkgo* is resistant to automobile exhaust fumes; Russian olive *(Elaeagnus angustifolia)* is salt-tolerant, whereas pines are not; and many conifers are drought-resistant and capable of growing on sites where water patterns have been upset. The experience with the American elm has taught us that diversity of species provides more safety than monoculture. Our experience with the chestnut should suggest caution against introduction of foreign species without long, intensive study. Our experience with maple decline has shown the need for interaction between the builder and the botanist. May we profit from these experiences!

ADDITIONAL READINGS

Agrios, G. N. *Plant Pathology.* 2nd Ed. New York: Academic Press, 1976.

Daubenmire, R. F. *Plants and Environment.* 2nd Ed. New York: Wiley, 1959.

Feininger, A. *Trees.* New York: Viking Press, 1968.

Gibbons, B. "Do We Treat Our Soil Like Dirt?" *National Geographic* 166, no. 3 (1984):351–399.

Harris, R. W. *Arboriculture: Care of Trees, Shrubs, and Vines in the Landscape.* Englewood Cliffs, N.J.: Prentice-Hall, 1983.

Li, H. L. *The Origin and Cultivation of Shade and Ornamental Trees.* University Park: Pennsylvania State University Press, 1963.

Menninger, E. A. *Fantastic Trees.* New York: Viking Press, 1967.

Pirone, P. P. *Diseases and Pests of Ornamental Plants.* 5th Ed. New York: Wiley Interscience, 1978.

Tropical Lands

. . . the sublimity of the primeval forests,
undefaced by the hand of man.

Notebooks of Charles Darwin, 1832

BIOLOGY AND GEOGRAPHY OF TROPICAL FORESTS

Flying over tropical forests at 35,000 feet is an awesome experience. As far as the eye can see, the land is green; villages and even towns are invisible from the air. Major tropical rivers, the Amazon, the Congo, the Mekong, and the Mahakan, are only occasionally glimpsed as silver streaks threading through overarching foliage. Nature is exuberant: of the 1.5 million named species of plants and animals in the world, 500,000 are tropical. Taxonomists conservatively estimate that this represents only a quarter of the tropical species. In contrast to temperate areas, where a single species may dominate a landscape, usually only one or a few individuals of a species are to be found in a hectare of tropical land surface. In the forests of Santo Domingo, for example, a tract of 170 hectares (420 acres) contains over 1000 named species of plants and at least 200 that are still imperfectly known. In one acre near the northern headwaters of the Amazon River, a botanical expedition in 1983 found that 90 percent of the species were previously unknown to science. One 25-acre (10-hectare) study area near Manaus, Brazil, has 350 tree species, compared with perhaps 20 in New England or Quebec. One hundred and fifty bird species are permanent residents, and over 250 species are migrants. Panama has as many plant species as all of Europe; there are as many different kinds of fish in the Amazon River as in the Atlantic Ocean.

This lavish growth is, however, most deceiving. The catch is in equating lush growth and species diversity with a highly productive soil. Tropical soils are, by their very nature, poor soils. Throughout geological time and in spite of being protected by their plant cover, the soils have been extensively weathered and leached by torrential rains that, on average, are five to ten times greater than those of temperate areas. The soils are primarily acidic, low-nutrient clays and sands with high concentrations of potentially toxic aluminum and iron. Under natural conditions, any nutrients in the soil must be immediately absorbed by plants or they will be lost. As debris and litter reach the forest floor, bacteria and fungi quickly degrade it to humus. This recycling is extremely efficient; it takes almost two years to degrade litter in the northern temperate zone and less than two months in the tropics. Little nutrient is lost. Root depth is shallow and root systems are laterally dispersed to absorb quickly any nutrients released during decomposition. In fact, almost all of the mineral nutrient in tropical forests is tied up in plants.

Basically, tropical forests are composed of thick stands of trees perched superficially on a thin film of rapidly decaying organic matter layered over almost sterile clay or sand. Left alone, tropical forests are in dynamic equilibrium with their environment and, barring severe climatic change or human intervention, will maintain their diversity and complexity.

The Amazon Basin includes the 3000-mile-long Amazon River and over a thousand tributaries, several being longer than the Mississippi River. The land area is over 3 million square miles. It includes a third of South America and has roughly 3 percent of the world's land. It is sparsely populated, mostly by preliterate Indian family groups, since the bulk of Brazil's population is in the cities and along the coast. As a potential resource, the lumber and minerals are estimated to be worth over a trillion dollars. The major port for the Amazon Basin is Belém, shipping out forest products, minerals, and some manufactured goods and receiving imports of all descriptions. It is also a major embarkation point for cocaine, marijuana, contraband birds, and skins of jaguars and alligators.

LAND USE AND MISUSE

It is obvious that these lands cannot be left alone. By the year 2000, of the 6.5 billion people on earth, three-fifths of them—almost 3.9 billion—will be living in countries between the Tropic of Cancer and the Tropic of Capricorn. Many women in these countries bear their first child by age 13. This growing population needs plants for food, shelter, and fuel and for industrial products involved in foreign exchange. These needs must be met from the lands upon which they live.

Until recently, Amazon agriculture was subsistence farming, in which isolated family groups used slash-and-burn (swidden) methods. They cleared a small area and planted corn, beans, manioc, and some upland rice for themselves, with a few banana or cacao trees as a cash crop (Figure 10.15). Within a few years, soil fertility was depleted and the farmer abandoned the clearing for another site. The forest quickly covered the clearing, leaving only an occasional coffee or cacao tree to mark the site. So long as there were few groups involved, the forest could recover with little permanent damage.

Exploitation of the Amazon Basin started when the countries of Europe were vying for control of sugar plantation land along the Atlantic coast south of the mouth of the river. Portugal eventually won out over Spain, England, and France. It established forts, imported

Figure 10.15 A field being cleared for crops by the destructive slash-and-burn procedure. Photograph by F. Botts, courtesy of the Food and Agriculture Organization, the United Nations. Photograph taken in Thailand.

blacks, impressed natives as slaves, and imposed its language on the country. As sugar cultivation moved to more hospitable climates, the Portuguese colonies were stagnant until the rubber boom of 1850–1900. When Asian rubber plantations proved to be more profitable, the boom collapsed. The people left for the cities or sank into abject poverty, leaving huge tracts of almost uninhabited land (Chapter 9). Small-scale swidden agriculture or minor crop farming—rice, bananas, Brazil nuts, and jute—did little damage to the forests, but did not provide the capital or the plant products needed to support the growing urban centers and their industrialization.

As the population shot up after 1950, the pressure to provide land and to exploit the resources in the Amazon Basin became overwhelming. In Latin America, 93 percent of all farmland is owned by 7 percent of the population. Birth rates currently exceed 3 percent a year, imposing unbearable pressures on governments. What else can be done when the population will double in less than 20 years except to make land available? In 1960, Brazil, financed by international capital, embarked on a massive industrialization plan that included exploitation of the Amazon tropics.

Clearing and lumbering these vast lands was the first order of business. The various states in Brazil had the power to sell any public land to individuals or corporations at a price defined as the value of the naked earth *(valor du terra nua),* and no value was assigned to trees. Land clearing and agricultural development were favored by government policy, with fiscal incentives including allowances of up to 50 percent of annual corporate taxes. As a condition for tax relief, at least half the land had to be cleared and, since there was a time limit for clearing, most of the trees were not harvested, but were simply burned; over 3 million hectares of usable lumber have gone up in smoke. Clearcutting could, if the trees were brought to market, produce a handsome profit, and the land could then be developed for grazing.

Mahogany and tropical cedar provided immediate returns, but the environmental costs were high. In contrast to more temperate regions, the desirable tree species do not grow in stands but as isolated specimens. Cutting down a forest giant crushes many other trees and opens up the canopy to the hot sun and the driving rains that quickly turn the soil into baked clay so that natural regeneration does not occur. Present logging operations are notably inefficient. Almost half of the remaining trees are damaged beyond recovery. The construction of skid trails and roads completes the ravaging of the forests.

Once road networks were established, landless peoples began to enter the previously inaccessible regions to take up farming activities. Such invasions, as evidenced by the push of the Americans westward, resulted in clashes with the native Indians, who were raped, exposed to the diseases of civilization, or killed outright. By 1970, fewer than 40 percent of the native Indians had survived and most had retreated further into the forests, where they occupied less productive lands. The road construction needed to bring cattle and timber to market made a bad situation worse. The Brazilian agency responsible for the protection of Indian life (FUNAI) is underfunded and is notably pliant to the demands of developers. Many people with many different goals have almost free access to these lands. In the western Pantanal region of Brazil, jaguars are being exterminated for their handsome spotted skins. Birds with bright feathers—egrets, macaws, and spoonbills—are now rarely seen as the traffic in living birds for North America and European collectors has increased.

Forest destruction is on an almost unimaginable level. The United Nations's Food and Agriculture Organization has estimated that 7.5 million hectares (18 million acres) of closed forest and 38 million hectares (95 million acres) of open forests are being cleared each year. Put another way, an area the size of Delaware is lost every week. Central and South America harvest 11 percent of the world's wood and use over 80 percent of it for fuel. The United States, on the other hand, produces 17 percent of the world's wood and uses less than 15 percent of it for energy production. Almost half of the world's tropical forests have already been converted into open land. If this rate is unaltered, tropical forests, and the plants and animals that are part of these forests, will be gone within 50–75 years. This projection is probably a bit pessimistic; cutting and land conversion go in spurts, as demand for forest products and for arable land wax and wane. More conservative estimates are closer to 90 years—as if this makes any real difference.

The destruction of tropical forests is not uniformly distributed. The Congo Basin, the interior of the Guyanas, and the western Brazilian Amazon are receiving less attention than lands of Central America, New Guinea, Southeast Asia, and the central Amazon Basin, where 30 acres are cut down every minute. El Salvador has destroyed over 80 percent of its forests and the Atlantic forests of northern and eastern Brazil are going fast.

Once a road network was established, additional forested areas were cut down to develop cattle ranches. Peasants were expelled and they, in turn, forced the Indians off lands that by legislation had been reserved in perpetuity for tribal life. Although initially they were small family operations, they have become large corporate enterprises. Volkswagen de Brasil owns 140,000 hectares (350,000 acres), Deltec International Packers—owned by Swift, Armour, and the King Ranch syndicate—has 72,000 hectares. Together with the Georgia-Pacific lumber company, these corporations are rapidly deforesting close to a half million acres. Almost without exception, cattle ranching is profitable for no more than seven to ten years. Overgrazing, soil erosion, and nutrient leaching are evident within five years, with consequent reduction in the quality and quantity of forage. The area needed per head of cattle increases from 1 to 5 hectares within five years, with a reduction in yield of beef to less than 25 pounds (10 kilograms) per hectare per year, a fraction of that in Florida or the Argentine pampa lands. This beef is too expensive for the people who raise it, but it represents a significant portion of Brazil's export economy. Subject to many insect pests, the cattle develop slowly, and the beef, of generally poor quality, becomes sausages or appears at your favorite fast food hamburger restaurant.

It would be nice to state that commercial exploitation of the Amazon Basin was successful, that jobs were created, and that the landless obtained the farms they needed. Unfortunately, this has not been the case. The well-publicized failure of Jari is the most outstanding example of that failure. Without doubt, the Jari plantation in northern Brazil is one of the most fantastic tales in the history of the tropics. In 1967, Daniel K. Ludwig, an American shipping magnate, purchased 6000 square miles (1.7 million hectares, or 380 million acres) at the confluence of the Jari and Amazon rivers. This is the size of Connecticut. It was purchased for $3 million, about $0.78 per acre. The tropical forests were leveled to build four cities, complete with 3000 miles of roads, an airport, a railroad, and all of the infrastructure needed for integrated commerce. The major project was a paper-pulp mill capable of producing 750 metric tons of pulp a day. The mill was powered by its own electric generating plant. Both the mill and the power plant had been made in Japan and towed halfway around the world and 300 miles up the Amazon River. Pulp, lumber, cattle, rice, and kaolin (porcelain clay) mining were each supposed to contribute to the financial success of Jari.

There was much less commercially valuable timber than expected, and Ludwig planted 490,000 acres of trees, expecting them to grow at rates comparable to those in their native habitats. His initial plantings were a Nigerian tree, the *Gmelina,* which can grow rapidly, but the trees died in the depleted sandy soil of Jari. A second planting of Caribbean pine did better, but the wood was of poor quality even for pulp. The rice operation was doomed by low prices due to surpluses of miracle rice, the cattle ranching lasted less than eight years, and the kaolin mining is barely in the black. By 1982, after over $1 billion had been invested, the entire complex was sold to Brazilian companies. The purchase was virtually mandated and financed by the Brazilian government. Isolation, poor management, clouded land titles, insects, malaria, yellow fever, and meningitis were contributing factors in the collapse of the Ludwig venture, but the major cause was failure to understand the nature of tropical ecosystems.

CONSEQUENCES OF FOREST DESTRUCTION

The ecological consequences of destruction of tropical forest ecosystems are profound, but not always immediately apparent. On a global level, we are confronted with the possibility that the rapid combustion of fossil fuels can cause an increase in the temperature of the earth by elevating the carbon dioxide concentration in the atmosphere, which prevents solar energy from escaping back into space. This is called the *greenhouse effect* by analogy to the heating up of a greenhouse through the failure of heat energy to escape back into the atmosphere through the glass.

Where there is dense plant cover, the carbon dioxide can be utilized through photosynthesis and converted into plant biomass. The rapid, year-round photosynthesis of tropical plants served this purpose in the past but may not in the future. The destruction, to date, of 30 percent of the earth's tropical biomass has been a factor in increasing the carbon dioxide concentration from 300 parts per million of air (ppm) at the beginning of this century to 350 ppm today and a projected increase to 450 ppm by the middle of the next century. At this level, the earth's average temperature may rise by several degrees Celsius, enough to cause melting of polar icecaps and an increase in the severity of droughts.

Conversion of forests to grazing land or even to young tree plantations also alters weather patterns. Many tropical forests are rightly

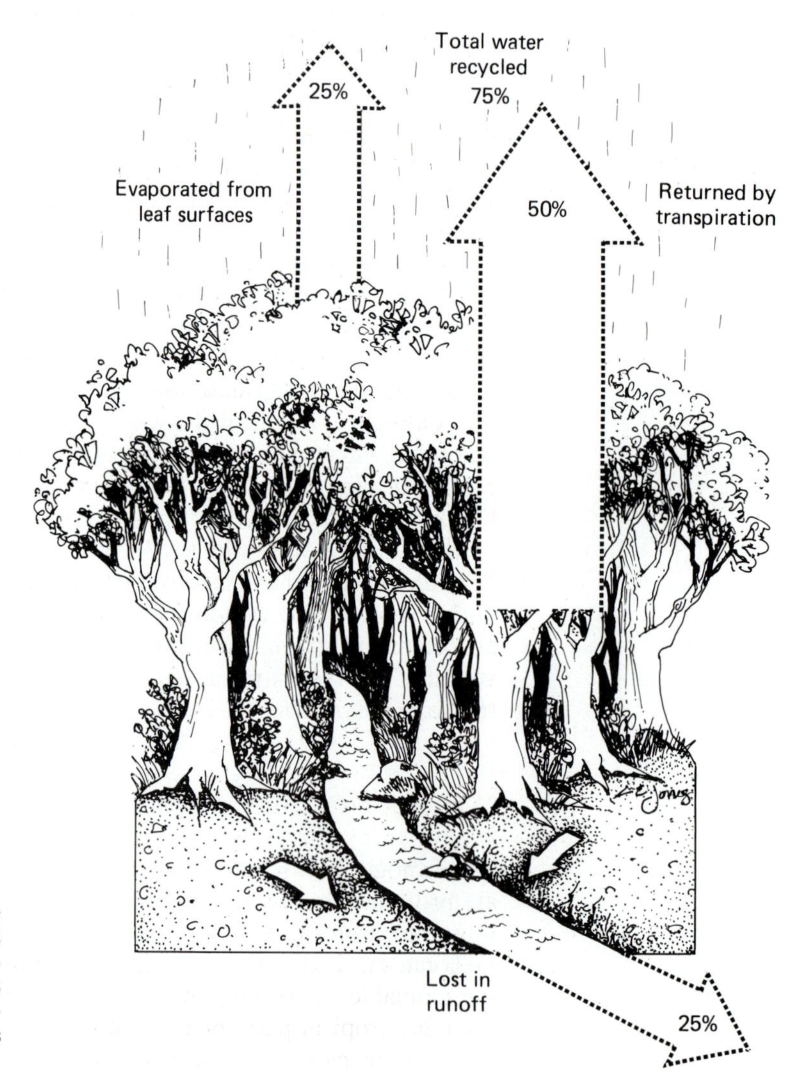

Figure 10.16 Recycling of water from a tropical forest. Close to 75 percent of incoming water is returned to the atmosphere, and 25 percent is lost as runoff. In deforested areas, these percentages are reversed.

called rain forests (jungles). Over 70 percent of the rainwater is, through evaporation from wetted leaf surfaces and the transpirational movement of water from soil out through the leaves, returned to the atmosphere, where it condenses into clouds, again to be returned as precipitation (Figure 10.16). When the trees have been removed, most of the water impinges on the soil to cause erosion; it flows into the rivers and moves uselessly out to sea. When the greenhouse effect is also factored in, the land can be expected to become dryer and hotter, severely limiting its ability to sustain plant life. Forage grasses will reduce erosion, but they cannot control water cycles as efficiently as forests. The altered weather patterns are not restricted to the denuded areas, since prevailing winds move moisture-laden clouds over lands that may be far removed from the lost forests.

The high humidity and temperatures of tropical forests promote the decomposition of plants and the recycling of the nutrients. With the trees removed, recycling of scarce nutrients does not occur, and these precious minerals are washed away. The trees, burned or removed, contain most of the forest's nutrients; their loss accelerates the depauperation of the soil. The remaining organic humus is decomposed, leaving the bare mineral soil exposed to torrential rains, periods of dry heat, and wind action. Clayey soil—acidic, low in silica, and high in aluminum and iron—is soon converted into a brick-hard *laterite* soil (Figure 10.17) that is so indestructible that it was used to construct Buddhist temples in Southeast Asia (Angkor Wat) and shrines in Latin America. Grasslands, agricultural fields, or open tree plantations do not greatly impede nutrient leaching and soil laterization. Only reforestation and time will serve this end, but few developers in Amazonia or elsewhere are prepared to wait 75–150 years to bring in a tree crop.

Many of the North American migratory birds utilize the Central and South American tropical forests during North American winters. Loss of habitat has already caused significant reductions in the numbers of some of the song birds upon which we depend for insect control.

The most serious consequence of unrestrained destruction of tropical forests has been and will continue to be the irretrievable loss of plant species. This collective simplification of the biosphere has no parallel in the history of the world. Professor E. O. Wilson has called the accelerated loss of species "the folly our descendants are least likely to forgive us." At present, close to half the medical prescriptions written in North America contain at least one plant product whose synthesis would be prohibitively expensive or downright impossible. Most of these "botanicals" are of ancient European or Asian origin. Pharmaceutical botanists have added important tropical plants to the medical armamentaria, but the pace of destruction has exceeded that of plant exploration. There are, in fact, fewer than 1500 knowledgeable botanists capable of naming and cataloguing tropical plants and no more than 200 tropical plant pharmacologists. It is likely that by the end of this century, close to a million species of tropical plants and animals will

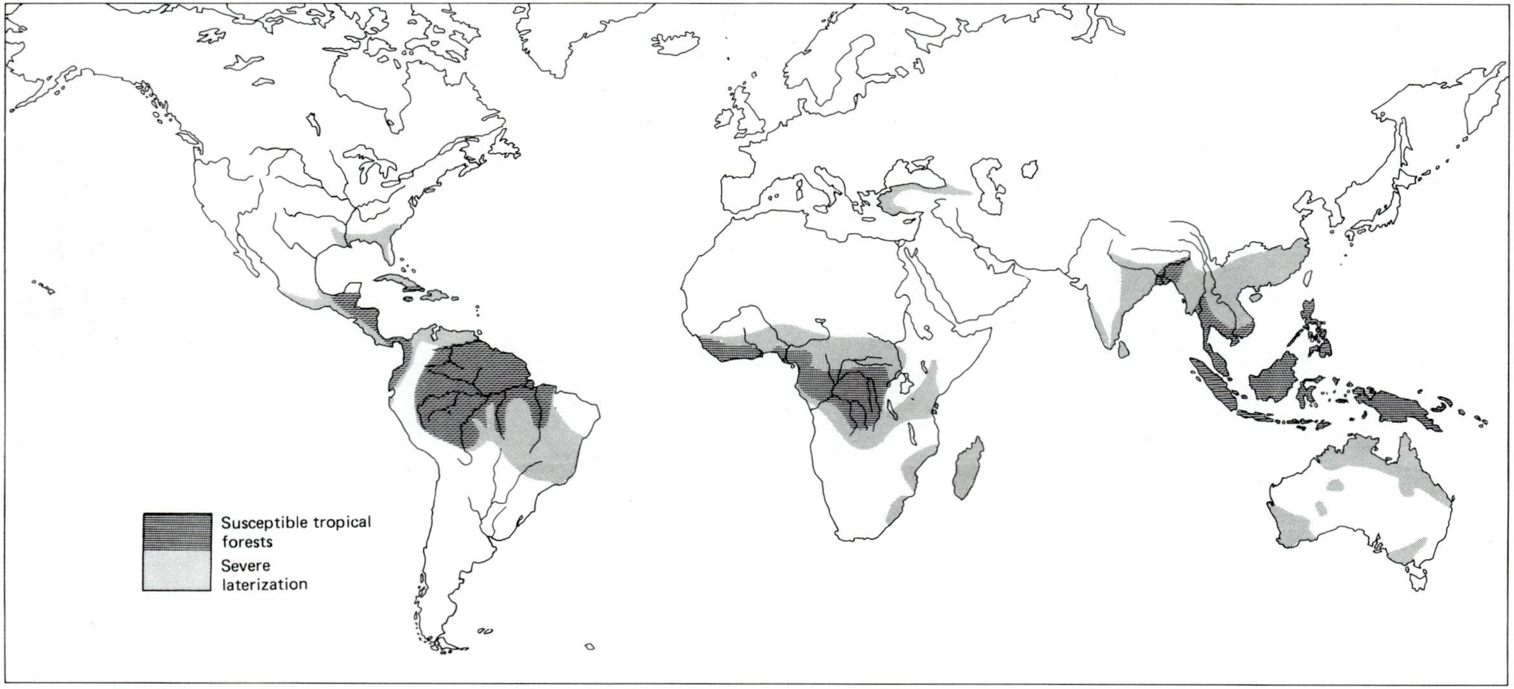

Figure 10.17 Areas of the world where soil laterization is severe. These are mainly in tropical regions. Redrawn from M. McNeil, "Lateritic Soils." *Scientific American* 211, no. 5 (1964):96–102.

become extinct, and we can only speculate on how many of these had economic or medical potential.

Because every species is, by definition, genetically unique, its germ plasm can never be duplicated. The insertion of hereditary information into our cultivated plants can confer disease resistance, more efficient growth, drought tolerance, or other desirable characteristics. Genetic engineering depends, however, entirely on the availability of wild plants that contain the hereditary information. Within the past decade, plants from tropical forests have been used as breeding stock to improve corn, sorghum, rice, and other fundamental food plants of tropical origin. It is ironic that the loss of species is accelerating at a time when techniques of genetic engineering have permitted the use of the vast gene pools of wild species. The disappearance of native tropical species is a matter of deep concern, but so far it has been impossible to do anything about it.

IS THERE ANY SOLUTION?

It is tempting for those of us in industrialized, wealthy countries to tell those of the developing world what to do and what not to do. In our smugness, we frequently fail to realize that for people without land, food, shelter, or much hope, conservation is just another pretty word. If we are to play any role in staying the destruction, it is important that our activities and recommendations have a rational basis; emotion and empathy are admirable, but policymakers and bankers think in dollars and the dispossessed and hungry think in terms of work and their children's supper.

There is no question that the rapidly increasing populations in the tropics must be provided with productive land and that the recovery of natural resources is necessary for the well-being of the people and their countries. But Amazonia and other tropical areas cannot be clearcut, burned, plowed, and grazed and then expected to become sustainable agricultural land. With such nonmanagement, lateritic semideserts form within a decade and no further productivity can be expected. Nor is there any justification for the disruption and eventual destruction of tribal life. Instead of greed and indifference, utilization of tropical lands must be based on sound ecological principles—both for wildlife and for people.

In spite of the efforts to develop beef cattle ranches, it is now obvious that tropical lands are usually unsuitable for this purpose. In areas where cattle ranching is dominant, soil fertility has decreased to the point where, even with fertilization, forage yields and hence cattle development has declined significantly. Identification of areas with more productive and sustainable soils for agriculture, greater emphasis on small landholdings capable of supporting individual families or village life, and modifications in lumbering—with due attention to retaining basic forest structure and nutrient cycling—are costly in the short run, but effective in the longer or even medium term. Road construction, now initiated at the whim of individual developers, can be incorporated into a national or even supranational plan to protect sensitive land and helpless people. Large forest reserves, embracing not just token acreages but hundreds if not thousands of square miles, will allow the Indians to maintain their own life styles, will protect animals, and will preserve for the world's future the irreplaceable germ plasm of countless plant species.

But you can't just put a fence around a piece of real estate. Poaching, as seen in the national parks of tropical Africa, and incursions onto the perimeter, like those next to the U.S. national parks, will occur. Tropical plant and animal species are very susceptible to extinction; all are found in small numbers; many have limited distribution ranges; and most have highly specialized ecological requirements and little adaptability to altered conditions. A concerted effort backed by the full and enthusiastic support of governments is the only hope for tropical forests. We in the temperate zones, who will benefit directly from such conservation efforts, must be prepared to help pay for their safety.

ADDITIONAL READINGS

Buschbacker, R. J. "Tropical Deforestation and Pasture Development." *BioScience* 36 (1986):22–28.

Carpenter, R. A. (ed.). *Assessing Tropical Forest Lands: Their Suitability for Sustained Uses*. New York: Unipub, 1982.

Ehrlich, P. and A. Ehrlich. *Extinction: The Causes and Consequences of the Disappearance of Species*. New York: Random House, 1981.

Forsyth, A., and K. Miyata. *Tropical Forests and Our Future*. New York: Norton, 1984.

Holzner, W., M. J. A. Werger, and I. Ikusima. *Man's Impact on Vegetation*. Hingham, Mass.: Kluwer, 1983.

Jordan, C. E. "Amazon Rain Forest." *American Scientist* 70, no. 4 (July–August 1982):394–401.

Kelly, B., and M. London. *Amazon*. New York: Harcourt Brace Jovanovich, 1983.

Moran, E. F. *Developing the Amazon: The Social and Economic Consequences of Government-directed Colonization Along Brazil's Transamazon Highway*. Bloomington: Indiana University Press, 1981.

Myers, N. *The Primary Source: Tropical Forests and Our Future*. New York: Norton, 1984.

Nir, D. *Man, A Geomorphic Agent*. Hingham, Mass.: Kluwer, 1983.

Prance, G. T., and T. S. Elias (eds.). *Extinction Is Forever*. Bronx: New York Botanical Garden, 1977.

Smith, N. J. H. *Rainforest Corridors: The Transamazon Colonization Scheme*. Berkeley: University of California Press, 1982.

Tangley, L. "Saving Tropical Forests." *BioScience* 36 (1986):4–8.

Index